Hydrothermal Vents and Processes

Geological Society Special Publications

Series Editor A.J. FLEET

GEOLOGICAL SOCIETY SPECIAL PUBLICATION NO. 87

Hydrothermal Vents and Processes

EDITED BY

L. M. PARSON
Institute of Oceanographic Sciences, Wormley, UK

C. L. WALKER
University of Leeds, UK

D. R. DIXON
Plymouth Marine Laboratories, UK

1995
Published by
The Geological Society
London

THE GEOLOGICAL SOCIETY

The Society was founded in 1807 as the Geological Society of London and is the oldest geological society in the world. It received its Royal Charter in 1825 for the purpose of 'investigating the mineral structure of the Earth'. The Society is Britain's national society for geology with a membership of 7500 (1993). It has countrywide coverage and approximately 1000 members reside overseas. The Society is responsible for all aspects of the geological sciences including professional matters. The Society has its own publishing house, which produces the Society's international journals, books and maps, and which acts as the European distributor for publication of the American Association of Petroleum Geologists, SEPM and the Geological Society of America.

Fellowship is open to those holding a recognized honours degree in geology or cognate subject and who have at least two years' relevant postgraduate experience, or who have not less than six years' relevant experience in geology or a cognate subject. A Fellow who has not less than five years' relevant postgraduate experience in the practice of geology may apply for validation and, subject to approval, may be able to use the designatory letters C Geol (Chartered Geologist).

Further information about the Society is available from the Membership Manager, The Geological Society, Burlington House, Piccadilly, London W1V 0JU, UK. The Society is a Registered Charity No. 210161.

Published by The Geological Society from:
The Geological Society Publishing House
Unit 7
Brassmill Enterprise Centre
Brassmill Lane
Bath BA1 3JN
UK
(*Orders*: Tel 01225 445046
      Fax 01225 442836)

First published 1995

The publisher makes no representation, express or implied, with regard to the accuracy of the information contained in this book and cannot accept any legal responsibility for any errors or omission that may be made.

British Library Cataloguing in Publication Data

A catalogue record for this book is available from the British Library

ISBN 1–897799–25–X

Typeset by Type Study, Scarborough, UK

Printed by Alden Press, Oxford, UK

Distributors

USA
  AAPG Bookstore
  PO Box 979
  Tulsa
  OK 74101–0979
  USA
  (*Orders*: Tel (918) 584–2555
      Fax (918) 584–0469)

Australia
  Australian Mineral Foundation
  63 Conyngham Street
  Glenside
  South Australia 5065
  Australia
  (*Orders*: Tel (08) 379–0444
      Fax (08) 379–4634)

India
  Affiliated East-West Press pvt Ltd
  G-1/16 Ansari Road
  New Delhi 110 002
  India
  (*Orders*: Tel (11) 327–9113
      Fax (11) 326–0538)

Japan
  Kanda Book Trading Co.
  Tanikawa Building
  3–2 Kanda Surugadai
  Chiyoda-Ku
  Tokyo 101
  Japan
  (*Orders*: Tel (03) 3255–3497
      Fax (03) 3255–3495)

Contents

PARSON, L. M. Introduction   1

GERMAN, C. R., BAKER, E. T. & KLINKHAMMER, G. Regional setting of hydrothermal activity   3

KRASNOV, S. G., POROSHINA, I. M. & CHERKASHEV, G. A. Geological setting of high-tempera-   17
ture hydrothermal activity and massive sulphide formation on fast- and slow-spreading ridges

MURTON, B. J., VAN DOVER, C. & SOUTHWARD, E. Geological setting and ecology of the   33
Broken Spur hydrothermal vent field: 29°10′N on the Mid-Atlantic Ridge

KRASNOV, S. G., CHERKASHEV, G. A., STEPANOVA, T. V., BATUYEV, B. N., KROTOV, A. G.,   43
MALIN, B. V., MASLOV, M. N., MARKOV, V. F., POROSHINA, I. M., SAMOVAROV, M. S.,
ASHADZE, A. M. & ERMOLAYEV, I. K. Detailed geographical studies of hydrothermal fields in
the North Atlantic

BAKER, E. T. Characteristics of hydrothermal discharge following a magmatic intrusion   65

EDMOND, J. M., CAMPBELL, A. C., PALMER, M. R., KLINKHAMMER, G. P., GERMAN, C. R.,   77
EDMONDS, H. N., ELDERFIELD, H., THOMPSON, G. & RONA, P. Time series studies of vent
fluids from the TAG and MARK sites (1986, 1990) Mid-Atlantic Ridge: a new solution
chemistry model and a mechanism for Cu/Zn zonation in massive sulphide orebodies

KLINKHAMMER, G. P., CHIN, C. S., WILSON, C. & GERMAN, C. R. Venting from the   87
Mid-Atlantic Ridge at 37°17′N: the Lucky Strike hydrothermal site

JAMES, R. H., ELDERFIELD, H., RUDNICKI, M. D., GERMAN, C. R., PALMER, M. R., CHIN, C.,   97
GREAVES, M. J., GURVICH, E., KLINKHAMMER, G. P., LUDFORD, E., MILLS, R. A., THOMSON, J.
& WILLIAMS, A. C. Hydrothermal plumes at Broken Spur 29°N Mid-Atlantic Ridge: chemical
and physical characteristics

PALMER, M. R., LUDFORD, E. M., GERMAN, C. R. & LILLEY, M. D. Dissolved methane and   111
hydrogen in the Steinahóll hydrothermal plume, 63°N Reykjanes Ridge

MILLS, R. A. Hydrothermal deposits and metalliferous sediments from TAG, 26°N   121
Mid-Atlantic Ridge

STUART, F. M., HARROP, P. J., KNOTT, R., FALLICK, A. E., TURNER, G., FOUQUET, Y. &   133
RICKARD, D. Noble gas isotopes in 25 000 years of hydrothermal fluids from 13°N on the East
Pacific Rise

DICKSON, P., SCHULTZ, A. & WOODS, A. Preliminary modelling of hydrothermal circulation   145
within mid-ocean ridge sulphide structures

PASCOE, A. R. & CANN, J. R. Modelling diffuse hydrothermal flow in black smoker vent fields   159

DUCKWORTH, R. C., KNOTT, R., FALLICK, A. E., RICKARD, D., MURTON, B. J. & VAN DOVER,   175
C. Mineralogy and sulphur isotope geochemistry of the Broken Spur sulphides, 29°N
Mid-Atlantic Ridge

SCOTT, S. D. & BINNS, R. A. Hydrothermal processes and contrasting styles of mineralization   191
in the western Woodlark and eastern Manus basins of the western Pacific

KNOTT, R., FALLICK, A. E., RICKARD, D. & BÄCKER, H. Mineralogy and sulphur isotope   207
characteristics of a massive sulphide boulder, Galapagos Rift, 85°55′W

CHERKASHEV, G. A. Hydrothermal input into sediments of the Mid-Atlantic Ridge   223

HODKINSON, R. A. & CRONAN, D. S. Hydrothermal sedimentation at ODP Sites 834 and 835 in   231
relation to crustal evolution of the Lau Backarc Basin

SUDARIKOV, S. M., DAVYDOV, M. P., BAZELYAN, V. L. & TARASOV, V. G. Distribution and transformation of Fe and Mn in hydrothermal plumes and sediments and the potential function of microbiocoenoses — 249

VAN DOVER, C. L. Ecology of Mid-Atlantic Ridge hydrothermal vents — 257

SHILLITO, B., LECHAIRE, J.-P., GOFFINET, G. & GAILL, F. Composition and morphogenesis of the tubes of vestimentiferan worms — 295

DANDO, P. R., HUGHES, J. A. & THIERMANN, F. Preliminary observations on biological communities at shallow hydrothermal vents in the Aegean Sea — 303

SUDARIKOV, S. M. & GALKIN, S. V. Geochemistry of the Snake Pit vent field and its implications for vent and non-vent fauna — 319

RIELEY, G., VAN DOVER, C. L., HEDRICK, D. B., WHITE, D. C. & EGLINTON, G. Lipid characteristics of hydrothermal vent organisms from 9°N, East Pacific Rise — 329

DIXON, D. R., JOLLIVET, D. A. S. B., DIXON, L. R. J., NOTT, J. A. & HOLLAND, P. W. H. Molecular identification of early life-history stages of hydrothermal vent organisms — 343

COWAN, D. A. Hyperthermophilic enzymes: biochemistry and biotechnology — 351

GERMAN, C. R. & ANGEL, M. V. Hydrothermal fluxes of metals to the oceans: a comparison with anthropogenic discharge — 365

SPEER, K. G. & HELFRICH, K. R. Hydrothermal plumes a review of flow and fluxes — 373

RUDNICKI, M. D. Particle formation, fallout and cycling within the buoyant and non-buoyant plume above the TAG vent field — 387

Index — 397

Hydrothermal vents and processes

L. M. PARSON[1], C. L. WALKER[2] & D. R. DIXON[3]

[1] Institute of Oceanographic Sciences, Deacon Laboratory, Wormley, Surrey GU8 5UB, UK

[2] Department of Earth Sciences, University of Leeds, Leeds LS2 9JT, UK

[3] Plymouth Marine Laboratory, Citadel Hill, Plymouth PL1 2PB, UK

Hydrothermal venting at mid-ocean ridges has become one of the fastest growing areas of interest in the marine sciences since its discovery in the late seventies. In tandem with the value of our increased knowledge of geological processes at actively spreading plate boundaries, we have been able to focus on the impact of the vent products on the global ocean chemical budget itself. The recognition that complex vent communities consisting of bizarre organisms dependent on geochemically-based chemosynthesis thrive in the extreme conditions of temperature, pressure and chemistry at the ridges, then go on to disperse and colonize new sites has undoubtedly led to novel and challenging research directions, but the very isolation of these communities over millions of years has led to suggestions that they may in themselves hold some of the keys to an understanding of the earliest evolutionary stages of life processes on the earth as a whole. Processes at mid-ocean ridge systems are four-dimensional in their character, with some events varying on a daily scale and others over a few hundreds or thousands of years. This volume represents the most recent reviews and reports of the latest advances in understanding of an area of marine science which we are only just beginning to recognize the scope and significance of.

The intense efforts which have been made over the past decade and a half have come some way to answering key questions which geologists, chemists and biologists have pointed to in the science of mid-ocean ridge hydrothermal activity. Despite these advances we still need to address critical areas of uncertainty. These include: What are the tectonic and volcanic controls on the location of, and evolutionary histories of hydrothermal sites? What are the dispersal mechanisms of hydrothermal vent products and what are the temporal and spatial variations in these? What is the composition of vent products and how do they vary according to the geological setting and maturity of the site? What are the separate contributions of the focused high temperature and the broad diffuse venting systems to thermal and chemical fluxes to the ocean? Is it possible to quantify the evolutionary history of sites from physical or chemical characteristics, or from the interrelationships of faunal assemblages on their genetics? What common parameters in hydrothermal activity can be found between sites occurring on fast- and slow-spreading ridges, and what reasons are there for the fundamental differences?

With the depth of knowledge of our best documented sites and the assessment of the steadily increasing number of newly discovered sites, we now know that some of our earliest preconceptions as to the location and geological setting of vent activity have to be revised. Hydrothermalism does not necessarily occur centrally to ridge segments where the thermal flux is reasoned to be at a maximum, rather it can equally be focused towards discontinuities where intersecting structural lineaments exert a controlling focus on fluid transport, at off-axis sites or at margins to axial zones where deep-seated flanking fault systems appear to act as conduits.

The latest techniques to identify the presence of hydrothermal activity involve real-time monitoring of chemical and/or physical signatures of plume activity, generally in its neutrally buoyant portion. Vertical profiles through the water column to detect absolute values in the marker anomalies in total dissolvable manganese, methane or Helium/temperature ratios can calibrate latest continuous surveying results.

An understanding of the physics of plume behaviour in its buoyant and non-buoyant phases is of critical importance to the reliability with which we can map the distribution of hydrothermal activity. We are reliant on the tracers within the plume to pinpoint venting, yet we must recognise that the complex dynamics of plumes as well as prevailing physical oceanographic conditions such as water mass movement, seafloor topography and diurnal and other temporal cycles can perturb our ability to repeat and verify our observations.

Each detailed investigation of communities associated with hydrothermal vents undertaken

From PARSON, L. M., WALKER, C. L. & DIXON, D. R. (eds), 1995, Hydrothermal Vents and Processes, Geological Society Special Publication No. 87, 1–2.

provides a richer catalogue of species than existed before. The tracing of biogeographical patterns, sensory physiology and determination of food web evolution depend on accurate identification of biological material, much of which is acquired under extreme sampling conditions. However, recent progress in molecular fingerprinting of species using DNA sequencing has maximised the use of the limited collections available. These state-of-the-art techniques will continue to strengthen our database and comprehension of vent ecosystem dynamics.

We are confident that this collection of papers represent a comprehensive synthesis of our current knowledge of vent processes, but are equally sure that we are only at the earliest stages of understanding the variability and interrelationships within these complex systems.

We are indebted to the sponsors of this meeting, namely the Geological Society of London, the BRIDGE initiative of the UK Natural Environment Research Council, the Challenger Society for Marine Sciences and Rio Tinto Zinc for their support for this publication. A list of referees who gave their time to review the papers appears below, and to each of them we extend our sincere thanks. We would also like to record our gratitude to numerous support staff at Burlington House, the Society Publishing House and at IOSDL, who all contributed in individual ways to the success and speedy delivery of this compilation, especially Angharad Hills, Annie Williams and Yvonne Baker. IOSDL Contribution No. 95016.

Gary Klinkhammer, Chris German, Harmon Craig, Ray Binns, Rachel Mills, Rob Ixer, Holger Jannasch, Rod Herbert, John Parkes, Keir Becker, Randy Koski, Toshitaka Gamo, Dave Coller, Martin Palmer, Mike Krom, Peter Herzig, Kevin Speer, Cara Wilson, Harry Elderfield, Eva Valsami-Jones, Annie Michard, Bob Nesbitt, Steve Scott, Bramley Murton, Ulrich von Stackleburg, Sven Patersen, Peter Rona, Roger Hekinian, John Edmond, José Honnorez, Ed Baker, Dave Scott, Karl Helfrich, Adam Schultz, Paul Dando, Paul Comet, Aline Fiala-Medioni, Fred Grassle, Judith Grassle, Steve Sparks, David Kadko, Ian Wright, Dave Rickard, Cindy van Dover, Eve Southward, Paul Tyler, Mark Rudnicki, Gerald Ernst, Philippe Jean-Baptiste, David Hilton, Jody Deming, Daniel Prieur, Michel Segonzak, Didier Jollivet, Francoise Gaill, Alick Jones, Michael Danson, Bambos Charalambous, Tony Williams, Alan Southward, Norma Sleep, John Ludden, Anne-Marie Karpoff.

Regional setting of hydrothermal activity

C. R. GERMAN[1], E. T. BAKER[2] & G. KLINKHAMMER[3]

[1]*Institute of Oceanographic Sciences Deacon Laboratory, Wormley, Surrey GU8 5UB, UK*

[2]*NOAA Pacific Marine Environmental Laboratory, Seattle, WA 98115–0070, USA*

[3]*COAS Oregon State University, Corvallis, OR97331–5530, USA*

Abstract: High-temperature hydrothermal venting, in which heat is transferred from the lithosphere to the oceans, is intimately associated with all forms of active plate boundaries – fast- and slow-spreading centres, fracture zones and even with subduction zones, in the form of back-arc spreading centres. A range of steady-state hydrothermal vent sites are described in various tectonic settings. A selection of the more unusual hydrothermal features and their geological environments are also considered, including hydrothermal 'events', gas-rich vents, and sites of phase separation. Modelling of hydrothermal plumes is introduced and the importance of focused high-temperature venting is reviewed with respect to global hydrothermal fluxes.

High temperature 'black smoker' hydrothermal activity was first discovered on the East Pacific Rise at 21°N in 1979 as part of a joint French–American submersible investigation of the ridge crest (Spiess *et al.* 1980). Since then, a wealth of further hydrothermal vent sites has been discovered throughout the world's oceans in a range of tectonic settings. The majority of

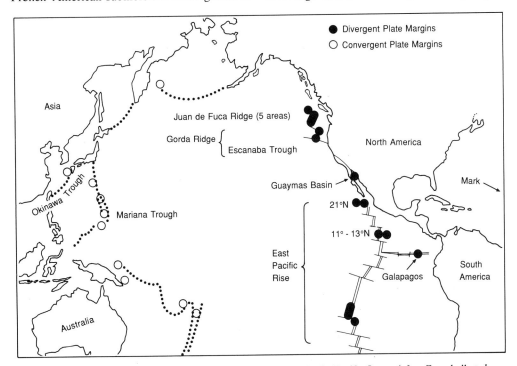

Fig. 1. Map of known high-temperature hydrothermal vent fields in the Pacific Ocean (after Campbell *et al.* 1994). Closed circles: divergent plate margin vent sites (Juan de Fuca Ridge, Gorda Ridge, East Pacific Rise, Galapagos Rift). Open circles: convergent plate margin vent sites (Lau Basin, Mariana Trough, Okinawa Trough, Aleutian Arc).

From PARSON, L. M., WALKER, C. L. & DIXON, D. R. (eds), 1995, *Hydrothermal Vents and Processes*, Geological Society Special Publication No. 87, 3–15.

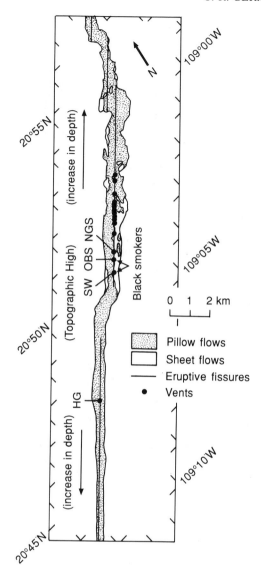

Fig. 2. Location of high-temperature 'black smoker' hydrothermal vents (solid circles) within the axial summit caldera of the East Pacific Rise (after Ballard & Francheteau 1982). Detailed investigations have been completed at the 'Hanging Gardens' (HG), 'Southwest' (SW), 'Ocean Bottom Seismometer' (OBS) and 'National Geographic Smoker' (NGS) vents (von Damm *et al.* 1985).

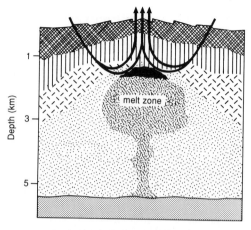

Fig. 3. Schematic cross-section of a fast-spreading section of mid-ocean ridge crest, such as the East Pacific Rise (after Sinton & Detrick 1992). Note the presence of a poorly developed axial summit caldera and an active melt zone beneath the spreading axis.

paper is to present some of these different types of hydrothermal activity in both regional and tectonic contexts and to introduce various aspects of hydrothermal activity and related processes which are currently receiving active investigation within the wider scientific community.

Fast-spreading ridges

The 'type' locality for hydrothermal activity on fast-spreading ridges is the 21°N vent sites on the East Pacific Rise, first discovered in 1979 (Fig. 2). Here, high temperature (350–360°C) fluids, rich in H_2S, Fe, Mn, Cu, Zn and Pb, erupting from the sea bed mix turbulently with cold ambient seawater to give rise to the thick clouds of metal-rich sulphide and oxide minerals which characterize 'black smokers' (Edmond *et al.* 1982; Von Damm *et al.* 1985; Campbell *et al.* 1988). Some of this material is entrained into the rising, buoyant plume and carried hundreds of metres from the seafloor (Lupton *et al.* 1985; Speer & Rona 1989), whereas more of the same material is precipitated in and around the vent orifices to produce large chimney-like constructs of Cu-, Zn-, Pb- and Fe-rich sulphide and Ca sulphate minerals in highly concentrated ore-grade deposits (Haymon & Kastner 1981; Haymon 1983).

On the East Pacific Rise, the spreading plate boundary is sharply defined by a shallow axial summit caldera which sits on top of the axial volcanic ridge and is typically only a few hundreds of metres wide (Fig. 3). Hydrothermally active sources along the East Pacific Rise are typically confined to this narrow band along

these sites of venting are associated with tectonic plate boundaries: fast- and slow-spreading mid-ocean ridges, fracture zones and back-arc spreading centres, although some low-temperature venting associated with oceanic seamounts such as Hawaii and the Society Islands has also been reported (Fig. 1). The purpose of this

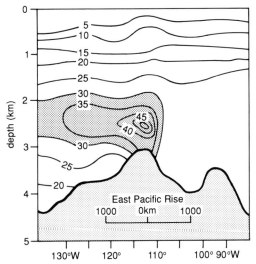

Fig. 4. Two-dimensional cross-section of the dissolved ^{3}He-enriched hydrothermal plume overlying the East Pacific Rise ridge crest at 15°S (after Lupton & Craig 1981). Contours represent ^{3}He:^{4}He ratios expressed as δ-^{3}He, which is a measure of ^{3}He enrichment in seawater relative to the atmosphere.

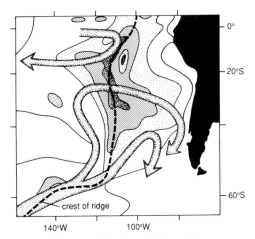

Fig. 5. Increase of hydrothermal Mn and Fe concentrations in uppermost sediments of the East Pacific Rise (after Boström *et al.* 1969) together with independently determined current flow across the ridge crest. Heaviest stipples indicate highest sediment (Al + Mn + Fe)/Al concentrations.

the trend of the ridge axis, although their effects can often be detected much further afield. Because the high temperature fluids erupting from a black smoker are hot, they are buoyant and continue to rise as they mix with local seawater until eventually a level of neutral buoyancy is attained (Lupton *et al.* 1985). The exact height of rise of a hydrothermal plume is a function of both the strength of the hydrothermal source itself and the intensity of the stratification of the water column into which it is emitted (Speer & Rona 1989). For hydrothermal vents along the East Pacific Rise, the height of rise is typically of the order of 100 m, whereas the sea bed within the axial summit caldera is typically only 10–20 m deeper than the shallowest point on the entire ridge crest locally. Thus hydrothermal plumes erupting from the East Pacific Rise typically rise clear of the surrounding, constraining topography and can be dispersed very widely across the entire Pacific Ocean.

Because hydrothermal fluids are extremely enriched in certain key tracers (e.g. dissolved Mn, CH$_4$, ^{3}He), relative to typical oceanic deep waters, chemical anomalies associated with hydrothermal plumes can often be detected at significant distances away from hydrothermal vent sites. One particularly good example is the hydrothermal plume at about 15°S on the East Pacific Rise. Lupton & Craig (1981) demonstrated that ^{3}He enrichments emitted from this portion of the East Pacific Rise could be traced over distances up to 2000 km away from the ridge axis (Fig. 4). Interestingly, the oceanic circulation patterns linked to this dispersion of dissolved hydrothermal ^{3}He coincided almost exactly with contours of anomalous metal enrichments in surface sediments from the flanks of the East Pacific Rise (Fig. 5). Dissolved manganese (Mn) and methane (CH$_4$) are enriched approximately 10^6-fold over ordinary seawater in high-temperature vent fluids (e.g. Welhan & Craig 1983; Von Damm *et al.* 1985) and so, even though these fluids undergo approximately 10^4-fold dilution in buoyant hydrothermal plumes, neutrally buoyant hydrothermal plumes directly above hydrothermal vent sites exhibit dissolved Mn and CH$_4$ concentrations which are approximately 100-fold enriched relative to typical oceanic deep water (Klinkhammer *et al.* 1986; Charlou *et al.* 1987). Because neutrally buoyant plumes overlie a much greater area of the mid-ocean ridge crest than is occupied by active hydrothermal chimneys and mounds, these water column enrichments give geochemists an important and valuable tool with which to prospect for and predict the occurrence of new hydrothermal vent sites.

A good example of the continuous nature of hydrothermal plume signatures along axis was presented by Bougault *et al.* (1990), who used

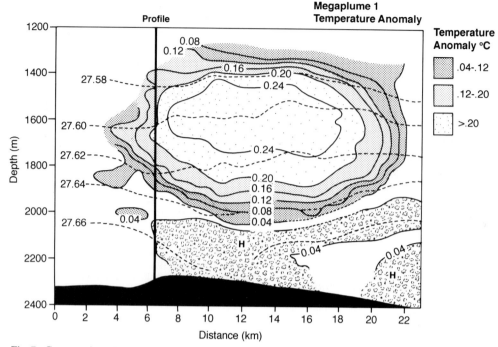

Fig. 7. Cross-section of temperature anomalies associated with MegaPlume I, Cleft Segment, Juan de Fuca Ridge, identified in 1986 (after Baker *et al.* 1987). Note that the maximum height of plume rise is extremely high, rising more than 1000 m above the sea bed.

IFREMER's novel survey sampling system, the Dynamic Hydrocast, to determine the extent of dissolved Mn and CH_4 anomalies along the ridge axis of the East Pacific Rise close to 13°N. As can be seen (Fig. 6), dissolved Mn and CH_4 anomalies were observable within the neutrally buoyant plume throughout the survey, but the highest concentrations coincided with sites of active venting in the region 12°40'–12°50'N.

Hydrothermal 'events'

Repeat study of the physical and chemical properties of hydrothermal vent fluids at both the East Pacific Rise, 21°N and the Mid-Atlantic Ridge, 23°N (see later) have demonstrated that the chemical composition of vent fluids at these sites can remain stable over time-scales of 5–10 years (e.g. Von Damm *et al.* 1985; Campbell *et al.* 1988). This need not always be the case, however. With the increase in experiments designed to identify and locate hydrothermal plumes, and with increasing sophistication in the nature of our detection techniques, a spate of 'new' evolving vent sites has been discovered in recent years.

The first such occurrences reported were the 'megaplumes' identified overlying the Cleft and Vance Segments of the northern Juan de Fuca ridge in 1986 and 1987 (Baker *et al.* 1987, 1989). These features took the form of unusually large hydrothermal plumes with temperature and light-attenuation anomalies which rose up to 1000 m above the sea bed (Fig. 7), i.e. much higher than the stable ('chronic') hydrothermal plumes which had been reported previously (Baker *et al.* 1985). The occurrence of such plumes, which were not continuous but were dissipated into the oceans over a time-scale of weeks to months, was attributed to some massive and sudden outpouring of fluids from a stable subsurface hydrothermal convection cell, perhaps due to tectonic fracturing of the host rock or related magmatic activity. This hypothesis was apparently confirmed by the work of Chadwick *et al.* (1991) and Embley *et al.* (1991), who re-surveyed the area using SeaMarc 1 sidescan sonar and sea bed photography in 1987–1989 and successfully identified areas of fresh volcanic lava flows close to the sites of the two located 'megaplumes' which had not appeared during a previous SeaBeam swath

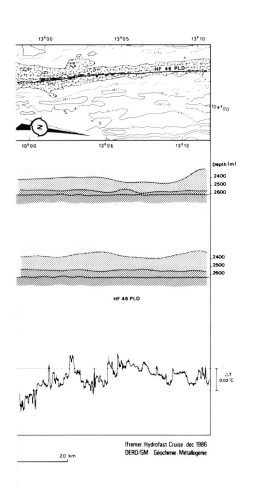

20'

10'

45°00'

50'

40'

44°30'

1). The
nd sites of fresh

bathymetry and SeaMarc 1 survey in 1981–1982 (Fig. 8).

A second recently identified 'hydrothermal event' site is the 9°N area on the East Pacific Rise. Haymon *et al.* (1991) completed a detailed sea bed survey of this section of ridge crest in 1989 using the ARGO deep-tow camera system. Upon their return with the submersible, *Alvin*, in 1991, entirely new lava flows were discovered, some of which were associated with freshly engulfed biological vent communities and an abundance of fresh immature systems in which high-temperature ($\geq 400°C$) vent fluids were erupting directly from holes, cracks and pits in fresh lava flows (Haymon *et al.* 1993). Subsequent radiometric dating of the lavas from one of these sites confirmed initial divers' impressions that these lava flows had only erupted a matter of days or, at most, weeks before the dive programme started (Rubin & MacDougall 1991). Intriguingly, return dives to the same area in March 1992 indicated that a more stable system was already being re-established with an abundant vent community and the construction of large, well-formed sulphide chimneys directly comparable with those surveyed in 1989, before the fresh volcanic eruptions (Haymon *et al.* 1992). This is an important result because it emphasizes the need for continuous monitoring of any particular vent site to ensure that any hydrothermal 'events' are detected. Over a time-scale of about two years between repeat visits (i.e. no return between 1989 and March 1992), the events of April 1991 at 9°N, East Pacific Rise would not have been recognized so readily. For comparison, the 'stable' black smoker vent fields identified at 21°N East Pacific Rise have only been visited in 1979, 1981 and 1985, whereas those at 23°–26°N Mid-Atlantic Ridge have been dived upon and/or imaged in 1985, 1986, 1990 and 1993.

The possibilities for success which might arise from continuous monitoring of any particular segment of ridge crest have been highlighted by the collaboration between NOAA and the US Navy, employing the Navy's SOSUS array to monitor part of the Juan de Fuca Ridge. A dramatic early result from that collaboration has been the seismic detection of a probable dyke intrusion along the CoAxial segment of the Juan de Fuca Ridge in June/July 1993 (Fox *et al.* 1995). Water column studies during July (i.e. during and immediately following the period of seismic activity) detected three separate event plumes, two of which were located directly above fresh sites of lava extrusion and one of which was evidently only a few days old (Baker *et al.* 1995). Increasing study and under-

standing of these remarkable features is rapidly enhancing our understanding of the stability and lifetimes of individual hydrothermal fields. It is to be hoped that similar, perhaps international, collaborations might be possible in future years along other similarly 'accessible' sections of ridge crest, e.g. the Reykjanes Ridge and/or the Azores Triple Junction area, Mid-Atlantic Ridge.

Slow-spreading ridges

Until 1984 it was widely predicted that hydrothermal activity might be restricted to fast-spreading ridges and that at slow-spreading ridges, such as the Mid-Atlantic Ridge, the heat flow would be insufficient to support active high-temperature black smoker vent fields. It is now clear that this is not the case. In 1984, Klinkhammer *et al.* (1985) detected the presence of abundant total dissolvable Mn (TDMn) anomalies along the Mid-Atlantic Ridge rift valley between 11°N and 26°N and argued that these anomalies provided evidence for high-temperature activity along the ridge axis. The following year, detailed water column surveys and deep-tow camera deployments identified the first ever high-temperature vent field on the Mid-Atlantic Ridge, the TAG hydrothermal field at 26°N (Rona *et al.* 1986). Subsequently further studies between the Kane and Atlantis fracture zones have confirmed the presence of at least two further high-temperature vent sites in the region, the 'Snakepit' hydrothermal field at 23°N (Detrick *et al.* 1986) and the 'Broken Spur' vent field at 29°N (Murton *et al.* 1994).

Unlike fast-spreading ridge axes such as the East Pacific Rise, the cross-sectional morphology of a slow-spreading ridge such as the Mid-Atlantic Ridge exhibits much more significant fracturing associated with its segmentation. The neovolcanic axis does not sit at an axial high but tends to be located, often asymmetrically, within a broad (5–10 km wide) axial rift valley which is bounded by steep fault-bounded valley walls which rise up to 1000–2000 m on either side (Fig. 9). The TAG hydrothermal field, 26°N Mid-Atlantic Ridge, is arguably one of the largest submarine hydrothermal fields known. The site comprises a huge sulphide mound approximately 200 m across which rises 50 m from the adjacent sea bed and sits astride a major fault zone, about 1.5 km east of the centre of the rift valley (Thompson *et al.* 1988). The site comprises an estimated 4.5×10^6 tonnes of massive sulphide and would almost certainly be an economically viable source of metal ores if it occurred in a land-based environment (Rona

Rift valley

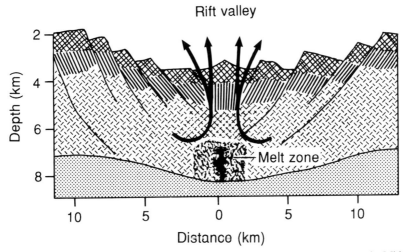

Fig. 9. Schematic cross-section of a slow-spreading section of mid-ocean ridge crest, such as the Mid-Atlantic Ridge (after Sinton & Detrick 1992). Note the well developed axial rift valley and poorly developed melt zone compared with fast-spreading ridge crests (Fig. 3).

et al. 1986). Within the limited field of submarine vent sites studied to date, however, TAG appears to be an exception rather than the rule.

The two other high-temperature vent sites identified in the Kane–Atlantis supersegment, at 23°N and 29°N, exhibit a tectonic setting more similar to the East Pacific Rise vent sites discussed previously. Although the Mid-Atlantic Ridge is characterized by a well-defined and deep axial rift valley, both the Snakepit and Broken Spur vent fields are associated with axial summit caldera fissures which sit on top of narrow neovolcanic ridge axes running parallel to the strike of the rift valley floor (Gente *et al.* 1991; Murton *et al.* 1994).

Whatever the fine detail of the tectonic setting of hydrothermal activity on the Mid-Atlantic Ridge, however, one general rule holds in all cases – the height of rise of any buoyant hydrothermal plume is always less than the bounding heights of the Mid-Atlantic Ridge rift valley. Consequently, any neutrally buoyant hydrothermal plumes overlying sites of active venting tend to be trapped within the Mid-Atlantic Ridge rift valley where it is relatively easy for characteristic hydrothermal anomalies (e.g. TDMn, CH_4, light-attenuation) to be detected (Klinkhammer *et al.* 1986; Nelsen *et al.* 1987; Charlou *et al.* 1991; Nelsen & Forde 1991; Murton *et al.* 1994). In this way, new sites of hydrothermal activity can be relatively easily prospected for along the Mid-Atlantic Ridge. Adopting this procedure, a survey of 17 new segments of the Mid-Atlantic Ridge between 32°

and 40°N in 1992 revealed evidence for up to seven additional, previously unknown, sites of high-temperature hydrothermal fields, including the 'Lucky Strike' field at 37°N (Klinkhammer *et al.* this volume). Combined with the more southerly work of Klinkhammer *et al.* (1985) and Murton *et al.* (1994), this yields an average of one hydrothermal field about every 175 km for the approximately 2500 km of ridge crest surveyed between 11°N and 40°N (Fig. 10).

Back-arc basins

In addition to fast- and slow-spreading mid-ocean ridges, high-temperature submarine hydrothermal activity has also been reported from a range of back-arc spreading centres. Briefly, as one oceanic plate is subducted beneath another, extension within the overriding plate occurs beyond the subduction trench and its associated chain of island arc volcanic rocks (Fig. 11). The first hydrothermal activity associated with such a back-arc spreading centre was reported from the Marianas back-arc system by Craig *et al.* (1987). Extensive submarine hydrothermal activity has also been reported from the Lau Basin, which plays host to one of the most active and intense hydrothermal fields yet found anywhere on the ocean floor (Fouquet *et al.* 1991*a, b*). The chemistry of these vent fluids is very different from that found in any other black smoker hydrothermal systems, with anomalously low pH values (pH 2) and high dissolved metal contents (Mn, Zn, Pb, As and

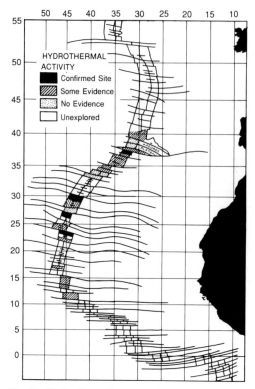

Fig. 10. Hydrothermally surveyed sections of the Mid-Atlantic Ridge crest between 11° and 40°N. Note that only five of the 25 segments examined between 1984 and 1993 have failed to yield any evidence for hydrothermal activity.

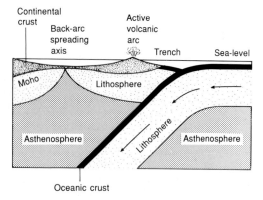

Fig. 11. Schematic cross-section of the subduction zone – island arc volcanic chain – back-arc spreading centre sequence typical of west Pacific marginal basins (e.g. Marianas Trough, Lau Basin). Hydrothermal circulation is frequently observed associated with the back-arc spreading axis.

Shallow vent sites and phase separation

Cu) relative to vent fluids sampled previously from mid-ocean ridge spreading centres. Much of this chemical variability has been attributed to the greenschist facies interactions of seawater with andesitic basement in the Lau Basin rather than the more typical interaction with basaltic basement at mid-ocean ridge spreading centres (Fouquet *et al.* 1991*a, b*).

Systematic plume sampling and near-bottom work is continuously giving rise to evidence for new hydrothermal vent sites throughout the marginal basins of the western Pacific Ocean, e.g. the Woodlark Basin (Binns *et al.* 1993), the North Fiji Basin (Craig & Poreda 1987; Auzende *et al.* 1988; Nojiri *et al.* 1989; Sedwick *et al.* 1990), the Manus Basin (Gamo *et al.* 1993) and the Okinawa and Mariana Troughs (Horibe *et al.* 1986; Horibe & Craig 1987; Ishibashi *et al.* 1988). A more detailed review of the hydrothermal processes active in these Pacific marginal basins than can be afforded here is presented elsewhere (Scott & Binns this volume).

Another important tectonic setting where vent fluid compositions show departures from 'typical' black smoker values is on particularly shallow sections of the mid-ocean ridge crest where confining pressures are no longer sufficient to maintain superheated fluids as a single phase. Crudely, at these reduced pressures phase separation of seawater occurs as the fluids become separated into a dense brine-type phase and a low-salinity, relatively metal-free, vapour-rich phase. To date, only one site of such activity has been unequivocally identified at the ASHES vent field, which sits on top of the Axial Volcano seamount, Juan de Fuca Ridge (Massoth *et al.* 1989). At this location, vent fluids collected in 1986, 1987 and 1988 ranged from a vent emitting relatively gas-rich, chloride-depleted and metal-depleted fluids (Virgin Mound) to a vent emitting relatively gas-depleted, chloride-rich and metal-rich fluids (Inferno Vent). The entire range of compositions observed, separated by only 50–100 m where they erupt from the sea bed, are best decribed by a single hydrothermal fluid undergoing phase separation followed by partial segregation of the vapour and brine phases in the subsurface (Butterfield *et al.* 1990). Although no further sites where phase separation definitely occurs have yet been confirmed, recent discoveries of hydrothermal plume signals overlying shallow areas of the Mid-Atlantic Ridge such as the Azores Triple Junction (Klinkhammer *et al.* this volume) and the Reykjanes Ridge (German *et al.* 1994; Palmer *et al.* this volume) suggest the presence of high-temperature venting in settings as shallow as, or shallower than, the ASHES vent site (i.e. ≤1700 m).

Fracture zones

In addition to hydrothermal discharge associated with mid-ocean ridge spreading centres, chemical anomalies in the water column overlying mid-ocean ridge fracture zones have also been reported. These anomalies typically take the form of large dissolved methane (CH_4) anomalies without any supporting TDMn or associated temperature anomalies. Characteristic sites where these features have been reported are at the 15°20'N fracture zone, Mid-Atlantic Ridge (Charlou et al. 1991) and in the FAMOUS area close to 38°N, Mid-Atlantic Ridge (Charlou et al. 1992). One possible interpretation is that these isolated dissolved CH_4 anomalies are derived not from conventional high-temperature fluid circulation through the basaltic upper layers of the oceanic crust, but are produced, instead, as a by-product from the serpentinization reactions which occur when seawater interacts with ultramafic, deep-crustal rocks – the Fischer Tropsch reaction (Charlou et al. 1991). Only at fracture zones should these rocks, understood to be formed a few kilometres below the seafloor, become exposed at the seawater–crustal interface. A necessary by-product of the hydrating reactions associated with the serpentinization of these rocks is dissolved CH_4 and it is certainly the case that ultramafic rocks are exposed at the sea bed in the immediate vicinity of the dissolved CH_4 anomalies identified at the 15°20'N fracture zone (Bougault et al. 1993).

Hydrothermal plumes and models

A recent and very welcome development in hydrothermal studies has been the application of plume theory to the understanding of the physical dispersion of hydrothermal products through the oceans. One of the most intriguing theories has been the proposal by Helfrich & Battisti (1991), Speer (1989) and Speer & Helfrich (this volume) that buoyant hydrothermal plumes should force a baroclinic vortex which is unstable and would be expected to give rise to isolated 'hydrothermal eddies' that propagate away from their source. A natural development of this theory is the proposal that oceanic hydrothermal plumes might shed isolated baroclinic vortices which are capable of transporting chemically and physically anomalous characteristics long distances away from individual vent sources.

Insufficient data yet exist for such models to be tested rigorously, but the discovery and preliminary characterization of the Steinahóll hydrothermal vent site on the Reykjanes Ridge close to Iceland (German et al. 1994) has provided an ideal natural laboratory where the required, detailed three-dimensional sampling coverage can be obtained. In 1993 a total of 32 stations were occupied for CTD deployments, along with sampling for TDMn, dissolved Si, CH_4 and H_2 analyses, in under 24 hours of ship-time. It is to be hoped, therefore, that the opportunity to exploit this unique environment will not be missed and that the development of hydrothermal plume theory and direct characterization and constraint of that new theory by pertinent field measurements will be able to progress, in concert, in future years.

High temperature 'black smoker' vents and global hydrothermal fluxes

In recent work, Stein & Stein (1994) have re-evaluated the magnitude and age variation of hydrothermal heat flow at mid-ocean ridges. Their study, based on an updated global heat flow model, predicts that no more than 30% (26 ± 4%) of hydrothermal heat flow occurs in crust younger than 1 Ma, whereas most occurs off-axis, away from active spreading centres, in crust up to 70 Ma. This is in agreement with the calculations of Mottl & Wheat (1994), who argue that high-temperature black smoker activity can only remove about 10–40% of the total riverine input of dissolved Mg to the oceans and that more than 70% of the total hydrothermal heat flux must occur on ridge flanks rather than at axial spreading centres. Assuming off-axis heat flow to result from the advection of seawater at low to moderate temperatures, an extremely large volume flux, much greater than that focused through high-temperature vents, would be required. At such a large volume flux (e.g. 2.5×10^{15} kg/a; Mottl & Wheat 1994), even scarcely detectable chemical changes in each litre of ridge flank fluid could dominate the global hydrothermal flux of chemicals to the oceans.

Furthermore, for the 25–30% of hydrothermal heat flow which is concentrated on young oceanic crust, on-axis, not all may be linked to focused 'black smoker' venting. Increasing attention has been paid to the importance of diffuse venting at axial hydrothermal fields (e.g. Rona & Trivett 1992; Schultz et al. 1992; Baker et al. 1993). Diffuse flow fluids may follow tortuous routes along networks of fractures through fresh or altered crust and/or hydrothermal sulphide structures. Consequently, the resultant fluids exiting the seafloor might be expected to exhibit a lower temperature than the primary high-temperature fluid, to be diluted with local (un)altered seawater and to exhibit a chemistry which is not a direct linear mixture of vent fluid and seawater because non-conservative mixing would occur for many of the dissolved species originally present in the high-temperature fluid.

In recent work from the Endeavour and Cleft Segments of the Juan de Fuca Ridge, Ginster *et al.* (1994) have compared summed hydrothermal heat fluxes from discrete black smoker sources with total heat fluxes estimated from measurements made in overlying neutrally buoyant plumes and of diffuse flow at the seafloor. Their estimated heat flux from summed black smoker discharge alone represents only 4–14% of their calculated *total* advective heat flux, but agrees well with the independent work of Bemis *et al.* (1993), who calculated black smoker vent fluxes in the same segments, based on *buoyant* plume investigations. Current best estimates therefore indicate that on-axis hydrothermal circulation may account for only 25% of the total hydrothermal heat flux which escapes through 0–65 Ma oceanic crust and, further, that of this 25%, only about 10% (i.e. just 2.5% of the global total) may be concentrated through focused 'black smoker' type chimneys – the remainder being dispersed through lower temperature, diffuse, on-axis venting.

Of course, studies to date on the partitioning between discrete and diffuse style venting in on-axis mid-ocean ridge settings have been restricted, almost entirely, to one particular subsection of the Juan de Fuca Ridge (Rona & Trivett 1992; Schultz *et al.* 1992; Baker *et al.* 1993; Bemis *et al.* 1993; Ginster *et al.* 1994). However, if the result obtained there proves to be the general case, then the common assumption of purely high-temperature on-axis discharge will need to be revised. In that case, the global fluxes of many non-conservative elements which have been calculated previously, assuming that all on-axis heat flow *is* derived from high-temperature 'black smoker' venting alone, will have been overestimated and those fluxes will need to be recalculated.

Summary

High-temperature hydrothermal venting occurs in a range of tectonic settings along plate boundaries, be they spreading centres, back-arc rifts or transform fracture zones. The regional setting of this high-temperature hydrothermal activity is becoming increasingly well understood as a result of a series of intelligently targetted, systematic, along-axis surveys. To date, however, techniques used in searching for hydrothermal venting have been focused almost exclusively on the detection of high-temperature 'black smoker' type activity. By contrast, global heat flow data suggest that lower temperature, off-axis flow probably constitutes >70% of the global hydrothermal heat and fluid flux (e.g. Stein & Stein 1994) and that even within on-axis flow, as much as 90% of hydrothermal heat flow (and an even greater proportion of volume flux) may be associated with diffuse rather than focused hydrothermal activity (e.g. Ginster *et al.* 1994). Because it is this lower temperature axial hydrothermal flow which gives rise to those environmental 'niches' most commonly associated with vent-specific organsims, the occurrence of axial diffuse-flow venting in the presence of or, perhaps more importantly, in the absence of, high-temperature focused flow may play an important part in the colonization of hydrothermal fields by vent-specific organisms which must migrate along-axis from one hydrothermal 'oasis' to another. This problem of colonisation is attracting the attention of many hydrothermal biologists today. Perhaps one of the most important challenges to be faced by those involved in the physico-chemical study of hydrothermal vents and processes in the immediate future, therefore, should be to understand: (i) what controls the composition and regional variability of lower temperature 'diffuse-flow' along axis and (ii) how to prospect for and unequivocally identify such diffuse flow.

Research for this manuscript was supported by the NERC through grant BRIDGE 15 (CRG), the NOAA Vents Program (ETB), NSF and ONR (GK) and by NATO Collaborative Research Grant CRG910921 (CRG/GK). We thank L. Parson and P. Herzig for their constructive comments during review. IOSDL Contribution No. 95009; Contribution No.1559 of the Pacific Marine Environmental Laboratory.

References

AUZENDE, J. M., EISSEN, J. P., LAFOY, Y., GENTE, P. & CHARLOU, J. L. 1988. Seafloor spreading in the North Fiji Basin (southwest Pacific). *Tectonophysics*, **146**, 317–351.

BAKER, E. T., LAVELLE, J. W., FEELY, R. A., MASSOTH, G. J., WALKER, S. L. & LUPTON, J. E. 1989. Episodic venting of hydrothermal fluids from the Juan de Fuca Ridge. *Journal of Geophysical Research*, **94**, 9237–9250.

——, —— & MASSOTH, G. J. 1985. Hydrothermal

particle plumes over the southern Juan de Fuca Ridge. *Nature*, **316**, 342–344.

——, MASSOTH, G. J. & FEELY, R. A. 1987. Cataclysmic hydrothermal venting on the Juan de Fuca Ridge. *Nature*, **329**, 149–151.

——, FEELY, R. A., THOMSON, R. E. & BURD, B. J. 1995. Hydrothermal event plumes from the CoAxial seafloor eruption site, Juan de Fuca Ridge. *Geophysical Research Letters*, **22**, 147–150.

——, ——, WALKER, S. L.& EMBLEY, R. W. 1993. A method for quantitatively estimating diffuse and discrete hydrothermal discharge. *Earth and Planetary Science Letters*, **118**, 235–249.

BALLARD, R. D. & FRANCHETEAU, J. 1982. The relationship between active sulphide deposition and the axial processes of the mid-ocean ridge. *Marine Technology Society Journal*, **16**, 8–22.

BEMIS, K. G., VON HERZEN, R. P. & MOTTL, M. J. 1993. Geothermal heat flux from hydrothermal plumes on the Juan de Fuca Ridge. *Journal of Geophysical Research*, **98**, 6351–6365.

BINNS, R. A., SCOTT, S. D. & 9 others 1993. Hydrothermal oxide and gold-rich sulfate deposits of Franklin Seamount, western Woodlark Basin, Papua, New Guinea. *Economic Geology*, **88**, 2122–2153.

BOSTRÖM, K., PETERSON, M. N. A., JOENSUU, O. & FISHER, D. E. 1969. Aluminum-poor ferromanganoan sediments on active oceanic ridges. *Journal of Geophysical Research*, **74**, 3261–3270.

BOUGAULT, H., CHARLOU, J. L., FOUQUET, Y. & NEEDHAM, H. D. 1990. Activité hydrothermale et structure axiale des dorsales Est-Pacifique et médio-Atlantique. *Oceanologica Acta Spécial*, **10**, 199–207.

——, —— & 8 others 1993. Fast and slow spreading ridges: structure and hydrothermal activity, ultramafic topographic highs, and CH$_4$ output. *Journal of Geophysical Research*, **98**, 9643–9651.

BUTTERFIELD, D. A., MASSOTH, G. J., McDUFF, R. E., LUPTON, J. E. & LILLEY, M. D. 1990. Geochemistry of hydrothermal fluids from Axial Seamount Hydrothermal Emissions Study vent field, Juan de Fuca Ridge: subseafloor boiling and subsequent fluid-rock interaction. *Journal of Geophysical Research*, **95**, 12 895–12 921.

CAMPBELL, A. C., BOWERS, T. S., MEASURES, C. I., FALKNER, K. K., KHADEM, M. & EDMOND, J. M. 1988. A time series of vent fluid compositions from 21°N, East Pacific Rise (1979, 1981, 1985) and the Guaymas Basin, Gulf of California (1982, 1985). *Journal of Geophysical Research*, **93**, 4537–4549.

——, GERMAN, C. R., PALMER, M. R., GAMO, T. & EDMOND, J. M. Chemistry of hydrothermal fluids from the Escanaba Trough, Gorda Ridge. *In*: MORTON, J. L. *ET AL*. (eds) *Geologic, hydrothermal and biologic studies at Escanaba Trough, Gorda Ridge, Offshore Northern California*. US Geological Survey Bulletin, 2022, 201–221.

CHADWICK, W. W., EMBLEY, R. W. & FOX, C. G. 1991. Evidence for volcanic eruption on the southern Juan de Fuca ridge between 1981 and 1987. *Nature*, **350**, 416–418.

CHARLOU, J. L., BOUGAULT, H., APPRIOU, P., NELSEN, T. & RONA, P. 1991. Different TDM/CH$_4$ hydrothermal plume signatures: TAG site at 26°N and serpentinized ultrabasic diapir at 15°05′N on the Mid-Atlantic Ridge. *Geochimica et Cosmochimica Acta*, **55**, 3209–3222.

——, ——, & FARANAUT 15°N Scientific Party 1992. Intense CH$_4$ plumes in seawater associated to ultramafic outcrops at 15°N on the Mid-Atlantic Ridge. *EOS, Transactions of the American Geophysical Union*, **73** (suppl.), 586.

——, RONA, P. & BOUGAULT, H. 1987. Methane anomalies over TAG hydrothermal field on Mid Atlantic Ridge. *Journal of Marine Research*, **45**, 461–472.

CRAIG, H. & POREDA, P. 1987. Papatua Expedition, Legs V and VI. *Scripps Institute of Oceanography Ref.*, **87–14**, 80 pp.

——, HORIBE, Y., FARLEY, K. A., WELHAN, J. A., KIM, K. R. & HEY, R. N. 1987. Hydrothermal vents in the Mariana Trough: results of the first Alvin dives. *EOS, Transactions of the American Geophysical Union*, **68**, 1531.

DETRICK, R. S., HONNOREZ, J. & ODP LEG 106 SCIENTIFIC PARTY 1986. Drilling the Snake Pit hydrothermal sulfide deposit on the Mid-Atlantic Ridge, lat 23°22′N. *Geology*, **14**, 1004–1007.

EDMOND, J. M., VON DAMM, K. L., McDUFF, R. E. & MEASURES, C. I. 1982. Chemistry of hot springs on the East Pacific Rise and their effluent dispersal. *Nature*, **297**, 187–191.

EMBLEY, R. W., CHADWICK, W., PERFIT, M. R. & BAKER, E. T. 1991. Geology of the northern Cleft segment, Juan de Fuca Ridge: recent lava flows, sea-floor spreading, and the formation of megaplumes. *Geology*, **19**, 771–775.

FOUQUET, Y., VON STACKLEBERG, U. & 9 others 1991*a*. Hydrothermal activity and metallogenesis in the Lau back-arc basin. *Nature*, **349**, 778–781.

——, —— & 9 others 1991*b*. Hydrothermal activity in the Lau back-arc basin: sulfides and water chemistry. *Geology*, **19**, 303–306.

FOX, C. G., RADFORD, W. E., DZIAK, R. P., LAU, T.-K., MATSUMOTO, H. & SCHREINER, A. E. 1995. Acoustic detection of a seafloor spreading episode on the Juan de Fuca ridge using military hydrophone arrays. *Geophysical Research Letters*, **22**, 131–134.

GAMO, T., SAKAI, H. & 7 others 1993. Hydrothermal plumes in the eastern Manus Basin, Bismark Sea: CH$_4$, Mn, Al and pH anomalies. *Deep-Sea Research*, **40**, 2335–2349.

GENTE, P., AUZENDE, J. M., KARSON, J. A., FOUQUET, Y. & MEVEL, C. 1991. An example of a recent accretion on the Mid-Atlantic Ridge: the Snake Pit neovolcanic ridge (MARK area, 23°22′N). *Tectonophysics*, **190**, 1–29.

GERMAN, C. R., BRIEM, J. & 10 others 1994. Hydrothermal activity on the Reykjanes Ridge: the Steinahóll vent-field at 63°06′N. *Earth and Planetary Science Letters*, **121**, 647–654.

GINSTER, U., MOTTL, M. J. & VON HERZEN, R. P. 1994. Heat flux from black smokers on the Endeavour and Cleft segments, Juan de Fuca

Ridge. *Journal of Geophysical Research,* **99**, 4937–4950.

HAYMON, R. M. 1983. Growth history of hydrothermal black smoker chimneys. *Nature,* **301**, 695–698.

—— & KASTNER, M. 1981. Hot spring deposits on the East Pacific Rise at 21°N: preliminary description of mineralogy and genesis. *Earth and Planetary Science Letters,* **53**, 363–381.

——, ——, EDWARDS, M., CARBOTTE, S. M., WRIGHT, D. & McDONALD, K. C. 1991. Hydrothermal vent distribution along the East Pacific Rise crest (9°09′–54′N) and its relationship to magmatic and tectonic processes on fast spreading mid-ocean ridges. *Earth and Planetary Science Letters,* **104**, 513–534.

——, FORNARI, D. & 16 others 1992. Dramatic short-term changes observed during March '92 dives to April '91 eruption site on the East Pacific Rise (EPR) crest, 9°45–52′N. *EOS, Transactions of the American Geophysical Union,* **73** (suppl.), 524.

——, —— & 13 others 1993. Volcanic eruption of the mid-ocean ridge along the East Pacific Rise crest at 9°45–52′N: direct submersible observations of seafloor phenomena associated with an eruption event in April, 1991. *Earth and Planetary Science Letters,* **119**, 85–101.

HELFRICH, K. R. & BATTISTI, T. M. 1991. Experiments on baroclinic vortex shedding from hydrothermal plumes. *Journal of Geophysical Research,* **96**, 12 511–12 518.

HORIBE, Y. & CRAIG, H. 1987. Papatua Expedition III: hydrothermal vents in the Mariana Trough and Kagoshima Bay (Sakurajima Volcano). *EOS, Transactions of the American Geophysical Union,* **68**, 100.

——, KIM, K. R. & CRAIG, H. 1986. Hydrothermal methane plumes in the Mariana back-arc spreading center. *Nature,* **324**, 131–133.

ISHIBASHI, J. I., GAMO, T., SAKAI, H., NOJIRI, Y., IGARASHI, G., SHITASHIMA, K. & TSUBOTA, H. 1988. Geochemical evidence for hydrothermal activity in the Okinawa Trough. *Geochemical Journal,* **22**, 107–114.

KLINKHAMMER, G., ELDERFIELD, H., GREAVES, M., RONA, P. & NELSEN, T. 1986. Manganese geochemistry near high-temperature vents in the Mid-Atlantic Ridge Rift Valley. *Earth and Planetary Science Letters,* **80**, 230–240.

——, RONA, P., GREAVES, M. J. & ELDERFIELD, H. 1985. Hydrothermal manganese plumes over the Mid-Atlantic Ridge rift valley. *Nature,* **314**, 727–731.

——, WILSON, C., CHIN, C. & GERMAN, C. R. Venting from the Mid-Atlantic Ridge at 37°17′N: the Lucky Strike hydrothermal site. *This volume.*

LUPTON, J. E. & CRAIG, H. 1981. A major helium-3 source at 15°S on the East Pacific Rise. *Science,* **214**, 13–18.

——, DELANEY, J. R., JOHNSON, H. P. & TIVEY, M. K. 1985. Entrainment and vertical transport of deep ocean water by buoyant hydrothermal plumes. *Nature,* **316**, 621–623.

MASSOTH, G. J., BUTTERFIELD, D. A., LUPTON, J. E.,

McDuff, R. E., LILLEY, M. D. & JONASSON, I. R. 1989. Submarine venting of phase-separated hydrothermal fluids at Axial Volcano, Juan de Fuca Ridge. *Nature,* **340**, 702–705.

MOTTL, M. J. & WHEAT, C. G. 1994. Hydrothermal circulation through mid-ocean ridge flanks: fluxes of heat and magnesium. *Geochimica et Cosmochimica Acta,* **58**, 2225–2237.

MURTON, B.J., KLINKHAMMER, G. & 11 others 1994. Direct evidence for the distribution and occurrence of hydrothermal activity between 27–30°N on the Mid-Atlantic Ridge. *Earth and Planetary Science Letters,* **125**, 119–128.

NELSEN, T. A. & FORDE, E. B. 1991. The structure, mass and interactions of the hydrothermal plumes at 26°N on the Mid-Atlantic Ridge. *Earth and Planetary Science Letters,* **106**, 1–16.

——, KLINKHAMMER, G. P., TREFRY, J. H. & TROCINE, R. P. 1987. Real-time observation of dispersed hydrothermal plumes using nephelometry: examples from the Mid-Atlantic Ridge. *Earth and Planetary Science Letters,* **81**, 245–252.

NOJIRI, Y., ISHIBASHI, J., KAWAI, T. & SAKAI, H. 1989. Hydrothermal plumes along the North Fiji Basin spreading axis. *Nature,* **342**, 667–670.

PALMER, M. R., LUDFORD, E. M., GERMAN, C. R. & LILLEY, M. D. Dissolved methane and hydrogen in the Steinahóll hydrothermal plume, 63°N, Reykjanes Ridge, this volume.

RONA, P. A. & TRIVETT, D. A. 1992. Discrete and diffuse heat transfer at ASHES vent field, Axial Volcano, Juan de Fuca Ridge. *Earth and Planetary Science Letters,* **109**, 57–71.

——, KLINKHAMMER, G., NELSEN, T. A., TREFRY, J. H. & ELDERFIELD, H. 1986. Black smokers, massive sulphides and vent biota at the Mid-Atlantic Ridge. *Nature,* **321**, 33–37.

RUBIN, K. H. & MacDOUGALL, J. D. 1991. Fine chronology of recent mid-ocean ridge eruptions on the southern JDF and 9°N EPR from ^{226}Ra–^{230}Th–^{238}U and ^{210}Po–^{210}Pb disequilibrium. *EOS, Transactions of the American Geophysical Union,* **72**, 231.

SCHULTZ, A., DELANEY, J. R. & McDUFF, R. E. 1992. On the partitioning of heat flux between diffuse and point-source seafloor venting. *Journal of Geophysical Research,* **97**, 12 229–12 314.

SCOTT, S. D. & BINNS, R. A. Hydrothermal processes and contrasting styles of mineralization in the western Woodlark and eastern Manos basins of the western Pacific, this volume.

SEDWICK, P. N., GAMO, T. & McMURTRY, G. M. 1990. Manganese and methane anomalies in the North Fiji Basin. *Deep-Sea Research,* **37**, 891–896.

SINTON, J. M. & DETRICK, R. S. 1992. Mid-Ocean Ridge magma chambers. *Journal of Geophysical Research,* **97**, 197–216.

SPEER, K. G. 1989. A forced baroclinic vortex around a hydrothermal plume. *Geophysical Research Letters,* **16**, 461–464.

—— & HELFRICH, K. R. Hydrothermal plumes: a review of flow and fluxes, this volume.

—— & RONA, P. A. 1989. A model of an Atlantic and

Pacific hydrothermal plume. *Journal of Geophysical Research,* **94**, 6213–6220.

SPIESS, F. N. & RISE PROJECT GROUP 1980. East Pacific Rise: hot springs and geophysical experiments. *Science,* **207**, 1421–1433.

STEIN, C. A. & STEIN, S. 1994. Constraints on hydrothermal heat flux through the oceanic lithosphere from global heat flow. *Journal of Geophysical Research,* **99**, 3081–3095.

THOMPSON, G., HUMPHRIS, S. E., SCHROEDER, B., SULANOWSKA, M. & RONA, P. A. 1988. Active vents and massive sulfides at 26°N (TAG) and 23°N (Snakepit) on the Mid-Atlantic Ridge. *Canadian Mineralogist,* **26**, 697–711.

VON DAMM, K. L., EDMOND, J. M., GRANT, B., MEASURES, C. I., WALDEN, B. & WEISS, R. F. 1985. Chemistry of submarine hydrothermal solutions at 21°N, East Pacific Rise. *Geochimica et Cosmochimica Acta,* **49**, 2197–2220.

WELHAN, J. A. & CRAIG, H. 1984. Methane, hydrogen and helium in hydrothermal fluids at 21°N on the East Pacific Rise. *In*: RONA, P. A. *ET AL.* (eds) *Hydrothermal Processes at Seafloor Spreading Centres*. Nato Conference Series, **4**. Plenum Press, New York, 391–409.

Geological setting of high-temperature hydrothermal activity and massive sulphide formation on fast- and slow-spreading ridges

SERGEY G. KRASNOV, IRINA M. POROSHINA &
GEORGIY A. CHERKASHEV

*Institute of Geology and Mineral Resources of the Ocean, 1 Angliysky Avenue,
190121 St Petersburg, Russia*

Abstract: Geological features which control massive sulphide formation on the fast-spreading East Pacific Rise (EPR) and slow-spreading Mid-Atlantic Ridge (MAR) can be specified based on data from the Sevmorgeologija Association (St Petersburg) research cruises and results of other studies. Wide crestal surfaces of undisturbed axial volcanoes and the presence of axial grabens that indicate voluminous subsurface magma chambers represent the sites most favourable for sulphide formation on the EPR. The elevation of rift segments and distance from major ridge axis discontinuities are less important for sulphide formation. Sites of localized magma delivery from subcrustal zones, as indicated by Mg anomalies in basalts, may be favourable. However, at least one site near 21°30'S on the EPR shows evidence of along-axial magma penetration from the central part of the rift segment to its tip, resulting in a lateral shift of hydrothermal activity with time. Higher crustal permeability for magma is required for the formation of subsurface chambers which initiate hydrothermal convection on the magmatically less active MAR. Rift valley marginal faults, and especially their intersections with minor transverse dislocations, locally control hydrothermal activity where magma laterally penetrates from the extrusive zones of the adjacent rift segments.

From 1985 to 1994, 14 research cruises to the East Pacific Rise (EPR) and Mid-Atlantic Ridge (MAR) were carried out by the Sevmorgeologija Association (St Petersburg), formerly part of the Ministry of Geology of the USSR. They were mostly devoted to prospecting for new hydrothermal fields and massive sulphide deposits. The EPR was studied between 13°N and the equator, and between 21°10' and 22°40'S (Fig. 1), and the MAR between 12° and 26°N.

In most of the cruises, the work included: (1) bathymetric (conventional single-beam echo-sounding) and magnetic surveys of the axial parts of ridges (30–100 km profiles across the ridge at intervals of 5–10 km); (2) hydrocast measurements (using a standard Neil Brown system) with water sampling over the ridge axis at 5–10 km intervals along the ridge, 300–600 m above the bottom, followed by onboard analysis of dissolved and particulate Fe, Mn, Zn and Cu by atomic absorption spectrometry; and (3) sediment sampling within 10 km of the EPR axis, or inside the MAR rift valley, at intervals of 5–10 km along the ridge, with subsequent geochemical and petrographic investigations.

More detailed sampling, sidescan sonar (30 and 100 kHz) and photo/TV surveys in zones of anomalous bottom water (increased temperature, light attenuation and heavy metal concentrations) led to the discovery of several previously unknown sites of hydrothermal activity and massive sulphide formation. The hydrothermal sites at the EPR near 13°N and 21°30'S studied in detail by Fouquet *et al.* (1988), Renard *et al.* (1985), Marchig *et al.* (1988) and Tufar & Näser (1991) were also revisited.

The data on the distribution of high-temperature hydrothermal fields, both previously known and newly found, allowed us to specify the geological settings most favourable for massive sulphide formation in zones of fast (EPR) and slow (MAR) spreading.

East Pacific Rise

The fast-spreading EPR is known to be an area of very high hyrothermal activity (Skornyakova 1965; Boström & Peterson 1966). High-temperature hydrothermal vents are probably the main source of Fe in metalliferous sediments (Von Damm *et al.* 1985). Spatial variations in the intensity of hydrothermal activity on the scale of the whole EPR are reflected in the distribution of these sediments (Fig. 1). The huge field of metalliferous sediments in the southern part of the EPR, corresponding approximately to the zone of Fe concentration above 10% in surficial sediments in Fig. 1, has previously been studied

From PARSON, L. M., WALKER, C. L. & DIXON, D. R. (eds), 1995, *Hydrothermal Vents and Processes*,
Geological Society Special Publication No. 87, 17–32.

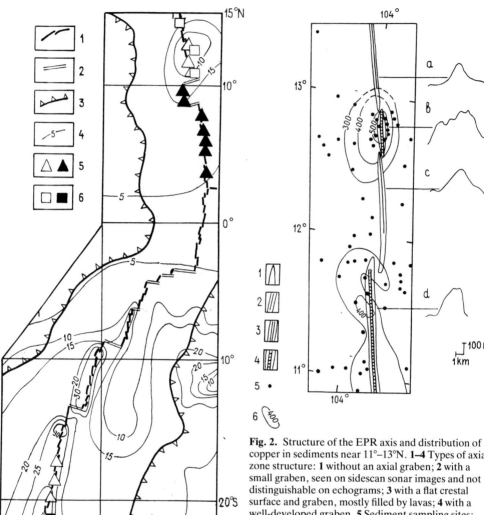

Fig. 2. Structure of the EPR axis and distribution of copper in sediments near 11°–13°N. **1–4** Types of axial zone structure: **1** without an axial graben; **2** with a small graben, seen on sidescan sonar images and not distinguishable on echograms; **3** with a flat crestal surface and graben, mostly filled by lavas; **4** with a well-developed graben. **5** Sediment sampling sites; **6** copper concentrations (in ppm) in surficial sediments, recalculated on carbonate-free and basalt glass-free basis; **a**, **b**, **c** and **d** examples of bathymetric cross-sections. Zones of copper enrichment in sediments of hydrothermal origin coincide with rift segments having a well-developed graben.

Fig. 1. Distribution of massive sulphide deposits and metalliferous sediments along the East Pacific Rise (modified from Krasnov *et al.* 1992a). **1** Segments of the EPR spreading axis; **2** transform faults; **3** outer contours of the EPR slopes; **4** total Fe+Mn concentrations (%) in surficial sediments, recalculated on the carbonate-free basis; **5** massive sulphide deposits in the EPR axis; and **6** massive sulphide deposits on off-axis seamounts. Sulphide deposits discovered in Sevmorgeologija research cruises are shown as closed symbols, those from other studies as open symbols.

in detail (Smirnov 1979; Dymond 1981). The smaller fields in the northern part of the EPR (between 9° and 15°N) were first described and mapped during the Sevmorgeologija cruises (Cherkashev 1990).

The higher hydrothermal activity of the southern part of the EPR compared with the northern part is probably related to the superfast spreading (up to 17 cm/a) of the southern EPR, resulting in more intense volcanism. The absence of metalliferous sediments near the equator may be an artifact of higher terrigenous

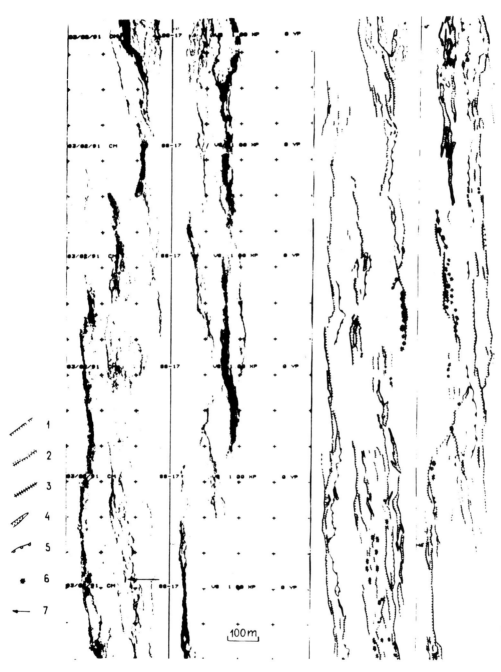

Fig. 3. Sidescan sonar images of the EPR axial zone between 12°46′ and 12°47′N and structural interpretations. **1** Main faults; **2** faults bounding tilted blocks; **3** crests of highs; **4** fissures; **5** concave bends; **6** hydrothermal structures; **7** acoustic shadow from a separate hydrothermal structure on the sidescan image.

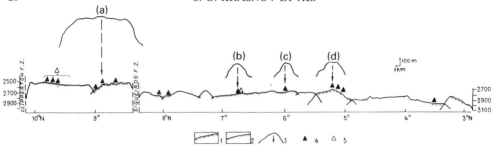

Fig. 4. Along-axis bathymetric profile of the EPR between 3° and 10°N and cross-sectional profiles of the axial zone (**a–d**). **1** Segments with a well-developed axial graben; **2** segments with a small graben; **3** cross-sectional profiles and their positions shown by arrows; **4** massive sulphide deposits discovered in the Sevmorgeologija cruises; and **5** massive sulphides known from other studies (see Krasnov *et al.* 1992*a* for review). Massive sulphides are present within the EPR axial zone segments with more or less pronounced axial grabens. General bathymetric levels of segments and distances from major ridge axis discontinuities are not important for sulphide localization.

sedimentation rates in the equatorial zone of the Pacific (Lisitsyn 1991), leading to increased dilution of the hydrothermal component in sediments by clays.

Several features of the tectonic control of hydrothermal fields on the EPR were formulated previously by Ballard *et al.* (1984), Macdonald *et al.* (1986) and Poroshina 1987. In general, the main sites of hydrothermal activity are known to be located at the shallowest ridge segments of the EPR axial zone, most often in their central parts, and often distant from the transform faults and other ridge axis discontinuities. In cross-section, such segments are wide trapeziform or dome-like. As seen from the results of seismic studies, the presence of a well-developed graben over these segments reflects the existence of a voluminous subsurface magma chamber (Macdonald *et al.* 1986).

The main features of the distribution of hydrothermal fields in the EPR axis between 11° and 13°N studied in detail in our cruises meet most of these criteria (Fig. 2). The area of maximum hydrothermal activity between 12°37′ and 12°54′N corresponds to a single rift segment, bounded by small overlapping spreading centres (OSCs). This segment is remarkable because of the structural uniformity of the axial rise and of its neovolcanic zone. In cross-section it forms a clearly distinguished asymmetrically uplifted axial block, 300 m high and 8 km wide at its base (Fig. 2b). An axial graben 300–500 m wide and up to 50 m deep is developed within its flattened 1 km wide crestal surface.

A high-frequency sidescan sonar survey of the crestal surface revealed the continuous development of the graben throughout the segment. The graben is cut by small (100–200 m amplitude) right-hand displacements into subsegments

3–4 km long (Fig. 3). No prominent along-axial variations were noticed in the inner structure of the graben. The sidescan sonar images show the continuous development of cone-shaped structures forming chains within each of the subsegments (Fig. 3). The subsequent photo/TV survey and dredging confirmed the hydrothermal origin of these structures.

The morphology of the next EPR axial zone segment, lying north of the 12°54′N OSC, is very different. The triangular-shaped axial volcano has no characteristic graben (Fig. 2a). South from the 12°37′N OSC the axial block also becomes narrower in cross-section (Fig. 2c).

Our data on the distribution of copper (Fig. 2) and other hydrothermal metals in sediments on the flanks of the axial zone indicate the decreased hydrothermal activity of segments north of 12°54′ and south of 12°37′N. The methane and dissolved manganese distributions in waters above the EPR axis also show sharp decreases in these hydrothermal components north of the 12°54′N OSC (Bougault *et al.* 1993). According to these workers, any sign of hydrothermal input into near-bottom waters disappears south of about 11°57′N. This latitude is the southern limit of a small graben developed within the southern segment.

According to our data on sediment composition (Fig. 2) and the results of temperature profiling (Crane *et al.* 1988), the segment lying south of 11°42′N, which has a trapeziform cross-section and a well developed graben, is again hydrothermally active.

The site of maximum hydrothermal activity at the northern EPR at 12°50′N coincides with a positive anomaly of Mg concentration in basalts with respect to the adjacent rift segments (Langmuir *et al.* 1986).

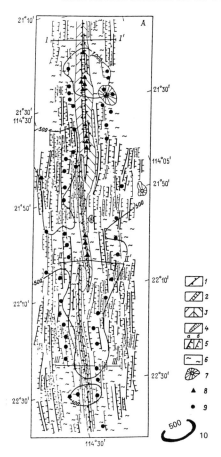

Fig. 5. Structural scheme of the EPR axial zone between 21°10′ and 22°40′S. **1** Crest of the axial high; **2** axial graben; **3** slopes of the linear axial volcano; **4** deeps in overlapping zones; **5** faults with amplitudes (**a** above 100 m; **b** below 100 m); **6** flattened surfaces; **7** off-axial volcanoes; **8** massive sulphide deposits; **9** sites of sediment sampling; **10** copper distribution in sediments (in ppm), recalculated on a carbonate-free and basalt glass-free basis. I–I′, II–II′, III–III′, positions of bathymetric profiles illustrated in Fig. 7.

Our surveys carried out between the 12°50′N field and the equator resulted in the discovery of previously unknown hydrothermal fields at 9°38′–9°44′, 8°53′, 7°53′–7°54′, 6°46′–6°56′, 6°13′–6°15′, 5°57′–6°05′, 5°13′–5°18′, 5°04′–5°10′ and 3°32′–3°37′N. The structural settings of these sites are diverse. Differences are revealed in general bathymetric levels of the EPR axial segments containing sulphides, in the morphology of their axial ridges and in the structure of their neovolcanic zones (Fig. 4). The common features of all the sites are the wide (more than 1 km) flattened crestal surface of the axial ridge and the presence of an axial graben.

The grabens are considerably different. At 5°59′–6°00′ and 6°46′–6°48′N they are depressions 10–15 m deep half-filled by lavas. Between 4°58′ and 5°22′N a caldera more than 50 m deep is bounded by well-developed bathymetric steps. Fresh lava flows are recognized on sidescan sonar images and lava lakes are ubiquitous. Sulphide deposits are often located close to fissures.

According to the results obtained in 1987–1988 during the RV *Geolog Fersman* Leg 4 (Krasnov *et al.* 1988, 1992*b*), the setting of hydrothermal fields between 21°20′ and 22°40′S also corresponds in general to those referred to above.

Of the three spreading segments studied (Figs 5 and 6), the northern, most hydrothermally active and sulphide-bearing segment (lying north of the 21°44′S overlapping spreading zone) is the most elevated and is relatively uniform throughout its length in terms of morphology. It is distinguished by the presence of a prominent, massive axial ridge, 250–400 m high and 5–7 km wide at its base, which is trapeziform in cross-section (Fig. 7), and a well-developed central graben 700–1000 m wide and 40–90 m deep (Fig. 8).

The axial part of the southern segment (lying southward of the 22°07′S OSC) is bathymetrically deeper and triangular in cross-section (Figs 6 and 7). The graben is 100–300 m wide but only 10–20 m deep and can only be locally traced. The segment lacks any sign of hydrothermal activity. Structurally, the segment between 21°44′ and 22°07′S combines elements of both its northern and the southern neighbours. The bottom waters over this segment show evidence of moderate hydrothermal activity and massive sulphides were discovered near 22°S (Fig. 6).

Variations exist between different parts of the northern segment that are mainly seen in the structural details of the neovolcanic zone as revealed by sonar studies and near-bottom echosounding. The southern part of the segment between 20°33′ and 21°42′S is characterized by the shallowest bathymetry and the deepest axial graben represented by a single uninterrupted depression (Fig. 8c).

North of 21°33′S, the crest of the axial ridge and the graben widen. The graben between 21°24′ and 21°33′S has the form of a shallow flat-bottomed depression cut by a narrow axial cleft (Fig. 8b). Detailed surveys carried out by French and German expeditions discovered extensive hydrothermal fields with massive sulphide deposits in the axial graben near 21°26′S (Renard *et al.* 1985; Tufar & Näser 1991).

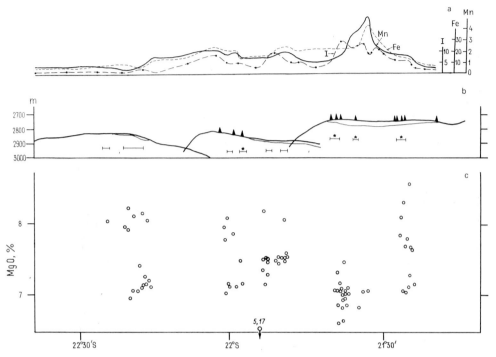

Fig. 6. Results of hydrological studies (**a**) above the EPR axis, (**b**) along-axial bathymetric profile and (**c**) Mg distribution in basalts between 21°15 and 22°40′S. (**a**) Distribution of dissolved Mn and particulate Fe concentrations in mkg/l average for the lowest 300 m of the water column; I is the maximum light attenuation (in m^{-1}) in the same water layer. (**b**) Bathymetric position of the EPR crest (thick lines) and graben floor (thin lines]. Closed triangles (**b**) are sulphide deposits from photo/TV profiling; horizontal bars and stars are lines of dredging and massive sulphide samples, respectively. (**c**) MgO concentrations in basalt samples illustrated with respect to latitude.

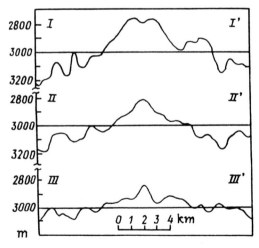

Fig. 7. Bathymetric profiles across different segments of the EPR axial zone between 21°S and 22°30′S. Positions of profiles given in Fig. 5.

The nearest flanks of the axial zone (up to 15 km from the axis) between 21°24′ and 21°33′S are also flattened. Few ridge-parallel faults have been observed here (Fig. 5). A large conical off-axis volcano was discovered at 21°30′S, 10 km east from the rift axis (Krasnov *et al.* 1988), which contains massive sulphides in its summit caldera.

Between 21°20′ and 21°24′S the graben floor is intensively fractured (Fig. 8a). North of 21°20′S, the axial ridge crestal surface and its central graben narrow again.

The most primitive Mg-rich basalts, typical of the EPR axial zone in the region studied (Fig. 6), were not found in the southern part of the segment but more Fe-rich varieties were sampled. According to temperature data, light attenuation and Fe and Mn distribution in the bottom waters (Krasnov *et al.* 1992*b*), this part of the northern segment is the site of the most intense hydrothermal activity (Fig. 6).

However, results of our photo/TV survey show the presence of massive sulphide deposits

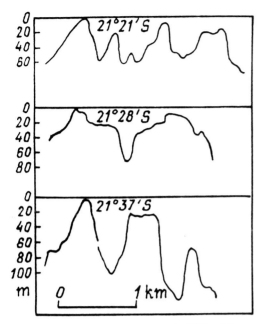

Fig. 8. Bathymetric profiles across the EPR axial graben near 21°30′S based on near-bottom echosounding results.

throughout the segment in the axial graben, where they are mostly confined to areas covered by collapsed sheet lava flows (Figs 9 and 10). The sediments sampled from the ridge flanks are homogenously rich in base metals throughout the region studied on a carbonate-free basis but recalculation on the basalt glass-free basis (Cherkashev 1990) allows us to define the zone of maximum metal concentration (Fig. 5). It is situated in the central part of the northern segment and does not coincide with the present site of maximum hydrothermal activity near its southern end (Fig. 6).

Mid-Atlantic Ridge

Until recently high-temperature vents and associated massive sulphides had been known on the MAR only within the TAG (26°N) and MARK (23°N) hydrothermal fields. In the course of the recent British (Duckworth *et al.*, German *et al.*, Murton *et al.* this volume), American (Langmuir *et al.* 1993) and Russian expeditions, hot vents and/or massive sulphides were also discovered at 37°N, 29°N, 24°30′N and 15°N (Fig. 11). The results of our studies at four of the six hydrothermal fields presently known on the MAR enables us to consider the general trends of their tectonic control.

The development of the axial part of the slow-spreading MAR is a less regular process than the development of the EPR axis. Segments of the MAR spreading zone appear morphologically as neovolcanic ridges representing clusters of small volcanoes (Smith & Cann 1990). These can develop at different parts of the rift valley floor. Jumps of the active spreading centres across the valley are common and result in the formation of new ridges parallel to the old ones within the same segments (Aplonov & Popov 1991).

Until recently, the MAR hydrothermal fields had only been found in two types of volcanic setting: (1) within well-developed neovolcanic ridges at the inner part of the rift valley floor (MARK, 29°N); and (2) at marginal parts of the rift valley at the base of its eastern wall – in these cases no clear neovolcanic ridge is developed, although young volcanics exist (TAG, 24°N).

The MARK hydrothermal field lies 30 km south of the Kane transform fault within an elongated graben-like structure in the crestal part of an unusually large, massive extrusive high in the central part of the rift valley (Karson & Brown 1988). The size of this high reflects exceptionally voluminous volcanism of the rift segment adjacent to the Kane Transform Fault. Aplonov *et al.* (1992) explained this redundant volcanism by damming of the northward directed subaxial asthenospheric flow by the transform fault. The setting of the MARK field in the graben of the high closely resembles that typical of EPR hydrothermal fields situated in crestal grabens of the axial rise.

The TAG field lies within the segment of the rift valley between small non-transform offsets at 26°03′ and 26°13′N (Fig. 12). It is confined to a bathymetric step, 6 km long and 3 km wide, at the base of the eastern wall of the valley. The most important features seen on sonar images within the step are several young lava fields. They represent volcanic centres of the relocated spreading zone which occupied its present position after an eastward jump 50 000 years ago (Aplonov & Popov 1991) and act as sulphide-controlling structures (Rona *et al.* 1993). The active TAG smoker is located within the southern volcanic centre.

The volcanic mounds lie in one line with a high, steep wall directly south of the TAG bathymetric step and corresponding to the rift valley marginal fault (Fig. 12). To the north, the group of young volcanic mounds which marks the present position of the spreading zone is continued outside the step by a line of more prominent volcanic mounds comprising the extrusive zone of the adjacent rift valley segment.

One more important feature controlling the

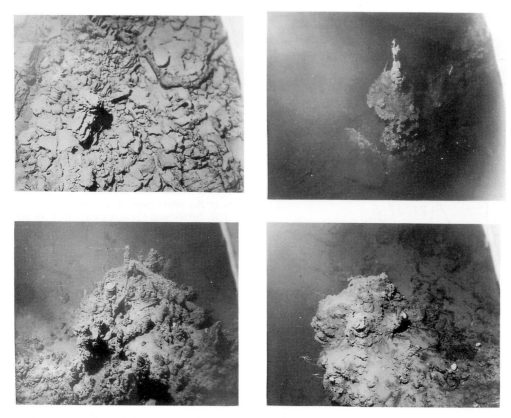

Fig. 9. Lava morphology typical of sites of hydrothermal activity and massive sulphide bodies in different stages of their degrading in the EPR axial graben near 21°30'S. An area of about 10 m^2 is illustrated in each photograph. (**a**) Platy lava talus and fragments of lava pillars on the bottom of a collapsed lava lake. A pillar in the centre of the picture can be identified by its shadow. (**b**) A sulphite chimney in the zone of continuing high-temperature hydrothermal activity. The chimney and surrounding basalts are covered by dark sulphide precipitate. (**c**) Fallen chimneys on the surface of a 3 m high non-active massive sulphide body. (**d**) A highly degraded massive sulphide body mostly covered with a light hydroxide crust. Disintegrating sulphides form dark patches on the seafloor at the base of the body.

hydrothermal activity is the zone of small transverse displacements passing directly through the TAG field at latitude 26°08'– 26°09'N, previously described by Karson & Rona (1990).

At 24°30'N, massive sulphides were dredged at the base of the eastern rift valley wall in a tectonic setting similar to that of the TAG field (Fig. 13). No constructional volcanic features can be seen in the relief at the site of sulphide formation. However, a neovolcanic high lies directly north of the site in line with the valley wall lying immediately south of the site.

The sulphide deposits discovered recently near 15°N (Krasnov *et al.* this volume) also fall into the second group. They are situated within an uplifted block of the rift valley adjacent to its eastern wall. The block causes a westward curve of the wall leading to a narrowing of the valley (Fig. 14).

The largest group of deposits is localized at the base of a narrow (200–250 m) meridional high within the uplifted block (Fig. 15). This high is directly continued by a normal extrusive zone of the valley floor southward of the uplifted block. Ultramafic rocks – peridotites (harzburgites), pyroxenites and serpentinites – are most common within the study area, especially in its eastern part, where they occupy the entire valley wall. Basalts are less common, developed mostly within the meridional high.

The second group of deposits is situated 5.5 km eastward at the base of the valley wall. Both sites are close to the zone of transverse

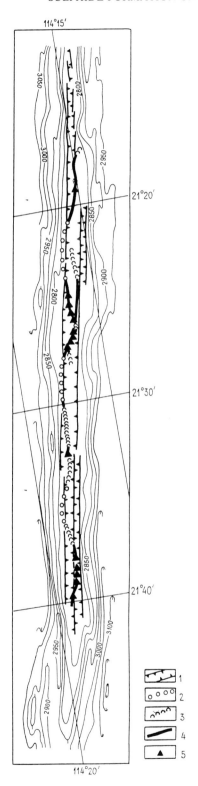

Fig. 10. Distribution of massive sulphide deposits and various morphological types of lavas along the axial part of the EPR spreading segment near 21°30'S according to TV/photo profiling. **1** Graben walls. **2–4** Dominant lava morphologies: **2** pillow; **3** lobate; **4** sheet flows and collapse structures; **5** sulphide deposits. Sulphides mainly associate with sheet flows and collapsed lava lakes within the axial graben.

linear structures, cutting the uplifted block in its central part and probably marking a high-order discontinuity.

The position of the Lucky Strike hydro-thermal field discovered on a seamount within the rift valley near 37°N (Langmuir *et al.* 1993) differs greatly from those of all the other vent fields known in the Atlantic. The features of its tectonic control may be atypical due to the influence of the Azores hot-spot, reflected in the compositions of some of the Lucky Strike seamount basalts (Langmuir *et al.* 1993). We therefore do not consider the Lucky Strike field further in this discussion.

Discussion

The results of our studies confirm the almost ubiquitous development of high-temperature sulphide-forming hydrothermal vents on the EPR. Taking into consideration the results of other surveys (Uchupi *et al.* 1988; Haymon *et al.* 1991), we conclude that massive sulphides are found within most of the EPR segments when studied in adequate detail. However, their quantities are very different within different segments.

Shallow, active magma chambers are needed to maintain high-temperature hydrothermal circulation in the upper crust (Cann & Strens 1982). Morphological features reflecting the presence of such large chambers at minimum depths beneath the ridge axis indicate the most hydrothermally active sites. The isostatic uplift of the whole axial rise above the zone of heated and partially melted rocks is reflected in trapezi-form or dome-like cross-sections (Macdonald 1989). The axial graben is a caldera-type structure forming during periods of emptying of the magma chamber. The largest ridge axis discontinuities separating different spreading segments act as heat sinks and cause gaps in the subaxial linear chamber or deepening of its roof.

We found no sign of hydrothermal activity in the OSCs or transform zones. However, our results do not prove any significant role of distance from major discontinuities in control-ling hydrothermal activity within the segments

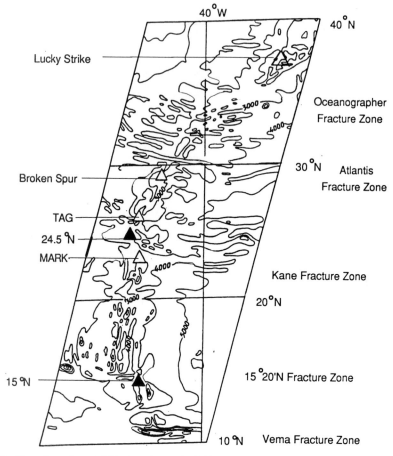

Fig. 11. Position of massive sulphide deposits on the Mid-Atlantic Ridge. Deposits discovered in Sevmorgeologija cruises are shown as closed triangles, the others as open triangles.

themselves. The smallest discontinuities (deviations in axial linearity, DEVALS) not associated with gaps in axial magmatic chambers may be favourable for high-temperature hydrothermal activity.

Hydrothermal vents and massive sulphides are not generally associated with basalts characterized by any definite compositional features. The positive Mg anomaly in basalts of the EPR axial zone at the site of intense hydrothermal activity at 12°50′N (Langmuir *et al.* 1986) probably reflects the rapid delivery of primitive magma to the surface from depth. The constant replenishment of the crustal chamber prevents the magma from extensive differentiation. The existence of the Mg anomaly in rocks and the position of this most active hydrothermal site of the northern EPR (Figs 1 and 2) can then be explained by the presence of a zone of enhanced magma generation.

There is a tendency, although not well pronounced, of Mg enrichment in basalts at sites of massive sulphide formation discovered at 3°35′, 8°53′ and 9°40′N on the EPR. However, this trend does not appear to be universal. Within the spreading segment between 21°20′ and 21°44′S studied in detail (Fig. 6), massive sulphides occur in association with both normal Mg-rich EPR basalts in the central part and with ferrobasalts in the southern part of the segment, which is most hydrothermally active at present.

The formation of ferrobasalts at the propagating tips of the EPR rift segments may be explained by advanced degrees of magma differentiation when it breaks through the colder crust surrounding the zone of propagation (Sinton *et al.* 1983).

The magma may be delivered to a propagating rift tip directly from the deep feeding zone or by

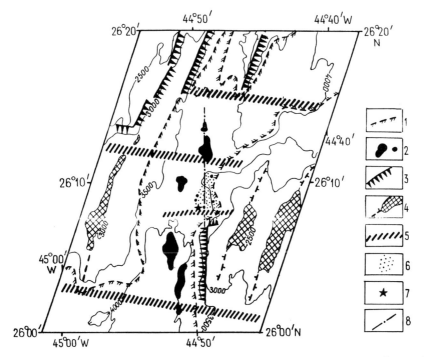

Fig. 12. Structural scheme of the TAG area. **1** Contours of the rift valley floor; **2** young volcanic mounds; **3** major faults; **4** crests and crestal surfaces of rift mountains; **5** zones of transverse dislocations; **6** bathymetric step, occupied by the TAG hydrothermal field; **7** active vent; **8** line of presumed magma penetration from the volcanic zone of the rift segment, lying north from 26°10′N, into the marginal fault of the southern segment.

an along-axis flow within a single chamber directed from the more mature part of the rift segment to the tip. In the latter instance, the process can be described by a model of centralized magma feeding of separate EPR spreading segments (Thompson *et al.* 1985). According to this model, sites of vertical magma delivery to subaxial chambers are mostly confined to the central uplifted parts of these segments. Later work proved, however, that the model of centralized magma feeding is valid only for some of the EPR segments (Thompson *et al.* 1989). The application of this model to the segment studied near 21°30′S seems to provide the best explanation for the whole set of data. The proposed existence of a zone of deep magma feeding in the central part of the segment and along-axis delivery of magma to its southern end is concordant with the following observations. (1) The most voluminous volcanism in the central part of the segment, reflected in the flattened profiles of the axial graben floor and rift flanks and in the presence of an off-axis volcano. Similar flattened characters of rift flanks were observed near 8°S (Macdonald *et al.*

1989) and 9°N (Fornari *et al.* 1992), where recent extensive lava fields have formed. Off-axis volcanoes are considered to act as 'safety valves' for redundant volcanism (Fornari *et al.* 1988). (2) The increasing rate of magma differentiation from the central part of the segment, where it could be delivered to the surface directly from the deep feeding channel to its southern end where basalts show evidence of evolved magma. (3) The presumed along-axis shift of the zone of most intense hydrothermal activity in time from the centre towards the southern part of the segment. The maximum metal enrichment in surficial sediments of the central part of the segment reflects the relatively recent position of this zone and hydrothermal plumes in near-bottom waters over the southern end mark its present position.

In contrast with the situation on the EPR, subsurface magma chambers are not directly observed under the MAR axial zone by seismic surveys (Detrick *et al.* 1989). However, the association of hot vents with the youngest sheet lavas observed both within the TAG and MARK fields suggests that hydrothermal systems may

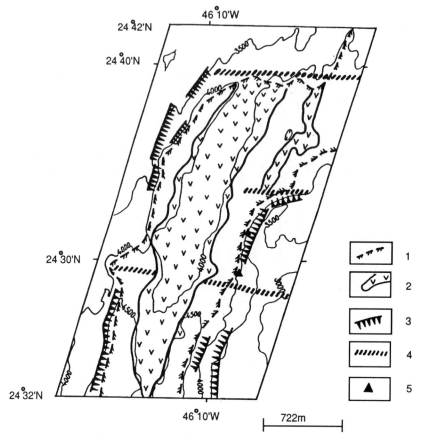

Fig. 13. Structural scheme of the MAR rift valley, near 24°30′N. **1** Contours of the rift valley floor; **2** extrusive highs; **3** major faults; **4** zones of transverse dislocations; and **5** sites of sulphide occurrence at the base of the valley wall directly continued northwards by an extrusive high.

also be developed over most shallow active crustal chambers. They are probably less voluminous than the EPR chambers as they cannot be resolved by existing geophysical methods, and may not be as long-lived. According to our TV studies of the TAG active smoker, hydrothermal activity is associated with sheet basalt flows as on the EPR. This lava type, atypical of the MAR, probably indicates the proximity of the magma chamber roof to the sea floor (Ballard *et al*. 1979).

The hydrothermal vents and massive sulphide deposits of the TAG and 24°30′N fields are controlled by magma penetration into zones of high permeability associated with marginal faults in the rift valley. These faults are very young (probably still active), steep (and presumably deep) and have large vertical displacements. The possibility of magma penetration into marginal faults is provided through the direct continuation of these faults to the extrusive zones of adjacent parts of the rift as a result of intersegmental lateral displacements by transverse faults.

Bettison-Varga *et al*. (1992) noted the important role of faults in localizing the near-surface intrusive bodies (penetrating into the dyke zone: 2B crustal layer) which control the zones of intense hydrothermal alteration (epidotization) of rocks, and massive sulphide deposits in the Troodos ophiolite (Cyprus).

At the EPR, ore-forming hydrothermal systems are almost entirely controlled by magmatic processes. Active extension and magma generation lead to the formation of the magma chambers necessary for the high-temperature hydrothermal circulation without any additional requirements. At major discontinuities where the crust permeability increases due to faulting, ore-forming processes are reduced as a result of

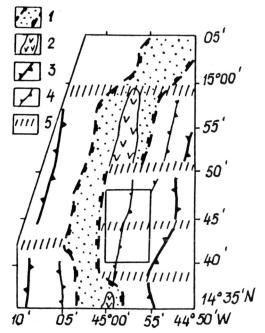

Fig. 14. Structural position of the study area near 14°45′N, MAR. **1** Rift valley floor; **2** extrusive zones; **3** crests of linear highs flanking the rift valley; **4** crests of linear highs on rift valley slopes; and **5** transverse dislocations. The rectangle shows the position of the area illustrated in Fig. 15.

the dilution of hot hydrothermal solutions with cold seawater.

Alternatively, minor transverse faults and, in particular, the most permeable zones of their intersection with rift valley marginal faults, control ore formation on the MAR in cases where they coincide with sites of magmatic activity. The most probable reason is that the formation of the subsurface magma chambers which are necessary for maintaining high-temperature convection (Cann & Strens 1982) can occur with less active magmatism only at sites of higher crustal permeability.

Conclusions

Data from the research cruises organized by the Sevmorgeologija Association (St Petersburg),

along with the results of other studies, allow us to formulate the tectonic settings favourable for hydrothermal activity and massive sulphide formation at fast- (EPR) and slow- (MAR) spreading centres.

Tectonic features directly connected with the volume and depth of the subaxial crustal linear magma chamber (such as the width of the crestal surface of the axial volcano and the presence of the graben) appear to be most significant for the EPR. The general bathymetric elevations of ridge segments are not always important and the distances from major discontinuities are not particularly important for the distribution of high-temperature hydrothermal activity.

The site of most intense hydrothermal activity in the northern part of the EPR near 12°50′N and some of the newly found sites of massive sulphide formation in the northern part of the EPR are marked by the development of the most primitive Mg-rich basalts. This phenomenon probably indicates direct vertical magma supply from its deep sources to the surface without significant differentiation in crustal chambers under hydrothermally active areas. It does not seem, however, to always be the case.

Various pieces of evidence support the hypothesis that an along-axis flow delivers magma to the site of present hydrothermal activity near 21°40′S, where ferrobasalts are developed. The site of relatively recent ore formation above the adjacent zone of localized magma supply within the same rift segment is less hydrothermally active at present.

In contrast with the EPR, where major ridge axis discontinuities are unfavourable for hydrothermal activity, the MAR hydrothermal sites studied are controlled by magma penetration into zones of increased crustal permeability. Transverse dislocations seem to play an important part in sulphide localization at the TAG and 15°N fields. Rift valley marginal faults are the main ore-localizing structures, especially at sites where their direct continuation by the extrusive rift zones of adjacent rift segments provide magma penetration into these faults (TAG, 24°30′N, 15°N).

Intense magmatic processes therefore entirely control the ore-forming hydrothermal activity of the fast-spreading EPR, whereas a combination

Fig. 15. (a) Geological setting of hydrothermal deposits within the uplifted block of the rift valley near 14°45′N, MAR. **1** Crests and crestal surfaces of linear highs; **2** bathymetric steps on the rift valley slope; **3** deeps between linear highs; **4** transverse tectonic dislocations; **5** outline of electric field anomalies; **6** outline of sulphide activity anomalies; and **7** hydrothermal deposits. (b) Bathymetric profile A–A′ across the study area. The figures on the profile show: **I** lower stage of the bathymetric step; **II** hydrothermal deposit; **III** meridional high with outcropping basalts; and **IV** rift valley wall.

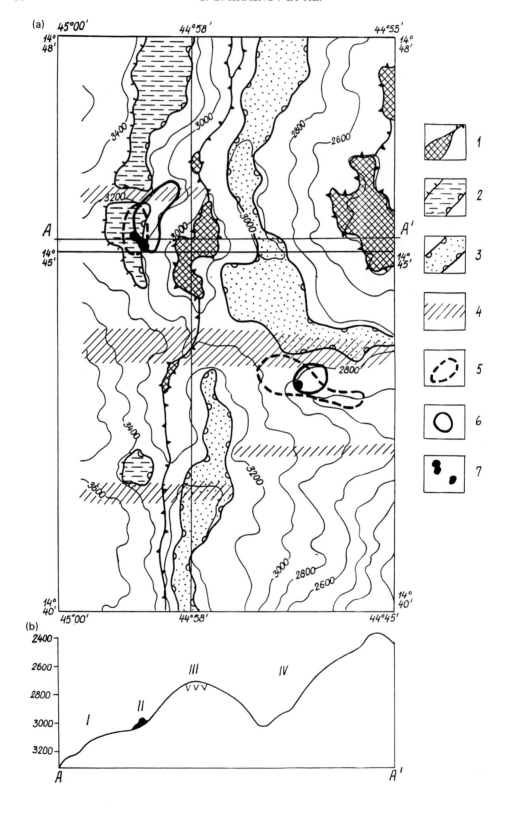

of magmatism and tectonics is important on the slow-spreading and less magmatically active MAR. High crustal permeability is probably required for the formation of subsurface magma chambers and is necessary for maintaining high-temperature hydrothermal convection under the MAR axis.

We thank all the participants of the numerous Sevmorgeologija research cruises on RV *Geolog Fersman*, who helped to obtain the data used in this study. We especially thank G. Glasby for his review of this paper. Critical reading of the manuscript by two anonymous reviewers was appreciated. This work was supported by Grant No. R1C000 from the International Science Foundation (Soros Foundation).

References

APLONOV, S. V. & POPOV, E. A. 1991. [Spreading instability and its reflection in the anomalous magnetic field]. *Izvestija Akademii Nauk SSSR, Ser. Phyzika Zemli*, **6**, 21–29 [in Russian].

——, ZAKHAROV, S. V., TRUNIN, A. A. & TIMOFEEV, V. I. 1992. [Migration of the area of spreading instability along the axis of a divergent boundary: Mid-Atlantic Ridge from 21° to 23°N]. *Izvestija Akademii Nauk SSSR, Ser. Phyzika Zemli*, **10**, 58–71 [in Russian].

BALLARD, R. D., HEKINIAN, R. & FRANCHETEAU, J. 1984. Geological setting of hydrothermal activity at 12°50'N on the East Pacific Rise: a submersible study. *Earth and Planetary Science Letters*, **69**, 176–186.

——, HOLCOMB, R. T. & VAN ANDEL, T. H. 1979. The Galapagos Rift at 86°W: 3. Sheet flows, collapse pits and lava lakes of the rift valley. *Journal of Geophysical Research*, **84**, 5407–5422.

BETTISON-VARGA, L., VARGA, R. J. & SHIFFMAN, P. 1992. Relation between ore-forming hydrothermal systems and extensional deformation in the Solea graben spreading center, Troodos ophiolite, Cyprus. *Geology*, **20**, 987–990.

BOSTRÖM, K. & PETERSON, M. N. A. 1966. Precipitates from hydrothermal exhalations on the East Pacific Rise. *Economic Geology*, **61**, 1258–1265.

BOUGAULT, H., CHARLOU, J.-L., FOUQUET, Y., APPRIOU, P. & JEAN-BAPTISTE, P. 1993. L'hydrothermalisme oceanique. *Societé Geologique de France Memoir*, **163**, 99–112.

CANN, J. R. & STRENS, M. R. 1982. Black smokers fuelled by freezing magma. *Nature*, **298**, 147–149.

CHERKASHEV, G. 1990. [Metalliferous sediments from sites of oceanic sulfide formation (example of the northern East Pacific Rise)]. PhD Thesis, VNIIOceangeologija, Leningrad [in Russian].

CRANE, K., AIKMAN, F. III & FOUCHER, J.-P. 1988. The distribution of geothermal fields along the East Pacific Rise from 13°10'N to 8°20'N: implications for deep seated origins. *Marine Geophysical Researches*, **9**, 211–236.

DETRICK, R. S., MUTTER, J. C. & BULL, P. 1989. Multichannel seismic data across the Snake Pit hydrothermal field (Mid-Atlantic Ridge, 23°N): no evidence for a crustal magma chamber. *EOS, Transactions of the American Geophysical Union*, **70**, 325–326.

DYMOND, J. 1981. Geochemistry of Nazca plate surface sediments: an evaluation of hydrothermal, biogenic, detrital and hydrogenous sources. *In*: LAVERNE, D. *et al.* (eds) *Nazca Plate: Crustal Formation and Andean Convergence*. Geological Society of America Memoir, **154**, 133–173.

FORNARI, D. J., PERFIT, M. R. & 8 others. 1988. Geochemical and structural studies of the Lamont seamounts: seamounts as indicators of mantle processes. *Earth and Planetary Science Letters*, **89**, 63–83.

——, ——, BATIZA, R. & EDWARDS, M. 1992. Submersible transects across the East Pacific Rise crest and upper flanks at 9°31'–32'N. 1. Observations of seafloor morphology and evidence for young volcanism off-axis. *EOS, Transactions of the American Geophysical Union*, **73**, 525.

FOUQUET, Y., AUCLAIR, G., CAMBON, P. & ETOUBLEAU, J. 1988. Geological setting and mineralogical and geochemical investigations of sulfide deposits near 13°N on the East Pacific Rise. *Marine Geology*, **84**, 143–178.

HAYMON, R. M., FORNARI, D. F., EDWARDS, M. H., CARBOTTE, S., WRIGHT, D. & MACDONALD, K. C. 1991. Hydrothermal vent distribution along the East Pacific Rise crest (9°09'–54'N) and its relationship to magmatic and tectonic processes on fast-spreading mid-ocean ridges. *Earth and Planetary Science Letters*, **104**, 513–534.

KARSON, J. A. & BROWN, J. R. 1988. Geological setting of the Snake Pit hydrothermal site: an active vent field on the Mid-Atlantic Ridge. *Marine Geophysical Researches*, **10**, 91–107.

—— & RONA, P. 1990. Block-tilting, transfer faults, and structural control of magmatic and hydrothermal processes in the TAG area, Mid-Atlantic ridge 26°N. *Bulletin of the Geological Society of America*, **102**, 1635–1645.

KRASNOV, S. G., CHERKASHEV, G. A. & 21 others. 1992a. [Hydrothermal Sulfide Ores and Metalliferous Sediments of the Ocean]. Nedra, St Petersburg, 279pp [in Russian].

——, KREYTER, I. I. & POROSHINA, I. M. 1992b. Distribution of hydrothermal vents on the East Pacific Rise (21°20'–22°40'S) based on a study of dispersion patterns of hydrothermal plumes. *Oceanology*, **32**, 375–381.

——, MASLOV, M. N., ANDREEV, N. M., KONFETKIN, V. M., KREYTER, I. I. & SMIRNOV, B. N. 1988. [Hydrothermal ore mineralization of the southern East Pacific Rise]. *Doklady Akademiji Nauk SSSR*, **302**, 161–164 [in Russian].

LANGMUIR, C., BENDER, J. & BATIZA, R. 1986. Petrological and tectonic segmentation of the East Pacific Rise, 5°30'–14°30'N. *Nature*, **322**, 422–429.

——, FORNARY, D. 1993. Geological setting and characteristics of the Lucky Strike Vent Field at 37°17'N on the Mid-Atlantic Ridge. *EOS,*

Transactions of the American Geophysical Union, **74**, 99.

LISITSYN, A. P. 1991. [*Processes of Terrigenous Sedimentation in Seas and Oceans*], Nauka, Moscow, 272pp [in Russian].

MACDONALD, K. C. 1989. Anatomy of the magma reservoirs. *Nature*, **336**, 178–179.

——, HAYMON, R. M. & PERRAM, L. J. 1986. 13 new hydrothermal vent sites found on the East Pacific Rise, 20–21°S. *EOS, Transactions of the American Geophysical Union*, **67**, 1231.

——, —— & SHOR, A. 1989. A 220 km recently erupted lava field on the East Pacific Rise near lat. 8°S. *Geology*, **17**, 212–216.

MARCHIG, V., GUNDLACH, H., HOLLER, G. & WILKE, M. 1988. New discoveries of massive sulfides on the East Pacific Rise. *Marine Geology*, **84**, 179–190.

POROSHINA, I. M. 1987. [Geomorphic features of sites of sulfide development on the East Pacific Rise]. *In*: [*Geomorphologic Studies of the Ocean Floor*]. PGO 'Sevmorgeologija', Leningrad, 73–85 [in Russian].

RENARD, V., HEKINIAN, R., FRANCHETEAU, J., BALLARD, R. D. & BACKER, H. 1985. Submersible observations at the ultra-fast spreading East Pacific Rise (17°30′ to 21°30′S). *Earth and Planetary Science Letters*, **75**, 339–353.

RONA, P. A., BOGDANOV, YU. A., GURVICH, E. G., RIMSKI-KORSAKOV, N. A., SAGALEVITCH, A. M., HANNINGTON, M. D. & THOMPSON, G. 1993. Relict hydrothermal zones in the TAG Hydrothermal Field, Mid-Atlantic Ridge 26°N, 45°W. *Journal of Geophysical Research*, **98**, 9715–9730.

SINTON, J. M., WILSON, D. S., CHRISTIE, D. M., HEY, R. N. & DELANEY, J. R. 1983. Petrologic consequences of rift propagation on oceanic spreading ridges. *Earth and Planetary Science Letters*, **62**, 193–207.

SKORNYAKOVA, N. S. 1965. Dispersed iron and manganese in Pacific Ocean sediments. *International Geology Review*, **7**, 2161–2174.

SMIRNOV, V. I. (ed.) 1979. [*Metalliferous Sediments of the South-Western Pacific*]. Nauka, Moscow, 280pp [in Russian].

SMITH, D. K. & CANN, J. R. 1990. Hundreds of small volcanoes on the median valley floor of the Mid-Atlantic Ridge at 24–30°N. *Nature*, **348**, 152–154.

THOMPSON, G., BRYAN, W., BALLARD, R., HAMURO, K. & MELSON, W. G. 1985. Axial processes along a segment of the East Pacific Rise, 10–12°N. *Nature*, **318**, 429–433.

——, —— & HUMPHRIS, S. E. 1989. Axial volcanism on the East Pacific Rise, 10–12°N. *In*: SAUNDERS, A. D. & NORRY, M. J. (eds) *Magmatism in Ocean Basins*. Geological Society, London, Special Publication, **42**, 181–200.

TUFAR, W. & NÄSER, G. 1991. *Forschungsfahrt, SONNE 63 – OLGA I. Lagerstattenkundliche Detailuntersuchungen eines aktiven Hydrothermalfeldes im Bereich eines Scheitelgrabens an divergierenden Plattenrandern mit hoher Divergenzrate im Ostpazifik*. Philipps-Universitat, Marburg.

UCHUPI, E., SCHWAB, W. C. & 6 others. 1988. An ANGUS/ARGO study of neovolcanic zone along the EPR from Clipperton fracture zone to 12°N. *Geo-Marine Letters*, **8**, 131–139.

VON DAMM, K., EDMOND, J. M., GRANT, B. & MEASURES, C. 1985. Chemistry of submarine hydrothermal solutions at 21°N, East Pacific Rise. *Geochimica et Cosmochimica Acta*, **49**, 2197–2220.

Geological setting and ecology of the Broken Spur hydrothermal vent field: 29°10′N on the Mid-Atlantic Ridge

BRAMLEY J. MURTON[1], CINDY VAN DOVER [2] & EVE SOUTHWARD[3]

[1]*Institute of Oceanographic Sciences, Wormley, Surrey GU8 5UB, UK*
[2]*Woods Hole Oceanographic Institute, MA 02543, USA*
[3]*Marine Biological Association, Citadel Hill, Plymouth PL1 2PB, UK*

Abstract: Deep-towed sidescan sonar and manned submersible studies have shown that hydrothermal activity within the Broken Spur vent field, located at 29°10′N on the Mid-Atlantic Ridge, is controlled by a combination of recent volcanic and tectonic activity. Three sulphide mounds, with high-temperature fluid vents, and two weathered sulphide mounds, with low-temperature fluid seeps, are aligned across an axial summit graben that lies along the crest of a neovolcanic ridge within the axial valley floor. The largest high-temperature venting sulphide mound, which is up to 40 m high, lies in the centre of the graben. Two further, but smaller, high-temperature sulphide mounds are located to the east and west of the larger mound. All three high-temperature venting mounds lie on an axis that strikes 115°, orthogonal to the trend of the axial summit graben and neovolcanic ridge. These directions are similar to the strike of two sets of faults that locally intersect at the vent site and which probably control the location of hydrothermal emission. The fauna colonizing the vents are distinct, at least at a species level, from those found at other hydrothermal sites on the Mid-Atlantic Ridge. New species of bresiliid shrimp, and a new genus of brittle star have been found along with other fauna in an ecosystem that is otherwise similar to those found at high-temperature hydrothermal sites elsewhere on the Mid-Atlantic Ridge. However, the populations of the bresiliid shrimp at Broken Spur are significantly lower in abundance than those of the same genus found elsewhere. The size, shape and state of alteration of the sulphide mounds and the extent of their oxyhydroxide sediments and weathered sulphide talus aprons, suggests that hydrothermal activity at the Broken Spur vent field has been long-lived, probably for several thousand years. This is supported by the unique speciation of fauna present, which also suggests that the Broken Spur vent field is isolated (in terms of faunal accessibility) from other vent sites on the Mid-Atlantic Ridge. The low population of the shrimp, in an otherwise active and long-established hydrothermal habitat, suggests that the hydrothermal activity at Broken Spur is in a state of change and may have been recently rejuvenated.

A combination of geophysical imaging and geochemical sensing techniques were used to assess the regional extent of hydrothermal activity along 300 km of the Mid-Atlantic Ridge (MAR) between 27°N and 30°N (Fig. 1) during cruise CD76 of the RRS *Charles Darwin* in February and March 1993 (Murton 1993). The study involved systematic acoustic imaging of the axial valley floor and sensing water column properties using the towed ocean bottom instrument (TOBI) (Murton *et al.* 1992). During the survey, the vehicle was towed at an altitude of between 200 and 300 m above the seafloor to allow it to encounter and measure the concentration of any hydrothermal plumes which might be present. Two survey tracks were occupied along the axial valley of the MAR, closely following the bathymetry (Purdy *et al.* 1990): a northerly pass parallel to the western valley wall, with a 180° clockwise turn in the western

nodal-deep basin of the Atlantis Fracture Zone, followed by a southerly pass parallel to the eastern valley wall.

As a result a site of hydrothermal activity was identified at 29°10′N, above which anomalies in the water column were detected over a distance of 14 km along-strike (Murton *et al.* 1994). These signals, which occurred at water depths between 2550 and 2650 m (Fig. 2), effected a decrease in light transmission of 0.17% and an increase in nephel concentrations of up to one hundred times above background. The presence of significant anomalous light attenuation and nephel concentrations, at a nearly constant depth in the water column over several kilometres along-strike, was taken as compelling evidence of a neutrally buoyant plume of the type generated by high-temperature metal-rich hydrothermal emissions. Anomalous increases in potential temperature at the site of the

From Parson, L. M., Walker, C. L. & Dixon, D. R. (eds), 1995, *Hydrothermal Vents and Processes*, Geological Society Special Publication No. 87, 33–41.

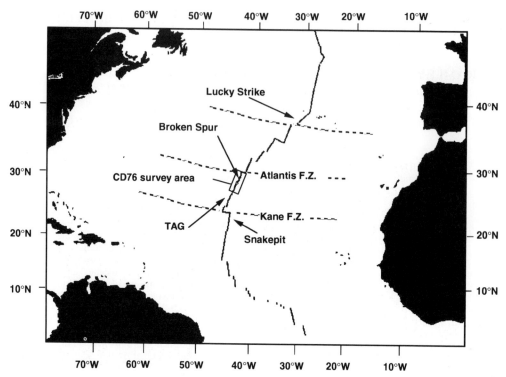

Fig. 1. Location of the area, between 27° and 30°N, surveyed during cruise CD76 of the RRS *Charles Darwin* and containing the Broken Spur hydrothermal vent field.

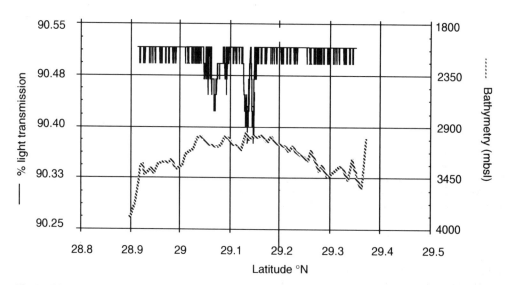

Fig. 2. TOBI transmissometer data (solid line) and bathymetry profile (broken line) showing the largest shifts in light attenuation at around 29.14°N, at a depth of about 2600 m, above the shallowest part of the second-order ridge segment. Other transmissometer signals, at 29.06°N and 29.08°N, suggest other sites of hydrothermal activity throughout the ridge segment.

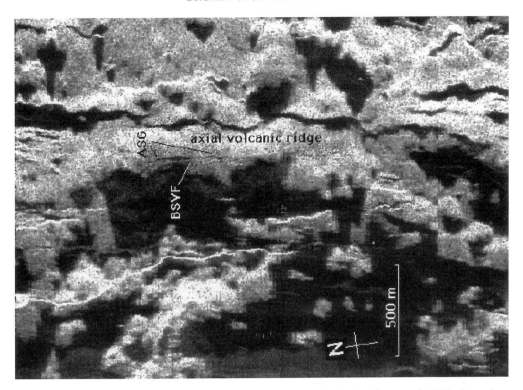

Fig. 3. TOBI sidescan sonar image (2 km by 3 km) of the location of the Broken Spur vent field. The image is similar to a monochrome photograph with bright areas being strong echoes and dark areas being weak echoes and shadows. The Broken Spur vent field (BSVF) lies at the top of an axial volcanic ridge and adjacent to an axial summit graben (ASG) running along the length of the ridge.

strongest plume signals were encountered along a 100 m transect between 8 and 10 m above the crest of a small neovolcanic ridge. These anomalies indicated several sources of hydrothermal activity at a site which has been named informally as the Broken Spur vent field (Murton *et al.* 1993, 1994).

The crest of the neovolcanic ridge, where the anomalies in particulate concentration and potential temperature were located, was examined during two dives using the DSV *Alvin* (dives 2624 and 2625 in June 1993). These studies confirmed the presence of a high-temperature hydrothermal field located within an axial summit graben (ASG) at 29°10′N, 43°10′W. Three discrete 'black smoker' sources of vent fluid in excess of 350°C were located, plus two weathered sulphide mounds emitting diffuse low-temperature fluids.

Geological setting of the vent field

Volcanic and tectonic environment

The Broken Spur vent field is only the fourth active hydrothermal system to be discovered on

the MAR, and is located within a basin in a strongly second-order segmented section of the MAR (Purdy *et al.* 1990). The vent field lies at 29°10′N, approximately 100 km south of the Atlantis Fracture Zone, on the crest of a 250 m high neovolcanic ridge. The neovolcanic ridge is parallel to the axis of the MAR and lies along the western side of the axial valley. The hydrothermal vents at the Broken Spur vent field are located within and on the walls of an ASG that lies along the crest of the neovolcanic ridge (Fig. 3). This 1.2 km long, 35–60 m wide and 30 m deep linear structure is asymmetrical with a single vertical wall forming its western side and a stepped wall forming its eastern side. The graben walls are covered by fresh and unbroken pillow lavas, suggesting that it is probably an axial summit caldera formed by subsidence during volcanic eruption. Elsewhere, the nearly flat-lying crest of the neovolcanic ridge at 29°10′N is covered with virtually sediment-free glassy pillow lavas with 2–3 m diameter, 1–2 m deep pits formed over collapsed lava tubes. Around the hydrothermal vents the lavas are dominantly lobate, suggesting more rapid or

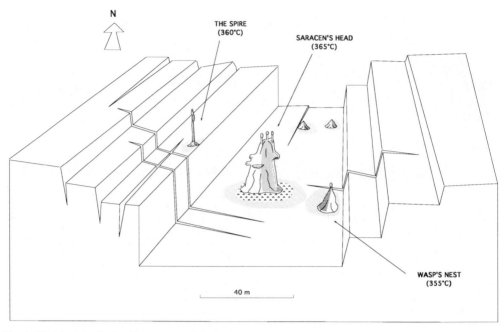

Fig. 4. Sketch of the Broken Spur vent field derived from a compilation of TOBI sidescan sonar imagery, seafloor photographs and observations from the DSV *Alvin*.

voluminous eruptions than for the peripheral pillow lava-dominated terrain (Fink and Griffiths 1992). Weathered hydrothermal sediments, first seen within 50 m of the vents, rapidly increase in thickness to at least several tens of centimetres at the base of the sulphide mounds.

The vent field lies at the intersection between two sets of cross-cutting fissures: one trends 028° and is parallel to the ASG, while the other trends 115° (orthogonal to the ASG) and is parallel to the alignment of three high-temperature venting mounds. The highest temperature venting was recorded at the largest sulphide mound, lying in the centre of the ASG. This suggests a structural control on the location of hydrothermal activity, probably by an increase in crustal permeability at the intersection of the two fault sets.

Sulphide mounds

At least five sulphide mounds occupy the Broken Spur vent field (Fig. 4). Two mounds, aligned 110°, vent fluids at less than 50°C. A further three mounds, aligned 115°, are located 150 m to the south and vent fluids up to 365°C.

The two lower temperature venting sulphide mounds are less than 5 m tall and about 3 m in diameter and are located on the floor of the ASG. An apron of sulphide rubble and hydrothermal sediments extends for 8 m from their periphery. The sediments are more than 30 cm deep within 2 m of the solid sulphide outcrop. The westernmost low-temperature venting sulphide mound (located at 29°10.095′N, 43°10.428′W, 3870 m bsl), which lies at the base of a 2 m high east-facing fault scarp, has shimmering water percolating from cracks in the sulphide outcrop. The easternmost mound (located at 29°10.098′N, 43°10.412′W, 3875 m bsl and marked with a tag K) is capped by branching sulphide chimneys, 0.3–0.5 m tall, and also has shimmering water escaping from cracks (Fig. 5). Although the presence of anhydrite deposits at the base of the lobate flanges indicates water temperatures in excess of 130°C (Haymon and Kastner 1981), *in situ* measurements only recorded temperatures up to 50°C.

Of the three high-temperature venting sulphide mounds, by far the most prominent is 35–40 m tall with a *c.* 20 m diameter base and an apron of sulphide rubble and soft oxyhydroxide sediments that extends for a further 15–20 m. This mound (located at 29°10.011′N, 43°10.437′W, 3057 m bsl and marked with a tag M), informally named the Saracen's Head, occurs in the middle of the ASG and is found

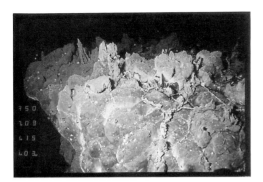

Fig. 5. Weathered sulphide mound (10.098'N, 43° 10.412'W, 3875 m bsl) capped by branching sulphide chimneys, 0.3–0.5 m tall, with shimmering water escaping from cracks.

Fig. 6. The top of the most prominent mound, the Saracen's Head (29°10.011'N, 43°10.437'W, 3057 m bsl), is rounded with a diameter of about 10 m. It comprises weathered sulphide rubble derived from fallen chimneys, solid, blocky sulphide outcrop and at least three spherical vent structures (beehives) exhaling black fluid.

mid-way between the two other high-temperature venting sulphide mounds. Its weathered sulphide apron has shimmering water escaping from between blocks. The sides of the mound are steep to vertical in places and also have shimmering water escaping from cracks. Flanges, several metres wide and 0.5–1 m thick, have rounded anhydrite encrusted bases. They jut out of the near-vertical wall of the mound at 5, 15 and 25 m above the surrounding seafloor. One flange has active sulphide chimneys on top venting black fluid and an anhydrite crust on its lower surface.

The top of the Saracen's Head mound is rounded with a diameter of about 5 m. It comprises weathered sulphide rubble derived from fallen chimneys, solid, blocky sulphide outcrop and at least three spherical vent structures exhaling black fluid (Fig. 6). These 0.5–1.5 m tall, 0.3–0.6 m diameter 'beehive'-like structures, composed of filigree-structured sulphide and anhydrite (see Duckworth *et al.* this volume, for a full description) are fed from below by open conduits, 3–5 cm in diameter, which vigorously vent black fluid. This fluid, which remains opalescent for 5–10 cm above the orifice of the conduit before rapidly turning black, has a maximum temperature of 361.7°C.

An 18 m tall sulphide chimney is located about 50 m to the west of the Saracen's Head mound. This feature (informally named The Spire) has a diameter of 1–2 m and protrudes from a 4 m tall, 10 m wide base of weathered sulphide rubble and oxyhydroxide sediments (located at 29°10.014'N, 43°10.459'W, depth 3050 m bsl and marked with a tag S). It occurs at the top of a 30 m high scarp that forms the western wall of the ASG. The solid sulphide outcrop forming The Spire is weathered, has diffuse fluid flow

from cracks halfway up and is capped by a 'beehive' structure of anhydrite and sulphide from which black fluid diffuses (Fig. 7). Fluid is supplied to the base of the 'beehive' through a single conduit, 5 cm in diameter, from where it escapes vigorously and is opalescent for up to 5 cm above the orifice before turning black. The temperature recorded in the opalescent fluid peaked at 359.9°C.

A third sulphide mound lies about 35 m to the east of the Saracen's Head. This feature is about 15–20 m tall and 8–10 m in diameter and has been informally named the Wasp's Nest. A weathered sulphide rubble and sediment apron, which extends for 5 m beyond the periphery of the solid sulphide outcrop, is colonized by fluted worm-tubes and anemones. The sulphide mound (located at 29°10.019'N, 43°10.422W, depth 3091 m bsl and marked by tag Q) lies at the base of the eastern wall of the ASG. Anhydrite crusts and bacterial mats are extensive on the surface of the mound 3–5 m below its top. The mound is capped by at least two anhydrite–sulphide 'beehive' structures and by superficial sulphide pipes, 50–70 cm long and 5 cm in diameter, venting black fluid. The fluids emerging from conduits supplying the 'beehives' have a maximum temperature of 357.5°C.

Ecology of the Broken Spur vent field

Faunal distribution

The ecology of the Broken Spur vent field is similar to the hydrothermal vent sites found elsewhere on the MAR (e.g. Snakepit at

Fig. 7. The solid sulphide outcrop forming the Spire (29°10.014′N, 43°10.459′W, depth 3050 m bsl) is weathered, has diffuse fluid flow from cracks halfway up and is capped by a 'beehive' structure of anhydrite and sulphide from which black fluid diffuses.

Fig. 9. Crabs are particularly populous at one high-temperature venting mound, the Wasp's Nest (located at 29°10.019′N, 43°10.422W, depth 3091 m bsl) where they have a population density of one per 10–15 cm^2.

Fig. 8. Aprons of sulphide sediment and rubble around the sulphide mounds are colonized by a high density of anemones and worm tubes.

23°23′N, TAG at 26°08′N (Grassle 1981, 1986) and Lucky Strike at 37°19′N), with shrimps dominating the proximal regions close to the hot vents, while worms, mussels, anemones and ophiuroids are more peripheral. Crabs and fish appear to be very mobile and are found both close to and far from the actively venting chimneys. Vestimentiferan tubeworms and alvinellid polychaetes, which characterize Pacific hydrothermal sites, are absent from the Broken Spur vent field.

The apron of sulphide sediment and rubble around the low-temperature venting mounds is colonized by a high-density of anemones, 15 mm long, 5–10 mm wide and one per 15 cm^2 (Fig. 8). Free-swimming polychaetes in the water column are also present, with a population density of one per 2–3 m^3. However, only dead mussels are present with partially dissolved shells (80 mm long by 40 mm wide).

The sediment apron around the high-temperature venting mounds, such as the Saracen's Head, is colonized by fluted worm-tubes, 20–30 mm long by 3–5 mm in diameter. They are attached to the substrate, oriented perpendicular to the seafloor and have an equidistant distribution of one per 5–6 cm^2. The fluted worm-tubes occupy local hollows, whereas the anemones colonize high-standing areas. The temperature within the sediment, at a depth of 5 cm, is 3.10°C (i.e. 0.21°C above ambient water temperature). Elsewhere white anemones (15 mm long, 5–10 mm wide), polychaetes and squat-lobsters occupy the sediment and polychaetes swim in the water column. The rubble aprons around the high-temperature venting mounds, which have a general angle of repose of *c.*50°, are colonized by fluted worm-tubes in sedimented hollows and anemones attached to high-standing sulphide blocks.

The solid surfaces of the high-temperature venting mounds are colonized by ophiuroids, brachyuran crabs, *Munidopsis* squat-lobsters and bresiliid shrimps. The crabs are particularly populous at the Wasp's Nest mound where they have a population density of one per 10–15 cm^2 (Fig. 9). Colonies of mussels on The Spire are located within crevices, with individuals 5–15 cm long, living but stained red by iron oxide, suggesting a very slow growth rate.

A flange, 5–6 m wide, 3–4 m deep and 2 m thick, located about 25 m above the base of the Saracen's Head mound, has a mat of filamentous bacteria covering its underside roof and walls. The water around the bacterial mat is at 3.96°C (i.e. 1.10°C above ambient). The outer surface of the flange is colonized by white anemones (5–10 mm long and 5 mm in diameter), living mussels with red-stained shells (25–40 mm long and 10–20 mm wide) and white ophiuroids up to 15 mm in diameter. The areas around the active vents on the Saracen's Head mound are inhabited by brachyuran crabs, bresiliid shrimps and ophiuroids. Temperatures for the ambient water around the shrimps range from 3.40–7.8°C (mean 4.0°C) and typically 3.8°C for the ophiuroids. The shrimp at Broken Spur have a population density of only one to four per 10 cm and, unlike at either the TAG or Snakepit hydrothermal sites on the MAR, were not observed in swarms.

Taxonomy

At the Broken Spur vent field, the large crabs are identical with *Segonzacia mesatlantica* known previously from the Snakepit vent field (M. Segonzac and D. Guille, pers. comm.; Guinot 1989), although the squat-lobster, *Munidopsis*, appears to belong to a hitherto undescribed species. The three species of shrimp (family *Bresiliidae*) collected from the high-temperature venting mounds all show differences from those described from either the Snakepit or TAG vent fields (Williams & Rona 1986). They belong to three genera, *Alvinocaris, Rimicaris* and possibly a hitherto undescribed genus. Three of the collected specimens of *Rimicaris* have been compared with specimens of *Rimicaris exoculata* (known from the TAG and Snakepit vent sites). The Broken Spur specimens have longer, thinner antennae and a less inflated carapace than *R. exoculata*. These and other differences indicate that the Broken Spur *Rimicaris* may belong to a new and hitherto undescribed species or subspecies. A single *Alvinocaris* collected from Broken Spur shows small differences from *Alvinocaris markensis*, as described from the Snakepit vent field. The most abundant shrimps present at Broken Spur are small (between 20 and 30 mm long) and appear to represent a hitherto undescribed species, or perhaps an undescribed genus. The mussels at Broken Spur, presumed to belong to the genus *Bathymodiolus* which is widespread at high-temperature vents on spreading ridges and low-temperature fluid seeps on continental shelves, have a different shell form and colour when compared with those seen elsewhere and in particular those from the Lucky Strike vent field. Polychaete worms of four species were collected at Broken Spur. Two were free-living and two were in tubes. One of the latter has wide flanges on its tubes and represents the fluted Ampharetidae and is similar to worms found at Snakepit and possibly TAG (D. Desbruyères,

pers. comm.). The small white ophiuroids collected at Broken Spur belong to a new genus and species of the family Ophiuridae which has also been found at Snakepit (Tyler *et al.* in press).

Discussion

The Broken Spur vent field lies within an ASG, which in turn is located along the crest of an axial volcanic ridge. The formation of an ASG, covered by fresh lava flows, suggests that this part of the ridge has been recently active. Similar features are common along the crest of the East Pacific Rise where they are interpreted as axial summit calderas, formed by subsidence of the seafloor above a linear magma reservoir (Haymon *et al.* 1991). At Broken Spur two fault sets, axis-parallel and axis-orthogonal, intersect in the vicinity of the vent field and may have localized hydrothermal activity by increasing the crustal permeability by fracturing and brecciation. The axis-parallel faults are normal to the plate separation direction and are thus probably related to simple extension. However, the axis orthogonal faults are not obviously related to crustal stretching and may instead be transfer faults linking differential extension between axis-parallel normal faults. Similar structures have been reported from 23°N on the MAR (Karson and Dick 1983).

There are three large high-temperature venting mounds, which in order from west to east are: The Spire (360°C, 18 m tall by 1.8 m wide), the Saracen's Head (365°C, 35–40 m tall by c.20 m wide) and the Wasp's Nest (357°C, 20 m tall and c.10 m wide). The east–west orientation of these three sulphide mounds (parallel to the axis-orthogonal faults) and the observation of the highest temperature fluids venting from the central and largest mound (the Saracen's Head), supports the interpretation that their location is controlled by a combination of enhanced crustal permeability within two local fault trends that intersect above an intra-crustal magma reservoir.

The size of the sulphide mounds at Broken Spur, with their aprons of weathered sulphide rubble and surrounding accumulations of oxyhydroxide sediments, show that it is a mature vent field which has probably been in existence for several thousand years, like the vent fields of TAG and Snakepit. Some of the animal species at Broken Spur appear to be distinct from those found elsewhere on the MAR, which suggests separation of the order of millions of years. The Broken Spur vent field thus appears to have a history of hydrothermal activity that has been isolated from other hydrothermal sites on the MAR for long enough for speciation of the fauna to occur. Isolation of the Broken Spur vent field may be the result of the strongly developed second-order segmentation of the MAR, which forms 50–100 km long basins (Purdy *et al.* 1990) that restrict the migration of vent fauna from one basin to another. This is compounded by the relative paucity of hydrothermal activity on the MAR, such that vent sites are separated on average by hundreds of kilometres, with the Broken Spur field being the only one within a 300 km section of the ridge (Murton *et al.* 1994).

There is, however, a disparity between the low population density, especially of bresiliid shrimp, found at Broken Spur and the apparently active hydrothermal habitat with accumulating sulphide structures and weathering talus aprons. Because the hydrothermal activity, sulphide mineralogy (Duckworth *et al.* this volume) and vent fluid geochemistry (James *et al.* this volume) are similar to those found at the TAG and Snakepit vent fields (Caye *et al.* 1988), we consider the comparatively low population density at Broken Spur to be the result of some instability, perhaps some recent and dramatic change in hydrothermal activity. This could have been a reactivation after a period of little activity, which has not yet given time for the population to recover. Alternatively, there may have been a recent catastrophic event which reduced the population, and again it has not had time to recover. The concept of short-period changes in hydrothermal activity has implications for the spatial and temporal frequency of magma supply, magma plumbing and magma residence at mid-ocean ridges. The response time for fauna recovery following such changes in hydrothermal activity, and hence the temporal and spatial scale of the causative volcanic processes, remains to be discovered.

The results reported here owe much to the efforts of many people, including the officers and crew of the RRS *Charles Darwin*, the RV *Atlantis II*, the IOSDL engineers who developed and ran the TOBI and WASP systems and the members of the Deep Submergence Group, Woods Hole Oceanographic Institution, for DSV *Alvin* operations. This work was supported by the Natural Environment Research Council, UK, the BRIDGE community research programme, UK and the National Sciences Foundation USA. We also are indebted to the helpful and constructive comments of reviewers of the early draft of this manuscript. IOSDL Contribution No. 95012.

References

CAYE, R., CERVELLE, B., CESBRON, F., OUDIN, E., PICOT, P. & PILLARD, F. 1988. The mineralogy of hydrothermal sulphide deposits. *Mineralogical Magazine*, **52**, 509–514.

DUCKWORTH, R. C., KNOTT, R., FALLICK, A. E., RICKARD, D., MURTON, B. J. & VAN DOVER, C. Mineralogy and sulphur isotope geochemistry of the Broken Spur sulphides, 29°N, Mid-Atlantic Ridge, this volume.

FINK, J. H. & GRIFFITHS, R. W. 1992. A laboratory analog study of the surface morphology of lava flows extruded from point and line sources. *Journal of Volcanology and Geothermal Research.* **54** 1/2, 19–32.

FU, L.-L., KEFFLER, T., NIILER, P. P. & WUNSCH, C. 1982. Observations of mesoscale variability in the western North Atlantic: a comparative study. *Journal of Marine Research*, **40**, 809–823.

GRASSLE, J. F. 1981. hydrothermal vent communities and biology. *Science*, **229**, 713–717.

—— 1986. The ecology of deep-sea hydrothermal vent communities. *Advances in Marine Biology*, **23**, 302–362.

GUINOT, D. 1989. Description de Segonzacia gen. nov. et remarques sur *Segonzacia mesatlantica* Williams: campagne HYDROSNAKE 1988 sur la dorsale medio-Atlantique Crustacea Decapoda Brachyura. *Bulletin Muséum Nationale d'Histoire Naturelle, Paris, Ser. II*, 130–145.

HAYMON, R. & KASTNER, M. 1981. Hot spring deposits on the East Pacific Rise at 21°N: preliminary description of mineralogy and genesis. *Earth and Planetary Science Letters*, **52**, 363–381.

——, FORNARI, D. J., EDWARDS, M. H., CARBOTTE, S., WRIGHT, D. & MACDONALD, K. C. 1991. hydrothermal vent distribution along the East Pacific Rise crest (9°09′–54′N) and its relationship to magmatic and tectonic processes on fast-spreading mid-ocean ridges. *Earth and Planetary Science Letters,* **104**, 513–534.

JAMES, R. H., ELDERFIELD, H. & 11 others. hydrothermal plumes at Broken Spur, 29°N Mid-Atlantic Ridge: chemical and physical characteristics, this volume.

KARSON, J. A. & DICK, H. J. B. 1983. Tectonics of ridge transform intersections at the Kane Fracture Zone. *Marine Geophysical Research*, **6**, 51–98.

MURTON, B. J. 1993. RRS *Charles Darwin* Cruise 76, geological and geochemical investigation between 27°N and 30°N of the Kane to Atlantis segment: the Mid-Atlantic Ridge. *IOSDL Cruise Report, No.* **236**, 34 pp.

——, KLINKHAMMER, G. & 11 others 1993. Direct evidence for the distribution and occurrence of hydrothermal activity between 27°N–30°N on the Mid-Atlantic Ridge. *EOS, Transactions of the American Geophysical Union*, **74**, 99.

——, —— & 11 others 1994. Direct evidence for the distribution and occurrence of hydrothermal activity between 27°N–30°N on the Mid-Atlantic Ridge. *Earth and Planetary Science Letters,* **125**, 119–128.

——, ROUSE, I. P., MILLARD, N. & FLEWELLEN, C. G. 1992. Deep-towed vehicle explores the ocean floor. *EOS, Transactions of the American Geophysical Union*, **73**, 1–4.

PURDY, G. M., SEMPÉRÉ, J.-C., SCHOUTEN, H., DuBois, D. L. & GOLDSMITH, R. 1990. Bathymetry of the mid Atlantic Ridge, 24°N–31°N: a map series. *Marine Geophysical Research,* **12,** 247–252.

TYLER, P., DESBRUYÈRES, D. & MURTON, B. J. A new genus of ophiuririd from hydrothermal vents on the Mid-Atlantic Ridge. *Journal of the Marine Biological Association,* in press.

WILLIAMS, A. B. & RONA, P. A. 1986. Two new caridean shrimps (Bresiliidae) from a hydrothermal field on the Mid-Atlantic Ridge. *Journal of Crustacean Biology,* **6,** 446–462.

Detailed geological studies of hydrothermal fields in the North Atlantic

S. G. KRASNOV[1], G. A. CHERKASHEV[1], T. V. STEPANOVA[1],
B. N. BATUYEV[2], A. G. KROTOV[2], B. V. MALIN[2], M. N. MASLOV[2],
V. F. MARKOV[2], I. M. POROSHINA[1], M. S. SAMOVAROV[2],
A. M. ASHADZE[2], L. I. LAZAREVA[2], I. K. ERMOLAYEV[2]

[1]*Institute of Geology and Mineral Resources of the Ocean, 1 Angliysky Avenue, 190121 St Petersburg, Russia.*

[2]*Polar Marine Geological Prospecting Expedition, Sevmorgeologija Association, 24 Pobedy Street, Lomonosov 189510, Russia*

Abstract: Sidescan sonar and TV/photo surveys, TV-controlled sulphide grab sampling and sediment sampling have been carried out at the TAG and MARK hydrothermal fields of the Mid-Atlantic Ridge during research cruises of the Sevmorgeologija Association, St Petersburg. Detailed mapping of the MARK field (23°N) allowed contouring of the massive sulphide bodies. The largest of them are Cu-enriched and the smallest zinc-enriched. Sulphides of the TAG field (26°N) active mound appear to be Cu-rich in its central and Zn-rich in its marginal parts. Sampling of the inactive Mir hydrothermal mound of the TAG field revealed an enrichment in Cu in its northern part and an abundance of silica in the south. This mound and especially the north hydrothermal zone, both presently inactive, determined the metal distribution in TAG sediments during the last 10 000–13 000 years. Extensive Cu-rich sulphide deposits were also discovered and sampled near 14°45′N. The dominance of Cu over Zn in mature Mid-Atlantic Ridge hydrothermal deposits, arising at late stages of their formation, can probably be explained by dominant leaching of Cu from basalts at advanced stages of development of the hydrothermal systems and/or by increasing the efficiency of Cu sulphide deposition in the course of formation of the ore body.

In the early 1990s, several research cruises were organized by the Sevmorgeologija Association (St Petersburg, Russia) to the hydrothermal fields of the Mid-Atlantic Ridge (MAR). The aims of our investigations were to map the sulphide deposits and obtain representative samples of massive sulphides and sulphide-bearing sediments to assess the deposits quantitatively.

Methods

The detailed work carried out by Sevmorgeologija's ships at the previously known TAG and MARK fields and at the newly found 15°N field (Fig. 1) included sidescan sonar and TV/photo surveys, conventional sediment sampling and massive sulphide sampling by a hydraulic grab equipped with a TV camera.

A towed sled equipped with a side-looking sonar (frequency 30 or 100 kHz), a TV and a 35 mm stereophoto camera system and an echosounder were used. The sonar surveys were carried out from 40 to 70 m above the seafloor at an average speed of 2 knots. The total swath coverage was about 250 m of seafloor on each side. Individual photos, taken at an average distance of 6 m above the seafloor, covered a bottom area of about 16.5 m².

The instrumented sled and the hydraulic grab sampler were acoustically navigated using systems with an array of up to seven transponders (mean square error of position determination less than 20 m). Ships were navigated by the GPS (mean square error less than 60 m).

The methods of chemical analysis used in our studies of hydrothermal aggregates have been detailed elsewhere (Krasnov *et al.* 1994).

Results

MARK field

The MARK (Snake Pit) hydrothermal field lies about 30 km south of the Kane Fracture Zone near 23°22′N and has previously been studied by

From PARSON, L. M., WALKER, C. L. & DIXON, D. R. (eds), 1995, *Hydrothermal Vents and Processes,* Geological Society Special Publication No. 87, 43–64.

43

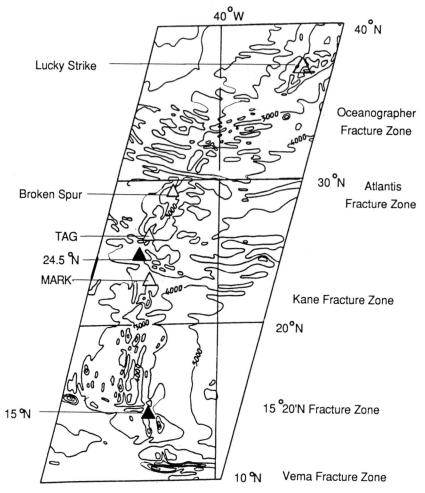

Fig. 1. Position of massive sulphide deposits of the Mid-Atlantic Ridge. Deposits discovered in Sevmorgeologija cruises are shown as solid triangles, the rest as open triangles.

drilling (Honnorez *et al.* 1990; Kase *et al.* 1990) and submersible dives (Thomson *et al.* 1988; Karson & Brown 1988; Mevel *et al.* 1989; Fouquet *et al.* 1993). The present work carried out during the third and fifth cruises of the R/V *Professor Logatchev* in 1991–1992 resulted in detailed mapping of the entire field at a scale of 1:2500, as well as sampling of the three major and several minor ore bodies and metalliferous sediments (Fig. 2).

The density of photo coverage obtained by towing along a systematic grid of profiles proved to be inadequate for studying such relatively small targets as the MARK sulphide deposits. Most of the photographic survey was therefore

carried out by positioning of the ship above certain sulphide bodies.

The general features of the geological setting of the MARK field are described by Krasnov *et al.* (this volume). The hydrothermal field located at the crest of the axial volcanic high is centred in a locally widened (up to 150 m) axial graben. Sulphide mounds are identified on the sonar images as chimney clusters.

Our data show that the MARK field is not as large (600 m long) as reported previously by Karson & Brown (1988). It appears to be 400 m long and 125–250 m wide oriented in an ENE direction (Fig. 2). The field may be contoured on the basis of the 30% areal coverage of the basalts

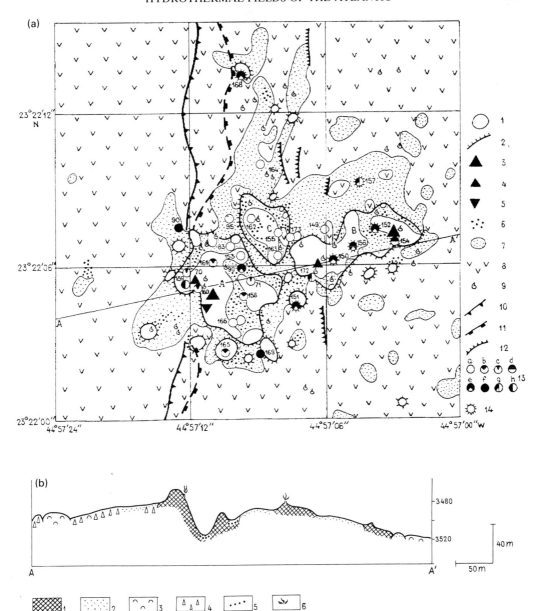

Fig. 2. Geological sketch (**a**) and cross-sectional profile (**b**) of the MARK hydrothermal field. (**a**) **1** outcrops of massive sulphides; **2** outlines of sulphide mounds (mounds 'A', 'B' and 'C' are described in the text); **3** black smokers; **4** white smokers; **5** fossil chimneys; **6** massive sulphide talus; **7** fields of metalliferous sediments; **8** basalt lavas and lava talus; **9** benthic communities; **10** upper rim of the western graben wall; **11** base of the western graben wall; **12** small faults, bounding the graben in the east; **13** sample sites (listed in Table 1) and dominant ore types, described in the text (**a** Mc–Py; **b** Bn–Cp; **c** Cp–Py; **d** Po–Sp–Cp; **e** Cp–Sp–Py; **f** Sp–Py; **g–h** combinations of different sulphide types in one grab sample: **g** Cp–Sp–Py and Mc–Py; **h** Sp–Py and Mc–Py); **14** chimney fragments in sulphide samples. A–A′ is the position of profile in Fig. 2b. (**b**) **1** Massive sulphides; **2** metalliferous sediments; **3** pillow lavas; **4** lava talus; **5** massive sulphide talus; and **6** chimney complexes and smokers. Position of the profile is shown in Fig. 2a.

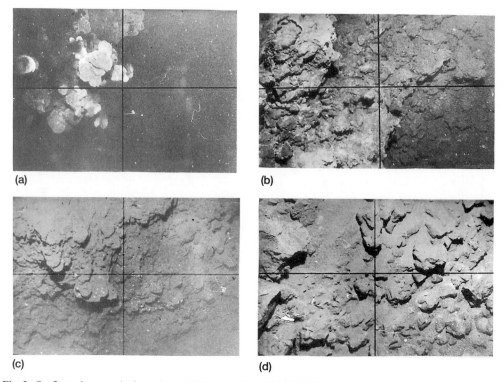

(a) (b) (c) (d)

Fig. 3. Seafloor photographs from the sulphide mound 'B', MARK field. (a) chimneys on top of the mound; (b) massive sulphides in the central part of the mound; (c) steep slope; and (d) sulphide talus at the base of the mound.

by hydrothermal sediments. Small massive sulphide deposits were discovered 50–100 m north of the main group of sulphide mounds and were surrounded by hydrothermal sediments. Single small sulphide deposits were also discovered 100 m to the west and 150 m to the south-east of the main field.

Eighteen massive sulphide deposits were identified by bottom photos (Fig. 3). Three of these are large mounds 20–50 m high. All the rest are small structures 5–10 m high and less than 25 m in diameter. The deposits are composed of accreted blocks of massive sulphides with metalliferous sediments occurring in the intervening areas. Sulphide and basalt talus cemented by hydrothermal sediments are developed at their base. The hydrothermal activity at some of the mounds is proved by the presence of 'smoke' and animal communities on photographs.

The largest, extreme western active massive sulphide deposit (marked 'A' on Fig. 2a) has a complex configuration and is about 100 m wide

and 50 m high. It runs 160 m along the western wall of the graben at a place where it is crossed by a north east-oriented depression. It forms a 15 m high cone above the graben wall (Fig. 2b) and is continuously traced across the wall with the meridional low at its base and eastward to the centre of the graben. In the low, massive sulphides outcrop from under sulphide talus, almost completely covering its floor. One more cone, 10 m high, exists in the eastern part of the deposit. A black smoker and a zone of white smoke emanating from fissures separating sulphide blocks were discovered on the eastern slope of the ore body corresponding to the graben wall.

The eastern active mound ('B' on Fig. 2a) is the second largest deposit. This is 150 m long, 30–60 m wide and 25 m high, trapezoidal in cross-section and strikes northeast. Ten chimneys more than 5 m high were identified in photographs on the flattened uplifted surface at the eastern margin. Many shrimps were observed in the clouds of black 'smoke' ejected by

Table 1. *Mineral and bulk chemical composition of massive sulphides from the MARK field*

Sample	Sample position within the mound	Sample weight kg	Ore types	Zn	Cu	Fe	Ag	Au	Si	S	Pb	Cd	As	Co	Sn	Bi	Mo	Mn	Sb
Mound A																			
70	Summit	110	Bn-Cp	0.53	45.04	20.44	90.00	0.30	0.55	32.87	55	30	<30	<5	31	0.5	<5	<100	<30
71-1	Upper slope	250	Bn-Cp	0.13	30.80	32.90	35.50	1.26	<0.05	35.86	3	<3	<30	360	29	1.4	78	<100	<30
160	Margin	1180	Cp-Py	2.33	25.00	29.05	79.53	7.10	5.47	30.88	32	39	200	1400	11	5.0	<5	<100	<30
158-1	Lower slope	120	Cp-Py	0.43	6.63	41.30	5.10	0.24	4.52	na	na	na	200	na	na	na	na	na	na
158-2	Lower slope	360	Cp-Py	1.49	11.00	36.75	22.16	0.74	4.52	34.94	64	25	300	1100	6	5.2	12	<100	<30
159-3	Summit	35	Sp-Py	22.13	0.88	18.55	173.60	1.10	19.63	25.61	1700	740	540	50	71	1.7	<5	<100	190
98	Margin	1260	Cp-Sp-Py	4.20	1.46	40.60	32.30	0.98	1.66	46.47	260	240	360	130	17	3.3	70	<100	<30
83	Lower slope	52	Mc-Py	0.08	0.17	45.85	12.00	0.16	<0.05	50.29	90	<3	350	330	5	1.7	72	<100	<30
153	Talus at the margin	16	Mc-Py	0.15	0.20	43.93	4.11	0.24	0.26	37.32	61	5	380	70	3	2.0	37	<100	<30
159	Summit	415	Mc-Py	2.33	0.84	39.03	23.40	1.90	3.69	30.14	76	32	380	250	10	3.4	14	<100	<30
161	Upper slope	100	Cp-Py	0.15	3.45	42.70	3.27	0.47	0.06	32.14	22	7	490	52	11	3.1	5	<100	<30
166	Upper slope	1300	Mc-Py	1.03	2.22	38.50	12.24	0.81	4.88	41.14	80	16	450	520	6	3.5	<5	<100	<30
Average A				3.09	10.79	35.65	46.32	1.28	3.70	36.19	222	102	346	387	18	2.8	27	<100	<30
Mound B																			
154	Margin	250	Po-Sp-Cp	8.00	11.38	29.75	59.00	0.18	6.78	31.43	180	140	350	720	22	8.6	15	<100	<30
150	Margin	700	Cp-Sp-Py	3.75	2.91	31.50	66.60	0.54	13.31	37.24	240	120	600	43	20	1.9	30	470	70
152	Margin	27.5	Cp-Sp-Py	6.44	5.07	30.19	46.30	0.81	10.62	30.76	240	270	300	270	15	9.3	13	<100	80
156	Margin	150	Cp-Sp-Py	4.82	2.36	25.81	20.69	9.10	20.69	31.34	270	800	220	82	25	8.5	30	420	100
172	Summit	700	Cp-Sp-Py	4.20	1.90	39.73	43.90	0.77	1.62	29.07	280	130	530	23	16	2.3	30	<100	<30
149	Talus at the base	450	Mc-Py	1.21	1.73	42.00	9.75	1.80	1.47	36.78	100	18	440	25	17	2.3	30	200	<30
172-2	Summit	300	Mc-Py	1.50	1.00	41.65	13.00	0.76	0.13	31.10	83	28	480	42	7	2.3	62	330	<30
173	Talus at the margin	26	Mc-Py	2.53	1.01	41.65	39.00	1.10	0.09	30.94	140	72	490	18	9	1.4	62	<100	<30
Average B				4.06	3.42	35.28	39.54	1.88	6.84	32.33	192	197	413	153	16	4.6	34		
Mound C																			
95	Talus at the base	670	Mc-Py	0.24	0.13	45.50	18.00	2.40	<0.05	53.15	130	<3	340	18	7	1.5	37	<100	<30
155	Margin	5	Mc-Py	0.22	0.19	44.98	1.91	1.30	0.25	34.17	64	<30	340	24	4	3.4	26	<100	<3
162-1	Summit	12	Mc-Py	0.25	0.90	43.22	4.83	0.43	0.06	49.34	54	7	360	20	7	1.5	15	<100	<30
163	Margin	57	Mc-Py	0.54	0.89	42.18	2.28	0.17	0.06	38.15	60	<3	440	5	6	3.2	13	<100	<30
Average C				0.31	0.53	43.97	6.75	1.08	0.10	43.70	77	3	370	61	6	2.4	23		
Small mounds																			
165	Margin	22	Cp-Sp-Py	11.66	6.25	31.50	93.00	2.00	0.66	39.30	200	280	600	170	35	3.3	10	<100	30
169	Margin	20	Sp-Py	17.19	0.44	21.70	233.60	4.50	10.42	31.11	650	540	1000	13	110	0.5	35	70	60
151-3	Margin	5	Cp-Sp-Py	5.62	1.55	12.69	161.30	2.10	36.72	16.32	930	250	690	70	13	12.0	40	800	130
168	Margin	60	Cp-Sp-Py	6.60	2.15	28.00	80.70	2.50	11.33	32.79	550	330	570	620	37	5.4	49	<100	100
90	Talus among sediments	580	Sp-Py	10.40	0.77	38.78	57.00	2.28	10.41	34.56	520	1000	260	27	28	0.6	27	<100	100
157-7	Talus among sediments	21	Mc-Py	1.11	0.90	43.05	10.93	0.12	na	na	na	na	na	na	na	na	na	na	na
157-8	Talus among sediments	45	Cp-Sp-Py	8.94	2.93	37.63	59.00	1.00	5.14	31.50	290	260	300	54	19	3.2	16	<100	50
164	Margin	48	Mc-Py	0.54	0.89	42.18	9.90	1.10	0.06	48.73	55	10	600	100	8	1.9	3	<100	<30
Average small mounds				7.27	1.98	31.94	88.18	1.95	10.68	33.47	456	381	574	150	36	3.8	26	<100	<30
Overall average				3.41	4.26	36.47	41.62	1.42	5.70	35.53	249	179	409	226	20	3.5	28		60

Note: Zn, Cu, Fe, Si and S are in wt. %, other elements in ppm.
S, Si and Al were determined by wet chemistry by O. Zaytzeva and T. Ivanova (Sevmorgeologija); atomic absorption spectrometry for Fe, Mn, Cu, Cd, Са, Mg, Zn and Pb was performed by N. Luneva (Sevmorgeologija)
and Au and Ag by M. Antonova and N. Smishlayeva (Sevzapgeologija).
Py, pyrite; Cp, chalcopyrite; Sp, sphalerite; Mc, marcasite; Bn, bornite.
Sampling sites are shown on Fig. 2.

Table 2. *Comparison of statistical data on bulk chemical compositions of MAR and EPR massive sulphides*

Element	TAG active mound				TAG inactive mound				MARK (surficial part)				MARK (core samples)				EPR			
	n	$\bar{x}$	$\bar{x}_g$	σ	n	$\bar{x}$	$\bar{x}_g$	σ	n	$\bar{x}$	$\bar{x}_g$	σ	n	$\bar{x}$	$\bar{x}_g$	σ	n	$\bar{x}$	$\bar{x}_g$	σ
Fe	41	30.19	28.74	9.18	137	27.87	25.34	9.69	152	33.60	32.08	9.31	14	37.23	37.17	2.42	360	30.98	28.44	10.72
Zn	39	5.24	2.83	12.25	136	8.69	3.36	9.63	152	4.59	1.20	6.67	14	4.65	3.34	2.63	364	6.33	2.13	8.28
Cu	39	9.25	1.66	7.22	134	5.01	1.38	9.12	152	9.03	2.43	13.07	14	10.16	9.54	3.48	348	9.50	3.05	11.84
Si	29	2.78	1.09	3.41	115	13.23	7.72	10.73	144	4.76	1.07	6.45	8	2.93	2.92	0.24	272	3.77	1.28	5.98
S	21	33.89	32.65	9.02	110	31.35	30.01	7.86	144	36.56	35.33	9.43	14	36.01	35.94	2.22	272	35.22	34.01	8.62
Ca	34	2.22	0.24	5.12	110	0.51	0.03	2.14	144	0.05	0.04	0.05	8	0.06	0.06	0.01	172	0.85	0.11	2.85
Pb	38	523	379	497	124	268	188	255	144	296	109	430	6	300	234	134	271	411	242	574
Au	30	2.08	1.35	3.29	126	4.20	2.25	5.50	152	1.77	0.86	2.16	8	0.60	0.58	0.14	114	0.46	0.30	0.52
Ag	38	70.45	35.74	77.91	129	112.58	59.20	140.61	151	61.99	34.15	62.65	6	15.83	14.14	6.64	215	50.74	21.27	65.46
Hg	na				98	5.20	2.50	8.40	5	16.10	15.20	5.60	na				47	11.30	7.10	11.60
Cd	26	144	66	179	124	212	47	400	144	175	35	256	6	221	192	94	206	259	82	349
Mn	na				111	150	76	229	144	132	75	197	na				77	212	89	540
Sb	9	12	9	9	122	54	24	57	152	43	26	50	8	20	20	0.20	62	7	3	11
As	12	83	70	64	124	144	114	85	152	365	277	213	8	113	93	62	150	217	128	225
Co	32	363	98	347	127	29	3	106	152	273	80	378	14	644	638	92	253	741	287	945
Sn	na				108	15	9	24	143	24	14	25	na				11	266	172	190
Bi	na				108	1.30	<0.5	2.70	144	2.9	1.70	2.80	na				na			
Mo	10	174	115	154	121	93	65	115	144	41	27	38	na				192	132	93	128

Note: Fe to Ca are in wt.%, Pb to Mo in ppm. n = quantity of analyses used. $\bar{x}$ = arithmetical mean, $\bar{x}_g$ = geometrical mean, σ = standard deviation. Data on the TAG active mound are from Rona *et al.* (1986), Lisitsyn *et al.* (1989) and from our samples; TAG inactive mounds from Lisitsyn *et al.* (1986), Rona *et al.* (1993) and our samples; MARK field (surficial part) from our samples; MARK field (core samples) from Honnorez *et al.* (1990) and Kase *et al.* (1990). Compositions of sulphides from the EPR are from Hekinian *et al.* (1980) and Bischoff *et al.* (1983) – 21°N; Hekinian and Fouquet (1985) and Fouquet *et al.* (1988) – 13°N; Bäcker *et al.* (1985) and Marchig *et al.* (1990) – 18°S and 21.5°S; our samples – 6°N, 9°N, 13°N and 21–22°S.

these chimneys. Another chimney cluster is situated at the southwestern flank of the mound. Five chimneys up to 1.5 m high emit white smoke. Dark sulphides are dusted by a white precipitate up to several metres from the chimneys.

The third largest mound ('C' on Fig. 2a) situated in the central part of the field is oval in shape, up to 70 m long, 50 m wide and 30 m high. It has steep slopes and a flattened top. We did not find any non-active chimneys as previously reported from the mound (Mevel et al. 1989).

More minor deposits are mostly conical agglomerations of sulphide blocks, partly covered by sulphide sediments. The unaltered chimney fragments found among hydrothermal sediments confirm the existence of other small sulphide deposits which are as yet unlocated.

Thirty-three large massive sulphide samples, between tens and hundreds of kilograms, were recovered by the heavy TV-equipped grab at 29 sites. Eight ore bodies were sampled. Three samples represent sulphide talus from outside the ore bodies. The element compositions of all these samples are given in Table 1. Statistical data on element concentrations in all the samples recovered during our expedition from surficial parts of the MARK field ore bodies, including the small samples of sulphide aggregates, are presented in Table 2.

Ore petrology and geochemistry were used to distinguish six basic ore types. Four belong mainly to zoned chimney fragments, namely (1) chalcopyrite–sphalerite–pyrite (Cp–Sp–Py), (2) bornite–chalcopyrite (Bn–Cp), (3) pyrrhotite–sphalerite–chalcopyrite, or isocubanite (Po–Sp–Cp, Ic) and (4) sphalerite–pyrite (Sp–Py). Two more types of ore material are massive sulphides without definite zonation, namely (5) marcasite–pyrite (Mc–Py) and (6) chalcopyrite–pyrite (Cp–Py).

The zonation of chimney fragments is shown by the transition from relatively low-temperature ore aggregates in their peripheral parts to high-temperature aggregates in central parts. Sp–Py (Fig. 4a and 4b) and Cp–Sp–Py ores are typical of the outer parts of chimneys. Subordinate amounts of gel-pyrite, mackinawite, isocunbanite, pyrrhotite and amorphous silica are present. At the chimney surfaces massive sulphides are characterized by highly porous, cavernous structures and colloform, dendritic textures with complex mineral intergrowth (Fig. 4a). Chalcopyrite increases towards the intermediate parts of the chimneys where mineral aggregates become more massive and globular in texture.

Massive Bn–Cp and Po–Sp–Cp, Ic aggregates

(Fig. 4c and 4d) are typical of the central parts of chimneys. Chalcocite, covellite, pyrite, marcasite, amorphous silica and quartz are minor minerals. The highest temperature phases, represented by chalcopyrite and isocubanite, are most enriched in the extreme inner parts of chimneys.

Quartz–hematite and amorphous silica–hematite mineralization was found in samples containing ores of Bn–Cp type.

Homogenous Mc–Py (Fig. 4e and 4f) and Cp–Py ores were recovered in irregular blocks and fragments, usually oxidized at the surface. Marcasite-bearing varieties are porous, others being more massive, recrystallized and often brecciated. Isocubanite is the only minor mineral found in these ore types.

The Sp–Py ores are most enriched in minor elements. The average concentrations in this ore type are 0.07% Pb, 0.05% As, 0.04% Cd, 0.009% Sb and 101 ppm Ag contents. The maximum enrichment of gold occurs in Cp–Sp–Py ores (average 2.9 ppm).

TAG field

The geological setting and structure of the TAG hydrothermal field (Fig. 5), situated within a bathymetric step at the base of the eastern rift valley wall near 26°N, has been described by many workers (Rona et al. 1986, 1993; Thomson et al. 1988; Zonenshain et al. 1989; Karson & Rona 1990; Krasnov et al. 1992). The main features of its tectonic control based on the results of our latest investigations are outlined by Krasnov et al. (this volume). Until recently, most effort in detailing the hydrothermal mineralization was concentrated on the active sulphide mound (Thomson et al. 1988; Lisitsyn et al. 1989; Lein et al. 1991), even though the existence of non-active mounds several kilometres northeast of the smokers had been known since 1985 (Rona et al. 1986; Lisitsyn et al. 1990). All the morphological elements of the mound (Fig. 6) were sampled during the Mir submersible dives of cruise 15 of R/V Akademik Mstislav Keldysh using the submersible arm manipulator. The four samples characterizing the different parts of the mound are described below and their composition is given in Table 3.

Sample #1 is from the mound outer slope. It is composed of chloride–oxide layered colloform aggregate. Thin layers of atacamite, ferrihydrite and hydrogoethite overgrow angular hyalobasalt fragments. The aggregate contains no relicts of primary sulphides or their pseudomorphs. According to our observations, the outer slope

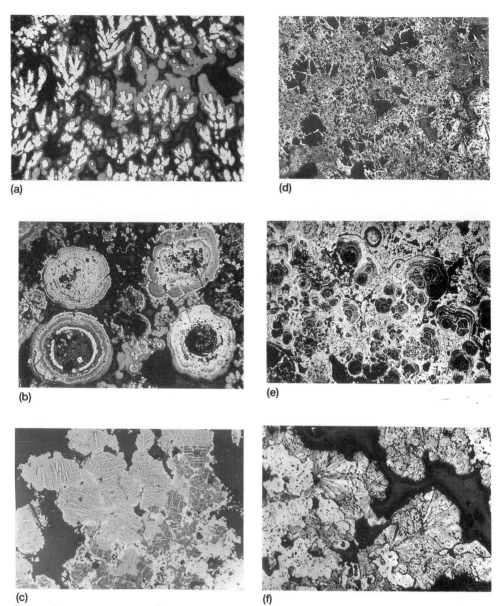

Fig. 4. Photomicrographs of polished sections from the MARK field. (**a**) Dendrites of marcasite (white) are overgrown by sphalerite (light grey) and amorphous silica (dark grey) in the shalerite–pyrite ore (×120). (**b**) Concentrically zoned pseudomorphic replacements of fauna in the sphalerite–pyrite ore. Marcasite (white), sphalerite (light grey), and amorphous silica show thin interlaying (×220). (**c**) Bornite–chalcopyrite ore. Bornite (light grey) with a thin network of platy chalcopyrite inclusions (white) is replaced by covellite (dark grey) and chalcopyrite (×120). (**d**) Platy aggregate of pyrrhotite [white with amorphous silica (grey)] in the pyrrhotite–sphalerite–chalcopyrite ore (×120). (**e**) Colloform rhythmically zoned aggregates of melnicovite–pyrite and amorphous silica (black interlayers) in the marcasite–pyrite ore (×120). (**f**) Overgrowth of pyrite (clear white) by radial aggregates of marcasite (light grey). Interstices are filled with amorphous silica (dark grey) (×220).

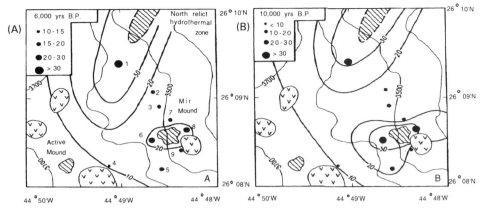

Fig. 5. Position of sediment cores (circles) from the TAG hydrothermal field relative to volcanic mounds (shown in v-hatching), sulphide mounds and the north hydrothermal zone of the TAG field (shown in diagonal hatching, after Rona *et al.* 1993). Core 1 from Metz *et al.* (1988), cores 2 to 4 from Lisitsyn *et al.* (1989), Core 5 from Shearme *et al.* (1983), the rest from this study. Circle sizes and heavy lines show Fe wt.% concentrations in sediments, recalculated on the carbonate-free basis; thin lines are isobaths from Rona *et al.* 1993. (**A**) – 6000 years BP and (**B**) 10 000 years BP.

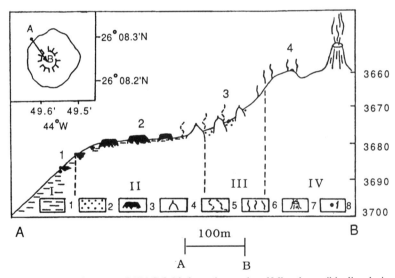

Fig. 6. Profile across the active mound, TAG field, from the results of Mir submersible dive during cruise 15 of the R/V *Akademik Mstislav Keldysh*. Upper left: plan view showing the outline of the mound, its summit surface and position of the profile A–B. **1** 'Ochres' (unlithified Fe-rich silica deposits with atacamite); **2** sulphide–sulphate talus; **3** blocky surface of massive sulphides; **4** anhydrite chimneys; **5** shimmering water; **6** black 'warm smokers'; **7** hot vent; **8** sampling sites (see Table 3 for sample description). I–IV: structural zones of the mound; **I** outer slope; **II** intermediate terrace; **III** inner slope; and **IV** summit surface.

of the mound contains fragments of basalt pillows.

Sample #2 was taken from the intermediate terrace. We did not observe any hydrothermal activity within the terrace. The sample represents Zn-rich massive sulphide (Fig. 7a). Sphalerite forms spherules 0.01–0.1 mm in diameter among coarse-grained marcasite interlayered with platy crystals with chalcopyrite emulsion.

Sample #3 is from the zone of shimmering

Table 3. *Mineral and bulk chemical composition of massive hydrothermal aggregates from the TAG active mound obtained in course of the submersible dive. Each of the four samples corresponds to the zone of the active mound of the same number*

Sample	Main minerals	Fe	Cu	Zn	S	SiO₂	Ca	Pb	Au	Ag	Sb	As	Co	Mo	Pt
1	At, FeH, Hy, AVG	14.03	35.33	0.40	4.07	7.20	0.31	150	0.57	4.2	0.99	170.0	34.0	47	na
2	Py, Mc, Sp	33.85	0.43	5.32	39.78	0.20	0.14	220	2.75	120.0	11.40	66.0	2.5	43	na
3	An, Cp, Py	12.20	3.10	0.05	29.56	0.23	20.12	55	0.15	1.2	<1	21.5	350.0	87	0.41
4	Py, Cp, AS	33.13	4.74	0.32	36.33	9.05	0.04	62	0.22	11.0	3.20	na	740.0	230	na

Note: Fe to Ca are in wt.%, Pb to Pt in ppm. Ca was determined by atomic absorption spectrometry at Sevmorgeologija, Pt by neutron activation analysis at the Geological Prospecting Institute of Rare and Noble Metals (Moscow). For methods of analysis of other elements, see Table 1.
Py, pyrite; Cp, chalcopyrite; Sp, sphalerite; Mc, marcasite; An, anhydrite; At, atacamite; Hy, hydrohematite; FeH, ferrihydrite; AS, amorphous silica; and AVG, altered volcanic glass.
Sample sites are shown on Fig. 6.

waters and black 'warm smokers' (below 100°) of the inner mound slope (Lisitsyn *et al.* 1990). Copper sulphides strongly dominate in this sample over Zn sulphides. All the sulphide minerals occur in interstices between anhydrite crystals (Fig. 7b). Chalcopyrite forms anhedral grains 0.02–0.06 mm in size clustered into groups no more than 0.15 mm in diameter. The single neutron activation analysis showed Pt concentrations previously unprecedented in oceanic massive sulphides (Table 3).

Sample #4 was taken from the upper mound surface near the hot vents. This sample is similar to sample #3 in its mineral composition, but amorphous silica is present instead of anhydrite. Chalcopyrite containing pyrite relicts replaces pyrite and forms granular aggregates and euhedral crystals. According to L.I. Bochek (pers. comm.) illite is formed during the alteration of volcanic glass and contains inclusions of palagonite, aragonite and plagioclase microlites.

The first data on the exact position and structure of a large non-active sulphide mound, referred to here as the Mir mound, were obtained during the R/V *Akademik Mstislav Keldysh* Cruise 23 in 1991 (Rona *et al.* 1993). In 1992–1993, during Cruise 6 of the R/V *Professor Logatchev*, detailed investigation and sampling of the Mir mound were carried out (Fig. 8).

Our photo/TV studies (Fig. 9) enabled the Mir mound to be mapped. It is a single mound about 400 m in diameter lying between 3420 and 3470 m water depth on the rim of a small terrace in the lower part of the rift valley wall above the level of the main TAG field. The non-active mound is partly covered by sediments. Outcrops which are probably small Fe–Mn oxide bodies with silica and separate clasts of the same composition appear to be covered by sediments at distances of up to 250 m from the mound.

Most of the massive sulphide mineral types found among our samples from the Mir mound (Fig. 7) are the same as described here for the MARK field. Cp–Py ores are most common (Figs 7c and 7d) comprising more than half of the ore material recovered. However, the hydrothermal deposits of the Mir mound differ from those of the MARK field in the abundance of silica-rich and hydroxide aggregates and the absence of pyrrhotite (only single relicts of its leached tabular crystals were seen in boxwork textures of silica). Amorphous silica or quartz are regular components of all ore types except Cp–Py.

Fig. 7. Photographs of polished sections from the active mound (**a, b**) and Mir mound (**c–h**) of the TAG field. Position of the samples from the active mound is shown in Fig. 6, from the Mir mound in Fig. 8. (**a**) Sphalerite–marcasite ore. Thin sphalerite interlayers (grey) are seen inside spherulitic marcasite aggregates (white). Sample 2 (×220). (**b**) Chalcopyrite (Cp) and pyrite (Py) in anhydrite. Sample 3 (×240). (**c**) Chalcopyrite–pyrite ore. Chalcopyrite (grey) overgrows and replaces recrystallized pyrite of the first generation (I) Pyrite of the second generation is marked (II). Sample 13 (×25). (**d**) Pyrite pseudomorphically replacing organic remains in chalcopyrite–pyrite ore. Sample 19 (×20). (**e**) Fragments of a melnikovite–pyrite layer at the contact of chalcopyrite–pyrite and marcasite–pyrite ores. Sample 20 (×30). (**f**) Sphalerite (grey) among thin pyrite dendrites in the sphalerite–pyrite ore. Sample 20 (×25). (**g**) Successive mineral formation in the sphalerite–chalcopyrite–pyrite ore: thin-grained chalcopyrite intergrown with amorphous silica (I), subhedral chalcopyrite (II), pyrite–marcasite intergrowth (III) and sphalerite (IV). Sample 20 (×25). (**h**) Spherulitic bornite aggregates (light grey) overgrowing amorphous silica (dark grey). Sample 5 (×30).

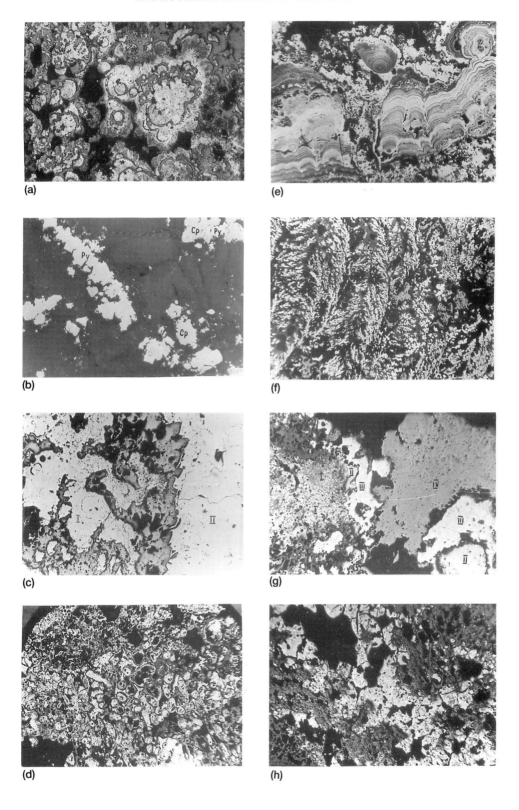

(a)

(e)

(b)

(f)

(c)

(g)

(d)

(h)

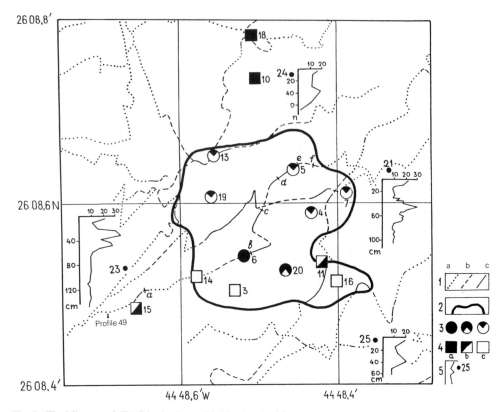

Fig. 8. The Mir mound, TAG hydrothermal field. **1** basalts (**a**), clasts of hydrothermal aggregates among sediments (**b**) and massive hydrothermal deposits (**c**) along the lines of TV/photo profiling (letters indicate the position of photographs in Fig. 9); **2** outline of the Mir mound; **3** massive sulphide samples, listed in Table 4, and dominant ore types (**a**) Sp–Py, (**b**) Cp–Sp, (**c**) Cp–Py; **4** low-temperature hydrothermal deposits, Mn-rich (**a**), Fe- and Mn-rich (**b**) and silica-rich (**c**); **5** sediment cores and vertical distribution of Fe in sediments (on a carbonate-free basis).

Owing to the absence of fresh chimneys, compositional and structural contrasts between chimney fragments and massive aggregates are not so distinctive as in the MARK field, although the recovered fragments of non-active chimneys are relatively enriched in chalcopyrite and sphalerite.

Among the large representative grab samples (Table 4), the samples enriched in Zn and silica also show the maximum enrichment of most of the minor elements. The average Pb concentration in Sp–Py ores reaches a maximum among all ore types of 0.03%. The same applies to Cd (0.05%), Sb (0.01%), Ag (0.02%) and Au (6.7 ppm). Some of the hydroxide aggregates are highly enriched in Co (Table 4).

The different types of hydrothermal material within the Mir mound show a well-defined zonation. Silica is the main component in the southern part of the mound, Zn sulphides in its central part and Cu sulphides in the northern part. Silica occurs mainly in the form of quartz in the western part of the mound and as amorphous silica in its eastern part.

Geochemical studies of four metalliferous sediment cores collected around the Mir mound at distances of up to 150 m from the mound (Table 5) showed similar downcore variations of metal concentrations in spite of the different density of sampling along the cores (Fig. 8). There are large peaks of Fe (Fig. 8), Cu and Zn enrichment at 20–50 cm below seafloor, where the sediments are of an unusual reddish colour. Metal concentrations also increase in surficial sediments.

15° Field

Anomalies of CH_4 and Mn in deep waters indicative of hydrothermal activity have been

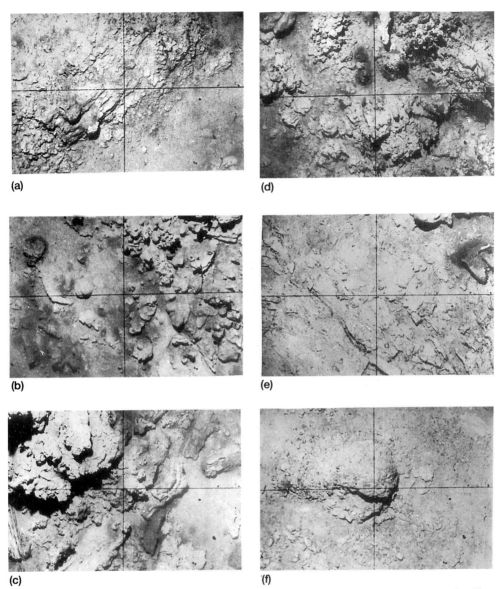

Fig. 9. Seafloor photographs along profile #49 across the Mir mound. Position of photographs located on Fig. 8. (a) Fe- and Mn-rich(?) hydrothermal deposits outcropping from under the sediments outside the mound; (b) surface of the mound partly covered by sediments; (c, d) inactive sulphide chimneys in the central part of the mound; (e) blocky surface of the hydrothermal deposit; (f) massive sulphide outcrops in the peripheral part of the mound. The field of view on each photograph is approximately 4.5 × 3.5 m.

known for some time above the MAR rift valley segment directly south of the 15°20'N fracture zone (Klinkhammer *et al.* 1985; Bougault *et al.* 1990). In addition, objects similar to sulphide deposits were photographed at 14°54'N near the base of the rift valley eastern wall (Eberhart *et al.* 1988). The existence of light attenuation,

temperature and dissolved Mn anomalies in the eastern part of the rift valley about 30 km south from the 15°20'N fracture zone was established during Cruises 10 and 12 of the R/V *Geolog Fersman* in 1991 and 1993.

Between November 1993 and February 1994, the RIFT towed system carrying temperature

Table 4. *Mineral and bulk chemical composition of massive hydrothermal aggregates from the 'Mir' mound, TAG field*

Sample	Type of material	Sample weight (kg)	Main minerals	Other minerals (rare in parentheses)	Fe	Cu	Zn	Si	S	Ca	Mn	Pb	Au	Ag	Hg	Cd	Sb	As	Co	Sn	Bi	Mo
13	Massive sulphide aggregate	150	Py,Cp	Mc (Sp)	34.12	18.10	0.20	2.43	40.00	0.06	0.02	100	0.67	9.90	0.78	12	<50	70	12	35	5.2	62
19	Chimneys and massive sulphide aggregate	1000	Py,Mc,Cp	(Sp,Is,Cv,Bn)	34.12	15.20	0.79	2.27	39.15	0.18	0.01	200	0.91	24.00	1.00	30	<50	70	<5	9	<1	62
7	Massive sulphide aggregate and ore breccia	150	Py,Cp	Cs,Ar	30.45	12.75	0.34	3.67	35.41	2.93	0.02	150	0.96	28.90	3.60	12	<50	90	160	23	4.2	130
5	Chimneys and massive sulphide–opaline aggregate	300	Py,Cp,AS	Mc,Bn (Cs, Cv,Id,Sp,At)	15.75	9.28	1.54	33.56	15.64	0.03	0.02	170	1.40	178.35	17.00	50	110	100	9	47	8.5	64
4	Sulphide breccia	100	Py,Cp,Ar	Cs,AS, (At)	19.60	11.75	1.53	10.53	18.05	8.61	0.06	120	1.14	41.20	2.60	20	<50	90	20	16	2.6	46
20-1	Massive aggregate	1200	Py,Mc,Sp,AS	Cp	27.65	2.40	7.55	12.80	37.35	0.01	0.02	600	5.05	225.15	9.20	270	120	150	<5	4	<1	36
20-2	Massive aggregate		Py,Mc,Cp,AS	Sp (Cv,Is)	32.55	4.75	3.22	10.24	37.32	0.01	0.01	200	2.70	69.65	3.60	95	100	170	<5	6	<1	54
6	Chimneys and massive sulphide aggregate	200	Sp,AS,Py	Mc,AS	19.42	1.21	15.05	19.47	26.51	0.18	0.07	250	7.50	239.35	17.00	550	150	260	29	30	<1	120
3	Massive quartz–sulphide aggregate	600	Q	Py,Sp,Mc (Cp)	12.95	0.81	1.90	40.79	15.24	0.06	0.09	100	1.05	34.35	2.00	12	<50	60	10	5	<1	51
16	Opaline aggregate	60	As	Sp,Py	04.10	0.12	0.96	54.24	4.09	0.06	0.06	600	3.37	440.00	16.00	12	140	<50	<5	3	<1	25
10	Oxide aggregate	30	MnHo	(FeHo)	2.24	1.30	0.67	4.80	0.06	1.20	41.20	<100	0.12	0.95	<0.01	12	400	70	50	<2	<1	1100
11-2	Opaline–oxide	40	As,FeHo		41.82	0.07	0.17	15.60	bd	0.14	2.31	200	0.08	2.68	0.09	<2	<50	90	<5	<2	4	90
11-1	Oxide aggregate	20	MnHo	FeHo, (At)	2.60	3.10	0.63	1.73	0.97	1.22	41.60	<100	0.21	1.80	<0.01	11	<50	<50	73	<2	<1	1080
15	Oxide aggregate	30	MnHo, FeHo	(At;As)	19.07	4.60	0.09	5.92	0.84	0.78	17.75	400	0.21	2.50	<0.01	5	<50	<50	>3000	<2	<1	940
18	Oxide aggregate	30	MnHo	FeHo,At	7.00	8.30	0.66	5.40	0.82	1.04	28.87	<100	0.20	1.17	<0.01	<2	<50	50	98	<2	<1	650
14	Opaline–oxide aggregate	50	AS,MnHo		2.00	0.45	0.14	45.28	1.10	0.59	6.46	<100	0.31	4.45	0.04	5	<50	55	280	13	<1	78

Note: Fe to Ca are in wt. %, all other elements in ppm.
For analytical methods, see Tables 1 and 3. Py, pyrite; Cp, chalcopyrite; Sp, sphalerite; Mc, marcasite; Bn, bornite; Ic, isocubanite; Cs, chalcocite; Cv, covellite; Id, idaite; At, atacamite; MnHo, Mn hydroxides; FeHo, Fe hydroxides; Q, quartz; AS, amorphous silica; and Ar, aragonite.
Sample sites are shown on Fig. 8.

Table 5. *Chemical composition of metalliferous sediments sampled near the 'Mir' mound, TAG field*

Station, sediment interval (cm)	CaCO₃	SiO₂ m.	Fe	Mn	Cu	Zn	Ni
St. 21							
0–4	71.43	0.02	6.27	1570	658	241	39
5–9	78.57	0.04	4.18	1450	306	182	23
10–14	79.46	0.04	2.22	1120	169	114	19
15–19	81.25	0.01	1.73	1040	154	129	23
20–24	79.02	0.01	2.07	1100	194	96	22
25–26	78.57	0.02	2.25	1050	178	106	22
27–28	78.57	0.26	2.16	1050	169	95	20
29–30	75.00	0.28	4.21	1280	288	162	23
31–32	69.64	0.33	4.05	1370	338	195	18
33–34	62.50	0.46	6.84	1670	528	301	33
35–36	57.14	0.37	7.72	2070	773	369	39
37–38	53.57	0.41	9.81	2250	852	457	49
39–40	48.21	0.50	11.25	2430	1181	509	48
41–42	43.30	0.14	13.97	2190	1851	624	52
43–44	43.30	0.21	17.05	2220	2364	721	42
45–46	45.98	0.21	13.41	2215	2248	714	58
47–48	47.30	0.21	13.35	2550	2449	683	45
49–50	48.04	0.16	11.45	2420	2091	643	70
51–52	51.30	0.10	9.07	2270	1671	552	46
53–54	54.64	0.16	8.10	2170	1422	524	47
55–56	55.96	0.08	6.70	1920	1246	449	42
57–58	57.30	0.09	6.36	1920	1090	451	39
59–60	56.64	0.10	6.47	1900	795	432	29
61–62	54.11	0.15	6.77	1550	758	456	29
63–64	49.98	0.14	9.00	1700	870	554	36
65–66	51.30	0.18	8.04	1540	779	495	33
67–68	56.64	0.14	7.97	1490	660	486	32
69–70	61.29	0.08	6.19	1470	322	301	21
71–74	66.96	0.20	2.97	1510	232	169	45
75–79	69.64	0.17	2.59	1210	199	137	40
80–84	71.43	0.20	2.24	1150	193	111	33
85–89	80.36	0.27	2.19	1135	165	96	18
90–94	69.64	0.45	2.55	1470	145	99	22
95–99	70.09	0.18	2.84	1475	162	102	20
100–104	69.64	0.22	2.76	1400	144	111	26
105–108	72.32	0.34	2.50	1300	182	86	17
St. 32							
0–6	63.39	0.08	7.60	1850	1087	392	27
6–16	75.45	0.17	2.42	1090	231	383	25
16–26	70.54	0.19	4.13	1500	506	653	37
26–31	27.68	0.30	22.60	3670	3364	3051	52
31–36	0.21	0.68	35.07	19 570	3279	4878	26
36–41	0.12	2.89	16.66	15 280	1467	2909	58
41–46	0.18	0.74	25.92	99 220	3235	5339	89
46–51	0.12	0.87	27.43	68 050	3765	4674	88
51–56	0.30	1.93	15.72	9910	1747	2018	63
56–61	19.87	0.57	8.05	3460	1060	1692	60
61–71	62.50	0.34	2.98	710	280	449	33
71–81	54.91	0.49	3.49	770	233	408	32
81–91	35.71	0.45	5.51	715	393	427	72
91–101	26.34	0.72	6.78	1750	479	527	64
101–106	17.14	0.77	6.97	1020	451	502	92
106–111	0.36	1.16	6.54	810	535	466	79
111–116	0.27	2.82	5.37	600	322	274	61
116–121	24.11	0.82	6.13	850	333	232	87
121–131	33.48	1.38	5.56	810	143	165	74
131–141	39.07	0.38	5.42	1150	197	138	76

continued

Table 5. *Continued*

Station, sediment interval (cm)	CaCO$_3$	SiO$_{2am.}$	Fe	Mn	Cu	Zn	Ni
141–151	34.82	0.38	5.03	980	190	141	102
151–156	34.82	0.31	4.64	1050	233	150	96
St. 24							
0–6	69.64	0.34	4.97	3540	691	243	37
7–12	75.89	0.62	2.91	1450	344	134	17
28–33	72.32	0.38	3.05	1260	387	140	27
34–39	51.79	0.47	8.98	2270	1415	441	44
53–57	67.86	0.54	3.50	1950	454	215	40
St. 25							
0–8	69.64	0.42	5.45	2080	790	259	25
10–20	79.46	0.40	1.87	1140	188	92	13
20–28	74.11	0.41	2.93	1400	337	138	25
28–33	62.50	0.32	6.18	1990	460	332	28
33–42	48.21	0.40	10.12	2690	1200	602	32
45–55	70.54	0.23	2.11	1130	177	145	14
55–65	73.21	0.41	1.96	1070	72	114	14

Note: CaCo$_2$, SiO$_{2am}$ and Fe are in wt.%, other elements in ppm.
CaCO$_3$ was determined by a volumetric method, amorphous silica by wet chemistry, other elements by atomic absorption spectrometry.
Sample sites are shown on Fig. 8.

and spontaneous electrical potential probes and potentiometric geochemical sensors was deployed for detailed studies of the same area during Cruise 7 of the R/V *Professor Logatchev*. The system operates between 30 and 40 m above the seafloor at a speed of 1 knot with an interval of 200–500 m between tracks. The results of the survey revealed two electrical potential and sulphide activity anomalies within a bathymetric step of the eastern wall.

Hydrothermal deposits were discovered during a subsequent TV/photo survey of the sites of these anomalies near 14°45'N (Fig. 10). The precise tectonic setting of the deposits is discussed by Krasnov *et al.* (this volume). Twelve sulphide mounds and two zones of sulphide talus developed among sediments were mapped (Fig. 11). The mounds are up to 20 m high. The largest is oval and 200 m long by 125 m wide. Two black smokers were observed at its crest, although no chimneys appear on the photographs and chimney fragments were rare among massive sulphide samples.

Samples from the mounds using the TV-grab included massive sulphides containing pyrite, marcasite, chalcopyrite, chalcocite, sphalerite and pyrrhotite as the principal minerals. Oxidized goethite–hydrogoethite aggregates with hematite were also present. Atacamite often covers the surfaces of sulphide samples. Element compositions of large sulphide samples

are given in Table 6. The highest Cu concentrations are connected with the presence of chalcocite in some samples.

The massive sulphide deposits and the volcanic rocks within 10–40 m are mostly covered by a thin layer of metalliferous sediments. Pelecypod and gastropod colonies and bacterial mats are present.

Hydrothermal deposits completely covered by a veneer of sediment were discovered 5.5 km east of the area described, although not sampled. Colonies of live pelecypods occur, probably indicating at least low-temperature activity at present.

Discussion

Comparison of hydrothermal fields

All three MAR hydrothermal fields studied are isometric in plan view, in contrast with the linear hydrothermal fields of the East Pacific Rise (EPR) axial graben (Bäcker *et al.* 1985; Fouquet *et al.* 1988; Krasnov *et al.* 1992). The MAR fields contain considerable quantities of massive sulphides occurring as a few, relatively large ore bodies.

The TAG field occupies the largest area (6 × 4 km). As far as we know, modern high-temperature sulphide-forming hydrothermal activity is confined to a single mound in the

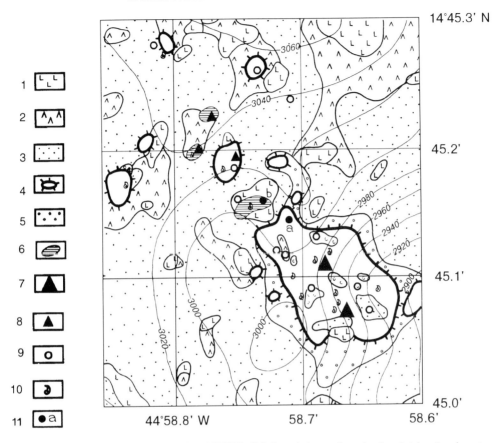

1

2

3

4

5

6

7

8

9

10

11

Fig. 10. Hydrothermal deposits at 14°45'N, 44°59'W. **1** Mafic and ultramafic rocks; **2** rock talus; **3** carbonate sediments; **4** sulphide mounds; **5** sulphide sediments; **6** low-temperature hydrothermal deposits; **7** high-temperature vents; **8** low-temperature vents; **9** hydrothermal benthos communities; **10** sample sites; **11** positions of photographs in Fig. 11.

southwestern corner of the field. This contrasts with the 'central-type' structure of the MARK field where the cluster of largest sulphide mounds is surrounded by smaller ore bodies.

The observed distributions of hydrothermal aggregates of various types within the active mound leads to the following conclusions.

1. The central part of the mound (upper surface and inner slope) is relatively enriched in Cu (samples ##3 and 4) and the peripheral part (intermediate step) in Zn (sample #2). The outer mound slope is mostly covered by low-temperature oxide, chloride and silicate aggregates (sample #1). The well-defined zonation of the active mound reflects its maturity (Krasnov *et al.* 1993).

2. The relative enrichment in Pb, As, Sb, Au and Ag is characteristic of the Py–Mc–Sp–

Cp association of the active mound peripheral zone. Co, Mo and Pt increase in the Py–Cp association of the central part.

3. The presence of basalt relicts in hydrothermal deposits of the active mound indicates complex spatial relations between massive sulphides and basalts. One possible explanation may be that ore material largely accumulates in the inner parts of massive sulphide bodies at the mature stage of their development (Krasnov *et al.* 1993). The ore is mainly deposited at this stage from fluids slowly moving through pores and cracks inside a sulphide mound. This results in 'swelling' of the mound. Basalt pillow fragments which could fall on the top of the ore body and become incorporated into its upper part at the early stages of formation may remain close to the surface, rising together with it.

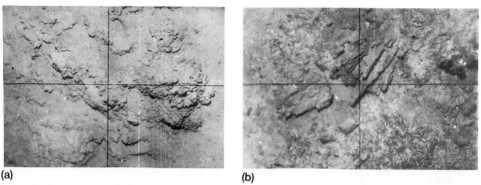

(a)   **(b)**

Fig. 11. Seafloor photographs from the 14°45′N hydrothermal field. (**a**) Surface of the hydrothermal deposit partly covered by sediments; (**b**) pelecypod covering the seafloor near an outcrop of ultramafic rocks. The field of view on each photograph is approximately 4.5 × 3.5 m. Positions of photographs located in Fig. 10.

Table 6. *Mineral and bulk chemical composition of massive sulphides from the 15°N field*

Sample	Sample wt (kg)	Main minerals	Other minerals	Cu	Zn	Pb	Fe	MnO	Cd	Co	Ag	Au
2	50	Cs,Bn,Cp,Ic	Bt	48.4	0.24	0.04	17.6	0.06	0.002	0.02	57	36.3
3	400	Po,Cp	AS,Ic,Sp	8.6	1.15	0.05	21.6	0.80	0.001	0.03	84	6.8
4	400	Cs,Bn,Cp,Sp,Py,As	Bt	5.2	5.75	0.06	33.1	0.20	0.007	0.06	72	3.2
6	30	Cs,Bn,Cp	–	33.2	0.20	0.02	21.6	0.06	0.007	0.02	89	0.3
9	250	AS,Bn,Po,Cp,Cs	Sp,Ic	8.4	2.10	0.05	27.0	0.03	0.004	0.06	65	8.2
12	400	Cv,At	–	22.2	0.11	0.04	6.5	0.08	0.001	0.02	84	6.1
13	500	Mc,Py,Fe,Ho,Cp	Cs,Bt,Sp	6.7	1.05	0.02	28.4	0.17	0.001	0.02	42	3.7

Note: Ag and Au are in ppm, other elements in wt.%.
Analyses were performed in the Geological Prospecting Institute of Rare and Noble Metals: Cu, Zn and Pb by atomic absorption spectrometry, other elements by wet chemistry. Po, pyrrhotite; Bt, barite; for other minerals see caption to Table 4. Sampling sites are shown on Fig. 10.

The Mir mound has been described by Rona *et al.* (1993) as subzone 2 of the whole Mir hydrothermal zone. The peaks of metal enrichment in the cross-section of sediments near the Mir mound (Table 5, Fig. 8) probably indicate two events of high-temperature activity. The synchronous sequence of events in all the four sites around the Mir mound makes mass wasting proposed by Metz *et al.* (1988) improbable as an origin for these layers. Metal-enriched layers are wholly composed of pelitic-size material. Hydroxide and sulphide clasts usual in hydrothermal turbidites of the same field (Bogdanov *et al.* 1994) are absent here. The layers could be formed by fallout of particulate material from the overlying hydrothermal plume. We have not dated these events directly in our cores but their ages (about 6000 and 10000 years) can be estimated from the results of dating other TAG cores and massive sulphides. Two metal-rich intervals from the core studied by Lisitsyn *et al.*

(1989) from the vicinity of the Mir mound (site 3 in Fig. 8) give ages of 6500 and 10000–13000 years, respectively. The youngest sulphides from the mound are dated at 9400 years (Rona *et al.* 1993).

The areal distribution of Fe in sediments formed about 10000 and 6000 years ago is shown in Fig. 5. It is evident that metal inputs from the north hydrothermal zone and the Mir mound were about equal during the first of these events, but the north zone input dominated during the second event. The presently active mound already existed at that time (Lalou *et al.* 1990, 1993), but its input did not significantly influence metal distribution in sediments on the scale of the whole TAG field.

The field discovered at 14°45′N is similar to the TAG field in its setting within a bathymetric step of the eastern rift valley wall. However, in the dimensions of sulphide deposits and in the character of their distribution (Fig. 10) it may be better compared with the MARK field, having the 'central-type' structure with the largest mound in the centre surrounded by smaller sulphide bodies. Although larger than the MARK field in diameter, it is hydrothermally less active. Sediments cover 63% of its surface. Chimneys are absent at most of the mounds, indicating the greater age of the sulphides.

The total volume of sulphide deposits from the MARK field was estimated by Karson & Brown (1988) to be 784 000 m^3. According to our estimates, the ore reserves of the MARK field are less than previously reported, about 270 000 m^3. The active TAG mound contains 1.5×10^6 m^3 of massive sulphides (Rona et al. 1986). We estimate the volume of the Mir mound, TAG field, to be 2.7×10^6 m^3. The largest ore body of the 15°N field contains about 205 000 m^3 of sulphides.

Chemical composition of sulphides

Element concentrations determined in ores of the TAG and MARK fields are close to those previously reported for these fields. Zinc and Cu concentrations in our large grab samples from the Mir mound (Table 4) do not reach the maximum values reported by Rona et al. (1993) who sampled separate Zn- and Cu-enriched hydrothermal spires from a submersible.

As seen from Table 2, the average concentrations of most elements in massive sulphides from the TAG and MARK fields are of the same order. Relatively high Ca concentrations in the TAG active mound samples may be at least partly due to the presence of anhydrite in chimneys. We did not sample active chimneys in the MARK field. The silica concentration increases in non-active TAG mounds in accordance with the observation that silicification is generally a late-stage process in the evolution of oceanic ore bodies (Krasnov et al. 1993). Concentrations of a number of elements (Sb, As, Au) that often closely associate with Si in oceanic sulphides (Krasnov et al. 1992) also increase in non-active mounds. The drill core samples from the MARK field (Honnorez et al. 1990; Kase et al. 1990) show lower concentrations of elements typical of surficial low-temperature aggregates (As, Sb, Ag, Au) compared with dredge samples or those obtained using the submersible. The most remarkable general feature of the core samples is the normal distribution of most elements (arithmetic and geometric means in Table 2 are about equal). The log-normal distribution of all elements except Fe and S in surficial samples is typical of oceanic massive sulphides (Krasnov et al. 1992).

Sulphides of the newly found hydrothermal field at 15°N are exceptionally rich in Cu (Table 6). Its maximum and average concentrations are higher than in any oceanic sulphide deposits sampled before. Zinc concentrations are relatively low. The most Cu-rich sample is also remarkably rich in Au (36.3 ppm).

Relations between sizes and compositions of sulphide deposits

According to the results of previous studies, statistical differences exist between the relative proportions of base metals in oceanic massive sulphide deposits of different sizes (Krasnov et al. 1993). The small ore bodies typical of the EPR axial graben are most often very rich in Zn, whereas the relative proportion of Fe and Cu increases in large sulphide mounds such as those of the MAR. Similar compositional relations exist between deposits of different sizes studied within the single extensive hydrothermal field near 13°N, EPR (Fouquet et al. 1988).

As a result of our study, we confirmed this trend. The extensive sulphide deposits from the TAG, MARK (Table 2) and probably also from the newly found 15°N fields are generally enriched in Cu compared with the usually smaller deposits of the EPR.

Different average compositions of massive sulphides characterize ore bodies of different sizes at the MARK field. Various compositional types of ores were sampled from the two largest active ore bodies (Table 1). Only Mc–Py ores were found in the large, but non-active 'C' sulphide body. Four of the five small ore bodies sampled appeared to be composed of Cp–Sp–Py ores. As seen from Table 1, only in these small deposits does Zn definitely dominate over Cu, whereas in large mounds Cu dominates, or is approximately equal to Zn.

According to the results of thermodynamic modelling performed by D.V. Grichuk and published by Krasnov et al. (1993), increasing rates of ore alteration in deep parts of the oceanic hydrothermal systems cause a decrease in Zn and an increase in Cu concentrations in ore-forming solutions in the course of evolution of these systems. This trend is reflected in the relative Cu enrichment in end-member solutions of the TAG field. Copper concentrations in these solutions are the highest measured for submarine vent fluids (Edmond et al. 1990), with approximately similar temperatures of about 350°C. On the axis of the EPR, hydrothermal systems usually do not reach the mature stages of their development because of frequent structural reorganizations associated with high tectonic and volcanic activity.

However, it is difficult to apply this explanation to processes at the MARK field where both large and small deposits fed by a single hydrothermal system have different Cu/Zn ratios. The differences in the efficiency with which metals deposit in ore bodies from solutions may provide an alternative explanation. At early ('black smoker') stages of the development of ore bodies, Zn and its associated elements carried by high-temperature solutions are much more efficiently deposited than Cu and Fe during the initial mixing of solutions with seawater (Krasnov *et al.* 1993). At late stages of sulphide mound growth, when the jet expulsion of fluids is replaced by slow advection through pores and cracks in sulphides, the metal ratios in sulphides become much closer to those of the hydrothermal solutions. In the MARK field, this results in the domination of Zn over Cu in most of the small sulphide deposits and of Cu over Zn in at least two of the three large examples sampled.

The waning stage of hydrothermal activity during which only the relatively low-temperature solutions reach the surface of ore bodies probably results in the covering of previously formed ores by low-temperature Fe sulphides. This may explain the absolute dominance of pyrite and marcasite in ores sampled from the surface of the non-active 'C' deposit, which has already passed through the whole cycle of development.

Conclusions

Detailed mapping of the MARK hydrothermal field and of the Mir non-active hydrothermal mound of the TAG field in the MAR rift valley was carried out with the help of TV/photo-profiling during the third, fifth and sixth cruises of the R/V *Professor Logatchev*. Large (tens to hundreds of kilograms) massive sulphide samples, taken by a TV-controlled hydraulic grab from different parts of ore bodies, and sediment cores allowed the main ore types to be identified and the zonation of the ores to be studied.

The MARK field appears to be 400 m long and up to 250 m wide, oriented in a ENE direction. Zinc sulphides dominate over Cu sulphides in the small ore bodies, but Cu sulphides either dominate in the larger mounds or the average concentrations of the two metals are about equal.

Submersible studies of the TAG active mound during Cruise 15 of the R/V *Akademik Mstislav*

Keldysh showed the enrichment of sulphides in its central part in Cu, Co, Pt and Mo and in the peripheral part in Zn, Pb, Sb, Ag and Au. This zoning reflects the relative maturity of the mound. The outer slopes are mostly covered by low-temperature hydrothermal aggregates including atacamite. The relatively high concentrations of Au in these aggregates supports the suggestion of its primary enrichment in low-temperature hydrothermal deposits (Krasnov *et al.* 1992).

The largest known Mir mound of the TAG field appears to be a single massive deposit 400 m in diameter, partly covered by sediments. Cu-enriched massive sulphides were recovered from its northern part, the central part is relatively Zn-enriched and quartz or amorphous silica dominate in the southern part. The mound is surrounded by a zone about 250 m wide of small low-temperature Fe and Mn oxide deposits with silica.

According to chemical analyses from the cores, the Mir mound and especially the north hydrothermal zone not investigated in our studies are both presently inactive, but controlled metal distribution in the TAG sediments during the last 10 000–13 000 years. The contribution from the presently active smoker was much lower.

The new hydrothermal field with extensive Cu-rich massive sulphide deposits was discovered at 14°45′N during the seventh cruise of the R/V *Professor Logatchev* of 1993–1994. It is comparable with the MARK field by its size and structure, but is evidently older and less hydrothermally active.

The study strongly suggested that large, mature, massive sulphide deposits formed on the MAR in long-lived hydrothermal systems in conditions of relatively stable volcano-tectonic development are mostly Cu-enriched. Two explanations may be offered (Krasnov *et al.* 1993). One of them, based on thermodynamic data of D.V. Grichuk, presumes that the domination of Cu over Zn in the MAR hydrothermal solutions and sulphides appears during the advanced stages of the development of convective systems, when most of the easier leachable Zn is already lost from the basalts. Alternatively, the relative efficiency of Cu sulphide deposition may increase at late stages of the formation of the ore bodies when the ejection of solutions from black smokers is replaced by diffusive penetration through previously deposited ores.

The authors thank all the participants of the R/V *Professor Logatchev* cruises who helped to obtain these data. We especially thank the head of Cruise 15 of the R/V *Akademik Mstislav Keldysh*, A.P. Lisitsyn, for providing to one of us the opportunity of submersible work and the expedition staff involved in the dive. We appreciate the efforts of G.P. Glasby who reviewed the initial version of this paper. S. Petersen and an anonymous reviewer provided extremely valuable comments on the manuscript. The study was partly supported by research grant # R1C000 of the International Science Foundation (Soros Foundation) to the first author.

References

BÄCKER, H., LANGE, J. & MARCHIG, V. 1985. Hydrothermal activity and sulfide formation in axial valleys of the East Pacific crest between 18 and 21°S. *Earth and Planetary Science Letters*, **12**, 9–22.

BISCHOFF, J. L., ROSENBAUER, R. J., ARUSCAVAGE, P. J., BAEDECKER, P. A. & CROCK, J. G. 1983. Sea-floor massive sulfide deposits from 21°N, East Pacific Rise, Juan de Fuca Ridge and Galapagos Rift: bulk chemical composition and economic implications. *Economic Geology*, **78**, 1711–1720.

BOGDANOV, YU. A., RONA, P. A., GURVICH, E. G., KUPTSOV, V. M., RIMSKI-KORSAKOV, N. A., SAGALEVITCH, A. M. & HANNINGTON, M. D. 1994. [Relict sulphide structures of the TAG hydrothermal field, Mid-Atlantic Ridge (26°N, 45°W)]. *Okeanologiya*, **34**, 590–599 [in Russian].

BOUGAULT, H., CHARLOU, J.-L., FOUQUET, Y., APPRIOU, P. & JEAN-BAPTISTE, P. 1993. L'hydrothermalisme oceanique. *Société Géologique de France Memoir*, **163**, 99–112.

EBERHART, G. L., RONA, P. A. & HONNOREZ, J. 1988. Geologic controls of hydrothermal activity in the Mid-Atlantic rift valley: tectonics and volcanics. *Marine Geophysical Researches*, **10**, 233–259.

EDMOND, J. M., CAMPBELL, A. C., PALMER, M. R. & GERMAN, C. R. 1990. Geochemistry of hydrothermal fluids from the Mid-Atlantic Ridge: TAG & MARK 1990. *EOS, Transactions of the American Geophysical Union*, **71**, 1650–1651.

FOUQUET, Y., AUCLAIR, G., CAMBON, P. & ETOUBLEAU, J. 1988. Geological setting and mineralogical and geochemical investigations of sulfide deposits near 13°N on the East Pacific Rise. *Marine Geology*, **84**, 143–178.

——, WAFIC, A., CAMBON, P., MEVEL, C., MEYER, G. & PASCAL, G. 1993. Tectonic setting and mineralogical and geochemical zonation in the Snake Pit sulfide deposit (Mid-Atlantic Ridge at 23°N). *Economic Geology*, **88**, 2018–2036.

HEKINIAN, R. & FOUQUET, Y. 1985. Volcanism and metallogenesis of axial and off-axial structures on the East Pacific Rise near 13°N. *Economic Geology*, **80**, 221–249.

——, FEVRIER, M., BISCHOFF, J. L., PICOT, P. & SHANKS, W. C. III 1980. Sulfide deposits from the East Pacific Rise near 21°N. *Science*, **207**, 1433–1444.

HONNOREZ, J., MEVEL, C. & HONNOREZ-GUERSTEIN,

B. M. 1990. Mineralogy and chemistry of sulfide deposits drilled from hydrothermal mound of the Snake Pit active field, MAR. *Proceedings, Ocean Drilling Program*, **106/109**, *Scientific Results*, 145–162.

KARSON, J. A. & BROWN, J. R. 1988. Geological setting of the Snake Pit hydrothermal site: an active vent field on the Mid-Atlantic Ridge. *Marine Geophysical Researches*, **10**, 91–107.

—— & RONA, P. 1990. Block-tilting, transfer faults, and structural control of magmatic and hydrothermal processes in the TAG area, Mid-Atlantic Ridge 26°N. *Bulletin of the Geological Society of America*, **102**, 1635–1645.

KASE, K., YAMAMOTO, M. & SHIBATA, T. 1990. Copper-rich sulfide deposits near 23°N, Mid-Atlantic Ridge: chemical composition, mineral chemistry, and sulfur isotopes. *Proceedings, Ocean Drilling Program*, **106/109**, *Scientific Results*, 163–177.

KLINKHAMMER, G., RONA, P., GLEAVER, M. & ELDERFIELD, H. 1985. Hydrothermal manganese plumes in the Mid-Atlantic Ridge valley. *Nature*, **314**, 727–731.

KRASNOV, S. G., CHERKASHEV, G. A. & 21 others 1992. [*Hydrothermal Sulfide Ores and Metalliferous Sediments of the Ocean*]. Nedra, St Petersburg, 279 pp [in Russian].

——, GRICHUK, D. V. & STEPANOVA, T. V. 1993. Evolutionary trends in composition of oceanic massive sulfide deposits. *Resource Geology Special Issue*, **17**, 173–179.

——, STEPANOVA, T. & STEPANOV, M. 1994. Chemical composition and formation of a massive sulfide deposit, Middle Valley, northern Juan de Fuca Ridge (Site 856). *Proceedings, Ocean Drilling Program*, **139**, *Scientific Results*, 353–372.

——, POROSHINA, I. M. & CHERKASHEV, G. A. Geological setting of high-temperature hydrothermal activity and massive sulphide formation on fast- and slow-spreading ridges. *This volume*.

LALOU, C., REYSS, J.-L., BRICHET, E., ARNOLD, M., THOMPSON, G., FOUQUET, Y. & RONA, P. 1993. New age data for Mid-Atlantic Ridge hydrothermal sites: TAG and Snakepit chronology revisited. *Journal of Geophysical Research*, **98**, 9705–9713.

——, THOMPSON, G., ARNOLD, M., BRICHET, E., DRUFFEL, E. & RONA, P. A. 1990. Geochronology of TAG and Snakepit hydrothermal fields, Mid-Atlantic Ridge: witness of a long and complex hydrothermal history. *Earth and Planetary Science Letters*, **97**, 113–128.

LEIN, A. YU., ULYANOVA, N. V., GRINENKO, V. A. & LISITSYN, A. P. 1991. [Geochemical features of hydrothermal sulfide ores from the Mid-Atlantic Ridge (26°N)]. *Geochymiya*, **3**, 307–319 [in Russian].

LISITSYN, A. P., BOGDANOV, Y. A., ZONENSHAIN, L. P., KUZMIN, M. I. & SAGALEVITCH, A. M. 1989. Hydrothermal phenomena in the Mid-Atlantic Ridge at latitude 26°N (TAG hydrothermal field). *International Geology Review*, **31**, 1183–1198.

——, SAGALEVITCH, A. M., CHERKASHEV, G. A. &

SHASHKOV, N. L. 1990. [Investigation of the hydrothermal vent in the Atlantic from 'Mir' submersible]. *Doklady Akademii Nauk SSSR,* **311**, 1462–1467 [in Russian].

MARCHIG, V., PUCHELT, H., ROSCH, H. & BLUM, N. 1990. Massive sulfides from ultra-fast spreading ridge, East Pacific Rise at 18–21°S: a geochemical stock report. *Marine Mining,* **9**, 459–493.

METZ, S., TREFRY, J. H. & NELSEN, T. A. 1988. History and geochemistry of a metalliferous sediment core from the Mid-Atlantic Ridge at 26°N. *Geochimica et Cosmochimica Acta,* **52**, 2369–2378.

MEVEL, C., AUZENDE, J.-M. & 9 others. 1989. La ride du Snake Pit (dorsale medio-Atlantique, 23°22′N): resultats preliminaires de la campagne HYDROSNAKE. *Comptes Rendus de l'Academie des Sciences Paris, Serie II,* **308**, 545–552.

RONA, P. A., BOGDANOV, YU. A., GURVICH, E. G., RIMSKI-KORSAKOV, N. A., SAGALEVITCH, A. M., HANNINGTON, M. D. & THOMPSON, G. 1993. Relict hydrothermal zones in the TAG Hydrothermal Field, Mid-Atlantic Ridge 26°N, 45°W. *Journal of Geophysical Research,* **98**, 9715–9730.

——, KLINKHAMMER, G., NELSEN, T. A., TREFRY, J. H. & ELDERFIELD, H. 1986. Black smokers, massive sulfides and vent biota at the Mid-Atlantic Ridge. *Nature,* **321**, 33–37.

SHEARME, S., CRONAN, D. S. & RONA, P. A. 1983. Geochemistry of sediments from the TAG hydrothermal field, MAR at latitude 26°N. *Marine Geology,* **51**, 269–291.

THOMPSON, G., HUMPHRIS, S., SCHROEDER, B., SULANOWSKA, M. & RONA, P. 1988. Active vents and massive sulfides at 26°N (TAG) and 23°N (Snake-pit) on the Mid-Atlantic Ridge. *Canadian Mineralogist,* **26**, 697–711.

ZONENSHAIN, L. P., KUZMIN, M. I., LISITSYN, A. P., BOGDANOV, Y. A. & BARANOV, B. V. 1989. Tectonics of the Mid-Atlantic rift valley between TAG and MARK areas (24–26°N): evidence for vertical tectonism. *Tectonophysics,* **159**, 1–23.

Characteristics of hydrothermal discharge following a magmatic intrusion

EDWARD T. BAKER

*Pacific Marine Environmental Laboratory, NOAA, 7600 Sand Point Way NE, Seattle,
WA 98115-0070, USA*

Abstract: Seafloor hydrothermal systems are profoundly altered by magmatic fluctuations,
which are inherently episodic and generally unpredictable. At present, three examples of
hydrothermal systems perturbed by magmatic intrusion have been identified and sampled:
1986 on the Cleft segment of the Juan de Fuca Ridge; 1991 at 10°N on the East Pacific Rise;
and 1993 on the CoAxial segment of the Juan de Fuca Ridge. From the fragmentary
observations at each site three trends can be identified that may be common to magmatically
perturbed hydrothermal systems. The flux of heat and mass can increase by orders of
magnitude virtually simultaneously with a magmatic intrusion by the sudden and short-lived
release of event plumes, while chronic discharge remains elevated for months or years
afterwards. Vent fluid composition is altered by at least two processes. Phase separation is
initiated or enhanced, producing fluids highly enriched in the vapour phase. The conjugate
brine-enriched fluid may be stored in the crust to be flushed months or years later by
convecting seawater. Magmatic degassing increases the flux of volatiles, temporarily
elevating ^{3}He/temperature ratios. Time series observations of magmatically altered vent
fields are vital, because chemical budget extrapolations and hypotheses derived only from
observations of stable hydrothermal discharge may be incomplete or unreliable.

Most of the available information about seafloor hydrothermal processes and fluxes has been obtained by measuring and sampling fluids and dispersed plumes from vent emissions that vary little in composition or magnitude between observations. Even though fluid chemistry may differ greatly within and between vent fields because of phase separation, sediment influence, rock composition and other factors, time series observations of individual vents commonly show stability on the order of years (e.g. Von Damm & Bischoff 1987; Campbell *et al.* 1988; Massoth *et al.* 1989; Butterfield *et al.* 1994). This image of stability, however, must be tempered with the realization that seafloor hydrothermal fields are hosted by the largest volcanic system on the planet. Volcanic systems are inherently episodic and generally unpredictable. The influence of episodic magmatic/volcanic events on hydrothermal flux and composition remains poorly understood.

Ridge-crest observations over the last few years have revealed that hydrothermal discharge occurs in two distinct styles. Discharge that continues over a period of months to years, as focused and/or diffuse flow, creates a chronic plume that may wax or wane but continually issues from the vent field. I prefer the term 'chronic' to 'steady state' because neither the discharge nor the resultant plume may be steady over this time frame. The frame of reference for chronic discharge is the vent field, because individual orifices may expire as others commence in response to changing circulation pathways in the crust. In contrast with the familiar chronic plumes are powerful and instantaneous event plumes that may release months or years worth of chronic discharge in a few hours or days. Even at the discovery of event plumes it was hypothesized that they were products of local magmatic/tectonic activity (Baker *et al.* 1987; Cann & Strens 1989), but recognition of the impact of such activity on subsequent chronic discharge has developed more slowly.

There are presently only three confirmed instances of ridge- crest eruptions that produced well-sampled examples of changes in hydrothermal discharge patterns and composition: August 1986, on the Cleft segment of the Juan de Fuca Ridge (JDFR); April 1991, on the East Pacific Rise (EPR) near 10°N; and July 1993, on the CoAxial segment of the JDFR (Fig. 1). Discovery techniques ranged from pure serendipity on the JDFR in 1986 to a sophisticated detection/response programme on the JDFR in 1993. Neither the circumstances nor the observations were similar in any of these cases, but together they provide an initial framework for evaluating the thermal and chemical effects

From PARSON, L. M., WALKER, C. L. & DIXON, D. R. (eds), 1995, *Hydrothermal Vents and Processes*,
Geological Society Special Publication No. 87, 65–76.

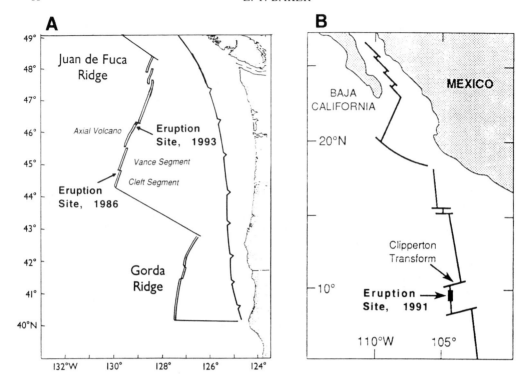

Fig. 1. Schematic maps of Pacific Ocean spreading centres showing location of eruption sites on (**A**) the Juan de Fuca Ridge in 1986 (Cleft segment) and 1993 (CoAxial segment) and (**B**) the northern East Pacific Rise in 1991.

associated with the creation or reinvigoration of hydrothermal discharge by a magmatic intrusion.

Three examples of disturbed systems

The discovery of 'event plumes' occurred in August 1986 during a systematic survey of hydrothermal plumes over the Cleft segment of the JDFR (Baker *et al.* 1987, 1989) (Fig. 2). A conductivity/temperature/depth/transmissometer (CTDT) package towed in a sawtooth pattern above the seafloor mapped a 700 m thick, 20 km diameter eddy-like megaplume (EP86) overlying a 200–300 m thick chronic plume from vent discharges that continue to the present day (Baker 1994). The hydrothermal origin of EP86 was confirmed by high concentrations of ^{3}He, Mn, Fe and Si, while the recovery from the plume centre of highly soluble anhydrite crystals (Feely *et al.* 1987) as large as 150 μm in length meant it was only a few days old when sampled (Baker *et al.* 1987, 1989). A plume model that relates heat flux to the observed plume rise found that about 10^8 m^3 of fluid with a total heat content of 10^{17} J were

necessary to generate the plume (Baker *et al.* 1989). Although EP86 was thoroughly sampled and annual sampling of the chronic plume continues to the present day, no vent fluid sample was collected until 1988 (Massoth *et al.* 1994; Butterfield & Massoth 1994). Chadwick *et al.* (1991) linked EP86 to a seafloor volcanic event by using repeat bathymetric surveys to identify a series of volcanic mounds totaling 0.05 km^3 and emplaced between 1981 and 1987 (Fig. 2). A second, smaller megaplume (EP87) was discovered in September 1987 over the Vance segment of the JDFR 45 km north of the first (Baker *et al.* 1989). This plume was 20 ± 9 days old when sampled, according to ^{222}Rn dating (Gendron *et al.* 1993). Its discharge location is unknown.

In April 1991, DSV *Alvin* conducted a dive series at a vent field on the EPR between 9°45′ and 9°52′N almost immediately after, if not simultaneously with, an extensive seafloor eruption (Haymon *et al.* 1993). A virtually complete video mapping of this vent field by the Argo imaging system in 1989 (Haymon *et al.* 1991) allowed precise pre- and post-eruption comparisons. Extensive vent fluid sampling over a

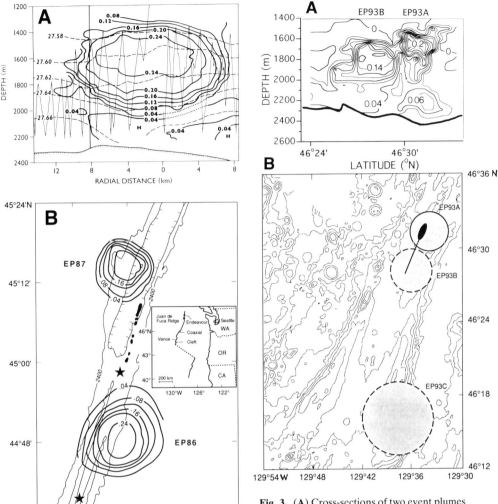

Fig. 2. (**A**) Cross-section of event plume EP86 (megaplume) and underlying chronic plume. Contours are isolines of hydrothermal temperature anomaly (°C). The temperature anomaly is the deviation along the potential temperature axis of the normally linear relationship between potential density and potential temperature. Potential density (kg m^{-3}) isolines shown as broken lines. Thin line in sawtooth pattern is track of the CTDT tow-yo path. (**B**) Plan view showing locations of EP86 and EP87 in relation to known hydrothermal fields (stars) and volcanic eruption mounds emplaced between 1981 and 1987 (solid beads). Megaplume contours in temperature anomaly. Origin of EP87 is unknown.

Fig. 3. (**A**) Cross-sections of two event plumes mapped over the CoAxial segment in July 1993. EP93B was an apparently mature, well-mixed plume at least several days old when discovered, whereas the shape and complex internal structure of EP93A suggests a very recent or still-forming event plume (Baker *et al.* 1995). Both plumes were smaller and had lower temperature anomalies than EP86 or EP87. (**B**) Plan view of all three CoAxial event plumes showing their relative locations when first discovered. EP93A and EP93B were found directly over the lava eruption mound (solid bead) and warm-water fissure (heavy line) near 46°30′N. EP93C was found two weeks after EP93A and EP93B; its discharge site is unknown. The size of EP93A was constrained by both along- and across-axis tows. The sizes of EP93B and EP93C are based on a single tow and an assumption of radial symmetry.

Table 1. *Hydrothermal heat supply*

Source	Annual heat supply (J)	Reference
Observed event plumes	$2 \times 10^{15} - 10^{17}$*	Baker *et al.* (1989), Baker *et al.* (1995)
Global ridge crest (<1 Ma)	$10^{20} \pm 0.1 \times 10^{20}$	Stein & Stein (1994)
Endeavour vent field, JDFR	$3-40 \times 10^{16}$	Baker & Massoth (1987), Rosenberg *et al.* (1988), Thomson *et al.* (1992)
North Cleft vent field, JDFR	$2-8 \times 10^{16}$	Baker (1994), Gendron *et al.* (1994), Baker *et al.* (1993)
TAG vent field, MAR	$2-4 \times 10^{16}$	Klinkhammer *et al.* (1986), Rudnicki & Elderfield (1992)
21°N vent field, EPR	7×10^{15}	Converse *et al.* (1984)
ASHES vent field, JDFR	$5-25 \times 10^{14}$	Rona & Trivett (1992)
Black smokers	$10^{13}-10^{15}$	Calculated from typical vent dimensions and discharge

JDFR, Juan de Fuca Ridge; MAR, Mid-Atlantic Ridge; and EPR, East Pacific Rise.
* Event plume annual heat supply is hydrothermal heat inventory of individual event plumes.

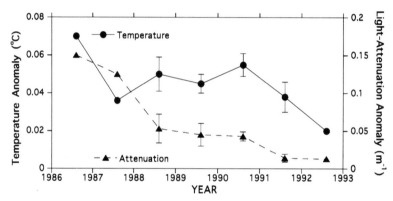

Fig. 4. Annual mean and standard deviation of temperature (solid circles) and light-attenuation (triangles) anomalies within the chronic plumes at the EP86 site on the Juan de Fuca Ridge. Error bars shown only for those years in which more than two CTDT tow- yos passed through the chronic plume. The chronic plume was more intense at the time of the EP86 discovery than at any subsequent time.

two-week period documented changes in temperature, chlorinity, gases and other constituents (Von Damm *et al.* 1991; Lilley *et al.* 1991; Lupton *et al.* 1991; Haymon *et al.* 1993). This initial sampling was followed a year later by another DSV *Alvin* programme (Von Damm *et al.* 1992; Lilley *et al.* 1992; Shanks *et al.* 1992). Vent sites outside the eruption area were also sampled for comparison. Unfortunately, plume sampling in April 1991 consisted only of CTD profiles on DSV *Alvin* descents, so hydrographic evidence for or against event plumes is inconclusive. A detailed plume survey was completed in November 1991 (Baker *et al.* 1994).

On June 22 1993, the NOAA/VENTS Program began real-time monitoring of seafloor acoustic events detected by the US Navy's northeast Pacific SOSUS hydrophone network (Fox *et al.* 1995). Between 26 June and 13 July, the network detected low- level seismic activity, with accompanying harmonic tremor, from the JDFR between 46°14′N, 129°50′W and 46°31′N, 129°36′W (Fox *et al.* 1995). More than 200 individual seismic events were recorded. Dziak *et al.* (1995) interpreted this seismic swarm as an along-axis dyke injection similar to those observed at Krafla (Iceland) and Kilauea (Hawaii) volcanos. A series of cruises to this area over the next four months found new volcanic flows, apparently new vent fields and several examples of event plumes with volumes up to 30% of the 1986 megaplume (Fig. 3)

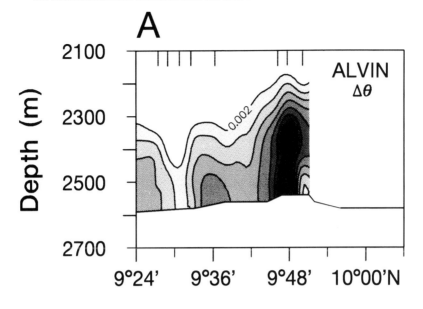

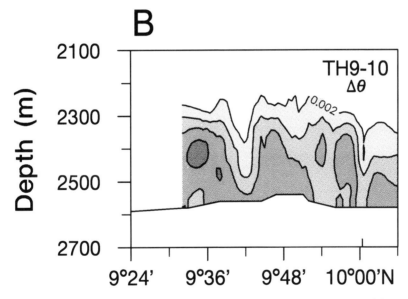

Fig. 5. Along-axis transects of temperature anomaly at the East Pacific Rise eruption site in (**A**) April 1991 and (**B**) November 1991. Contour interval is 0.002°C. Maximum anomaly in core of plume at 9°48′N in April was about 0.03°C. Ticks along the top axis of (**A**) mark the location of CTD profiles used to construct the transect; transect (**B**) is a continuous CTDT tow-yo. The April maximum was located directly over the eruption site (9°45′–9°52′N).

(Embley *et al.* 1995; Chadwick *et al.* 1995; Baker *et al.* 1995). Event plumes were observed only during the first few weeks of the response effort. Chronic plumes were most intense immediately after the eruption and measurably declined over a four month observation period. Systematic vent fluid sampling was not conducted until a DSV *Alvin* cruise in October 1993, and results are not yet available.

Because of the logistical complexities involved in oceanographic sampling, especially sampling at the seafloor, each of these three

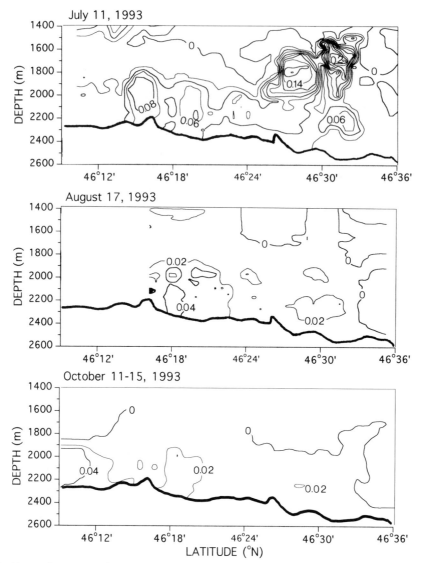

Fig. 6. Along-axis transects of temperature anomaly in chronic plumes along the CoAxial segment of the Juan de Fuca Ridge from CTDT tow-yos on 11 July, 17 August and 15 October. Contour interval is 0.02°C in each case. Two event plumes were detected near 46°30′N on 11 July. Intensity and rise height of plumes over the two main discharge sites, 46°30′N and 46°16′–19′N, decreased markedly in the three months after the seismic eruptions. Plumes at the 46°10 N site were not discovered until 15 October; the history of venting at this site before 15 October is uncertain.

eruptions received a response with a different sampling emphasis. Although these differences make for imperfect comparisons, we can use parts of each to construct a hypothetical history of the discharge following an eruption event that encompasses both fine-scale sampling immediately after an eruption and annual-scale sampling lasting for nearly a decade.

Temporal patterns of hydrothermal flux and fluid composition

Flux enhancement–event and chronic plumes

Individual event plumes surveyed along the JDFR since 1986 incorporated hydrothermal heat equivalent to the total annual supply of a typical vent field, and 10–1000 times that of a single black smoker (Table 1). The horizontal and vertical symmetry characteristic of event plumes (Figs 2 and 3) implies formation by a single, continuous and brief fluid release.

In addition to the pulse of fluid discharge from event plumes, it also appears that chronic discharge from low- and high-temperature orifices is enhanced by an eruption. The chronic plume at the EP86 site has been mapped each year since 1986, and the mean intensity was greatest at the time of the megaplume discovery (Fig. 4). Concentrations of hydrothermal Mn, Fe and ^{3}He in the chronic plume were also greatest in 1986 (Massoth *et al.* 1994). Plume intensity at this site has declined significantly, though not continuously, between 1986 and 1991 (Baker 1994).

Plume mapping was cursory at the time of the 1991 EPR eruption, but Macdonald *et al.* (1992) constructed an along-axis transect of hydro-thermal temperature anomaly from profiles recorded by a CTD mounted on DSV *Alvin* during eight descents (Fig. 5). Plume intensity directly above the eruption site was several times higher in April than during a detailed plume survey seven months later in November (Baker *et al.* 1994).

Because of the location and timing of the 1993 CoAxial event (just days before the first of several scheduled cruises to the JDFR), we can assemble a detailed history of plume intensity during the four months immediately following the eruption. CTDT tow-yo transects along the axial strike of the presumed dyke injection were carried out on 11 July, 17 August and 11–15 October (Fig. 6). A 1991 pre-eruption tow that reached as far north as 46°17′N along this same transect found no hydrothermal indicators (E.T. Baker, unpubl. data). On 11 July, 14 days after the first seismic events, plume maxima with temperature anomalies of 0.06 to 0.08°C were centred at 46°31′N, 46°23′N, 46°19′N and 46°16′N lying below the event plumes. No event plumes were found along the axis after 15 July. By 17 August the plume at 46°23′N had dissipated and the intensity of plumes at the 46°31′N, 46°19′N and 46°16′N sites had decreased by more than half. Two months later plumes at all three sites had weakened even further, while a previously undiscovered and substantial plume was found farther south along the same trend, at 46°10′N. Any connection between the 46°10′N source and the seismic events is uncertain, but the nearby 1991 tow found no evidence of a plume in that area.

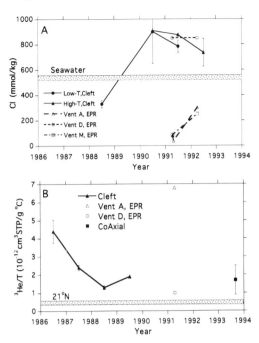

Fig. 7. (**A**) Variability of Cl concentration in vent fluids from the Cleft segment (solid lines, closed symbols) and East Pacific Rise (EPR) (broken lines, open symbols) eruption sites. Horizontal bar marks average seawater value. On the EPR, vents A and M were in the eruption zone, while vent D was about 30 km south. Error bars on the Cleft segment data are ±1 standard deviation based on measurements at several sites (Butterfield & Massoth 1994). The EPR data are single measurements (Shanks *et al.* 1992; Von Damm *et al.* 1992). (**B**) ^{3}He/*T* ratio in chronic plumes at the Cleft segment (solid line, closed triangles) and CoAxial segment (closed square) eruption sites, and in vent fluids from EPR eruption site (open symbols). Horizontal bar marks 'normal' value typical of stable vent fields such as at 21°N on the EPR (Lupton *et al.* 1980). Error bars on the Cleft and CoAxial segment data are ±1 standard deviation (Baker & Lupton 1990; Massoth *et al.* 1994; Lupton *et al.* 1995). ^{3}He/*T* ratios from the EPR estimated from total He concentration (Lupton *et al.* 1991) and a typical vent fluid ^{3}He/^{4}He ratio of 1.1×10^{-5} (Lupton *et al.* 1980).

Compositional evolution of chronic vent fluids and plumes

Chronic plumes from magmatically disturbed vent fields show several characteristics consistent with magmatic degassing and/or phase separation of hydrothermal fluids. Several hydrothermal tracers can illustrate these processes; we here limit the discussion to two of the most diagnostic: chlorinity and ^{3}He/temperature ratio.

Chlorine concentration is a sensitive indicator of the extent of phase separation by vent fluids into vapour and brine components (e.g. Von Damm 1988; Cowen & Cann 1988). The presence of a low- chlorinity vapour phase in a hydrothermal system could reflect subcritical and/or supercritical phase separation when seawater contacts or very hot rock, as during a dyke injection. Alternatively, or additionally, low-chlorinity vapour could arise directly from magma degassing during depressurization and cooling, followed by exsolution of brine as crystallization proceeds. The presence of vent fluids with chlorinity significantly different from seawater ($540 \, \mathrm{mmol \, kg^{-1}}$) does not, however, require a recent and local injection of magma. Vent fields with stable (over several years) chlorinities below and/or above seawater (typically 50–200% seawater chlorinity) are common (summarized by Von Damm 1990). A systematic temporal change in chlorinity is the indicator of recent magmatic activity.

Temporal changes in chlorinity have been observed at both the Cleft segment and the EPR at 10°N (Fig. 7). At the Cleft segment, low-temperature fluids were first sampled in 1988 and high-temperature fluids were not found until 1990, so the chlorinity trend is only weakly constrained for the first four years after the 1986 megaplume. Butterfield & Massoth (1994) interpret the observed trend as one of large-scale but temporary (1986 to 1988 or 1989) venting of low-chlorinity fluids produced by phase separation during a dyke intrusion associated with the megaplume. A transition to brine-enriched fluids that had been accumulating in the crust occurred by 1990. Dilution of this brine by non-phase-separated hydrothermal fluid has evidently lowered the chlorinity since 1990 or 1991.

Sampling of vent fluids during the first year after the 1991 eruption on the EPR supports the pattern deduced from the Cleft segment data (Fig. 7). During the two week period immediately following discovery of the eruption, chlorinity (and dissolved silica) at vent A within the eruption zone decreased while temperature increased (Fig. 8). These changes suggest an increasing proportion of phase-separated vapour in the emissions and a readjustment of the fluid–quartz equilibrium to conditions consistent with a fluid circulation depth of only 200 m (Shanks et al. 1992; Von Damm et al. 1992; Haymon et al. 1993). Approximately a year later, the chlorinity of vents A and M (also within the eruption zone) had risen from lows of 38 and $85 \, \mathrm{mmol \, kg^{-1}}$ in 1991 to 300 and $250 \, \mathrm{mmol \, kg^{-1}}$, respectively, in March 1992.

Dissolved silica and temperature at vent A underwent concomitant changes (Fig. 8). These trends indicate that one year after the eruption a lower proportion of the discharge was vapour and that the depth of circulation had increased to about 1500 m (Shanks et al. 1992; Von Damm et al. 1992), approximately the depth of the seismically identified magma chamber (Detrick et al. 1987). Vent D , about 30 km south of the eruption zone, remained stable at $850 \, \mathrm{mmol \, kg^{-1}}$ from 1991 to 1992.

Helium-3, derived only from mantle magma, unequivocally indicates magma degassing and hydrothermal activity. The ratio of ^{3}He to temperature was initially assumed to be uniform in submarine hydrothermal systems (Edmond et al. 1982), but later observations revealed that it can vary significantly between systems depending on the dynamic conditions of the ridgecrest at the vent site (Lupton et al. 1989). The first observation of a temporal change in the ^{3}He/T ratio within a hydrothermal system was made at the EP86 site (Baker & Lupton 1990) (Fig. 7). The ^{3}He/T trend in the chronic plume was interpreted as arising from an initial surge of ^{3}He from degassing magma, followed by a decline as the residence time of fluids in the circulation system increased and the extraction rates for heat and He converged. Poreda & Arnórsson (1992) similarly argue that preferential loss of ^{3}He relative to heat in young magmas causes initially high but eventually decreasing ^{3}He/enthalpy ratios in Icelandic geothermal systems.

Time-series data on post-eruption ^{3}He/T trends are not yet available at either 10°N on the EPR or the CoAxial segment on the JDFR. However, samples collected from vents and chronic plumes do reveal elevated ^{3}He/T ratios immediately after both eruptions (Fig. 7). At the EPR, the ^{3}He/T ratio in vent A fluids (about $6.9 \times 10^{-12} \, \mathrm{cm^3 \, STP \, g^{-1} °C^{-1}}$) (Lupton et al. 1991) was similar to that sampled in 1986 in the chronic plume at the EP86 site (about $4.4 \times 10^{-12} \, \mathrm{cm^3 \, STP \, g^{-1} °C^{-1}}$) (Baker & Lupton 1990) (Fig. 7). Vent D , outside the eruption zone, yielded a 'normal' value of about $1 \times 10^{-12} \, \mathrm{cm^3 \, STP \, g^{-1} °C^{-1}}$, similar to stable vent fields such as 21°N on the EPR (Lupton et al. 1980). At the CoAxial eruption site, 50 samples collected from the chronic plume in July and August, 1993, had a mean ^{3}He/T ratio of 1.7 ± 0.8 (1σ) $\times 10^{-12} \, \mathrm{cm^3 \, STP \, g^{-1} °C^{-1}}$ (Lupton et al. 1995). This value is modest relative to that associated with the 1986 Cleft segment eruption, perhaps because of the comparatively small size of the CoAxial eruption. Chadwick et al. (1995) report that the CoAxial eruption produced only $5.4 \times 10^6 \, \mathrm{m^3}$ of lava, compared with $51 \times 10^6 \, \mathrm{m^3}$

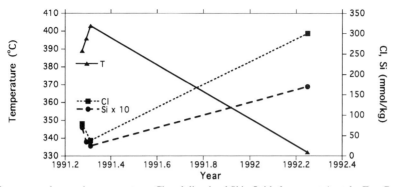

Fig. 8. Short-term changes in temperature, Cl and dissolved Si in fluids from vent A at the East Pacific Rise eruption site (Von Damm *et al.* 1992; Shanks *et al.* 1992; Haymon *et al.* 1993). Increasing *T* and decreasing Cl and Si after the March 1991 eruption indicates that at least several weeks were required to reach the maximum physical and chemical perturbation of hydrothermal fluids after a magmatic intrusion.

emplaced between 1983 and 1987 at the EP86 site.

Discussion

The piecemeal observations presently available from axial eruption sites indicate that seafloor spreading events perturb hydrothermal systems in several typical ways. Firstly, heat and mass flux are increased virtually instantaneously by the release of event plumes. Chronic flux is initiated or invigorated and declines over months or years depending on the amount of new magmatic heat available. Secondly, post-eruptive discharge has a high proportion of phase-separated vapour, with measured chlorinity as low as 7% of seawater. The proportion of vapour also declines over months to years. Thirdly, post-eruptive discharge has a high concentration of volatiles produced by phase separation and exsolution of gases from the intruded magma. Volatile concentrations in the discharge should decline faster than the vapour concentrations owing to the relatively brief effect of magma degassing. Other effects exist as well.

These perturbations are idealized in Fig. 9, a hypothetical summary of the trends of heat flux, ^{3}He/T ratio and Cl concentration in hydrothermal discharge after a magma intrusion. The magnitude of changes in Cl and ^{3}He/T is derived from actual plume and vent fluid measurements at the three eruption sites, but the magnitude of flux changes is conjectural owing to the difficulty of quantifying total vent field flux. Time and relative intensity values are keyed principally to EP86, which has the longest sampling history; both will probably vary according to the size of

the magma intrusion. The most uncertain of the three trends is that of Cl, particularly in the light of multi-year observations of stable chlorinity values both higher and lower than seawater (e.g. Von Damm 1990; Butterfield *et al.* 1994). Temporal changes in Cl at specific vent orifices are well documented, however (Fig. 8); at the Cleft eruption site variations of Li, B and Mn relative to Cl suggest that variations in the high-temperature Monolith vent discharge between 1990 and 1992 can be explained by mixing a previously formed brine with convecting seawater (Butterfield and Massoth 1994).

A conceptual model formulated to explain event plume discharges and their possible traces in the geological record supports the trends hypothesized in Fig. 9 (Cathles 1993). Cathles (1993) postulates that hydrothermal discharge originates in a narrow (*c.* 3 m wide) zone separated from 1200°C magma by a few hundred metres thick thermal boundary layer. A tectonic or magmatic disturbance (e.g. deflation of the deep magma reservoir during a shallow dyke intrusion) can increase the permeability of this zone enough to briefly increase the volume flux to megaplume levels, followed by a return to pre-megaplume flux. Cracking of the high-temperature thermal boundary layer accelerates the leaching of volatiles and hence increases the ^{3}He/T ratio. The permeability increase also allows seawater to invade further into the thermal boundary layer and undergo phase separation. The vapour phase is expelled into the upflow zone to produce an immediate low-Cl discharge, and the brine is only later flushed out to produce a subsequent high-Cl discharge. Cathles (1993) further argues that petrological observations in the Bushveld gabbro (Schiffries

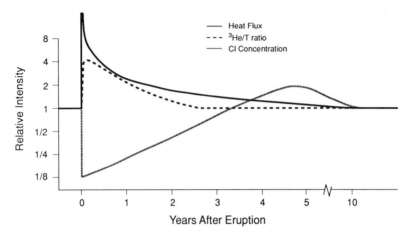

Fig. 9. Hypothetical trends of heat flux, ^{3}He/T ratio and Cl concentration in vent fluids following a magmatic intrusion. Time and heat flux scales are arbitrary and probably depend on the size of the intrusion; the scales here are based on changes following the 1986 cleft segment eruption. Pre-eruptive values of ^{3}He/T and Cl concentrations are assumed to be about 0.5×10^{-12} cm^3 STP g^{-1}°C^{-1} and 540 mmol kg^{-1}, respectively, although these values are known to vary along the mid-ocean ridge.

& Skinner 1987) describe alterations consistent with fractures filled first with brine and vapour and later flushed of the remaining brine.

Conclusions

Sampling of hydrothermal plumes and vent fluids in the days and years after a local magmatic intrusion is logistically complex and dependent to a greater degree than desirable on luck. Nevertheless, in the past eight years there have been three opportunities for such sampling, enough to at least pose some hypotheses about the effect of magmatic intrusions and seafloor eruptions on hydrothermal systems. The discovery of a 'megaplume' over the Cleft segment of the JDFR in 1986 and the suggestion of its link with a contemporaneous seafloor eruption instantly secured an appreciation of the importance of hydrothermal events. The fortunate encounter of DSV *Alvin* with a seafloor eruption at 10°N on the EPR in 1991, together with extensive follow-up work, is providing detailed temporal and spatial sampling of hydrothermal fluids altered by a sudden magmatic intrusion. Extensive water column and seafloor mapping and sampling in response to a seismically detected dyke intrusion on the CoAxial segment of the JDFR in 1993 confirmed the connection between event plumes and magmatic intrusions and is providing a second opportunity for post-eruptive time series sampling.

Soon after the discovery of seafloor vents it became clear that hydrothermal discharge is a major or substantial source of elements such as Mn, Ca, Si and others (Edmond *et al.* 1979). Global budget calculations originally assumed a steady state and globally invariant discharge. Continued observations have discredited both these assumptions. Although the importance of hydrothermal circulation as an oceanic chemical regulator remains incontestable, budget specifics for many elements are uncertain. The high flux and extreme compositional variations characteristic of magmatically disturbed hydrothermal systems may significantly affect the global budgets of some elements to a degree not yet appreciated.

This research is supported by the VENTS Program of the National Oceanic and Atmospheric Administration. G. Massoth, R. Feely, G. Cannon, S. Walker, D. Tennant and others have been invaluable in helping collect plume data over the last decade. I thank J. Lupton for his conception of an early version of Fig. 9. Contribution No. 1522 from NOAA/Pacific Marine Environmental Laboratory.

References

BAKER, E. T. 1994. A six-year time series of hydrothermal plumes over the Cleft segment of the Juan de Fuca Ridge. *Journal of Geophysical Research*, **99**, 4889–4904.
—— & LUPTON, J. L. 1990. Changes in submarine hydrothermal helium-3/heat ratios as an indicator of magmatic/tectonic activity. *Nature*, **346**, 556–558.
—— & MASSOTH, G. J. 1987. Characteristics of hydrothermal plumes from two vent fields on the

Juan de Fuca Ridge, northeast Pacific Ocean. *Earth and Planetary Science Letters*, **85**, 59–73.

——, FEELY, R. A., MOTTL, M. J., SANSONE, F. J., WHEAT, C. G., RESING, J. A. & LUPTON, J. E. Hydrothermal plumes along the East Pacific Rise, 8°40′ to 11°50′N: Plume distribution and relationship to the apparent magmatic budget. *Earth and Planetary Science Letters*, **128**, 1–17.

——, LAVELLE, J. W., FEELY, R. A., MASSOTH, G. J., WALKER, S. L. & LUPTON, J. E. 1989. Episodic venting of hydrothermal fluids from the Juan de Fuca Ridge. *Journal of Geophysical Research*, **94**, 9237–9250.

——, MASSOTH, G. L., WALKER, S. L. & EMBLEY, R. W. 1993. A method for quantitatively estimating diffuse and discrete hydrothermal discharge. *Earth and Planetary Science Letters*, **118**, 235–249.

——, FEELY, R. A., EMBLEY, R. W., THOMSON, R. E. & BURD, B. J. 1995. Hydrothermal event plumes from the CoAxial seafloor eruption site, Juan de Fuca Ridge. *Geophysical Research Letters*, **22**, 141–150.

——, —— & —— 1987. Cataclysmic hydrothermal venting on the Juan de Fuca Ridge. *Nature*, **329**, 149–151.

BUTTERFIELD, D. A. & MASSOTH, G. J. 1994. Geochemistry of north Cleft segment vent fluids: temporal changes in chlorinity and their possible relationship to recent volcanism. *Journal of Geophysical Research*, **99**, 4951–4968.

——, MCDUFF, R. E., MOTTL, M. J., LILLEY, M. D., LUPTON, J. E. & MASSOTH, G. J. 1994. Gradients in the composition of hydrothermal fluids from the Endeavour segment vent field: phase separation and brine loss. *Journal of Geophysical Research*, **99**, 9561–9583.

CAMPBELL, A. C., BOWERS, T. S., MEASURES, C. I., FALKNER, K. K., KHADEM, M. & EDMOND, J. H. 1988. A time series of vent fluid compositions from 21°N, East Pacific Rise (1979, 1981, 1985), and the Guaymas Basin, Gulf of California (1982, 1985). *Journal of Geophysical Research*, **93**, 4537–4549.

CANN, J. R. & STRENS, M. R. 1989. Modeling periodic megaplume emission by black smoker systems. *Journal of Geophysical Research*, **94**, 12 227–12 237.

CATHLES, L. M. 1993. A capless 350°C flow zone model to explain megaplumes, salinity variations, and high-temperature veins in ridge axis hydrothermal systems. *Economic Geology*, **88**, 1977–1988.

CHADWICK, JR, W. W., EMBLEY, R. W. & FOX, C. G. 1991. Evidence for volcanic eruption on the southern Juan de Fuca Ridge between 1981 and 1987. *Nature*, **313**, 212–214.

——, —— & —— 1995. SeaBeam depth changes associated with recent lava flows, CoAxial segment, Juan de Fuca Ridge: Evidence for multiple eruptions between 1981–1993, *Geophysical Research Letters*, **22**, 167–170.

CONVERSE, D. R., HOLLAND, H. D. & EDMOND, J. M. 1984. Flow rates in the axial hot springs of the East Pacific Rise (21°N): implications for the heat budget and the formation of massive sulfide deposits. *Earth and Planetary Science Letters*, **69**, 159–175.

COWEN, J. & CANN, J. R. 1988. Supercritical two phase separation of hydrothermal fluids in the Troodos ophiolite. *Nature*, **333**, 259–261.

DETRICK, R. S., BUHL, P., VERA, E., MUTTER, J., ORCUTT, J., MADSEN, J. & BROCHER, T. 1987. Multi-channel seismic imaging of a crustal magma chamber along the East Pacific Rise. *Nature*, **326**, 35–41.

DZIAK, R. P., FOX, C. G. & SCHREINER, A. E. 1995. The June–July seismo-acoustic event at CoAxial segment, Juan de Fuca Ridge: evidence for a lateral dike injection. *Geophysical Research Letters*, **22**, 135–138.

EDMOND, J. M., MEASURES, C., MCDUFF, R. E., CHAN, L. H., COLLIER, R., GRANT, B., GORDON, L. I. & CORLISS, J. B. 1979. Ridge crest hydrothermal activity and the balances of major and minor elements in the ocean: the Galapagos data. *Earth and Planetary Science Letters*, **46**, 1–18.

——, VON DAMM, K. L., MCDUFF, R. E. & MEASURES, C. I. 1982. Chemistry of hot springs on the East Pacific Rise and their effluent dispersal. *Nature*, **297**, 187–191.

EMBLEY, R. W., CHADWICK JR, W. W., JONASSON, I. R., BUTTERFIELD, D. A. & BAKER, E. T. 1995. Initial results of the rapid response to the 1993 CoAxial event: relationships between hydrothermal and volcanic processes. *Geophysical Research Letters*, **22**, 143–146.

FEELY, R. A., LEWISON, M., MASSOTH, G. J., ROBERT-BALDO, G., LAVELLE, J. W., BYRNE, R. H., VON DAMM, K. L. & CURL, JR, H. C. 1987. Composition and dissolution of black smoker particulates from active vents on the Juan de Fuca Ridge. *Journal of Geophysical Research*, **92**, 11 347–11 363.

FOX, C. G., RADFORD, W. E., DZIAK, R. P., LAU, T.-K., MATSUMOTO, H. & SCHREINER, A. E. 1995. Acoustic detection of a seafloor spreading episode on the Juan de Fuca Ridge using military hydrophone arrays. *Geophysical Research Letters*, **22**, 131–134.

GENDRON, J. F., COWEN, J. P., FEELY, R. A. & BAKER, E. T. 1993. Age estimate for the 1987 megaplume on the southern Juan de Fuca Ridge using excess radon and manganese partitioning. *Deep-Sea Research*, **40**, 1559–1567.

——, TODD, J. F., FEELY, R. A., BAKER, E. T. & KADKO, D. 1994. Excess ^{222}Rn over the Cleft segment, Juan de Fuca Ridge. *Journal of Geophysical Research*, **99**, 5007–5016.

HAYMON, R. M., FORNARI, D. J., EDWARDS, M. H., CARBOTTE, S., WRIGHT, D. & MACDONALD, K. C. 1991. Hydrothermal vent distribution along the East Pacific Rise crest (9°09′–54′N) and its relationship to magmatic and tectonic processes on fast-spreading mid-ocean ridges. *Earth and Planetary Science Letters*, **104**, 513–534.

——, ——, VON DAMM, K. L., LILLEYM, M. D., PERFIT, M. R., EDMOND, J. M., SHANKS, W. C. III,

LUTZ, R. A., GREBMEIER, J. M., CARBOTTE, S., WRIGHT, D., MCLAUGHLIN, E., SMITH, M., BEEDLE, N. & OLSON, E. 1993. Volcanic eruption of the mid-ocean ridge along the East Pacific Rise crest at 9°45′–52′N: direct submersible observations of seafloor phenomena associated with an eruption event in April, 1991. *Earth and Planetary Science Letters*, **119**, 85–101.

KLINKHAMMER, G., ELDERFIELD, H., GREAVES, M., RONA, P. & NELSEN, T. 1986. Manganese geochemistry near high-temperature vents in the Mid- Atlantic rift valley. *Earth and Planetary Science Letters*, **80**, 230–240.

LILLEY, M. D., LUPTON, J. E. & VON DAMM, K. 1992. Volatiles in the 9°N hydrothermal system: a comparison of 1991 and 1992 data. *Eos, Transactions of the American Geophysical Union*, **73** (Fall Meeting Supplement), 524.

——, OLSON, E. J., MCLAUGHLIN, E. & VON DAMM, K. 1991. Methane, hydrogen, and carbon dioxide in vent fluids from the 9°N hydrothermal system, *Eos, Transactions of the American Geophysical Union*, **72** (Fall Meeting Supplement). 481.

LUPTON, J. E., BAKER, E. T. & MASSOTH, G. J. 1989. Variable ³He/heat ratios in submarine hydrothermal systems: evidence from two plumes over the Juan de Fuca Ridge. *Nature*, **337**, 161–164.

——, ——, ——, THOMSON, R. E., BURD, B. J., BUTTERFIELD, D. A., EMBLEY, R. W.& CANNON, G. A. 1995. Variations in water-column ³He/heat ratios associated with the 1993 CoAxial event, Juan de Fuca Ridge. *Geophysical Research Letters*, **22**, 155–158.

——, KLINKHAMMER, G. P., NORMARK, W. R., HAYMON, R., MACDONALD, M. C., WEISS, R. F. & CRAIG, H. 1980. Helium-3 and manganese at the 21°N East Pacific Rise hydrothermal site. *Earth and Planetary Science Letters*, **50**, 115–127.

——, LILLEY, M., OLSON, E. & VON DAMM, K. 1991. Gas chemistry of vent fluids from 9°–10°N on the East Pacific Rise. *Eos, Transactions of the American Geophysical Union*, **72** (Fall Meeting Supplement), 481.

MACDONALD, K. C., BAKER, E. T., HAYMON, R., LILLEY, M. D., PERFIT, M. R., FORNARI, D. J. & VON DAMM, K. 1992. Time-series water column measurements and submersible observations at EPR 9°30′N–10°02′N from 4/91 to 3/92: evidence for changing hydrothermal activity. *Eos, Transactions of the American Geophysical Union*, **73** (Fall Meeting Supplement), 530.

MASSOTH, G. J., BAKER, E. T., LUPTON, J. E., FEELY, R. A., BUTTERFIELD, D. A., VON DAMM, K., ROE, K. K. & LEBON, G. T. 1994. Temporal and spatial variability of hydrothermal manganese and iron at Cleft segment, Juan de Fuca Ridge. *Journal of Geophysical Research*, **99**, 4905–4924.

——, BUTTERFIELD, D. A., LUPTON, J. E., MCDUFF, R. E., LILLEY, M. D. & JONASSON, I. R. 1989. Submarine venting of phase-separated hydrothermal fluids at Axial Volcano, Juan de Fuca Ridge. *Nature*, **340**, 702–705.

POREDA, R. J. & ARNÓRSSON, S. 1992. Helium isotopes in Icelandic geothermal systems: II. Helium–heat relationships. *Geochimica et Cosmochima Acta*, **56**, 4229–4235.

RONA, P. A. & TRIVETT, D. A. 1992. Discrete and diffuse heat transfer at ASHES vent field, Axial Volcano, Juan de Fuca Ridge. *Earth and Planetary Science Letters*, **109**, 57–71.

ROSENBERG, N. D., LUPTON, J. E., KADKO, D., COLLIER, R., LILLEY, M. D. & PAK, H. 1988. Estimation of heat and chemical fluxes from a seafloor hydrothermal vent field using radon measurements. *Nature*, **334**, 604–607.

RUDNICKI, M. D. & ELDERFIELD, H. 1992. Theory applied to the Mid-Atlantic Ridge hydrothermal plumes: the finite difference approach. *Journal of Volcanology and Geothermal Research*, **50**, 161–172.

SCHIFFRIES, C. M. & SKINNER, B. J. 1987. The Bushveld hydrothermal system: field and petrologic evidence. *American Journal of Science*, **287**, 566–595.

SHANKS III, W. C., BÖHLKE, J. K. & VON DAMM, K. 1992. Stable isotope variations in vent fluids on short time scales: observations at the 1991 eruption site on the East Pacific Rise at 9°33′–52′N. *Eos, Transactions of the American Geophysical Union*, **73** (Fall Meeting Supplement), 524.

STEIN, C. A. & STEIN, S. 1994. Constraints on hydrothermal heat flux through the oceanic lithosphere from global heat flow. *Journal of Geophysical Research*, **99**, 3081–3095.

THOMSON, R. E., DELANEY, J. R., MCDUFF, R. E., JANECKY, D. R. & MCCLAIN, J. S. 1992. Physical characteristics of the Endeavour Ridge hydrothermal plume during July 1988. *Earth and Planetary Science Letters*, **111**, 141–154.

VON DAMM, K. L. 1988. Systematics of and postulated controls on submarine hydrothermal solution chemistry. *Journal of Geophysical Research*, **93**, 4551–4561.

—— 1990. Seafloor hydrothermal activity: black smoker chemistry and chimneys. *Annual Reviews of Earth and Planetary Science*, **18**, 173–204.

—— & BISCHOFF, J. L. 1987. Chemistry of hydrothermal solutions from the southern Juan de Fuca Ridge. *Journal of Geophysical Research*, **92**, 11 334–11 346.

——, COLODNER, D. C. & EDMONDS, H. N. 1992. Hydrothermal fluid chemistry at 9°–10°N EPR '92: big changes and still changing. *Eos, Transactions of the American Geophysical Union*, **73** (Fall Meeting Supplement), 524.

——, GREBMEIER, J. M. & EDMOND, J. M. 1991. Preliminary chemistry of hydrothermal vent fluids from 9°–10°N on the East Pacific Rise. *Eos, Transactions of the American Geophysical Union*, **72** (Fall Meeting Supplement), 480.

Time series studies of vent fluids from the TAG and MARK sites (1986, 1990) Mid-Atlantic Ridge: a new solution chemistry model and a mechanism for Cu/Zn zonation in massive sulphide orebodies

J. M. EDMOND[1], A. C. CAMPBELL[1], M. R. PALMER[1],
G. P. KLINKHAMMER[1], C. R. GERMAN[1], H. N. EDMONDS[1],
H. ELDERFIELD[2], G. THOMPSON[3] & P. RONA[4]

[1]Department of Earth, Atmospheric and Planetary Sciences, Massachussetts Institute of Technology, E34-201, Cambridge, MA 02139, USA

[2]Department of Earth Sciences, University of Cambridge, Cambridge, UK

[3]Department of Marine Chemistry, Woods Hole Oceanographic Institution, Woods Hole, MA 02543, USA

[4]NOAA/AOML, 4301 Rickenbacker Causeway, Miami, FL 33149, USA

Abstract: The hot springs at TAG and MARK on the Mid-Atlantic Ridge have been resampled after an interval of four years (1986–1990). The fluid compositions show the same temporal stability as observed elsewhere, e.g. 21°N, East Pacific Rise (EPR) and the Guaymas Basin. Although the MARK fluids have no chemical characteristics that would distinguish them from those on the faster spreading ridges of the Pacific, TAG has pronounced differences. The depletion in B and the small shift in $\delta^{11}B$ indicate a substantial degree of reaction at intermediate temperatures along the recharge path that is unique to TAG. In addition, the TAG mound contains a large cluster of sphalerite-rich 'onion domes' that have been formed from an approximately 5:1 mixture of the primary hydrothermal end-member fluid and seawater that is formed within the deposit. This has resulted in the extensive precipitation of FeS within the mound and a resulting decrease in pH to values below 3. The low pH causes the large-scale remobilization of Zn from the interior of the deposit and its reprecipitation on the surface as the domes. Such compositional zoning is a common feature of ophiolite-type massive sulphide ore bodies and probably results by the same mechanism. The end-member data from both hydrothermal areas fall on the mixing planes defined by the EPR and Juan de Fuca data in a new three-component mixing model, indicating the presence of a phase-separated brine pool at depth under these Mid-Atlantic Ridge systems which is very similar in composition to those on the Pacific ridges.

Since their original discovery in 1977 (Corliss *et al.* 1979), active hydrothermal systems have been found at numerous locations along the various intermediate and fast-spreading ridges of the Eastern Pacific (Von Damm 1990). The continuing exploration efforts have been invariably successful (e.g. Haymon *et al.* 1993). On slow-spreading ridges the modelled thermal regime is such that permanent magma chambers cannot be sustained over the long intervals between volcanic events (Sleep & Wolery 1978). It has been commonly assumed that active hydrothermal systems would therefore be rare on slow-spreading ridges. Consistent with this the very low regional ^{3}He anomaly in the Atlantic (Ostlund *et al.* 1987), an ocean dominated by slow-spreading ridges, could be ascribed to a very low rate of hydrothermal supply of the isotope in addition to the diluting effects of the relatively vigorous circulation. However, hydrothermal systems on slow ridges are of great potential interest as the tectonic, structural and thermal characteristics of this environment are markedly different from those at faster spreading rates.

The first active oceanic hydrothermal systems on a slow-spreading centre were discovered on the Mid-Atlantic Ridge (MAR) at the TAG and MARK sites (26 and 23°N) in 1985 (Rona *et al.* 1986; Detrick *et al.* 1988). New sites, Lucky Strike and Broken Spur (37° and 29°N), have since been discovered (Klinkhammer *et al.* 1993; Murton *et al.* 1993). The first sampling attempts at TAG and MARK were made in 1986 during a multidisciplinary cruise using *Alvin* (Karson *et al.* 1987); four dives were devoted to TAG and one to MARK (Campbell *et al.* 1988a).

TAG is located in carbonate sedimented

From PARSON, L. M., WALKER, C. L. & DIXON, D. R. (eds), 1995, *Hydrothermal Vents and Processes*, Geological Society Special Publication No. 87, 77–86.

Table 1. *Time series of Mid-Atlantic Ridge vent fluids*

	MARK 1990	MARK 1986	TAG 1990 black smokers	TAG 1990 white smokers	TAG 1986 seeps
Temperature (°C)	335–356	335–350	360–366	273–301	290–320
pH (NBS, 23°C)	3.7	3.7–3.9	3.35	3	–
Alkalinity (med./l)	−0.25	−0.06 to −0.24	−0.45	−1.1	–
$Si(OH)_4$(mmol/l)	19.2	18.2	20.75	19.1	22
H_2S(mmol/l)	6.1	5.9	2.5–3.5	0.5	–
Cl(mmol/l)	563	559(563)	636	–	659
Na(mmol/l)	520	510	557	–	584
K(mmol/l)	23	23.6	17.1	17.1	17
Rb(μmol/l)	11.3	11	9.1	9.4	10
Cs(nmol/l)	181	180	108	113	100
B(μmol/l)	562	520	356	388	–
$\gamma^{11}B$	26.3	26.8	30.9	35	–
Ca(mmol/l)	9.96	10	30.8	27	26
Sr(μmol/l)	48	50	103	91	99
$^{87}Sr/^{86}Sr$	0.7039	0.7028	0.7038	0.7046	0.7029
Mn(μmol/l)	451	491	680	750	1000
Fe(μmol/l)	2560	2180	5590	3830	1640
Cu(μmol/l)	15–18	10–17	120–150	3	–
Zn(μmol/l)	52	47–50	46	300–400	–

pillows on approximately 100 000 year old crust at the base of the eastern wall of the rift valley in 3700 m water depth (Eberhart *et al.* 1988). It was found to consist of a large mound, about 100 m in diameter and 30–50 m high, composed of hydrothermally deposited sulphide minerals and their oxidation products (Thompson *et al.* 1988). This mound was dominated by a large cluster of tall anhydrite chimneys surrounded by numerous black smoker seeps. The smoke from the seeps completely enveloped the submersible before the primary chimney orifices could be sampled. Samples from the seeps themselves were considerably diluted by entrained ambient seawater (Campbell *et al.* 1988a). However, estimates could be made of the end-member major ion compositions. These were remarkably similar to those observed on the intermediate and fast-spreading ridges of the Eastern Pacific (Von Damm 1990; Bowers *et al.* 1988). There was no indication of a markedly deeper reaction zone, even though the site was on old crust on a slow-spreading ridge.

These observations were confirmed by the data from the excellent samples collected at MARK (Campbell *et al.* 1988a). Here the hydrothermal fields are more typical, located on the glassy basalt ridges that define the neovolcanic zone along the axis of the rift valley (Kong *et al.* 1988; Karson *et al.* 1987). Samples were collected from the chimneys and from an onion-shaped structure on the flanks of one of

the ridges. Again, the complete data set generated from these samples was closely comparable with those from vent fields on the much faster spreading ridge segments on the East Pacific Ridge (EPR) (Campbell *et al.* 1988a).

A return cruise to the MAR sites in 1990 was dedicated to the study of the geology, geochemistry and biology of the hydrothermal systems themselves. A particular objective was the thorough sampling of the TAG mound. Time series sampling was also initiated at MARK to complement the long-term observations of the fields at 21°N on the EPR (Campbell *et al.* 1988b).

Chemistry of hydrothermal fluids at the TAG and MARK sites

End-member fluids

Although the overall configuration of the MARK field was unchanged in 1990, that of TAG had been greatly altered. The steep sulphide cliffs surrounding parts of the mound had collapsed, as had the massive central cluster of chimneys, presumably as a result of local earthquake activity (Huang *et al.* 1986; Kong *et al.* 1992). The vent orifices were now easily accessible for sampling. As a result dive time was also available for excavating and sampling an unusual cluster of constructional features

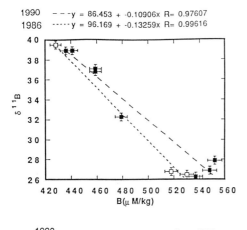

1990 − − −y = 86.453 + -0.10906x R= 0.97607
1986 ·····y = 96.169 + -0.13259x R= 0.99616

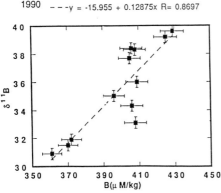

1990 − − −y = -15.955 + 0.12875x R= 0.8697

Fig. 1. Boron isotope systematics for MARK (upper panel) and TAG (lower panel). The 1990 data for MARK are indicated by the closed squares.

shaped like vertically elongate onion domes which had been seen on the mound in 1986. Sampling and analytical methods were as summarized in Campbell *et al.* (1988a). The complete data set for the two expeditions is given in Table 1.

The repeated sampling of the MARK site shows generally minor variations in the chemical composition of the vent fluids over the four-year interval. The largest differences are for silica, with an increase of 1 mmol/l ($\approx$5%) and the Sr isotope ratio, where the increase of 0.001 is substantial. The Sr data imply an increase in the integrated water to rock ratio. This is at variance with the behaviour of other elements, e.g. Li, that are equally if not more sensitive to this parameter. At TAG the estimated composition of the 1986 end-member, based on lengthy extrapolations, is consistent with the much better data from 1990.

One striking feature in the TAG data is the behaviour of B (Fig. 1). In all other systems analysed to date, including MARK, the concentration and isotope systematics of B can be interpreted as a simple mixing process between seawater B (430 μmol/l, $\delta^{11}B = 39.5‰$) and a component extracted quantitatively from the basalt (c. 0.5 ppm, $-3.5‰$) at high temperature with no isotope frationation (Spivack & Edmond 1987). This results in a concentration increase of 15–20% and an isotopic shift of -10 to $-15‰$ relative to the original seawater. These results also indicate low water to rock ratios (0.5–1.0) and no resolvable uptake of B at low and intermediate temperatures with the attendant fractionation. At TAG the latter process is clearly important (Fig. 1). The end-member concentrations are c. 70 μmol/l lower than seawater and the isotopic depletion is only c. 10‰.

These results can be interpreted more quantitatively using the simple model of Spivack & Edmond (1987). It is assumed that the uptake occurs as a simple Rayleigh fractionation process at temperatures of $\approx$150°C and constant fractionation factor, α, of 0.967, calculated from the observed $\delta^{11}B$ values for seawater and serpentinites and metabasalts. This evolved fluid then reacts with pristine tholeiite at 350°C and quantitatively extracts boron with no frationation. At TAG the fraction of seawater B remaining after the first step is $\approx$0.55 with an isotopic composition of c. 58‰. High-temperature reactions then occur with a water to rock ratio of about one-third. Application of this model to the MARK and EPR data (Campbell *et al.* 1988a; Spivack & Edmond 1987) constrains the extent of the low-temperature removal in these systems to less than 10% of the seawater boron, only slightly greater than the resolution of the measurements. The effect at TAG is much larger.

As observed in 1986 (Palmer & Edmond 1989), the data for the heavy alkalis, Rb and Cs, indicate the involvement of the pre-existing country rock, altered greenstones, in the hydrothermal reactions at both TAG and MARK. These elements are strongly enriched during the low-temperature alteration of basalts; Cs is taken up preferentially. In the various EPR hot spring systems the Cs to Rb ratio is slightly higher than for fresh tholeiite, but lower than for hydrothermal smectites (Fig. 2). At TAG and especially MARK (and the Marianas; Campbell *et al.* 1994) the ratio approaches that for weathered basalts. This requires that altered basalts are reacting with the recharging fluids at intermediate temperatures ($\approx$150°C) consistent with the B data.

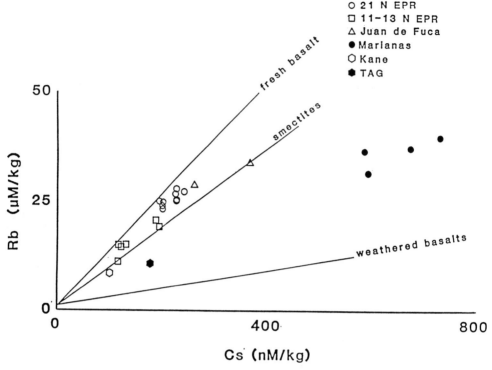

Fig. 2. Compositional relationships between Cs and Rb in hydrothermal fluids, tholeiitic basalts and greenstone alteration minerals and basalt weathered at low temperature.

By contrast with the alkalis and B, Li shows no evidence for reaction at intermediate temperatures (Chan *et al.* 1993). The concentrations, ≈850 µmol/l, are within the range observed on the EPR, as are the isotopic compositions of about −8.5‰. Lithium has an alteration chemistry similar to boron, with large-scale uptake and fractionation at low and intermediate temperatures (Chan *et al.* 1992). Its lack of remobilization possibly reflects the ability of Li, with its small ionic radius, to replace Mg and Fe in octahedral sites in layered silicates and mafic minerals, i.e. it is structurally bound rather than absorbed in ion-exchange sites (Chan *et al.* 1993).

Both the unique behaviour of boron and the remobilization of the heavy alkalis require that the fluids recharging the reaction zone warm up much more slowly than in other systems. In general, the fluid chemistries, the water isotopes and the relative abundances of temperature-dependent rock alteration facies in basalts recovered from the seafloor by dredging and drilling argue for rapid heating of the circulating fluids with no significant water–rock interactions at intermediate temperatures (Mottl 1983;

Bowers & Taylor, 1985; Bowers *et al.* 1988). Thus the isotherms must be bunched tightly around the heat source. Relaxation of these temperature gradients, i.e. lateral conductive heating, requires diffuse flow, an adjustment in the balance between inward advection and outward conduction of heat. TAG is unusual in being located in pillowed terrain mantled by carbonate sediments (Eberhart *et al.* 1988). It is possible that these act as a seal against rapid local recharge along fissures and intra-pillow channels, thus allowing the conductive heating of a substantial volume of the country rock.

Onion domes

An unusual cluster of dome-shaped constructional features occupies about 10% of the upper surface of the TAG mound. These vertically elongate sphalerite-rich domes range up to a few metres tall and about a metre in diameter; they emit a diffuse flow of white 'smoke' from their upper surfaces. A large example, on the outer edge of the cluster, was removed to gain access to the feeder conduit for sampling. The fluid chemistries (Table 1) show unequivocal evidence for the entrainment of ambient seawater into the mound and mixing-induced reaction with the end-member fluid. The lowest Mg value was 12 mmol/l in what appears to be a pair of excellent samples. This value is consistent with the measured temperatures which, between several domes, fall in the range 301–273°C, indicative of between 15 and 20% entrained seawater. For comparative purposes the concentration data for the white smokers in Table 1 are calculated by extrapolation to zero Mg values. The actual measurements are shown in Fig. 3, along with those for the primary fluids at TAG and MARK. It is seen that for elements that might be expected to behave conservatively during this mixing process, e.g. K, Rb and Cs, the extrapolated values are indistinguishable from those measured in the black smokers. There may be a slight loss of Si, as opal or quartz, and Ca and Sr are definitely depleted by anhydrite precipitation. Sulphide is largely removed from solution along with Fe in approximately the stoichiometric proportions indicative of the formation of pyrrhotite, FeS (the end-member Fe:S ratio is c. 1.85). This reaction causes the pH (25°C) to drop from 3.35 to slightly less than 3, one of the lowest values yet observed in oceanic hot springs. Copper is also strongly depleted. However, the Zn concentrations increase by an order of magnitude, indicating remobilization by the increased acidity. From these observations it is clear that the circulation of seawater within a hydrothermal mound greatly increases the efficiency of ore localization compared with free venting of the end-member fluid, and, by causing lowering of the temperature and pH, promotes the differential precipitation and remobilization of particular elements, e.g. Cu and Zn. Therefore this process also contributes to the large-scale compositional zonation observed in Cyprus-type ore bodies (Coleman 1977; Franklin et al. 1981). Copper is retained with Fe in the body of the deposit, whereas Zn is stripped from these buried chimney fragments due to the lowering of the pH, and concentrated at the seawater interface.

The effectiveness of this 'leaky mound' model for ore localization depends on the relative abundances of Fe and H_2S. The Fe:H_2S ratios in all the end-member hydrothermal fluids measured to date are a linear function of the chloride concentration, even though the relation of the individual species concentrations is much more scattered (Fig. 4). The transition to Fe:S ratios greater than unity occurs at a chloride concentration of c. 650 mmol/l, which is c. 20%

above the average seawater value. Hence at high chloride levels the efficiency of ore metal precipitation by the in situ mixing process is limited by the availability of sulphide. Therefore any such fluid subjected to this process, either to produce the mixtures seen at the onion domes or the more common, very dilute Galapagos-type diffuse flow (Edmond et al. 1979), will emerge strongly depleted in sulphide. This explains the occurrence of the 'azoic zones' around the high-chloride vent fields that occur between 11 and 13°N on the EPR. There the fields are devoid of megafauna; the basalts are mantled with a dark brown slime, presumably formed by Fe–Mn oxidizing bacteria. It can be speculated that the macrofauna, dependent as they are on chemosynthetic sulphide oxidizing bacteria, cannot utilize reduced Fe and Mn as energy sources as their oxidation products are very insoluble compared with the oxyanions produced from hydrogen sulphide. It is remarkable that a comparable number of high-chloride systems (Fe:S > 1) have been found as of low-chloride (Fig. 4), despite the fact that the latter can support a Galapagos-type 'halo' of megafauna that greatly facilitates their identification either photographically or from submersibles.

The linear relationship between the ratio Fe:H_2S and Cl cannot be explained thermodynamically; chloro-complexing of Fe^{2+} produces $FeCl_2^0$ as the major species, leading to a higher order dependence. However, the relationship is unavoidable if, in fact, the vent fluid compositions are produced by the three-component mixing of a phase-separated, volatile-poor brine, the associated dilute, volatile-rich phase and a partially reacted (Mg = 0) seawater ([Fe + Cl; $H_2S = 0$], [H_2S; Fe = Cl = 0], [Cl; Fe = $H_2S = 0$]). This is best illustrated on a three-dimensional diagram of Fe versus H_2S versus Cl (Fig. 5). Data from all the hot springs on sediment-starved ridge axes define a mixing plane with a scatter compatible with the assigned analytical error. There is no systematic deviation with relative composition, confirming the overall integrity of the samples, or with spreading rates or depth, suggesting that the process is general. Presumably the accumulation of a brine phase occurs early in an eruptive–hydrothermal cycle, as observed at 9°N on the EPR, where a dilute phase (<0.3‰) with a temperature of 405°C was observed to vent for several days (Von Damm et al. 1991). Qualitative versions of this mixing model have been discussed by Bischoff & Rosenbauer (1989) and by Von Damm (1990). A quantitative analyses is presented by Edmonds & Edmond (in press).

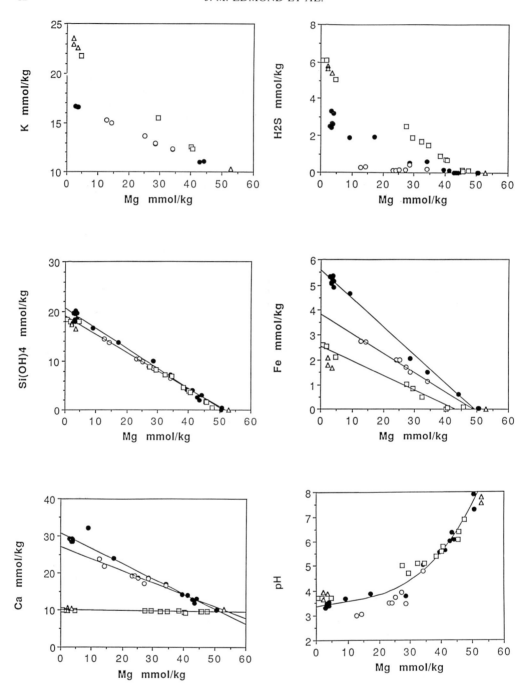

Fig. 3. Data plots for the black (●) and white (○) smokers at TAG and for the black smokers at MARK 90 (□) and MARK 86 (△).

Ironically, this model explains the large range observed in the chloride concentrations in the vent fluids, it is not so obvious how long-term temporal stability can be maintained in such a mixing process. The thermodynamic models (Bowers *et al.* 1988) explain the stability

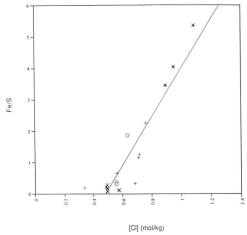

Fe/S

[Cl] (mol/kg)

Fig. 4. Relationship between the ratio Fe : H₂S and Cl in hydrothermal fluids from sediment starved ridge axes. ×, 21°N, EPR; +, 11–13°N, EPR; *, southern Juan de Fuca Ridge; o, TAG and MARK.
A quadratic or other higher order fit to the data does not improve upon the linear treatment shown here.

but not the chloride systematics. One of these, the two outstanding problems in hydrothermal fluid chemistry, can be explained, but at the expense of the other.

Conclusions

The chemical data from the TAG and MARK sites show the same general temporal stability observed over longer periods at 21°N, EPR and at the sediment covered system in Guaymas Basin, Gulf of California (Campbell *et al.* 1988*b*). In addition, the compositional data, with the exception of B, fall within the range observed on the shallower, faster spreading segments of the EPR (Bowers *et al.* 1988). However, the wide ranges in chloride characteristic of the EPR between 9 and 11°N and of the Juan de Fuca (Von Damm & Bischoff 1987) are not observed. Taken together this argues against any significant contribution of fluids from reaction zones deep in the crust of the MAR. The estimated reaction depths on the EPR, based on the silica geobarometry, are in the upper crust above the seismically imaged magma sill. No such feature has been resolved on the MAR (Detrick *et al.* 1990). A plausible scenario involves a deep reaction zone on the MAR with the fluids ascending several kilometres to the seafloor along the boundary faults of the rift valley. This is certainly consistent with the location of the TAG field (Eberhart *et al.* 1988).

However, the fluid chemistries argue the contrary. At the much greater hydrostatic pressures the reaction temperatures would certainly be higher, leading to phase separation during ascent. There is no evidence of this. In addition, the equilibrium thermodynamic calculations are consistent with a shallow reaction zone (Campbell *et al.* 1988*a*).

The inferred location of the high temperature water–rock reaction zone in the upper crust makes the nature of the heat source at TAG especially problematic. There is little geophysical information available for the immediate area; from studies of microearthquakes, a low velocity zone is inferred at mid- to deep crustal levels immediately below the axis itself (Kong *et al.* 1992). Given the enormous size of the TAG mound, the heat source driving it must be large and long-lived. Small, young volcanic features have been reported along the eastern wall associated with other large sulphide deposits that are extinct (P. Rona, unpublished data). Presumably, multiple episodes of off-axis intrusive activity have occurred that drive the hydrothermal systems.

Multi-channel seismic lines in the MARK area did not detect either a magma lens or a shallow low velocity zone (Detrick *et al.* 1990). The hydrothermal deposits at MARK are similar in scale to those on the EPR (Thompson *et al.* 1988). It is likely that the hydrothermal activity is driven by heat from the feeder dykes responsible for the recent volcanism around the field.

Hydrothermal deposits formed in the neovolcanic zone, e.g. MARK, EPR, must be susceptible to destruction in subsequent eruptive events, i.e. their lifetime as recognizable geological entities is limited to the period between volcanic episodes. Off-axis locations in the rift valleys on slow-spreading ridges appear to be volcanically and tectonically more quiescent. Hence, given a suitable heat source, these locations may be more conducive to the formation of major hydrothermal ore deposits. However, to be preserved they must survive dismemberment during tectonic uplift in the walls of the rift valley.

Results to date show no relationship between fluid chemistry and spreading rate despite the large tectonic, magmatic and thermal variations associated with the latter variable. There is a greater involvement of altered country rock in the reaction zone, at slow-spreading rates, but the overall effects are not dramatic at MARK, the on-axis site. At TAG there is unique evidence from the isotope systematics of B that extensive reactions occur at intermediate temperatures during recharge. This may be an

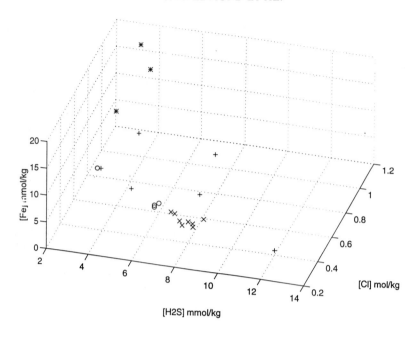

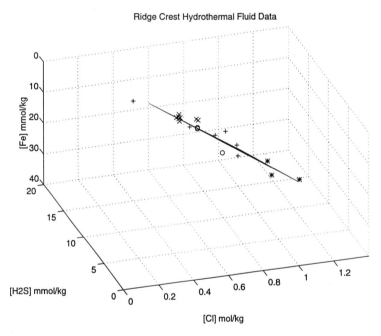

Fig. 5. Data from Fig. 4 plotted on an Fe–H₂S–Cl three-dimensional diagram. The data define a mixing plane. For clarity, the upper panel shows an oblique view to indicate the range in the data; the lower panel is a rotated 'edge on' view. Sample identification as in Fig. 4.

artifact of TAG's off-axis location in pillow terrain, largely sealed by carbonate pelagic sediments.

The data from TAG and MARK fall precisely on the three-component mixing plane, brine, dilute phase, altered seawater, of Edmonds and

Edmond (in press). Thus the compositions of these three components, the primary reactants, must be fairly constant regardless of the spreading rate. The reasons for the stability of the mixing process remain unknown.

It is a pleasure to acknowledge the energy, enthusiasm and expertise of the ALVIN Group and the dedicated support of the captain, officers and crew of the *Atlantis II*, even under very dry conditions. H. Elderfield and his colleagues from Cambridge were a source of stimulation. We thank S. Hart for his continued generous provision of access to his mass spectrometry laboratory. This work was supported by the National Science Foundation and NOAA as part of the RIDGE/FARA initiative.

References

BISCHOFF, J. L. & ROSENBAUER, R. J. 1989. Salinity variations in submarine hydrothermal systems by layered double-diffusive convection. *Journal of Geology*, **97**, 613–623.

BOWERS, T.S. & TAYLOR, H. P. 1985. An integrated chemical and stable-isotope model of the origin of mid-ocean ridge hot spring sytems. *Journal of Geophysical Research*, **90**, 12 583–12 606.

——, CAMPBELL, A. C., MEASURES, C. I., SPIVACK, A. J., KHADEM, M. & EDMOND, J. M. 1988. Chemical controls on the composition of vent fluids at 11–13°N and 21°N, East Pacific Rise. *Journal of Geophysical Research*, **93**, 4522–4536.

CAMPBELL, A. C., BOWERS, T. S., MEASURES, C. I., FALKNER, K. K., KHADEM, M. & EDMOND, J. M. 1988*b*. A time-series of vent fluid compositions from 21°N, East Pacific Rise (1979, 1981, 1985) and Guaymas Basin (1982, 1985). *Journal of Geophysical Research*, **93**, 4537–4549.

——, GERMAN, C. R., PALMER, M. R., GAMO, T. & EDMOND, J. M. 1994. Chemistry of hydrothermal fluids from Escanaba Trough, Gorda Ridge. *In:* MORTON, J. L. *ET AL*. (eds) *Geologic, Hydrothermal and Biologic Studies at Escanaba Trough, Gorda Ridge, Offshore Northern California. US Geological Survey Bulletin*, **2022**, 201–222.

——, PALMER, M. R. & 9 others. 1988*a*. Chemistry of hot springs on the Mid-Atlantic Ridge. *Nature*, **335**, 514–519.

CHAN, L. H., EDMOND, J. M. & THOMPSON, G. 1993. A lithium isotope study of hotsprings, and metabasalts from mid-ocean ridge hydrothermal systems. *Journal of Geophysical Research*, **98**, 9653–9659.

——, ——, —— & GILLIS, K. 1992. Lithium isotope composition of submarine basalts: implications for the lithium cycle in the oceans. *Earth Planetary Science Letters*, **108**, 151–160.

COLEMAN, R. G. 1977. *Ophiolites: Ancient Oceanic Lithosphere?* Springer-Verlag, New York, 299 pp.

CORLISS, J. B., DYMOND, D. & 9 others. 1979. Submarine thermal springs on the Galapagos Rift. *Science*, **203**, 1073–1083.

DETRICK, R. S., FOX, P. J., SCHULZ, N., POCKALNY, R.,

KONG, L., MAYER, L. & RYAN, W. B. F. 1988. Geologic and tectonic setting of the MARK area. *Proceedings of the ODP, Initial Reports (Pt A)*, **106/109**, 15–20.

——, MUTTER, J. C., BUHL, P. & KIM, I. I. 1990. No evidence from multichannel reflection data for a crustal magma chamber in the MARK area on the Mid-Atlantic Ridge. *Nature*, **347**, 61–63.

EBERHART, G. L., RONA, P. A. & HONNOREZ, J. 1988. Geologic controls of hydrothermal activity in the Mid-Atlantic Rift Valley: tectonics and volcanics. *Marine Geophysical Research*, **10**, 233–259.

EDMOND, J. M., MEASURES, C. I. & 6 others. 1979. Ridge crest hydrothermal activity and the balances of the major and minor elements in the ocean: the Galapagos data. *Earth and Planetary Science Letters*, **46**, 1–18.

EDMONDS, H. N. & EDMOND, J. M. A three component mixing model for ridge crest hydrothermal fluids. *Earth and Planetary Science Letters*, in press.

FRANKLIN, J. M., SANGSTER, D. M. & LYDON, J. W. 1981. Volcanic-associated massive sulphide deposits. *In:* SKINNER, B. J. (ed.) *Economic Geology 75th Anniversary Volume*, 485–627.

HAYMON, R. M., FORNARI, D. J., VON DAMM, K. L. *et al.* 1993. Volcanic eruption of the mid-ocean ridge along the East Pacific Rise crest at 9°45–52′N: I. Direct submersible observations of seafloor phenomena associated with an eruption event in April 1991. *Earth and Planetary Science Letters*, **119**, 85–101.

HEKINIAN, R., AUZENDE, J. M., FRANCHETEAU, J., GENTE, P., RYAN, W. B. F. & KAPPEL, E. S. 1985. Offset spreading centers near 12°53′N on the East Pacific Rise: submersible observations and compositions of the volcanics. *Marine Geophysical Research*, **7**, 359–377.

HUANG, P. Y., SOLOMON, S. C., BERGMAN, E. A. & NABELEK, J. L. 1986. Focal depths and mechanisms of Mid-Atlantic Ridge earthquakes from body waveform inversion. *Journal of Geophysical Research*, **91**, 579–598.

KARSON, J. A., THOMPSON, G. & 12 others. 1987. Along-axis variations in seafloor spreading in the MARK area. *Nature*, **328**, 681–685.

KONG, L. S. L., DETRICK, R. S., FOX, P. J., MAYER, L. A. & RYAN, W. B. F. 1988. The morphology and tectonics of the MARK area from SeaBeam and SeaMARC 1 observations (Mid-Atlantic Ridge 23°N). *Marine Geophysical Research*, **10**, 59–90.

——, SOLOMON, S. C. & PURDY, G. M. 1992. Microearthquake characteristics of a Mid-Ocean Ridge along-axis high. *Journal of Geophysical Research*, **97**, 1659–1685.

MOTTL, M. J. 1983. Metabasalts, axial hot springs and the structure of hydrothermal systems at mid-ocean ridges. *Geological Society of America Bulletin*, **94**, 161–18.

MURTON, B. J., KLINKHAMMER, G. & 12 others. 1993. Direct measurements of the distribution and occurrence of hydrothermal activity between 27°N and 30°N on the Mid-Atlantic Ridge. *EOS*,

Transactions of the American Geophysical Union, **74**, 99.

OSTLUND, H. G., CRAIG, H., BROECKER, W. S. & SPENCER, D. 1987. *GEOSECS Atlantic, Pacific and Indian Ocean Expeditions: Shorebased Data and Graphics*, Vol. 7. US Govt Printing Office, Washington DC, 200 pp.

PALMER, M. R. & EDMOND, J. M. 1989. Cesium and rubidium in hydrothermal fluids: evidence for recycling of alkali elements. *Earth and Planetary Science Letters*, **95**, 8–14.

RONA, P. A., KLINKHAMMER, G., NELSEN, T. A., TREFRY, J. H. & ELDERFIELD, H. 1986. Black smokers, massive sulphides and vent biota at the Mid-Atlantic Ridge. *Nature*, **321**, 33–37.

SLEEP, N. H. & WOLERY, T. J. 1978. Egress of hot water from mid-ocean ridge hydrothermal systems: some thermal constraints: *Journal of Geophysical Research*, **83**, 5913–5922.

SPIVACK, A. J. & EDMOND, J. M. 1986. Boron isotope exchange between seawater and the oceanic

crust. *Geochimica et Cosmochimica Acta*, **51**, 1033–1043.

THOMPSON, G., HUMPHRIS, S. E., SCHROEDER, B., SULANOWSKA, M. & RONA, P. 1988. Active vents and massive sulphides at 26°N (TAG) and 23°N (SNAKEPIT) on the Mid-Atlantic Ridge. *Canadian Mineralogist*, **26**, 697–711.

VON DAMM, K. L. 1990. Seafloor hydrothermal activity: black smoker chemistry and chimneys. *Annual Reviews of Earth and Planetary Sciences*, **18**, 173–204.

——, GREBMEIER, J. M. & EDMOND, J. M. 1991. Preliminary chemistry of hydrothermal vent fluids from 9–10°N, East Pacific Rise. *EOS, Transactions of the American Geophysical Union*, **72**, 480.

WILSON, C., KLINKHAMMER, G., SPEER, K. & CHARLOU, J-L. 1993. Hydrothermal venting at the Lucky Strike segment (37°17'N) of the Mid-Atlantic Ridge. *EOS, Transactions of the American Geophysical Union*, **74**, 360.

Venting from the Mid-Atlantic Ridge at 37°17′N: the Lucky Strike hydrothermal site

G. P. KLINKHAMMER[1], C. S. CHIN[1], C. WILSON[1] & C. R. GERMAN[2]

[1]*College of Oceanic and Atmospheric Sciences, Oregon State University, Corvallis, OR 97331, USA*

[2]*Institute of Oceanographic Sciences Deacon Laboratory, Brook Road, Wormley, Godalming, Surrey GU8 5UB, UK*

Abstract: The Lucky Strike hydrothermal field on the Mid-Atlantic Ridge (MAR) at 37°17′N was discovered in 1992 during dredging operations carried out as part of the FARA Program (French–American Ridge Atlantic). The vent field at Lucky Strike lies in a depression between three cones that form the summit of an axial seamount. The summit rises 400 m above the rift valley floor to a water depth of 1570 m. Hydrothermal fluids stream from several vent complexes with exit temperatures up to 333°C. This paper presents the first observations of hydrothermal plumes in the water column over Lucky Strike, including anomalies in light scattering with a nephelometer and dissolved Mn with the zero-angle photon fibre optic spectrometer. Plumes at Lucky Strike are small compared with those at other MAR hydrothermal sites such as TAG (26°N) and MARK (23°N). These small plumes are consistent with the relatively low Fe and Mn concentrations found in associated vent waters. The Sr/Ca ratios of these fluids are low, suggesting that the basaltic substrate below Lucky Strike is highly altered, but rare earth element concentrations are also low, indicating reduced mobilization. These observations together with the low fluid temperatures are suggestive of an old and waning hydrothermal system with extensive zones of subfloor alteration.

Hydrothermal fluxes from mid-ocean ridges affect ocean chemistry, the geology and mineralization of oceanic crust, the circulation of ocean water, the heat budget of the Earth, the mechanical and physical properties of the lithosphere and the speciation and biogeography of deep-sea animals. Quantifying the extent of these effects and exploring for new hydrothermal sources are areas of increasing interest as we strive to understand the systematics of ridges.

Venting of metal-rich, anoxic fluids from the seafloor produces plumes of dissolved and particulate matter in the overlying water column. Such anomalies can be used to backtrack to previously undiscovered vent sites. This technique has proved to be especially useful in the vast rift valley of the slow-spreading Mid-Atlantic Ridge (MAR) (Murton *et al.* 1994). Plumes are also diagnostic of vent conditions. Soon after the first deep-sea hydrothermal vents were discovered it was recognized that the low-temperature hot springs at Galapagos (Klinkhammer *et al.* 1977) produce distinctly different plumes from the high temperature jets at 21°N (Lupton *et al.* 1980). More recently, we have learned that episodic, intrusive magmatic activity can create event megaplumes with a distinctly different morphology and chemistry from the chronic plumes associated with steady-state venting (Baker, this volume). As we continue to find new vent sites, survey their associated plumes and visit the vents with submersibles, we continue to unravel relationships between venting style and plume signatures in the water column.

Being able to relate hydrothermal plume distributions and characteristics to the type and extent of venting has important implications for the global exploration of the ridge system. As it will not be feasible to undertake detailed submersible or remotely operated vehicle studies at every vent site along the ridge, we must be able to estimate fluxes of geothermal heat and mass from simple observations made from a surface ship. Likewise, as it will not be possible to drill every hydrothermal system visited by submersible, it would be an advantage to be able to distinguish the sub-seafloor characteristics of different vent systems from more easily accessible information, such as vent fluid chemistry. This information could then be used to establish drilling priorities, for example. This paper explores these ideas by examining plume and fluid data from the Lucky Strike hydrothermal field.

From PARSON, L. M., WALKER, C. L. & DIXON, D. R. (eds), 1995, *Hydrothermal Vents and Processes*, Geological Society Special Publication No. 87, 87–96.

FAZAR study region

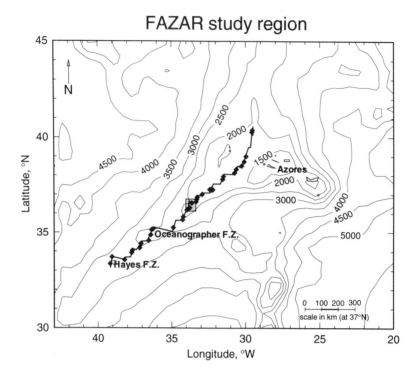

Fig. 1. Regional scale bathymetric map of the Mid-Atlantic Ridge between 30°N and 45°N. The Lucky Strike segment is located within the shaded box, just south of the Pico Offset. Water sampling stations occupied during FAZAR are shown as black diamonds.

Setting and methods

Lucky Strike lies 600 km south of the Azores between the Oceanographer Fracture Zone and Pico Offset (Fig. 1). The site was discovered during the FAZAR cruise, a sampling leg of the FARA Program. FAZAR was a systematic survey of water and rock chemistry along the MAR from 15°N to 40°N. Rock sampling was accomplished with a combination of wax coring and dredging. The water survey consisted of water collection with a rosette system (IFREMER) and *in situ* geochemical measurements using an instrument package designed and fabricated at Oregon State University (OSU). FAZAR included a hydrographic survey of the Lucky Strike segment using the IFREMER rosette system and OSU package.

One of the optical instruments in the OSU package was the zero-angle photon fibre optic spectrometer (ZAPS). During the FAZAR expedition, ZAPS was deployed as a chemical sensor for the detection of dissolved Mn. In this technique seawater is pumped through two cartridges: the first contains calcium periodate

crystals and the second retained *N, N'*-diethy-laniline (DEA). Reduced forms of manganese have a catalytic effect on the oxidative coupling of the fluorophor DEA producing a non-fluorescent product. ZAPS monitors the Mn concentration of the sample stream by following changes in the fluorescence of DEA; see Klinkhammer (1994) for further details of the ZAPS instrument.

A vent site at Lucky Strike was discovered when sulphides were dredged from the large axial volcano at 37°17'N. This discovery was made in autumn 1992; vent fluids were collected during a visit by the submersible *Alvin* the following spring (Langmuir *et al.* 1993a). *Alvin* work at Lucky Strike included fluid sampling from the six vents shown in Fig. 2. The axial seamount rises 400 m above the floor of the rift valley to a water depth of 1570 m. The vents visited by *Alvin* lie in a depression between three cones at the summit. The vent creatures at Lucky Strike include numerous mussels, sea urchins and some shrimps, but fewer than at other MAR sites. Hydrothermal deposits on the seafloor consist mainly of the high-temperature

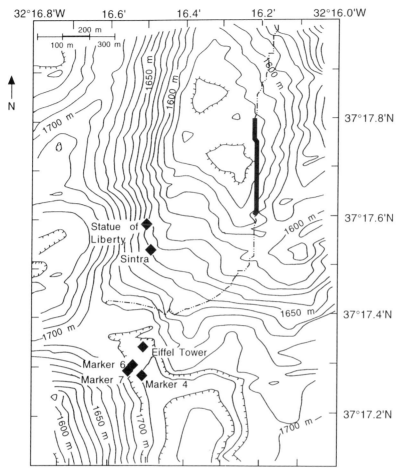

Fig. 2. Vent sites at Lucky Strike (diamonds) placed on the bathymetry at Lucky Strike. The broken line is the track of the tow-yow that produced the Mn anomaly shown in Fig. 3. The thick solid line is the calculated positions of the sled when it was recording the anomaly. The map is adapted from the *Alvin* cruise report (LDEO-93-2) (Langmuir *et al.* 1993*b*).

sulphidic minerals pyrite, sphalerite and chalcopyrite, but lower temperature minerals such as barite and anhydrite are also abundant. The vent structures are constructed on altered basalt and sulphidic rubble with little associated sediment.

Fluid samples were taken with *Alvin* using standard Ti samplers. An aliquot from the sampler was drawn as soon as possible after collection and acidified to prevent sulphide precipitation, then filtered and subsampled for rare earth elements (REE). There was no visible precipitate in the REE samples when they were analysed at OSU about one year after the cruise.

In the laboratory, fluid aliquots of approximately 8 ml volumes were poured into sample tubes of an inductively coupled plasma mass spectrometer (ICPMS, Fisons PlasmaQuad 2 Plus) autosampler. This aliquot was used to flush and fill a 5-ml sample loop inside a Dionex ion chromatograph. The Dionex chelation module separated the transition elements, including the REE, from anions and the alkalis using in-line resin cartridges; elutions were tightly controlled with gradient pumps. The eluent from the ion chromatograph was shunted automatically to the ICPMS with a peristaltic pump and VGS 100 flow valve. The ICPMS scanned the mass spectrum from 139 to 175 AMU 480 times. The REE peaks developed during these scans were normalized to [115]In to cancel out fluctuations in the plasma. Concentrations were calculated by comparing these normalized values with similar

Table 1. *Rare earth element (REE) data for vent fluids from the Lucky Strike Hydrothermal Field*

Site: sample(s):	Mkr-4/5 2608-14		Mkr-6 2606-15 2607-9		Mkr-7 2607-14		SL 2605-15		ET-A 2608-4		Sintra 2606-14		2500 m Seawater	
Temperature (°C):	292		298		297		197		316		207			
Sr/Ca × 1000 (mol/l):	2.25		2.74		2.4		2.78		2.65		2.61		8.53	
Total + 3 REE (nmol/kg):	13.69		8.40		12.46		1.155		11.59		5.59		0.100	
REE (pmol/kg)	EM	NORM	EM	NORM	EM	NORM	EM	NORM	EM	NORM	EM	NORM	SW	NORM
La	5540	340	3890	238	6520	400	292	18	3070	188	1190	73	8.57	0.526
Ce	5760	115	3030	61	4610	92	351	7.0	4910	98	2180	44	3.50	0.070
Pr	415	48	233	27	253	29	43	5.0	526	61	257	30	3.42	0.398
Nd	920	20	690	15	644	14	206	4.5	1960	43	1010	22	20	0.429
Sm	154	10	106	7.1	98	6.6	81	5.5	386	26	238	16	7.67	0.515
Eu	550	96	375	66	564	99	946	166	621	109	1300	228	3.06	0.537
Gd	146	7.5	107	5.5	96	4.9	75	3.8	280	14	238	12	11	0.567
Tb	19	5.5	16	4.4	10	2.8	11	3.3	42	12	32	9.0	2.12	0.606
Dy	110	4.8	95	4.1	72	3.1	64	2.8	223	10	141	6.1	15	0.635
Ho	20	4.1	15	3.1	9.5	1.9	7.6	1.5	40	8.1	29	6.0	3.41	0.696
Er	71	5.1	64	4.6	21	1.5	19	1.4	99	7.1	49	3.5	10	0.714
Tm	8.7	3.5	8.8	3.5	3.5	1.4	4.1	1.7	10	4.1	11	4.2	1.88	0.750
Yb	38	2.8	46	3.4	20	1.5	18	1.3	54	4.0	41	3.0	11	0.830
Lu	2.3	1.2	7.20	3.6	2.4	1.2	3.7	1.9	7.9	4.0	7.1	3.6	1.73	0.880

Data from six vent sites are given: the Marker sites (Mkr), Statue of Liberty (SL), Eiffel Tower (ET) and Sintra. Individual fluid samples are designated by an *Alvin* dive number followed by a bottle number. Duplicate samples from Mkr-6 agreed to within 10% and were averaged together to get these results. EM are end-member concentrations assuming 0 Mg. NORM are fluid concentrations normalized to 'average typical MORB' taken from Sun (1980) times 10[6]. Seawater data (SW) were taken from Elderfield & Greaves (1982); underlined data were interpolated or extrapolated from their data set.

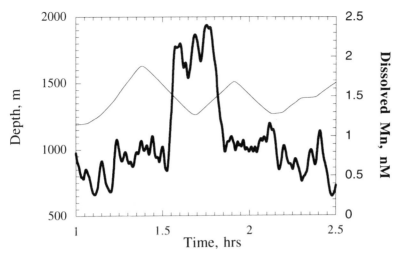

Fig. 3. Depth (light line) and dissolved Mn (heavy line) from the N–S tow-yow shown in Fig. 2. The anomaly in dissolved Mn was detected just east of the summit at depths between 1380 and 1520 m. These plume heights are consistent with vent depths of 1600–1750 m assuming normal fluid buoyancy, which is in turn consistent with the depths of the vent sites discovered with *Alvin* (Fig. 2).

results from a standard. The composition of the standard was determined independently by highly accurate thermal ionization mass spectrometry (Klinkhammer *et al.* 1994*b*). This ion chromatography–ICPMS technique had the triple benefit of increasing the effective detection limit of transition elements by concentrating them into a smaller volume, avoiding major isobaric interferences from barium oxides by separating barium from the REE, and decreasing ion suppression in the plasma by reducing total dissolved solids in the sample stream. Detection limits were 0.5 pmol/kg, five times higher than the lowest concentration in these fluids (Table 1). Precision varied with concentration, ranging from ±2% for the light REE to ±4% for the heavy REE. Accuracies were of the order of ±5% for the REE levels in these fluids.

Hydrothermal plumes at Lucky Strike

The OSU instrument package consisted of a fibre optic spectrometer (ZAPS, Klinkhammer 1994) for measuring dissolved Mn, and a nephelometer and transmissometer for detecting particles. The package was designed to detect vent sites emanating metal-rich fluids, areas that support complex ecosystems. The OSU package was lowered in 11 of the 19 segments making up the FAZAR section (Fig. 1). Plumes of anomalously high Mn and particles were detected in seven of the 11 segments. This

frequency should be considered as an underestimate of total venting for two reasons: firstly, carrying out one or two lowerings in a segment, as during FAZAR, can miss plume areas; and secondly, the OSU package may not detect CH_4-rich fluids with low metal concentrations (Charlou *et al.* 1988).

The water column at Lucky Strike contains plumes of CH_4, particles and Mn, but these plumes are small compared with other MAR locations. A Mn trace from the ZAPS probe recorded during a tow-yow operation on the axial seamount at Lucky Strike is shown in Fig. 3. This anomaly was detected east of the summit, 460 m northeast of the Statue of Liberty and Sintra vents subsequently visited by *Alvin* (Fig. 2). The maximum Mn anomaly observed was about 2.0 nmol/l; this is small compared with anomalies of >30 nmol/l at TAG (Klinkhammer *et al.* 1986) and >200 nmol/l on the Juan de Fuca Ridge (Coale *et al.* 1991). The weak Mn plumes at Lucky Strike are in line with the small nephel anomalies (Fig. 4). Not only are the plumes at Lucky Strike small, but they are also broad and often occur at more than one depth (Fig. 4).

The low levels of Mn and particles observed at Lucky Strike during FAZAR suggested to some of us on the cruise that the fluids feeding the water column were chemically deficient. On the other hand, the observation that the plumes occurred well above the seafloor and surrounding terrain indicated that these anomalies

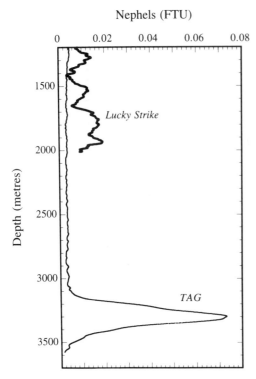

Fig. 4. Light scattering profiles over Lucky Strike compared with similar profiles over the TAG hydrothermal mound at 26°N. Both traces were generated with a Chelsea Aquatracka Mk III Fluorimeter deployed as a nephelometer (420 nm). The upper two anomalies in the Lucky Strike trace are consistent with the water depths of the vents visited by *Alvin* in 1993 (Fig. 2). The deepest anomaly requires additional venting at deeper water depths assuming normal fluid buoyancy (perhaps near the base of the seamount).

originated as buoyant, high-temperature jets and not low-temperature hot springs or seeps. These initial conclusions were verified during the *Alvin* campaign in 1993. High-temperature vents were found at Lucky Strike (Fig. 2), but the Mn and Fe concentrations in these fluids were about five times lower than the most enriched fluids from TAG (Colodner *et al.* 1993).

The relatively low metal concentrations of the Lucky Strike fluids are unusual for high-temperature, bare rock systems. There are a number of possible explanations: fluid–rock interactions are anomalous at Lucky Strike because of the shallow water depth and plume–basalt chemistry associated with the Azores Platform (Schilling 1975); the fluids have low metal-carrying capacities; the Lucky Strike

hydrothermal cells are in a waning, cool-down stage with reduced mobilization in the high-temperature reaction zone; or there is significant removal of metals from fluids during mineralization. Of course, these mechanisms are not exclusive and may be linked. An examination of REE data for Lucky Strike fluids can begin to address this problem.

Rare earth element geochemistry

Rare earth elements in fluids as indicators of alteration

During nuclear synthesis REE with even atomic numbers are more stable than REE with odd atomic numbers. Even REE are thus more abundant in nature. When examining systematic trends in REE concentrations across the series it is convenient to eliminate this overriding odd–even effect through normalization. For hydrothermal data it is reasonable to normalize against the average composition of a normal mid-ocean ridge basalt (N-MORB). Patterns from this normalization are shown in Fig. 5.

The REE patterns of hydrothermal fluids are affected by dissolution and recrystallization reactions in basalt as well as retrograde reactions in secondary alteration assemblages (Klinkhammer *et al.* 1994a). The REEs are not particularly chalchophilic (Morgan and Wandless 1980), but are removed effectively from solution by coprecipitation with Fe, as when hydrothermal fluids mix with seawater (Klinkhammer *et al.* 1983; German *et al.* 1990).

Hydrothermal waters from high-temperature vents are enriched in light REE and possess a strikingly positive Eu anomaly relative to N-MORB (Klinkhammer *et al.* 1994a). Fluids from Lucky Strike also possess this pattern (Table 1, Fig. 5). One interesting fact about the deep-sea fluid pattern in general is that it looks like neither MORB nor seawater. Klinkhammer *et al.* (1994a) have argued that this pattern is fixed by dissolution and recrystallization reactions of hot seawater with the plagioclase component of MORB and modified by ion exchange with Ca and Sr during alteration of MORB plagioclase to hydrothermal plagioclase.

When MORB is leached by circulating fluids, plagioclase feldspars are altered from An-rich (Ca-feldspar) to Ab-rich (Na-feldspar). Berndt *et al.* (1988) have shown that the Sr/Ca ratio of hydrothermal fluid is an effective indicator of the extent of hydrothermal plagioclase formation. This reaction sequence also affects the partitioning of REE between fluid and rock

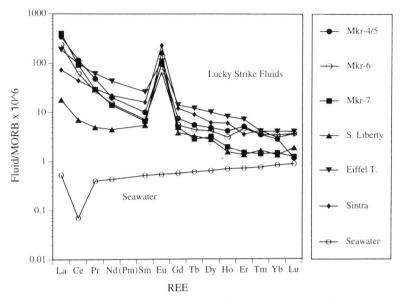

Fig. 5. Rare earth element (REE) concentrations in vent fluids from Lucky Strike (Table 1) and Atlantic seawater at 2500 m (Elderfield and Greaves 1982) normalized to 'average typical mid-ocean ridge basalt (MORB)' (Sun 1980). The REE patterns in these fluids are similar to those from other sites (Klinkhammer *et al.* in press) but concentrations are atypically low and related to temperature (Table 1).

because trivalent REE in VIII coordination (Er 1.004 Å to La 1.16 Å) have effective ionic radii (IR; Shannon 1976) that overlap with the ionic radii of Ca in plagioclase (VI 1.00 Å, VIII 1.12 Å; Bragg *et al.* 1965). As a result of this crystochemical exchange +3 REE are preferentially retained by the plagioclase in An-rich systems but given up more readily to the fluid in Ab-rich systems. Just the opposite is true for Sr (IR = 1.26 Å) which fits more readily into the Ab lattice (Blundy & Wood 1991) and therefore +2 Eu which has an essentially identical IR (1.25 Å). Divalent Eu also partitions into plagioclase preferentially during magma segregation for the same reason; it exists as the +2 cation at low oxygen fugacities (Weill & Drake 1973), which fits readily into the Sr lattice site. In this view, the Eu anomaly of hydrothermal fluid is a product of the crystallization history of the basalt (Humphris *et al.* 1978) and subsequent alteration, whereas the + 3 REE concentration of pristine fluid is controlled by partitioning during dissolution and recrystallization of MORB plagioclase.

In this model, variations in the ratio of An to Ab in the hydrothermal substrate produce the inverse relationship we observe between the Sr/Ca ratios of different fluids and their total +3 REE contents, as illustrated in Fig. 6. The field in Fig. 6 contains data from the East Pacific Rise

for bare rock (e.g. 21°, 11° and 13°N) as well as sediment-hosted systems (e.g. Guaymas Basin and Escanaba Trough). The MAR fluids from the MARK site at 23°N (triangle) and TAG mound at 26°N (diamond) fall in this same field, but Lucky Strike fluids are distinctly different.

Rare earth elements in fluids from Lucky Strike

As mentioned earlier in the discussion, fluid compositions at Lucky Strike (Colodner *et al.* 1993) are consistent with the small plumes found during FAZAR. Mn and Fe concentrations at Lucky Strike are five- to eight-fold lower than fluids at TAG or MARK, and it seems clear that the metal-poor quality of these fluids is responsible for the modest plumes in the water column above Lucky Strike. It is less clear what causes these fluids to be chemically depleted. The REE systematics of the fluids can shed some light on this problem.

Fluid temperatures at Lucky Strike ranged between 197 and 316°C (Table 1) and there is a weak but positive correlation between these temperatures and +3 REE concentrations, with a correlation coefficient of 0.77. This relationship can be seen in Table 1 and Fig. 5; the lowest temperature (197°C) and +3 REE content

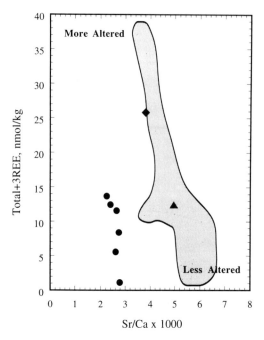

Fig. 6. Total +3 REE concentrations (Table 1) plotted against Sr/Ca ratios (Von Damm and Colodner, unpublished data). The field includes data from the East Pacific Rise for bare-rock and sediment-hosted systems (Klinkhammer *et al.* in press). The triangle is MARK (Mid-Atlantic Ridge at 23°N) and the diamond TAG (Mid-Atlantic Ridge at 26°N). Data from Lucky Strike (closed circles) are subparallel to the field for typical fluids.

(1.16 nmol/kg) were recorded at the Statue of Liberty vent complex, whereas the highest +3 REE level (11.6 nmol/kg) and temperature (316°C) were found at Eiffel Tower. These exit temperatures are lower than the 350°C fluid typical of most vent areas (Bowers *et al.* 1988). If we extrapolate linearly along the temperature–REE trend of these samples, the predicted +3 REE concentration for a 350°C fluid at Lucky Strike would be 17.5 nmol/kg. This value lies between the +3 REE contents at the other MAR sites, MARK and TAG. From this analysis it seems that an end-member fluid at Lucky Strike contains a typical amount of REEs for its temperature. Based on this result it seems likely that the metal-carrying capacity of this fluid is normal, and that the fluid is not responding to the enriched REE content of plume basalts (Schilling 1975) or to reduced water pressure. Further considerations of these processes will be possible when the chemical compositions of the fluids (Colodner *et al.* 1993) and rocks (Langmuir *et al.* 1993) from this site become available.

Low REE concentrations in these fluids may result from either reduced mobilization or removal within the cell. It seems unlikely that this removal occurs during mixing of seawater with fluid near the seafloor as the samples have low Mg contents (Colodner *et al.* 1993). Similarly, removal of REE with lower temperature minerals such as anhydrite and barite does not seem to fit as the fluid temperatures are too high. Also, in the case of anhydrite, the Sr/Ca ratios in these fluids are lower than normal, not higher (Fig. 6) as might be expected if there was extensive precipitation of anhydrite below the seafloor.

It seems more likely that reduced metal levels in the fluids at Lucky Strike result from reactions deep in the system. As pointed out earlier, low Sr/Ca ratios in these fluids (Fig. 6) suggest that the basaltic substrate is highly altered. Moreover, the fact that the Sr/Ca ratios are low because of unusually low Sr concentrations is what would be expected for fluids reacting with Ab-rich plagioclase (Blundy & Wood 1991). The inverse, linear relationship between +3 REE levels and Sr/Ca ratios (Fig. 6) are also consistent with extensive alteration.

However, alteration alone cannot explain the low REE concentrations in these fluids. The observation that these low REE concentrations are related to exit temperatures and that these temperatures are lower than those for other systems suggests that there is reduced mobilization at the fluid–basalt boundary, either because of a temperature-related shift in crystochemical exchange equilibria or because the substrate is highly leached. The former seems more likely as the relationship with Sr/Ca predicted for exchange is preserved in these samples (Fig. 6). This situation contrasts sharply with the chemistry of fluids from the southern Juan de Fuca Ridge (SJFR) where Sr/Ca ratios are also low (Von Damm and Bischoff 1987) and similar to those at Lucky Strike. Rare earth element patterns have not been published for the SJFR. However, end-member Nd concentrations for this system are 18 nmol/kg (Hinkley and Tatsumoto 1987) so that the total +3 REE content at the SJFR is likely to be >100 nmol/kg and more in line with other systems (Fig. 6), compared with <14 nmol/kg for Lucky Strike (Table 1). If the REE levels were unrelated to temperature, then we could invoke REE removal in secondary mineral phases, but this process is harder to rationalize with the linear Sr/Ca relationship. In either case, low temperatures, low REE concentrations and low Sr/Ca ratios are indicative of a waning system with zones of extensive alteration.

Conclusions

Predictions of the nature of venting at Lucky Strike based on an instrumental survey of the water column during the FARA Program are consistent with observations made with the submersible *Alvin*. Plumes at Lucky Strike rise above the local topography, but have low levels of light-scattering particles and dissolved Mn concentrations. Fluids from Lucky Strike are buoyant but metal-poor compared with other MAR sites. This ground-truthing exercise reinforces the idea that simple observations of hydrothermal plumes from a surface ship can be used to formulate a first-order geochemical picture of venting from the seafloor.

Although it is clear that reduced metal levels in the vent fluids at Lucky Strike are responsible for the small plumes we observe in the overlying water column, it is less clear why metal levels in these fluids are low. We can gain some insight into this problem by looking at REE data for the fluids. In general, fluid REE patterns at Lucky Strike are similar to those from other deep-sea systems: light REE enriched with a positive Eu anomaly. However, REE concentrations at Lucky Strike are relatively low and directly related to temperature. These observations are consistent with limited metal mobilization in a mature and waning hydrothermal system. Moreover, a regular relationship between total +3 REE concentrations and low Sr/Ca ratios indicates extensive alteration of the basaltic substrate below Lucky Strike.

C. Langmuir (LDEO) was chief scientist for FARA and the *Alvin* dives on Lucky Strike. K. Von Damm (University of New Hampshire) and D. Colodner (LDEO) processed the fluid samples and graciously gave us permission to use preliminary data. A. Ungerer (OSU) was mainly responsible for the ICP-MS analyses. This work was supported by NSF grant OCE–9105177 to GK and NATO Collaborative Research grant CRG 910921 to CRG and GK. IOSDL Contribution no. **95001**.

References

BAKER, E. T. Characteristics of hydrothermal discharge following a magmatic intrusion, this volume.

BERNDT, M. E., SEYFRIED, W. E. JR & BECK, J. W. 1988. Hydrothermal alteration processes at mid-ocean ridges: experimental and theoretical constraints from Ca and Sr exchange reactions and Sr isotopic ratios. *Journal of Geophysical Research,* **93**, 4573–4583.

BLUNDY, J. D. & WOOD, B. J. 1991. Crystal-chemical controls on the partitioning of Sr and Ba between plagioclase feldspar, silicate melts, and hydrothermal solutions. *Geochimica et Cosmochimica Acta,* **55**, 193–209.

BOWERS, T. S., CAMPBELL, A. C., MEASURES, C. I., SPIVACK, A. J., KHADEM, M. & EDMOND, J. M. 1988. Chemical controls on the composition of vent fluids at 13°–11°N and 21°N, East Pacific Rise. *Journal of Geophysical Research,* **93**, 4522–4536.

BRAGG, L., CLARINGBULL, G. F. & TAYLOR, W. H. 1965. *The Crystalline State, Vol. IV,* Cornell University Press.

CHARLOU, J.-L., DMITRIEV, L. & BOUGAULT, H. 1988. Hydrothermal CH$_4$ between 12°N and 15°N over the Mid-Atlantic Ridge. *Deep Sea Research,* **35**, 121–131.

COALE, K. H., CHIN, C. S., MASSOTH, G. J., JOHNSON, K. S. & BAKER, E. T. 1991. In situ chemical mapping of dissolved iron and manganese in hydrothermal plumes. *Nature,* **352**, 325–328.

COLODNER, D., LIN, J. & 6 others 1993. Chemistry of Lucky Strike hydrothermal fluids: initial results. *EOS, Transactions of the American Geophysical Union,* **74**, 99.

ELDERFIELD, H. & GREAVES, M. J. 1982. The rare earth elements in seawater. *Nature,* **296**, 214–219.

GERMAN, C. R., KLINKHAMMER, G. P., EDMOND, J. M., MITRA, A. & ELDERFIELD, H. 1990. Hydrothermal scavenging of rare-earth elements in the ocean. *Nature,* **345**, 516–518.

HINKLEY, T. K. & TATSUMOTO, M. 1987. Metals and isotopes in Juan de Fuca Ridge hydrothermal fluids and their associated solid materials. *Journal of Geophysical Research,* **92**, 11 400–11 410.

HUMPHRIS, S. E., MORRISON, M. A. & THOMPSON, R. N. 1978. Influence of rock crystallisation history upon subsequent lanthanide mobility during hydrothermal alteration of basalts. *Chemical Geology,* **23**, 125–137.

KLINKHAMMER, G. P. 1994. Fiber optic spectrometers for in-situ measurements in the oceans: the ZAPS Probe. *Marine Chemistry,* **47**, 13–20.

——, ELDERFIELD, H., EDMOND, J. M. & MITRA, A. 1994*a*. Geochemical implications of rare earth element patterns in hydrothermal fluids from mid-ocean ridges. *Geochimica et Cosmochimica Acta,* **58**, 5105–5113

——, ELDERFIELD, H., GREAVES, M., RONA, P. & NELSEN, T. A. 1986. Manganese geochemistry near high-temperature vents in the Mid-Atlantic Ridge rift valley. *Earth and Planetary Science Letters,* **80**, 230–240.

——, ELDERFIELD, H. & HUDSON, A. 1983. Rare earth elements in seawater near hydrothermal vents. *Nature,* **305**, 185–188.

——, GERMAN, C. R., ELDERFIELD, H., GREAVES, M. J. & MITRA, A. 1994*b*. Rare earth elements in hydrothermal fluids and plume particulates by inductively coupled plasma mass spectrometry. *Marine Chemistry,* **45**, 179–186.

——, WEISS, R. F. & BENDER, M. L. 1977. Hydrothermal manganese in the Galapagos Rift. *Nature,* **269**, 319–320.

LANGMUIR, C. H. & LUCKY STRIKE TEAM 1993.

Geological setting and characteristics of the Lucky Strike vent field at 37°17′N on the Mid-Atlantic Ridge. *EOS, Transactions of the American Geophysical Union*, **74**, 99.

——, *ET AL.* 1993. *Lucky Strike/ALVIN expedition cruise report*. Technical Report LDEO-93-2, Lamant-Doherty Earth Observatory.

LUPTON, J. E., KLINKHAMMER, G., NORMARK, W. R., HAYMON, R., MACDONALD, K., WEISS, R. & CRAIG, H. 1980. Helium–3 and manganese at the 21°N East Pacific Rise hydrothermal site. *Earth and Planetary Science Letters*, **50**, 115–127.

MORGAN, J. W. & WANDLESS, G. A. 1980. Rare earth element distribution in some hydrothermal minerals: evidence for crystallographic control. *Geochimica et Cosmochimica Acta*, **44**, 973–980.

MURTON, B. J., KLINKHAMMER, G. & 11 others 1994. Direct evidence for the distribution and occurrence of hydrothermal activity between 27–30°N on the Mid Atlantic Ridge. *Earth and Planetary Science Letters*, **125**, 119–128.

SCHILLING, J.-G. 1975. Azores mantle blob: rare-earth evidence. *Earth and Planetary Science Letters*, **25**, 103–115.

SHANNON, R. D. 1976. Revised effective ionic radii and systematic studies of interatomic distances in halides and chalcogenides. *Acta Crystallography*, **A32**, 751–767.

SUN, S.-S. 1980. Lead isotopic study of young volcanic rocks from mid-ocean ridges, ocean islands and island arcs. *Philosophical Transactions of the Royal Society of London*, **297**, 409–445.

WEILL, D. F. & DRAKE M. J. 1973. Europium anomaly in plagioclase feldspar: experimental results and semiquantitative model. *Science*, **180**, 1059–1060.

Hydrothermal plumes at Broken Spur, 29°N Mid-Atlantic Ridge: chemical and physical characteristics

R. H. JAMES[1], H. ELDERFIELD[1], M. D. RUDNICKI[1,6], C. R. GERMAN[2], M. R. PALMER[3], C. CHIN[4], M. J. GREAVES[1], E. GURVICH[5], G. P. KLINKHAMMER[4], E. LUDFORD[3], R. A. MILLS[1,7], J. THOMSON[2] & A. C. WILLIAMS[1]

[1]*Department of Earth Sciences, Cambridge University, Downing Street, Cambridge CB2 3EQ, UK*

[2]*Institute of Oceanographic Sciences, Wormley GU5 8UB, UK*

[3]*Department of Geology, University of Bristol, Bristol BS8 1RJ, UK*

[4]*College of Oceanography, Oregon State University, Corvallis, OR 97331, USA*

[5]*P.P. Shirshov Institute of Oceanology, Moscow 117218, Russia*

[6]*Present address: College of Oceanography, Oregon State University, Corvallis, OR 97331, USA*

[7]*Present address: Department of Oceanography, University of Southampton, Highfield, Southampton SO9 5NH, UK*

Abstract: Two distinct hydrothermal plumes have been identified and sampled in the vicinity of active hydrothermal vents recently discovered at a site named Broken Spur, 29°10.15′N 43°10.28′W, on the Mid-Atlantic Ridge. The northern plume (29°10.13′N 43°10.61′W) most closely associated with Broken Spur consists of a weak plume (heat flux *c.* 5 MW), whereas plumes detected to the south (29°09.11′N 43°10.54′W) are much stronger (heat flux *c.* 57 MW). The plumes are characterized by dissolved ^{222}Rn, Fe and Mn anomalies of up to 20 dpm/100 kg, 32 and 14 nmol/l, respectively, and enhanced concentrations of Fe, Mn, Cu, Zn, Cd, Co and Pb in the particulate phase. Comparison of Fe:Mn ratios in the plume (3.7) to Fe:Mn in vent fluids collected at Broken Spur (7.9) suggests that *c.* 53% of hydrothermal Fe is removed during plume evolution.

Submarine hydrothermal activity has been recognized and sampled in a wide variety of oceanographic conditions in the Pacific Ocean (Edmond *et al.* 1979; Von Damm *et al.* 1985a, b; Von Damm and Bischoff 1987; Baker *et al.* 1987; Massoth *et al.* 1989; Grimaud *et al.* 1991; Sedwick *et al.* 1992; Stuben *et al.* 1992), whereas investigation of hydrothermal activity in the Atlantic is comparatively sparse. To date, five sites of hydrothermal activity have been discovered; all are mid-ocean ridge axis systems. The Trans-Atlantic Geotraverse (TAG) area at 26°N on the Mid-Atlantic Ridge (MAR) and the MAR at Kane (MARK; subsequently referred to as Snakepit) site at 23°N MAR were both discovered in the summer of 1985 and are the best studied sites in the Atlantic (Rona *et al.* 1986, Detrick *et al.* 1986). Both sites are at depths of *c.* 3500–3700 m. Two shallower sites of hydrothermal activity have been discovered more recently at Steinahóll, 63°N MAR (water

depth *c.* 600 m; Olafsson *et al.* 1991; German *et al.* 1994) and Lucky Strike, 37°N MAR (water depth *c.* 1600 m; Langmuir *et al.* 1993). Vent fluids have been sampled from TAG and Snakepit (Campbell *et al.* 1988) and Lucky Strike (Colodner *et al.* 1993), but have yet to be sampled from Steinahóll. The fifth site, Broken Spur (29°N MAR) is the subject of this paper.

Evidence of a further site of hydrothermal activity at the MAR was first obtained during February/March 1993 (Murton *et al.* 1994). A systematic survey of the axial valley of the MAR was carried out between 27°N and 30°N using TOBI, a towed package equipped with a 30 kHz sidescan sonar array and a variety of real-time sensors. These sensors included a temperature probe, transmissometer and zero-angle photon spectrometer (ZAPS) which measures dissolved oxidizable manganese (Klinkhammer 1994). Dissolved Mn, temperature and particle anomalies in the water column suggested that an active

From PARSON, L. M., WALKER, C. L. & DIXON, D. R. (eds), 1995, *Hydrothermal Vents and Processes*, Geological Society Special Publication No. 87, 97–109.

97

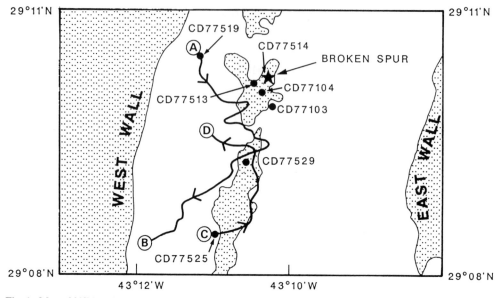

Fig. 1. Map of 29°N study area. The shaded areas are <3100 m and denote the west and east walls of the median valley and the axial volcanic ridge. The location of Broken Spur is 29°10.15′N, 43°10.28′W.

vent was located at approximately 29°10.13′N 43°10.28′W, and this site was subsequently named Broken Spur. RRS *Charles Darwin* cruise 77 (CD77; March/April 1993) carried out a survey in the 29°N area to map the extent of hydrothermal activity defined by particle, temperature, salinity and chemical anomalies in the water column. During the course of the cruise samples of plume waters, particles and sulphide chimney material were collected. Hydrothermal activity in the area was confirmed in June 1993 by the discovery (and sampling) of active vents by dives performed by DSRV *Alvin* (Murton & Van Dover 1993).

Hydrothermal plumes integrate regional venting; their distinct thermochemical signature reflects processes contributing to their formation. The principal purpose of this paper is to document the physical and chemical characteristics of hydrothermal plumes at Broken Spur and to attempt to relate these characteristics to seafloor sources. We begin by summarizing the background temperature and salinity structure at 29°N, then investigate how these signals are modified in the vicinity of a plume. This information is used to calculate the heat flux associated with venting at Broken Spur. Fe, Mn and ^{222}Rn have been used as chemical tracers of hydrothermal activity; their concentrations in the plume are shown to reflect chemical and physical processes occurring during plume evolution. Insight into the modification of hydrothermal chemical fluxes within the plume is

gained from investigation of particulate mineral phases (essentially Fe oxyhydroxides) in the plume.

Study area and methods

Figure 1 shows the study area of cruise CD77 within the median valley of the MAR at 29°N. The position of Broken Spur is shown and is located at the crest of an axial volcanic ridge, similar to the setting of the Snakepit hydrothermal vent site at 23°N (Detrick *et al.* 1986). Work was carried out within a transponder net fixed by GPS navigation. The hydrothermal plume was surveyed using a CTD equipped with a nephelometer, transmissometer, relay transponder and a rosette of 12 10-l Go-Flo bottles. On recovery of the package, samples were withdrawn for shipboard analyses of ^{222}Rn (Mathieu 1977) and Si and PO_4 (Hydes 1984). Samples for nitrate analysis were preserved by poisoning with $HgCl_2$ and analysed on return following the method of Hydes (1984). The remainder was filtered under clean conditions through acid-cleaned 0.4 μm Nuclepore filters and stored in acid-clean 1-l high density polyethylene or polypropylene bottles and acidified with 1 ml sub-boiling quartz-distilled HNO_3. The filters were stored in Petri dishes and refrigerated at 4°C. Upon return, samples were analysed for dissolved Mn and Fe using the solvent extraction/ graphite furnace atomic absorption spectrometry (GF-AAS) technique

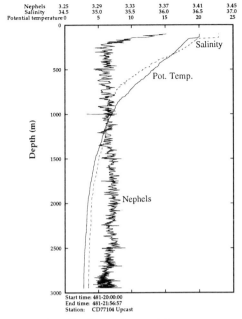

Fig. 2. Plot of background potential temperature (°C), salinity and nephels (raw volts) versus depth in the vicinity of Broken Spur (CD77104).

of Danielsson *et al.* (1978) as modified by Statham (1985). Particulate trace metals were extracted with HNO_3 and HF following the method of Trefry & Trocine (1991) and also analysed by GF-AAS. Analytical precision was better than 8% (2σ) for all analyses.

Hydrological and chemical structure of the Atlantic in the vicinity of 29°N

Figure 2 (CD77104) shows the hydrological structure of the western Atlantic at 29°N. North Atlantic Central Water, from the surface to a depth of 1000 m, is underlain by North Atlantic Deep Water (Tchernia 1980). Higher salinity Mediterranean outflow water intrudes the water column at approximately 1000 m, producing a pronounced discontinuity in the plot of potential temperature–salinity (Fig. 3).

These hydrological observations are also to some degree reflected in the chemical characteristics of the water column at 29°N. Figure 4 illustrates the vertical distributions of phosphate, silicate, nitrate and nitrite determined in this study (CD77104). The profiles are similar to those obtained by Bruland & Franks (1983) and can largely be interpreted by biogeochemical cycling. The discontinuity in the silicate profile at around 1000 m reflects the intrusion of relatively nutrient-rich Mediterranean water; this may also enhance the nitrate and phosphate maximum.

Hydrothermal plumes at Broken Spur

Physical characteristics

During cruise CD77 two long survey lines and 21 vertical profiles from six stations were obtained during the plume survey (Fig. 1). Plume signals were recognized on 16 of the lowerings. Figure 5 shows two examples. The record in Fig. 5a

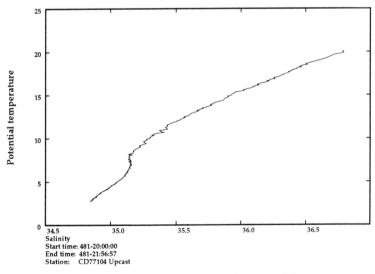

Fig. 3. Plot of background (CD77104) potential temperature (°C) versus salinity.

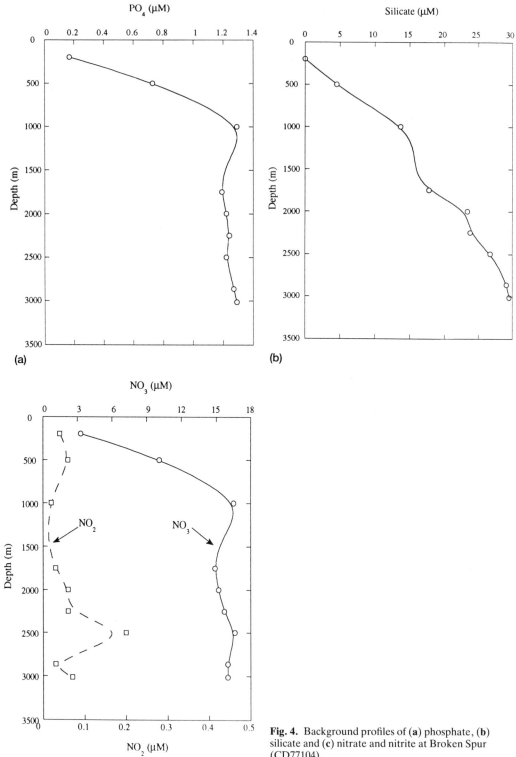

Fig. 4. Background profiles of (**a**) phosphate, (**b**) silicate and (**c**) nitrate and nitrite at Broken Spur (CD77104).

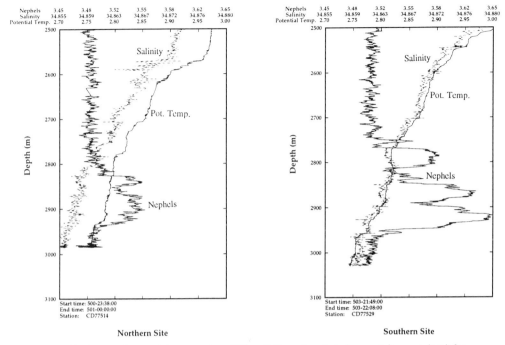

Fig. 5. Plot of background potential temperature (°C), salinity and nephels (raw volts) versus depth for (**a**) northern site CD77514-dl and (**b**) southern site CD77529-dl.

(CD77514) was obtained at 29°10.13′N 43°10.61′W, close to the location of the Broken Spur hydrothermal site (29°10.15′N 43°10.28′W) and consists of a weak plume that may be recognized from the nephelometry profile. The plume extends over 2830–2930 m compared with a bottom depth of 3050 m. It is layered and comprises three main maxima at 2840, 2885 and 2920 m. The plume record shown in Fig. 5b (CD77529) was obtained from 29°09.11′N 43°10.54′W, 1.9 km to the south of the Broken Spur site. It shows an intense multiple plume over 2750–2950 m with maxima at 2750, 2785, 2830, 2870, 2920 and 2945 m. Associated with the nephelometry profiles of the plume are anomalies in potential temperature and salinity. The salinity profile of Fig. 5b (CD77529) clearly shows a reduction in salinity in particle-rich plume layers. This is a result of the entrainment of bottom waters of lower salinity and temperature than ambient seawater at plume height, a characteristic signal of mid-Atlantic hydrothermal plumes (Speer & Rona 1989; Rudnicki & Elderfield 1992a).

Evidence for strong plume signals at the southern site may also be seen from the two nephelometry sections shown in Fig. 6, the tracks of which are marked on Fig. 1. CD77519 is a N–S section and shows an intense plume just

north of 29°09′N. CD77525 is a S–N section and also shows an intense plume very close to the site of the CD77529 vertical profile shown in Fig. 5b. The regional extent of the plume signals coupled with the intensity of the signals to the south, away from Broken Spur, strongly suggests that there must be more than one hydrothermal vent site in this ridge segment.

An estimate of the heat flux from venting may be obtained by application of a plume model (Rudnicki & Elderfield 1992a). The background salinity and temperature gradients in the area are 2.7×10^{-5} m^{-1} and 3.8×10^{-4}°C m^{-1}, respectively. Given a venting depth of 3050 m for the northern site derived from Broken Spur, the plume layers are approximately 130, 165 and 210 m off-bottom, which correspond to a total heat flux of 5 MW. The plume layers for the southern site shown in Fig. 5b are approximately 155, 180, 230, 270, 315 and 350 m off-bottom and, assuming the vents are at 3100 m depth, correspond to a total heat flux of 57 MW (a venting depth of 3050 m gives a calculated heat flux of 28 MW). The order-of-magnitude difference in heat fluxes between the two sites means (a) it is unlikely that the source of the southern plume is also the Broken Spur hydrothermal site and (b) the southern plume results from a larger vent complex than that at Broken Spur. As vents

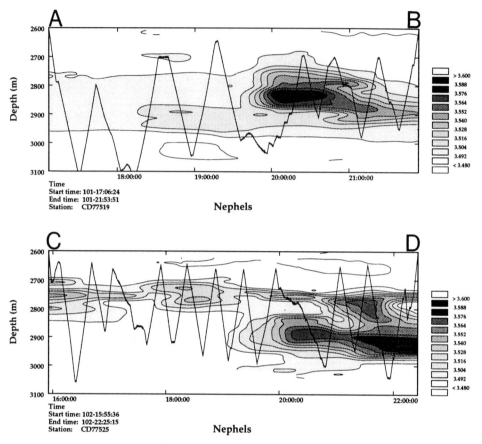

Fig. 6. Sections showing particle concentrations measured by nephelometry defining dispersion of neutrally buoyant plumes in the 29°N area. The horizontal scale is time and is approximately linear. Solid lines show the tracks through the water column of the towed instrument package. Contours show the nephelometry output in volts (background values <3.5). The track lines of the sections are shown in Fig. 1: CD77519 is a N–S track (A–B) and shows a large plume signal where the track is closest to the location of CD77529 (see Fig. 5b); CD77525 is a S–N track (C–D) and shows a northern plume signal, presumably from Broken Spur (see Fig. 5a) and an intense southern plume signal over the NE–SW segment of the track line.

sampled by Murton & Van Dover (1993) were located close to the site of Broken Spur as identified by the TOBI/ZAPS survey of Murton *et al.* (1994), this has interesting implications for the discovery of further vents.

Chemical characteristics

Figure 7 shows elevated ^{222}Rn activities 100–200 m off-bottom in the region of the hydrothermal plume defined by nephelometry anomalies (Fig. 5). The anomalies, compared with the background station, are 10–20 dpm 100 kg^{-1}. These are much too high to be accounted for by the dilution of hydrothermal vent fluids, for which typical values are 300–900 dpm kg^{-1} for two Pacific sites (Kim & Finkel 1980; Kadko & Moore 1988) and 100–200 dpm kg^{-1} for two Atlantic sites (Rudnicki & Elderfield 1992b). The total Mn concentration of these samples implies a vent fluid dilution of $1–2 \times 10^{4}$, giving a calculated ^{222}Rn vent fluid end-member of 1000–4000 dpm kg^{-1}. The discrepancy between the calculated end-member and measured values has previously been explained by entrainment of ^{222}Rn-enriched bottom waters in the rising plume (Rosenberg *et al.* 1988; Rudnicki & Elderfield 1992b); indeed, the profiles in Fig. 7 all show indications of Rn-enriched bottom waters. However, without the collection of gas-tight vent fluid samples at Broken Spur, the

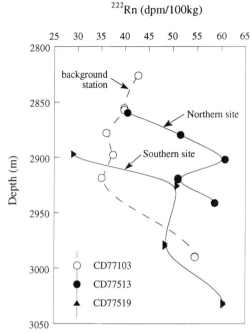

^{222}Rn (dpm/100kg)

Fig. 7. ^{222}Rn profiles from three stations in median valley at 29°N: CD77103 (background station); CD77513 (near Broken Spur); CD77519 (southern site).

Fuca Ridge are reported to reside almost exclusively in colloidal-size oxyhydroxide phases (Feely *et al.* 1987; Walker & Baker 1988). As no sample was taken within the larger, lower plume at the southern site, it is difficult to assess whether differences exist in the dissolved Mn and Fe concentrations between the northern and southern sites. Dissolved Fe and Mn are correlated in the neutrally buoyant plume (Fig. 10); the Fe:Mn ratios for the southern (CD77525) and northern (CD77514) sites are, respectively, 2.0 and 2.5.

Chemistry of hydrothermal plume particles

Metal concentrations of particles from plumes at sites CD77514 and CD77525 are presented in Table 1 and compared with typical background values in the Atlantic. Generally, particulate Fe is used as a conservative element against which to plot particulate trace metal data (Trocine & Trefry 1988; German *et al.* 1991). However, as a high proportion of Fe is measured in the 'dissolved' fraction, we have plotted particle data for Fe, Mn, Cd, Co, Zn, Pb and Cu versus dissolved Mn (Fig. 11). In addition, Fig. 11 distinguishes between particles from the northern (CD77514) and southern (CD77525) sites, and, where appropriate, outlines a first- or second-order differential relationship between dissolved Mn and particulate metal concentration.

Although concentrations of all metals in the particulate phase increase with increasing dissolved Mn, the relationship between the trace elements and Mn concentration in the plume particles at Broken Spur appears to be characterized by two distinct patterns.

(1) Particle concentrations of Cu, Mn and Zn (northern site) and Cd, Fe and Pb (southern site) increase linearly with increasing dissolved Mn. The result for Mn at the northern site contrasts with previous studies which suggest that Mn concentrations do not vary in the plume (Trocine & Trefry 1988), or vary only at high particulate Fe concentrations (>100 nmol/l; German *et al.* 1991).

(2) Particle concentrations of Cd, Co, Fe and Pb (northern site) and Cu and Zn (southern site) show non-conservative behaviour with respect to dissolved Mn; there is a negative departure from the notional mixing line. The implications of this relationship are discussed in the following.

Negative departure from linear mixing is most notable for particles from the northern site where concentrations of dissolved Mn and particulate Fe are higher. German *et al.* (1991)

origin of the neutrally buoyant plume ^{222}Rn must remain in doubt.

Other chemical indicators of hydrothermal activity include elevated levels of dissolved Mn and Fe. Figure 8 shows Mn and Fe profiles in the region of the plume defined by nephelometry anomalies [Fig. 9; note Fig. 9 (left-hand panel) depicts a different cast from Fig. 5a but is at the same station]. Background dissolved Mn and Fe profiles are also given. Maximum dissolved Fe and Mn concentrations clearly coincide with the particle maximum at the northern site. No sample was collected from the lower (2890–2970 m) plume at the southern site; hence this plume is not reflected in the chemical profiles. The upper (2740–2700 m) plume was sampled and is related to the particle anomaly. Figure 8 shows Mn and Fe anomalies of up to 14 and 32 nmol/l, respectively, 100–200 m above bottom. Iron is removed into particulate phases within hours to days after discharge (Rudnicki & Elderfield 1993; Chin *et al.* 1994). Hence this is an extremely high level of apparently dissolved Fe; the 'dissolved' fraction is operationally defined as Fe passing through a 0.4 μm filter and may include Fe in colloidal form. Particulate Fe within plumes at the Cleft segment, Juan de

Table 1. *Chemical data for hydrothermal plume particles and dissolved Mn from Broken Spur*

Sample	Depth (m)	Fe (nmol/l)	Co (pmol/l)	Pb (pmol/l)	Cd (pmol/l)	Cu (pmol/l)	Mn (pmol/l)	Zn (pmol/l)	Dissolved Mn (nmol/l)
CD77514-1	2962	5	2.5	5.8	0.52	120	110	270	3.23
CD77514-2	2951	6	2.7	7.3	0.86	230	140	210	5.1
CD77514-6	2941	16	4.3	14	2.6	790	180	1030	9
CD77514-4	2936	14	3.7	21	2.8	620	170	970	9.42
CD77514-5	2934	14	3.6	11	3.5	750	150	630	11.2
CD77514-7	2925	11	3.9	14	4	690	150	510	10.4
CD77514-8	2914	19	5.5	26	4.2	1030	170	720	11.2
CD77514-9	2876	22	5.1	36	7.8	990	170	1080	14.3
CD77514-11	2839	8	3.5	13	1.3	370	130	650	6.09
CD77525-1	2730	4		4.3	0.36	200	140	490	0.98
CD77525-2	2758	12	4.1	11	1.7	620	170	740	7.16
CD77525-5	2778	8	3	7.6	0.66	280	170	530	6.38
CD77525-8	2888	17	3.3	9.8	1.5	660	130	1310	10.8
CD77525-10	3001	8	2.9	10	1.4	170	130	590	5.23
CD77525-11	3000	14	4	7.7	0.62	230	190	–	5.68
Background*		3	3	1	0.5	20	150	40	0.68

* German *et al.* (1991); dissolved Mn this study.

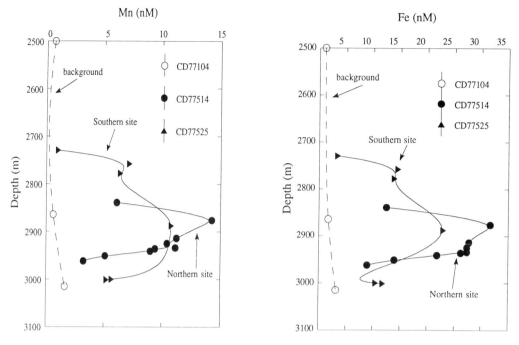

Fig. 8. (**a**) Dissolved Mn and (**b**) dissolved Fe profiles from three stations in the median valley at 29°N: CD77104 (background station); CD77514 (near Broken Spur); CD77525 (southern site).

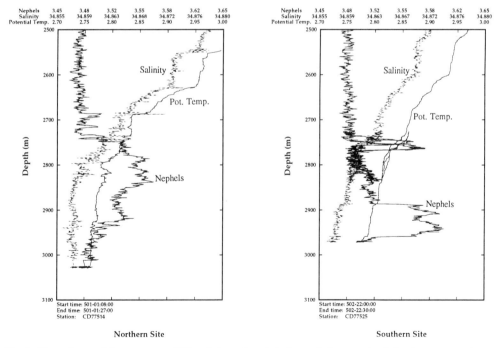

Fig. 9. Plots of potential temperature (°C), salinity and nephels (raw volts) versus depth for the trace metal casts in Fig. 8a and b: (**a**) CD77514u-1 and (**b**) CD77525.

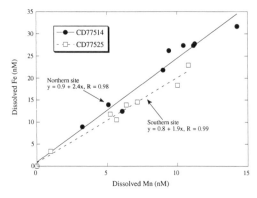

Fig. 10. Plot of dissolved Mn versus dissolved Fe at plume height.

plots particulate metal concentrations against particulate Fe for particles collected from the TAG hydrothermal plume. A relationship of this type is observed for Cu, Zn, Co and Pb. The greatest departure from linearity occurs at particulate Fe concentrations of c. 100 nmol/l; highest particulate Fe concentrations recorded in this study are c. 20 nmol/l.

This behaviour indicates that there is removal of these elements from the neutrally buoyant plume. These chalcophile elements appear to be largely associated with sulphides transported with the plume and this removal pattern reflects either their oxidative dissolution or their fallout by gravitational settling (Trocine & Trefry 1988; Metz & Trefry 1993). These processes may therefore be detected even at relatively low plume dissolved Mn and particulate Fe concentrations.

Although concentrations of particulate Co, Pb, Cu and Mn are of the same magnitude as observed at Axial Volcano on the Juan de Fuca Ridge (Feely *et al.* 1990) and the MAR at 26°N (Trocine & Trefry 1988), the elements Cd and Zn appear to exhibit an order of magnitude enrichment. This may reflect the vent fluid composition; in addition to the trace element composition itself, concentrations are also controlled by the pH, chlorinity and H_2S content of the fluid. The trace metal content of vent fluids at Broken Spur is summarized in Table 2. The Zn composition of vent fluids at Axial Volcano varies from an order of magnitude smaller to roughly the same as the concentration observed at Broken Spur (Massoth *et al.* 1989). The Cd content of Axial Volcano fluids appears to be slightly greater than that found at Broken Spur (Trefry *et al.* 1994); no Cd or Zn data is available for the MAR at 26°N.

Modification of vent fluid fluxes during plume evolution

Hydrothermal vent fluids have been collected at the Northern site (Broken Spur) close to CD77514. The chemistry of the vent fluids is reported in full by James *et al.* (in press). Maximum concentrations of Co, Pb, Cd, Cu and Zn in the vent fluids are compared with maximum particulate concentrations observed at plume height in Table 2. Iron is treated separately as a large fraction is observed to reside in the dissolved (<0.4 μm) phase. The maximum dissolved Mn concentration at plume height is c. 14 nmol/l; the measured hydrothermal end-member at Broken Spur is c. 250 μmol/l. Manganese has been shown to persist in hydrothermal plumes with a removal time in excess of two weeks (Kadko *et al.* 1990) or even two years (Lavelle *et al.* 1992), hence the concentration of Mn recorded at plume height reflects dilution of the vent fluid by a factor of 2×10^4, close to model predictions (Rudnicki & Elderfield 1993). Plume height concentrations of Co, Pb, Zn, Cu and Cd projected from 2×10^4 dilution are given in the final column in Table 2. Assuming all vent fluid injected metal is present in particle form at plume height (dissolved Co, Pb, Cd, Cu and Zn concentrations at plume height appear to be within analytical error of background concentration; R. James, unpublished data), the data suggest that up to c. 70% of hydrothermally injected Co, Cu and Zn is lost during transit to plume height, presumably as sulphides. Particulate Cd and Pb concentrations in the plume appear too large to be accounted for by vent fluid input. Trefry *et al.* (1994) report that up to 58% of vent fluid Cd and up to 38% of vent fluid Pb may be present in vent fluid samples as residual precipitates; hence their concentration in the vent fluid may be underestimated.

Maximum particulate Fe concentrations observed at Broken Spur are c. 20 nmol/l with a corresponding dissolved Fe concentration of 32 nmol/l. The measured hydrothermal end-member Fe is 2.0 nmol/l. A 2×10^4 dilution of the vent fluid generates an Fe anomaly of c. 100 nmol/l at plume height. As Fe in the dissolved and particulate phases is significantly above background, the total Fe anomaly at plume height is c. 50 nmol/l, indicating removal of c. 50% of hydrothermal vent Fe, presumably deposited as large sulphide particles close to the vent site. This is similar to results from the East Pacific Rise, near 21°N, where 50% of hydrothermally injected Fe is thought to form primary sulphides immediately after venting (Mottl & McConachy 1990), and consistent with particle

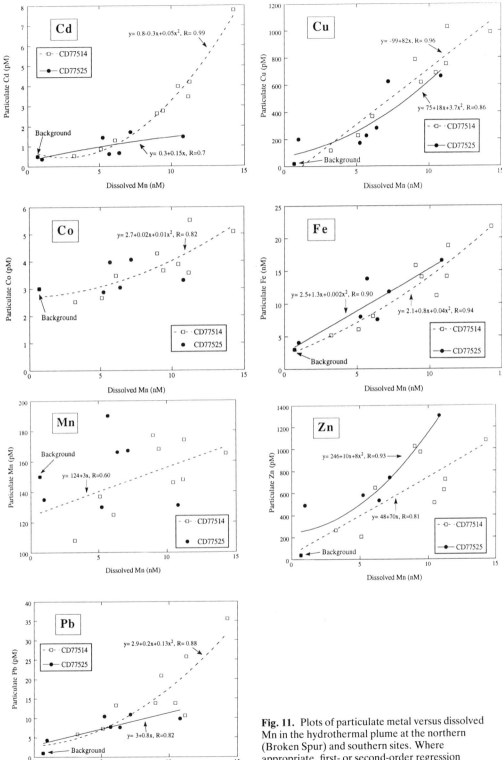

Fig. 11. Plots of particulate metal versus dissolved Mn in the hydrothermal plume at the northern (Broken Spur) and southern sites. Where appropriate, first- or second-order regression relationships are indicated.

Table 2. *Metal anomalies at plume height predicted from 2×10^4 dilution of vent fluids*

Element	Maximum vent fluid concentration*	Particle anomaly at plume height	Projected vent fluid concentration at 2×10^4 dilution
Co	420 nmol/l	5.5 pmol/l	21 pmol/l
Pb	380 nmol/l	36 pmol/l	19 pmol/l
Cd	150 nmol/l	7.8 pmol/l	7 pmol/l
Cu	65 nmol/l	1 nmol/l	3 nmol/l
Zn	90 μmol/l	1.3 nmol/l	4.5 nmol/l
Fe	2 mmol/l	50 nmol/l[†]	100 nmol/l

* Data from James *et al.* (in press).
† Total Fe (see text for details).

formation in the TAG buoyant plume where oxyanion/Fe ratios in plume particles are consistent with *c.* 50% removal of Fe from vent fluids as sulphides (Rudnicki & Elderfield 1993).

Conclusions

An evaluation of samples and data from the MAR at 29°N shows that hydrothermal plumes exist both at the Broken Spur hydrothermal site and also further to the south. The southerly plume appears more intense, which suggests the existence of vents in addition to those sampled by Murton & Van Dover (1993). Plumes show multiple layering as at TAG, with temperature and salinity anomalies resulting from the entrainment of bottom waters rising in the plume. ^{222}Rn anomalies at plume height also reflect this process. In addition to entrainment, concentrations of Fe and Mn at plume height are also perturbed by chemical processes occurring during plume evolution, such as the removal of Fe as sulphides. The plume particles at Broken Spur are enriched in Fe, Mn, Cu, Cd, Co, Zn and Pb.

We are grateful to the master, officers and crew of RRS *Charles Darwin* and the technical support of D. Dunster, H. Evans, A. Lord, S. Watts and G. White (RVS *Barry*) for their assistance during cruise CD77. This work was supported by NERC grants GR9/513 (H.E., C.R.G. & M.R.P.) and GR3/8596 (H.E.). R.H.J. was supported by NERC studentship GT4/92/128/G. IOSDL Contribution No. 95013.

References

BAKER, E. T., MASSOTH, G. J. & FEELY, R. A. 1987. Cataclysmic hydrothermal venting on the Juan de Fuca Ridge. *Nature*, **329**, 149–151.

BRULAND, K. W. & FRANKS, R. P. 1983. Mn, Ni, Cu, Zn and Cd in the western north Atlantic. *In*: C. S. WONG, E. BOYLE, K. W. BRULAND, J. D. BURTON & E. D. GOLDBERG (eds) *Trace Metals in Seawater*. Plenum Press, New York, 395–414.

CAMPBELL, A. C., PALMER, M. R. & 9 others 1988. Chemistry of hot springs on the Mid-Atlantic Ridge. *Nature*, **335**, 514–519.

CHIN, C. S., COALE, K. H., ELROD, V. A., JOHNSON, K. S., MASSOTH, G. J. & BAKER, E. T. 1994. In situ observations of dissolved iron and manganese in hydrothermal vent plumes, Juan de Fuca Ridge. *Journal of Geophysical Research*, **99**, 4969–4984.

COLODNER, D., LIN, J., VON DAMM, K., BUTTERMORE, L., KOZLOWSKI, R., CHARLOU, J.-L., DONVAL, J. P. & WILSON, C. 1993. Chemistry of the Lucky Strike hydrothermal fluids: initial results. *Eos, Transactions of the American Geophysical Union*, **74**, 99.

DANIELSSON, L. G., MAGNUSSON, B. & WESTERLUND, S. 1978. An improved metal extraction procedure for the determination of trace metals in seawater by atomic absorption spectrometry with electrothermal atomisation. *Analytica Chimica Acta*, **98**, 47–57.

DETRICK, R. S. & THE SCIENTIFIC PARTY, LEG 106 1986. Drilling the Snakepit hydrothermal sulphide deposits on the Mid-Atlantic Ridge, Lat. 23°22'N. *Geology*, **14**, 1004–1007.

EDMOND, J. M., MEASURES, C., MCDUFF, R. F., CHAN, L.-H., COLLIER, R., GRANT, B., GORDON, L. I. & CORLISS, J. B. 1979. Ridge crest hydrothermal activity and the balances of the major and minor elements in the ocean: the Galapagos data. *Earth and Planetary Science Letters*, **46**, 1–18.

FEELY, R. A., GEISELMAN, T. L., BAKER, E. T. & MASSOTH, G. J. 1990. Distribution and composition of hydrothermal plume particles from the ASHES vent field at Axial Volcano, Juan de Fuca Ridge. *Journal of Geophysical Research*, **95**, 12 855–12 873.

——, LEWISON, M. & 6 others 1987. Composition and dissolution of black smoker particulates from active vents on the Juan de Fuca Ridge. *Journal of Geophysical Research*, **92**, 11 347–11 363.

GERMAN, C. R., BRIEM, J. & 9 others 1994. Hydrothermal activity on the Reykjanes Ridge: the Steinahóll vent field at 63°06'N. *Earth and Planetary Science Letters*, **121**, 647–654.

——, CAMPBELL, A. C. & EDMOND, J. M. 1991.

Hydrothermal scavenging at the Mid-Atlantic Ridge: modification of trace element dissolved fluxes. *Earth and Planetary Science Letters*, **107**, 101–114.

GRIMAUD, D., ISHIBASHI, J.-T., LAGABRIELLE, Y., AUZENDE, J. M. & URABE, T. 1991. Chemistry of hydrothermal fluids from the 17°S active site on the North Fiji Basin Ridge (SW Pacific). *Chemical Geology*, **93**, 209–218.

HYDES, D. J. 1984. A manual of methods for the continuous flow determination of ammonia, nitrate-nitrite, phosphate and silicate in seawater. *IOS Report*, **177**, 37pp.

JAMES, R. H., ELDERFIELD, H. & PALMER, M. R. 1995. The chemistry of hydrothermal fluids from the Broken Spur site, 29°N Mid-Atlantic Ridge. *Geochimica et Cosmochimica Acta*, **59**, 651–659.

KADKO, D. & MOORE, W. 1988. Radiochemical constraints on the crustal residence time of submarine hydrothermal fluids: Endeavour Ridge. *Geochimica et Cosmochimica Acta*, **53**, 659–668.

KADKO, D. C., ROSENBERG, N. D., LUPTON, J. E., COLLIER, R. W. & LILLEY, M. D. 1990. Chemical reaction rates and entrainment within the Endeavour Ridge hydrothermal plume. *Earth and Planetary Science Letters*, **99**, 315–335.

KIM, K. & FINKEL, R. 1980. Radioactivities on 21°N hydrothermal systems. *EOS, Transactions of the American Geophysical Union*, **61**, 995.

KLINKHAMMER, G. Fiber optic spectrometers for in-situ measurements in the oceans: the ZAPS probe. *Marine Chemistry*, in press.

LANGMUIR, C. H., FORNARI, D. & 17 others 1993. Geological setting and characteristics of the Lucky Strike vent field at 37°17′N on the Mid-Atlantic Ridge. *Eos, Transactions of the American Geophysical Union*, **74**, 99.

LAVELLE, J. W., COWEN, J. P. & MASSOTH, G. J. 1992. A model for the deposition of hydrothermal manganese near ridge crests. *Journal of Geophysical Research*, **97**, 7413–7427.

MASSOTH, G. J., BUTTERFIELD, D. A., LUPTON, J. E., McDUFF, R. E., LILLEY, M. D. & JONASSON, I. R. 1989. Submarine venting of phase-separated hydrothermal fluids at Axial Volcano, Juan de Fuca Ridge. *Nature*, **340**, 702–705.

MATHIEU, G. G. 1977. Rn^{222}–Ra^{226} techniques of analysis. *Annual Technical Report COO-2185-0 to ERDA, Lamont-Doherty Geological Observatory*.

METZ, S. & TREFRY, J. H. 1993. Field and laboratory studies of metal uptake and release by hydrothermal precipitates. *Journal of Geophysical Research*, **98**, 9661–9666.

MOTTL, M. J. & McCONACHY, T. F. 1990. Chemical processes in buoyant hydrothermal plumes on the East Pacific Rise near 21°N. *Geochimica et Cosmochimica Acta*, **54**, 1911–1927.

MURTON, B. J. & VAN DOVER, C. 1993. ALVIN dives on the Broken Spur hydrothermal vent field at 29°10′N on the Mid-Atlantic Ridge. *BRIDGE Newsletter*, **5**, 11–14.

MURTON, B. J., KLINKHAMMER, G. & 11 others. 1994.

Direct measurements of the distribution and occurrence of hydrothermal activity between 27°N and 30°N on the Mid-Atlantic Ridge. *Earth and Planetary Science Letters*, **125**, 119–125.

OLAFSSON, J., THORS, K. & CANN, J. 1991. A sudden cruise off Iceland. *Ridge Events*, **2**, 35–38.

RONA, P. A., KLINKHAMMER, G., NELSEN, T. A., TREFRY, J. H. & ELDERFIELD, H. 1986. Black smokers, massive sulphides and vent biota at the Mid-Atlantic Ridge. *Nature*, **321**, 33–37.

ROSENBERG, N. D., LUPTON, J. E., KADKO, D., COLLIER, R., LILLEY, M. D. & PAK, H. 1988. Estimation of heat and chemical fluxes from a seafloor hydrothermal vent field using radon measurements. *Nature*, **334**, 604–607.

RUDNICKI, M. D. & ELDERFIELD, H. 1992a. Theory applied to the Mid-Atlantic Ridge hydrothermal plumes: The finite-difference approach. *Journal of Volcanology and Geothermal Research*, **50**, 161–172.

—— & —— 1992b. Helium, radon and manganese at the TAG and Snakepit hydrothermal vent fields, 26°N and 23°N, Mid-Atlantic Ridge. *Earth and Planetary Science Letters*, **113**, 307–321.

—— & —— 1993. A chemical model of the buoyant and neutrally buoyant plume above the TAG vent field, 26°N, Mid-Atlantic Ridge. *Geochimica et Cosmochimica Acta*, **57**, 2939–2957.

SEDWICK, P. N., McMURTRY, G. M. & MacDOUGALL, J. D. 1992. Chemistry of hydrothermal solutions from Pele's Vents, Loihi Seamount, Hawaii. *Geochimica et Cosmochimica Acta*, **56**, 3643–3667.

SPEER, K. G. & RONA, P. A. 1989. A model of an Atlantic and Pacific hydrothermal plume. *Journal of Geophysical Research*, **94**, 6213–6220.

STATHAM, P. J. 1985. The determination of dissolved manganese and cadmium in seawater at low nmol/litre concentrations by chelation and extraction followed by electrothermal atomic absorption spectrophotometry. *Analytica Chimica Acta*, **169**, 149–159.

STUBEN, D., STOFFERS, P. & 6 others 1992. Manganese, methane, iron, zinc and nickel anomalies in hydrothermal plumes from Teahita and MacDonald volcanoes. *Geochimica et Cosmochimica Acta*, **56**, 3693–3704.

TCHERNIA, P. 1980. *Descriptive Regional Oceanography*. Pergamon Marine Series, Vol. 3. Pergamon Press, Oxford, 253 pp.

TREFRY, J. H. & TROCINE, R. P. 1991. Collection and analysis of marine particles for trace elements. *In*: HURD, D. C. & SPENCER, D. W. (eds) *Marine Particles: Analysis and Characterisation*. AGU Geophysical Monograph Series **63**, 311–316.

——, BUTTERFIELD, D. B., METZ, S., MASSOTH, G. J., TROCINE, R. P. & FEELY, R. A. 1994. Trace metals in hydrothermal solutions from the Cleft Segment on the southern Juan de Fuca Ridge. *Journal of Geophysical Research*, **99**, 4925–4935.

TROCINE, R. P. & TREFRY, J. H. 1988. Distribution and chemistry of suspended particles from an active hydrothermal vent site on the Mid-Atlantic Ridge at 26°N. *Earth and Planetary Science Letters*, **88**, 1–15.

Von Damm, K. L. & Bischoff, J. L. 1987. Chemistry of hydrothermal solutions from the southern Juan de Fuca Ridge. *Journal of Geophysical Research,* **92**, 11334–11346.

——, Edmond, J. M., Grant, B., Measures, C. I., Walden, B. & Weiss, R. F. 1985*a*. Chemistry of submarine hydrothermal solutions at 21°N, East Pacific Rise. *Geochimica et Cosmochimica Acta,* **49**, 2197–2220.

——, ——, Measures, C. I. & Grant, B. 1985*b*. Chemistry of submarine hydrothermal solutions at Guaymas Basin, Gulf of California. *Geochimica et Cosmochimica Acta,* **49**, 2221–2237.

Walker, S. L. & Baker, E. T. 1988. Particle-size distributions within hydrothermal plumes over the Juan de Fuca Ridge. *Marine Geology* **78**, 217–226.

Dissolved methane and hydrogen in the Steinahóll hydrothermal plume, 63°N, Reykjanes Ridge

M. R. PALMER[1], E. M. LUDFORD[1], C. R. GERMAN[2]
& M. D. LILLEY[3]

[1]Department of Geology, University of Bristol, Wills Memorial Building, Queen's Road, Bristol BS8 1RJ, UK

[2]Institute of Oceanographic Sciences, Wormley, Surrey GU8 5UB, UK

[3]School of Oceanography, University of Washington, Seattle, WA 98195, USA

Abstract: Dissolved concentrations of CH_4, H_2 and Mn have been measured in a hydrothermal plume above the shallow water (c. 250 m) Steinahóll vent site at 63°N on the Reykjanes Ridge. Samples were taken in October 1990 and June 1993. In 1990 dissolved CH_4 levels of up to 108 nmol/l and dissolved H_2 levels of up to 42 nmol/l were recorded. By 1993 these had declined to a maximum of 18 nmol/l for CH_4 and 30 nmol/l for H_2. Re-occupation of the site in 1993 included a higher density of sampling, suggesting that the fall in measured levels of these dissolved gases was not a sampling artefact. The high dissolved CH_4 and H_2 concentrations reported for 1990 were measured immediately after an episode of earthquake activity which may have resulted in the injection of magmatic gases into the local hydrothermal system. By 1993 the dissolved CH_4 levels had returned to values more typical of submarine hydrothermal systems elsewhere on the Mid-Atlantic Ridge. Dissolved H_2 levels remained high at Steinahóll, relative to other mid-ocean ridge sites, and it is believed that this is due to a combination of inorganic and biogenic processes. The gas chemistry of the Steinahóll hydrothermal site bears some similarities to that of the geothermal waters on the nearby Reykjanes Peninsula, Iceland, whereas the high levels of dissolved Mn in the plume are more typical of a submarine hydrothermal system. It appears, therefore, that the Steinahóll hydrothermal site may represent a transition between submarine and subaerial geothermal systems in its gas chemistry.

The detection of plumes in the water column enriched in Mn, particulates and CH_4 is often the first indication of hydrothermal activity on the seabed below (Klinkhammer et al. 1985; Nelsen et al. 1986; Charlou et al. 1987). The chemical composition and physical characteristics of the plume can provide information about the nature of the hydrothermal system responsible for the plume (e.g. Charlou et al. 1991), the dispersal of heat and chemical effluents from the underlying vent system (Cann & Strens 1989; German et al. 1991), the age of the hydrothermal plume and the role of biological processes in modifying the plume signature (Kadko et al. 1990).

Most of the mid-ocean ridge submarine hydrothermal systems studied to date have been from depths greater than 2000 m and most of these studies have been conducted in the Pacific Ocean (Von Damm 1990). In this paper we present results of a preliminary survey of the gas geochemistry of the Steinahóll hydrothermal plume above a shallow (c. 250 m) vent site on the Reykjanes Ridge at 63°N on the Mid-Atlantic Ridge (Fig. 1).

Analytical methods

The first sampling cruise was in November 1990 and immediately followed the detection of a swarm of earthquakes in the Steinahóll area less than a week previously (Olafsson et al. 1991). Sampling was performed in the same area in June 1993. In both cases the ship was the R/S *Bjarni Saemundsson*, and similar sampling equipment was deployed, namely, a Seabird 911 plus CTD (conductivity, temperature and depth) sensor equipped with a 25 cm pathlength SeaTech transmissometer and a Chelsea Instruments AquaTracka III nephelometer. The instruments were mounted in a frame with a rosette of 12 1.71 TPN Hydrobios water samplers.

In 1990 water samples from four stations at the Steinahóll site (Fig. 1) were drawn into (250 ml) tubular flasks fitted with PTFE vacuum valves. The samples were poisoned with $HgCl_2$ and held at 2°C until analyses of H_2 and CH_4 were made by gas chromatography using a modified version of the method described by Lilley et al. (1983).

From PARSON, L. M., WALKER, C. L. & DIXON, D. R. (eds), 1995, *Hydrothermal Vents and Processes*, Geological Society Special Publication No. 87, 111–120.

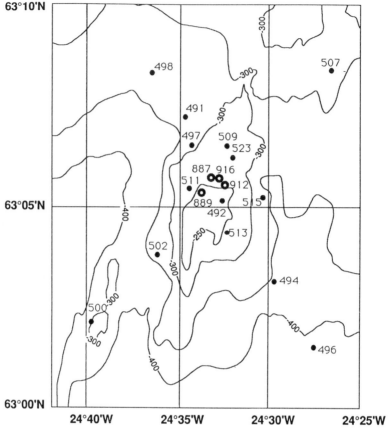

Fig. 1. Map of Steinahóll site with sample locations. Open circles indicate stations occupied in 1990 and closed circles are stations occupied in 1993.

In 1993 dissolved CH_4 and H_2 were determined in samples from 14 stations at the Steinahóll site (Fig. 1). The CH_4 and H_2 concentrations were determined on board ship by headspace analysis using a portable gas chromatograph and the method described by Dando *et al.* (1991). Depth profiles of these data were presented in German *et al.* (1994) as a comparison with the lack of hydrothermal activity elsewhere on the Reykjanes Ridge, but no interpretation of the Steinahóll data was made. Average background values during the 1993 cruise (including the total blank) for CH_4 and H_2 were 1.4 and 0.6 nmol/l, respectively, based on analyses of samples collected at similar depths from stations occupied along the Reykjanes Ridge away from areas of obvious hydrothermal activity (German *et al.* 1994). These background values are comparable with CH_4 and H_2 analyses of other relatively shallow water sites (e.g. Bullister *et al.* 1982). The precision of the analyses is $\pm 10\%$ based on analyses of

samples taken from different water sampling bottles tripped together at the same depth. The major source of potential error in sampling on both cruises was the problem of leaking water bottles. The background (non-hydrothermally influenced) dissolved Si concentrations are relatively uniform with depth in this area so there was no independent method for quantifying this problem, apart from cases where seawater was obviously leaking from the sample bottles when they were brought on board. However, the generally good correlation between dissolved CH_4 and Mn concentrations (Fig. 5a) suggests that preferential leaking of gas from the sample bottles was not a significant problem in this study. The data listed in Table 1 have not been corrected for background concentrations, but the samples that may have been taken from leaking sample bottles are denoted with an asterisk. Dissolved Mn and Si concentrations were determined on-board ship using standard spectrometric techniques.

Table 1. *Dissolved CH₄ and H₂ concentrations in the Steinahóll hydrothermal plume*

Station	1990			Station			
	Depth	CH₄	H₂				
887	250	21.1	7.8	492	150	7.2	15.8
					200*	1.6	2.7
					240	8.0	4.2
889	212	23.5	7.1				
	222	33.0					
912	175	21.6	4.4	494	100	1.3	0.6
	195	35.4			200	8.4	17.1
	203	22.0	12.1		250	10.3	23.0
	211	18.6	4.2		300	16.8	21.5
	211	20.2	7.9		350	11.4	12.6
916	100	18	5.5	509	100	3.8	13.4
	145	39.9	11.2		150	10.8	25.2
	170	67.6	33.7		200	9.0	13.4
	200	108	42.0		250	6.2	12.3
	207	44.4	33.8		260	7.0	11.5
	221	50.4	23.4	515	150	5.1	4.8
					200	7.1	5.4
					250	14.3	19.8
					280	17.4	23.7
					296	16.9	18.8
				523	150	17.4	12.6
					150	16.4	10.4
					170	18.4	16.0
					170	16.4	11.9
					182	14.3	27.2
					200	9.2	12.3
					200	8.2	10.2
					230	16.4	13.1
					275	16.4	30.2

All depths are in metres and all concentrations are in nmol/l. Data are only shown for the 1993 stations with concentrations above background values.
* possible leaking bottle.

Results

During the first sampling period in 1990, measured dissolved CH₄ and H₂ concentrations reached maximum values of 108 and 42 nmol/l, respectively. The highest concentrations were observed at a depth of 200 m, although the coverage of samples was not sufficient to delineate the plume precisely (Fig. 2a).

During the 1993 cruise measured dissolved CH₄ and H₂ concentrations only reached maximum values of 18.4 and 30.2 nmol/l, respectively, despite a considerably higher sampling density. When all the 1993 Steinahóll CH₄ and H₂ depth profiles are plotted together (Fig. 2b) the highest concentrations are observed at a depth of 150–300 m. When the data are plotted separately for the stations showing CH₄ and H₂ levels above background (Fig. 3), elevated concentrations are observed close to the bottom (in general the deepest sample was taken c. 10 m

above the seafloor). The water column anomalies close to the seafloor may be generated by a diffuse flow of gas, whereas the elevated dissolved gas concentrations higher in the water column may result from focused, high-temperature venting or entrainment of bottom water by gas bubble-rich plumes (German *et al.* 1994).

The lateral dispersion of the hydrothermal plume cannot be constrained using the 1990 data, but when the CH₄ and H₂ concentrations for the 1993 samples taken from 150–250 m are plotted on the sampling grid (Fig. 4) they show a sharp boundary along the northern and western margin of the plume. The extent of dispersion to the east is less clearly resolved because of the absence of sampling in this area. The highest concentrations are observed above the topographic high at the centre of the sampling grid, although station 494, to the southeast, also has high CH₄ and H₂ levels. This may indicate that the hydrothermal plume extends along a

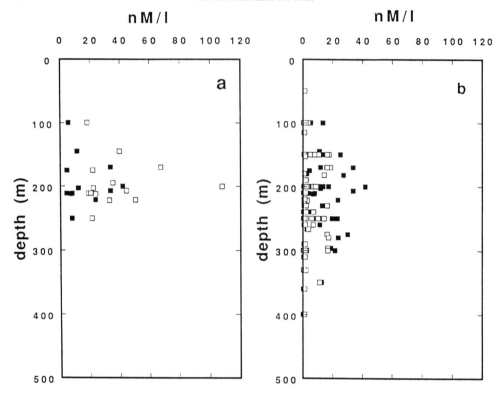

Fig. 2. Integrated depth profiles of dissolved CH_4 and H_2. (**a**) 1990 data. (**b**) 1993 data. Open squares are CH_4 concentrations and closed squares are H_2 concentrations.

NW–SE trend, or there may be at least two sites of hydrothermal discharge in the area.

The CH_4 and H_2 concentrations in all the samples are well below saturation with respect to their pure phases in seawater (Wiesenberg & Guinasso 1979; Millero 1983). Therefore the bubble plumes in the area (German *et al.* 1994) are most probably largely composed of CO_2 as this is the most abundant gas at most hydrothermal sites (e.g. Evans *et al.* 1988).

Discussion

The dissolved CH_4 and H_2 in the Steinahóll plume could be derived from abiogenic water–rock interactions (Welhan & Craig 1983), bacteriological activity (Baross *et al.* 1982) or a combination of both processes.

There are several lines of evidence which indicate that at least some of the CH_4 and H_2 were produced by hydrothermal activity. During the 1983 cruise, dissolved Mn and Si concentrations in subsamples from the same water bottles were enriched above background levels by up to 60 nmol/l and 2 µmol/l respectively,

yielding a $\Delta Mn/\Delta Si$ ratio of 30×10^{-3} (German *et al.* 1994). This compares with values of $27–45 \times 10^{-3}$ measured in hydrothermal springs elsewhere on the Mid-Atlantic Ridge (Campbell *et al.* 1988). Good correlations exist between dissolved CH_4 and both Mn and Si concentrations for the 1993 cruise (Fig. 5). During the previous cruise to the Steinahóll area elevated Mn and Si levels of a similar magnitude were also recorded (Olafsson *et al.* 1991). In addition, the $\Delta Mn/\Delta CH_4$ molar ratio in the Steinahóll plume of 3.5 (German *et al.* 1994) is virtually identical to that measured in the hydrothermal plume from the TAG hydrothermal site at 26°N on the Mid-Atlantic Ridge of 3.4 (Charlou *et al.* 1991).

Underwater video images of the Steinahóll site have shown areas of shimmering water flowing into overlying seawater (J. Olafsson, pers. comm. 1993), which may represent discharge sites of warm water. These video images show that the seafloor at Steinahóll consists of blocky basalt with a light dusting of sediment and bacterial mats in some areas. Culture experiments of hydrothermal waters and chimney material at temperatures of 100°C and 1 atm

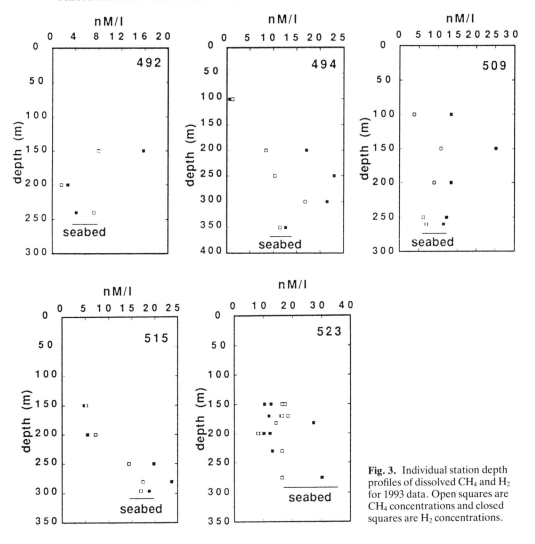

Fig. 3. Individual station depth profiles of dissolved CH$_4$ and H$_2$ for 1993 data. Open squares are CH$_4$ concentrations and closed squares are H$_2$ concentrations.

pressure have shown that thermophilic bacteria are capable of producing significant levels of H$_2$ and CH$_4$ utilising inorganic energy sources (Baross *et al.* 1982). In all these experiments H$_2$ was produced in equal or greater abundances than CH$_4$ and both were produced at rates of tens of nanomoles per millilitre of culture per hour. The bacteria in the experiments grew very slowly at temperatures of 80–86°C and ceased growth entirely when the temperature was lowered to 70–75°C (Baross *et al.* 1982).

Data from four Pacific hydrothermal sites have CH$_4$ and H$_2$ levels in end-member vent fluids that range from 52 to 118 μmol/kg and 143 to 1610 μmol/kg, respectively (Welhan & Craig 1983; Kim *et al.* 1984; Evans *et al.* 1988). The H$_2$/CH$_4$ ratios are between 3 and 5 at southern Juan de Fuca (Evans *et al.* 1988), 6 and 38 at

21°N EPR (Welhan & Craig, 1983) and 3 and 7 at 11–13°N (Kim *et al.* 1984). These compare with values of 3–30 in MORB glasses (Welhan 1988) and suggest that a significant proportion of CH$_4$ and H$_2$ in hydrothermal fluids is derived from the oceanic crust.

Dissolved CH$_4$ concentrations have been measured in plumes and vent fluids from a number of mid-ocean ridge hydrothermal sites in the Atlantic and Pacific Oceans (e.g. Kadko *et al.* 1990; Charlou *et al.* 1991), but the only detailed hydrothermal plume dissolved H$_2$ data are for the Endeavour segment of the Juan de Fuca Ridge (Kadko *et al.* 1990). In that study, CH$_4$ and H$_2$ were both shown to be rapidly consumed by bacteria within the hydrothermal plume, with the half-life of CH$_4$ being ten days compared with only ten hours for H$_2$. This was

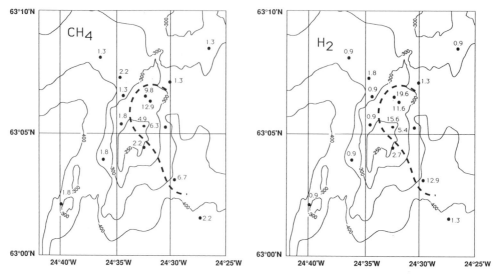

Fig. 4. Maps of (**a**) dissolved CH_4 and (**b**) dissolved H_2 concentrations between depths of 150 and 250 m for 1993 data. Broken line outlines area of samples with CH_4 and H_2 concentrations above background.

reflected in the H_2/CH_4 levels of the Endeavour plume samples which were only 0.034 in the sample with the highest hydrothermal component, whereas most of the samples had considerably lower ratios (Kadko *et al.* 1990). The average H_2/CH_4 ratios for the Steinahóll plume were 0.55 in 1990 and 1.14 in 1993. The 1993 Steinahóll samples have similar CH_4 levels to those observed in other plume studies, but they are marked by considerably higher H_2 concentrations. It is possible that these relatively high H_2/CH_4 ratios are because the Steinahóll samples were taken from younger portions of the plume before significant bacterial oxidation of the H_2. Evidence for bacterial oxidation of CH_4 and H_2 in the plume is given by plots of CH_4 and H_2 versus dissolved Mn concentrations (Fig. 5). Within close proximity of hydrothermal vents dissolved Mn can essentially be considered to be a conservative tracer (Klinkhammer *et al.* 1985). The correlation coefficient between the CH_4 and Mn data ($r^2 = 0.866$) is significantly better than between the H_2 and dissolved Mn data ($r^2 = 0.723$), which is consistent with the more rapid bacterial consumption of H_2 (Kadko *et al.* 1990). There is also considerable scatter in the CH_4 versus Mn plot, which suggests that oxidation of CH_4 has also occurred within the plume.

Bacterial activity within the plume decreases the H_2/CH_4 ratio through more rapid oxidation of H_2, whereas thermophilic bacterial activity at the source of hydrothermal fluids can lead to an increase in H_2/CH_4 ratios by preferential production of H_2 (Baross *et al.* 1982). In addition to these biogenic influences the H_2/CH_4 ratio can also be affected by the inorganic reaction

$$CO_2 + 4H_2 = CH_4 + 2H_2O \qquad (1)$$

The equilibrium position of this reaction is pressure and temperature dependent (with CH_4 being consumed at higher temperatures) and the relative proportions of the gas phases have been used as a thermometer in geothermal systems (Arnorsson & Gunnlaugsson 1985). The composition of the gas phase of the seawater-dominated geothermal systems on the Reykjanes Peninsula (southwest Iceland) have H_2/CH_4 ratios of 9–27. The carbon isotope fractionation between CH_4 and CO_2 in these geothermal waters yields an equilibrium temperature of *c.* 350°C for the above reaction, compared with a reservoir temperature of 285°C (Poreda *et al.* 1992). The carbon isotope equilibrium temperature for mid-ocean ridge geothermal systems has been calculated to be >650°C, compared with vent water temperatures of *c.* 350°C (Welhan & Craig 1983; Evans *et al.* 1988; Poreda *et al.* 1992). These studies illustrate that dissolved gases are slow to re-equilibrate at lower temperatures. It is possible, therefore, that the relatively high H_2/CH_4 ratios in the Steinahóll plume are the result of the incorporation of a gas phase into the hydrothermal fluids with a chemistry defined at high (magmatic?) temperatures.

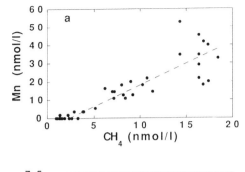

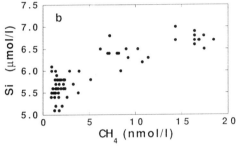

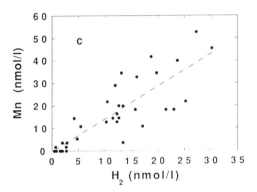

Fig. 5. (**a**) Dissolved CH₄ versus dissolved Mn.
(**b**) Dissolved CH₄ versus dissolved Si. (**c**) Dissolved
H₂ versus dissolved Mn. 1993 data for all figures.

No direct measurements have been made of
the temperatures of the shimmering water
observed at Steinahóll, or of any other submarine hydrothermal fluids in the area. Reservoir temperatures of 269–283°C have been
measured in the Reykjanes Peninsula geothermal site at a subsurface depth of 1600 m
(Arnorsson 1978). Video images of the Steinahóll area have revealed the presence of diffuse
shimmering water (of unknown temperature)
and do not show evidence for focused, high-
temperature venting, but more detailed coverage would be required before the existence of
high-temperature fluids could be discounted. It

is possible that the hydrothermal fluids at
Steinahóll were heated to temperatures similar
to the reservoir temperatures on the Reykjanes
Peninsula, but have been cooled by subsurface
mixing with seawater, as observed at Galapagos
(Edmond *et al*. 1979). This process results in the
production of clear, diffuse, low-temperature
fluids which preserve the Si and Mn anomalies of
the original high-temperature hydrothermal
fluids (Edmond *et al*. 1979). The major element
composition of the Reykjanes Peninsula geothermal fluids is similar to submarine hydrothermal fluids from the Mid-Atlantic Ridge, but
they have much lower H_2S levels (Arnorsson
1978; Campbell *et al*. 1988) due to the precipitation of Fe sulphides (Gunlaugsson & Arnorsson 1982), so if similar fluids are present at
Steinahóll then black smoker-type vents would
not be expected at this site.

The Mn concentration of the 1993 Steinahóll
plume reaches values of up to 60 nmol/l (German *et al*. 1994), but comparison with other
hydrothermal systems suggests that the concentration of Mn in the undiluted hydrothermal
fluids is likely to be several orders of magnitude
higher than this value (Klinkhammer *et al*. 1985;
Campbell *et al*. 1988). In contrast, the maximum
dissolved Mn concentration observed in the
undiluted high-temperature (285°C) reservoir
waters of the Reykjanes Peninsula is only
56 nmol/l (Kristmannsdottir & Olafson 1989),
compared with Mn levels of 0.5–1.0 mmol/l in
Mid-Atlantic Ridge hydrothermal fluids
(Campbell *et al*. 1988). Both Mn and Si are
mobile at relatively low temperatures (70°C) in
seawater–basalt hydrothermal experiments, but
the concentration of dissolved Mn in these
experiments only exceeds values measured in
the Steinahóll plume at temperatures >150°C
(Seyfried & Bischoff 1979). Again, this suggests
that the fluids at the Steinahóll site have reached
elevated temperatures even though they may
have been subsequently cooled by subsurface
seawater entrainment.

Comparison of the data from 1990 and 1993
show that there has been an evolution in the H_2
and CH₄ geochemistry of the hydrothermal
plume at the Steinahóll site over this interval. In
1990 the maximum CH₄ concentration was
108 nmol/l compared with a maximum of only
18 nmol/l in 1993, whereas maximum dissolved
H_2 concentrations dropped from 42 to 30 nmol/l
over the same period. There are a number of
possible hypotheses that might explain these
observations.

It is possible that the 1993 cruise did not
sample the main portion of the hydrothermal
plume. However, the sampling density was far

higher in 1993 than in 1990 making this unlikely. In addition, the maximum dissolved Mn and Si levels were very similar in both cruises (Olafsson *et al.* 1991; German *et al.* 1994) which would again argue that a similar portion of the plume was sampled during both cruises.

The lower dissolved CH_4 and H_2 levels may reflect a decrease in the level of hydrothermal activity between the cruises. However, the relatively constant Mn and Si levels do not support this hypothesis. Si concentrations in hydrothermal fluids are sensitive to temperature (e.g. Von Damm *et al.* 1985) and would be expected to drop significantly in a waning hydrothermal system. In addition to a fall in gas concentrations, the CH_4/H_2 ratios in the most enriched samples fell between 1990 and 1993. Again, this argues against a fall in the temperature of the system because falling temperatures favour higher CH_4/H_2 ratios (Arnorsson & Gunlaugsson 1985).

As noted above, the dissolved CH_4 levels and the $\Delta Mn/\Delta CH_4$ molar ratios in the Steinahóll plume in 1993 (German *et al.* 1994) are very similar to those observed in hydrothermal plumes elsewhere on the Mid-Atlantic Ridge (Charlou *et al.* 1991). This suggests that the CH_4 and H_2 concentrations observed in 1990 are anomalously high. It is significant, therefore, that the 1990 cruise immediately followed the detection of a swarm of micro-earthquakes in the Steinahóll area. Studies in the Pacific have shown that megaplumes can be released into the water column by bursts of tectonic activity due to a sudden increase in the permeability of rock overlying a pre-existing hydrothermal field (Baker *et al.* 1989; Cann & Strens 1989). It has been suggested that this increase in permeability may result from hydrofracturing associated with the rapid injection of magmatic gas into the hydrothermal circulation cell (Cann & Strens 1987, 1989). Such an event would not lead to a large change in the Mn and Si concentrations of the hydrothermal fluid, but could result in a large increase in the CH_4 and H_2 levels and may, therefore, explain the elevated gas concentrations in the 1990 plume. Elevated concentrations of volatile species (such as CH_4 and H_2) in hydrothermal plumes has also been linked with recent magmatic activity on the East Pacific Rise (Lupton *et al.* 1993).

A number of questions remain to be answered. It is likely that entrainment of water by rising bubbles at the Steinahóll site (German *et al.* 1994) play a significant part in affecting the distribution of dissolved species within the plume. The physics of this process is not well understood, but we have recently started a research project into the theoretical and experimental aspects of this problem. The question as to whether or not high-temperature hydrothermal fluids are present at Steinahóll will only be resolved by a more intensive study of the area. In addition, the relative importance of biogenic versus inorganic processes in controlling the CH_4 and H_2 levels in the Steinahóll hydrothermal system require a detailed biogeochemical study of the site. To answer some of these questions we hope to visit the area for a more extended geochemical and physical survey in the near future.

Summary

1. Dissolved CH_4 concentrations in the Steinahóll plume declined from maximum levels of 108 nmol/l in 1990 to 18 nmol/l in 1993. Over the same interval maximum dissolved H_2 concentrations dropped from 42 to 30 nmol/l.

2. The high CH_4 and H_2 concentrations in the 1990 plume may be the result of injection of magmatic gas into the hydrothermal system that coincided with increased earthquake activity in the area immediately prior to the 1990 sampling expedition.

3. The CH_4 levels in 1993 are similar to those observed in hydrothermal fluids from other mid-ocean ridge sites, whereas the H_2 concentrations are considerably higher than those measured in the only other study of H_2 in submarine mid-ocean ridge hydrothermal plumes. The high H_2/CH_4 ratios in the 1993 Steinahóll plume may arise from inorganic re-equilibrium at higher temperatures and pressures. Alternatively, thermophilic bacterial activity may serve as a source of the H_2-rich gas.

4. The concentrations of dissolved H_2 and CH_4 are both well below saturation with respect to their pure gaseous phases, so it is probable that the bubble plumes at Steinahóll are dominated by CO_2.

5. The gas chemistry of the Steinahóll hydrothermal system shares some affinities with Icelandic subaerial geothermal systems, but other aspects of the chemical composition of the hydrothermal plume at Steinahóll are more typical of submarine hydrothermal fluids.

This work was supported by the Natural Environment Research Council's BRIDGE and ODP Community Research Projects. We thank Captain I. Lárusson and the officers and crew of R/S *Bjarni Saemundsson* Cruise B8/1993 and V. Thoroddsen (Icelandic Marine

Research Institute) for support in the planning and execution of the cruise. We also thank P. Dando (PML, UK) for loan of the gas chromatograph. Support was also received from an NERC post-graduate studentship (EML) and the Royal Society (MRP). 105DL Contribution No. 95014.

References

ARNORSSON, S. 1978. Major element chemistry of the geothermal sea-water at Reykjanes and Svartsengi, Iceland. *Mineralogical Magazine*, **42**, 209–220.

—— & GUNNLAUGSSON, E. 1985. New gas thermometers for geothermal exploration – calibration and application. *Geochimica et Cosmochimica Acta*, **49**, 1307–1325.

BAKER, E. T., LAVELLE, W., FEELY, R. A., MASSOTH, G. J., WALKER, S. L. & LUPTON, J. E. 1989. Episodic venting of hydrothermal fluids from the Juan de Fuca Ridge. *Journal of Geophysical Research*, **94**, 9237–9250.

BAROSS, J A., LILLEY, M. D. & GORDON, L. I. 1982. Is the CH_4, H_2 and CO venting from submarine hydrothermal systems produced by thermophilic bacteria? *Nature*, **298**, 366–368.

BULLISTER, J. L., GUINASSO, N. L. & SCHINK, D. R. 1982. Dissolved hydrogen, carbon monoxide and methane at the CEPEX site. *Journal of Geophysical Research*, **87**, 2022–2034.

CAMPBELL, A. C., PALMER, M. R. & 9 others 1988. Chemistry of hot springs on the Mid-Atlantic Ridge. *Nature*, **335**, 514–519.

CANN, J. R. & STRENS, M. R. 1987. Venting events in hot water. *Nature*, **329**, 104.

—— & —— 1989. Modelling periodic megaplume emission by black smoker systems. *Journal of Geophysical Research*, **94**, 12 227–12 237.

CHARLOU, J. L., BOUGAULT, H., APPRIOU, P., NELSEN, T. & RONA, P. 1991. Different TDMn/CH_4 hydrothermal plume signatures: TAG site at 26°N and serpentinised ultrabasic diapir at 15°05′N on the Mid-Atlantic Ridge. *Geochimica et Cosmochimica Acta*, **55**, 3209–3222.

——, RONA, P. A. & BOUGAULT, H. 1987. Methane anomalies over TAG hydrothermal field on Mid Atlantic Ridge. *Journal of Marine Research*, **45**, 461–472.

DANDO, P. R., AUSTEN, M. C. & 8 others 1991. Ecology of a North Sea pockmark with an active methane seep. *Marine Ecology*, **70**, 49–63.

EDMOND, J. M., MEASURES, C. & 6 others 1979. Ridge crest hydrothermal activity and the balances of the major and minor elements in the ocean: the Galapagos data. *Earth and Planetary Science Letters*, **46**, 1–18.

EVANS, W. C., WHITE, L. D. & RAPP, J. B. 1988. Geochemistry of some gases in hydrothermal fluids from the Southern Juan de Fuca Ridge. *Journal of Geophysical Research*, **93**, 15 305–15 313.

GERMAN, C. R., BRIEM, J. & 10 others 1994. Hydrothermal activity on the Reykjanes ridge:

the Steinahóll vent-field at 63°06′N. *Earth and Planetary Science Letters*, **121**, 647–654.

——, CAMPBELL A. C. & EDMOND, J. M. 1991. Hydrothermal scavenging at the Mid-Atlantic Ridge: modification of dissolved trace element fluxes. *Earth and Planetary Science Letters*, **107**, 101–114.

GUNNLAUGSSON, E. & ARNORSSON, S. 1982. The chemistry of iron in geothermal systems in Iceland. *Journal of Volcanology and Geothermal Research*, **14**, 281–289.

KADKO, D. C., ROSENBERG, N. D., LUPTON, J. E., COLLIER, R. W. & LILLEY, M. D. 1990. Chemical reaction rates and entrainment within the Endeavour Ridge hydrothermal plume. *Earth and Planetary Science Letters*, **99**, 315–335.

KIM, K. R., WELHAN, J. A. & CRAIG, H. 1984. The hydrothermal vent fluids at 13°N and 11°N on the East Pacific Rise: Alvin 1984 results (abstract). *Eos, Transactions of the American Geophysical Union*, **65**, 973.

KLINKHAMMER, G., RONA, P. A., GREAVES, M. & ELDERFIELD, H. 1985. Hydrothermal plumes in the Mid-Atlantic Ridge rift valley. *Nature*, **314**, 727–731.

KRISTMANNSDOTTIR, H. & OLAFSSON, M. 1989. Manganese and iron in saline groundwater and geothermal brines in Iceland. *In*: MILES, D. L. (ed.) *Water–Rock Interactions 6*. Balkema, Rotterdam, 393–396.

LILLEY, M. D., BAROSS, J. A. & GORDON, L. I. 1983. Reduced gases and bacteria in hydrothermal fluids: the Galapagos spreading center and 21°N East Pacific Rise. *In*: RONA, P. A., BOSTRÖM, K., LAUBIER, L. & SMITH, K. L. (eds) *Hydrothermal Processes at Seafloor Spreading Centers*. Plenum, New York, 411–450.

LUPTON, J. E., BAKER, E. T. & 7 others 1993. Chemical and physical diversity of hydrothermal plumes along the East Pacific Rise, 8°45′N to 11°50′N. *Geophysical Research Letters*, **20**, 2913–2916.

MILLERO, R. J. 1983. Influence of pressure on chemical processes in the sea. *In*: RILEY, J. P. & CHESTER, R. (eds) *Chemical Oceanography*, Vol. 8. Academic Press, London, 1–88.

NELSEN, T. A., KLINKHAMMER, G. & TREFRY, J. 1986/1987. Real-time observation and tracking of dispersed hydrothermal plumes using nephelometry: examples from the Mid-Atlantic Ridge. *Earth and Planetary Science Letters*, **81**, 245–252.

OLAFSSON, J., THORS, K. & CANN, J. 1991. A sudden cruise off Iceland. *RIDGE Events*, **2**, 35–38.

POREDA, R. J., CRAIG, H., ARNORSSON, A. & WELHAN, J. A. 1992. Helium isotopes in Icelandic geothermal systems: I: ^{3}He, gas chemistry, and ^{13}C relations. *Geochimica et Cosmochimica Acta*, **56**, 4221–4228.

SEYFRIED, W. E. & BISCHOFF, J. L. 1979. Low temperature basalt alteration of seawater: an experimental study at 70°C and 150°C. *Geochimica et Cosmochimica Acta*, **43**, 1937–1947.

VON DAMM, K. L. 1990. Seafloor hydrothermal activity: black smoker chemistry and chimneys.

Annual Review of Earth and Planetary Science, **18**, 173–204.

——, EDMOND, J. M., GRANT, B., MEASURES, C. I., WALDEN, B. & WEISS, R. F. 1985. Chemistry of submarine hydrothermal fluids at 21°N, East Pacific Rise. *Geochimica et Cosmochimica Acta*, **49**, 2197–2220.

WELHAN, J. A. 1988. Methane and hydrogen in mid-ocean-ridge basalt glasses: analysis by vacuum crushing. *Canadian Journal of Earth Sciences*, **25**, 38–48.

—— & CRAIG, H. 1983. Methane, hydrogen and helium in hydrothermal fluids at 21°N on the East Pacific Rise. *In*: RONA, P. A., BOSTRÖM, K., LAUBIER, L. & SMITH, K. L. (eds) *Hydrothermal Processes at Seafloor Spreading Centers*. Plenum, New York, 391–409.

WIESENBERG, D. A. & GUINASSO, N. L. 1979. Equilibrium solubilities of methane, carbon monoxide, and hydrogen in water and seawater. *Journal of Chemical Engineering Data*, **24**, 356–360.

Hydrothermal deposits and metalliferous sediments from TAG, 26°N Mid-Atlantic Ridge

RACHEL A. MILLS

Department of Oceanography, University of Southampton, Southampton SO17 1BJ, UK

Abstract: The Trans-Atlantic Geotraverse (TAG) hydrothermal site at 26°N on the Mid-Atlantic Ridge has been the focus of many studies since the early 1970s. The hydrothermal field includes a wide variety of deposits ranging from high-temperature sulphide chimneys, low-temperature Mn deposits, diffuse flow to relict, inactive mounds. These deposits reflect the different styles of venting and the evolution of the hydrothermal system over time. Metalliferous sediments in this area are derived from both mass wasting of mound features and fall out of plume particulates. The active TAG mound was drilled by the Ocean Drilling Programme in October/November 1994, which provided the first samples from beneath the surface at this site. The current state of knowledge of processes occurring within TAG metalliferous deposits are reviewed here.

Metal-rich ore bodies have been found throughout the geological record and many of the deposits are associated with the transport of ore-forming elements by fluids in tectonically active environments. The possibility of widespread hydrothermal circulation at mid-ocean ridges (MORs) was recognized in the early 1970s (Wolery & Sleep 1976), later than the recognition of ophiolites as remnants of oceanic crustal material (Gass 1968). Much of our present understanding of MOR processes and structure comes from studies of ophiolites such as Troodos (e.g. Robertson & Xenophontos 1993); however, the discovery of high-temperature venting at MORs (Corliss *et al.* 1979) has revolutionized our understanding of how metalliferous deposits form on the ocean floor.

The present state of knowledge of metalliferous deposit formation, history and fate at one site on the Mid-Atlantic Ridge (MAR) is reviewed here. The Trans-Atlantic Geotraverse (TAG) site is named after a North Atlantic basin study carried out in the late 1960s (Rona 1970). This site has been studied intensively since the early 1970s and is comparatively well characterized (Scott *et al.* 1974, 1978; Rona 1980; Shearme *et al.* 1983; Rona *et al.* 1993a). Although the TAG site is not typical of MOR deposits because of its large size and long history, study of this site has given an insight into hydrothermal processes at slow-spreading centres. The TAG mound was a target for drilling by the Ocean Drilling Program (ODP) in October/November 1994, which provided information on the subsurface structure, composition and processes, but may well have altered the current circulation and depositional characteristics of the site.

Mineralization at mid-ocean ridges

Cold seawater penetrates into the oceanic crust through fissures and cracks which develop as the crust cools (e.g. Alt in press). As the seawater permeates downwards into the crust it is heated and reactions occur with the basalt (e.g. Alt in press). The extent of these chemical reactions increases with temperature and pressure until the rock becomes too impermeable and the fluid rises back to the seafloor. Seawater penetrates 2–4 km into the oceanic crust at MORs and the resultant fluid is acidic, reducing, sulphide-rich and rich in ore-forming metals (e.g. Von Damm 1990). As this fluid mixes with cold, oxidizing seawater, ore-forming metals precipitate both as chimney structures and dispersed sulphides and oxides in the close vicinity of the hydrothermal vents. The buoyant plume rises and entrains ambient seawater until it reaches neutral buoyancy about 200 m above the site of venting (Speer & Rona 1989). The neutrally buoyant plume is then dispersed laterally over large distances and the Fe oxyhydroxide particles fall out to form metal-rich sediments (Bostrom *et al.* 1969).

Many mineral deposits have been found at MORs (Rona 1988). These deposits are found in a variety of tectonic settings (German *et al.* this volume), but in general there appear to be fewer occurrences of mineral deposits on slow-spreading ridges than on fast-spreading ridges. The size of the mineral deposit must depend, among

From PARSON, L. M., WALKER, C. L. & DIXON, D. R. (eds), 1995, *Hydrothermal Vents and Processes*, Geological Society Special Publication No. 87, 121–132.

121

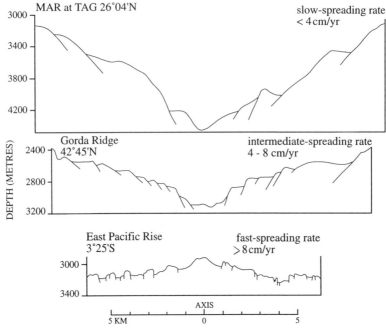

Fig. 1. Bathymetric profiles of slow-, intermediate- and fast-spreading oceanic ridges. Vertical exaggeration 4:1.

other things, on the persistence of hydrothermal activity at any one site and in general the deposits along slow-spreading ridges appear to be larger in size than along fast-spreading ridges (Rona 1988).

Mid-Atlantic Ridge

The MAR is a slow-spreading centre (full spreading rate 1–4 cm a^{-1}) and has a distinct median valley running along the ridge axis. The median valley is 10–20 km wide with steep, normal faulted, scarp walls rising 1–2 km above the valley floor. The ridge topography is distinct from that of fast-spreading ridges such as the East Pacific Rise (EPR) (full spreading rate up to 18 cm a^{-1}) and has implications for the transport and deposition of metal-rich deposits. A deep median valley will act to limit the dispersion of the neutrally-buoyant plume away from the ridge axis, whereas the smooth topography of fast-spreading ridges allows dispersion of the neutrally buoyant plume over large distances (Lupton & Craig 1981). Figure 1 demonstrates the different ridge topography at three sites with a range of spreading rates; a plume rising 250 m above the site of venting will be limited to the median valley at the TAG site.

There are five known sites of high-temperature hydrothermal activity on the MAR: Menez Gwen (38°N), Lucky Strike (37°N), Broken Spur (29°N), TAG (26°N) and Snake Pit (23°N) (Fig. 2). Menez Gwen is the shallowest hydrothermal site in the Atlantic (c. 700 m) and was discovered in 1994 (H. Bougault pers. comm.). Lucky Strike was sampled from submersible in May 1993 and the active venting occurs in discrete, dispersed sites on a seamount within the median valley (Langmuir et al. 1993). Broken Spur hosts three discrete high-temperature sources and a number of inactive sites of diffuse flow that are situated within an axial graben (Murton et al. 1993). Snake Pit was discovered in 1986 (Detrick et al. 1986) and is situated on a neovolcanic ridge within an axial graben (Thompson et al. 1988). There is evidence at Lucky Strike and Snake Pit that hydrothermal activity has been intermittent at these sites and has a long and complex history (Lalou et al. 1990; Humphris et al. 1993).

TAG hydrothermal site

The TAG site shares few of the characteristics of the Atlantic sites other than evidence for persistent and intermittent hydrothermal

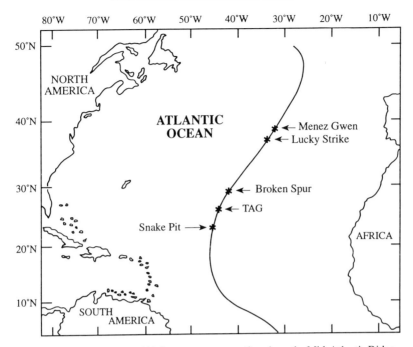

Fig. 2. Locations of the five known sites of high-temperature venting along the Mid-Atlantic Ridge.

activity (Lalou *et al.* 1990). There are a number of different components present including the active site of venting discovered in 1985 (Rona *et al.* 1986) and these are shown in Fig. 3. The active mound lies to the eastern edge of the median valley floor at a depth of 3680 m and is approximately 200 m in diameter and 50 m high (Thompson *et al.* 1988; Tivey *et al.* in press). A cross-section of the active TAG mound is shown in Fig. 4. A number of inactive mounds lie to the north of the active site (Rona *et al.* 1993*a*); two of these (Mir and Alvin) have been sampled from submersible (Lisitsyn *et al.* 1989; Rona *et al.* 1993*a*; Von Herzen *et al.* 1993). These inactive zones lie close to the foot of the eastern valley wall. The Mir and Alvin zones are large (1–2 km in length) and have undergone extensive alteration and mass wasting, though relict chimney features are still apparent (Rona *et al.* 1993*b*). There is an area of Mn-rich, low-temperature deposits to the east of the active mound between depths of 2500 and 3500 m (Thompson *et al.* 1985).

Fluid flow at TAG

There is a wide variety of styles of hydrothermal venting and fluid flow within the TAG field and this is reflected in the range of deposits observed. Three main styles of venting occur: black

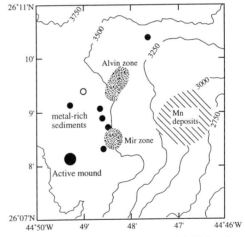

Fig. 3. Map of the TAG hydrothermal field showing the site of active venting, the low-temperature Mn deposits and the inactive, relict mounds (adapted from Rona *et al.* 1993*a*). Core positions are shown as small circles. Bathymetric contour intervals are in metres.

smoker, white smoker and diffuse lower temperature. High-temperature fluids vent from the apex of the active TAG mound at temperatures of up to 366°C (Edmond *et al.* 1990). The centre of the active TAG mound is shrouded in black

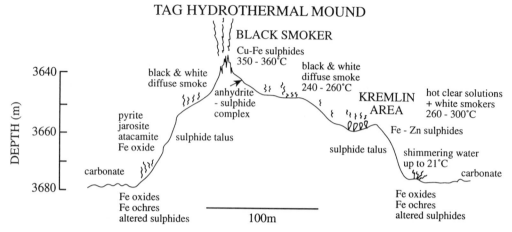

Fig. 4. Cross-section of the active TAG hydrothermal mound with the different deposits and styles of venting shown (adapted from Thompson *et al.* 1988).

'smoke' which is vigorously emitted from a cluster of chimneys. Black sulphides precipitate as the hydrothermal fluid mixes with cold seawater (Edmond *et al.* 1990) and this dense black 'smoke' shrouds the individual chimneys and rises rapidly, mixing turbulently to coalesce and form a buoyant plume that rises into the water column. The dominant particulate phase in the buoyant plume is 'nano-phase' Fe oxide with some larger chalcopyrite and pyrite grains (Campbell 1991). The temperatures measured in TAG vents appear to be relatively stable over the five year period that the active TAG mound has been studied (Von Herzen *et al.* 1993).

At the base of the black smoker complex are 1–5 m blocks of massive anhydrite mixed with sulphide undergoing dissolution in the ambient seawater (Thompson *et al.* 1988; Tivey *et al.* in press). This anhydrite is interspersed with plate-like layers of sulphides, some of which are coated with Fe oxide (Tivey *et al.* in press). Black smoke and hot shimmering water emanates from fissures in this surface and is entrained upwards into the main buoyant plume (Tivey *et al.* in press). These anhydrite deposits precipitate from a mixture of seawater and hydrothermal fluid and dissolve in the cold ambient seawater which bathes the mound (Tivey *et al.* in press). The isotope systematics of the anhydrite suggest that the fluid has undergone conductive heating and mixing with seawater, suggesting that circulation of fluid within the mound is more complex than mixing of fluid with cold seawater (Mills & Elderfield 1992; Tivey *et al.* in press).

Black smokers are one end-member of a spectrum of venting styles at TAG. To the

southeast of the black smoker complex is the area dubbed the 'Kremlin' owing to the bulbous shape of the sphalerite-rich chimneys (Thompson *et al.* 1988; Tivey *et al.* in press). These chimneys host lower temperature fluids ($\leqslant 300°C$) which are depleted in H_2S, Fe and Cu relative to the black smoker fluids (Edmond *et al.* 1990). The white smoker particles are predominantly Fe-coated amorphous silica with occasional Fe sulphide grains (M. Cooper, pers. comm.).

Low-temperature diffuse flow emanates from a great proportion of the mound surface. The zones of diffuse flow are delineated by the presence of anemones and a heat flow survey demonstrates that the spatial distribution is variable (Becker & Von Herzen 1993). Areas of upwelling of hot (up to 45°C) fluids have been identified at the northwestern and southern extremes of the active TAG mound (Becker & Von Herzen 1993). A swath of anomalously low heat flow has been identified along the western margin of the mound surface, suggesting that some recharge of ambient seawater may be occurring (Becker & Von Herzen 1993). The inactive Mir zone also exhibits above ambient heat flow values (Rona *et al.* 1993*b*)

The inferred fluid flow within the TAG mound involves strongly directed flow from depth discharging from the black smoker chimneys, with pooling and mixing of some of the fluid within the mound (Tivey *et al.* in press). Amorphous Si will only precipitate if the hydrothermal fluid undergoes conductive cooling during circulation within the mound (Tivey *et al.* in press). The white smoker fluid chemistry suggests that a combination of conductive cooling, precipitation and dissolution of various phases and mixing with seawater has taken place (Tivey *et al.* in press; Mills & Elderfield in press). Low-temperature pore fluids sampled from areas of upwelling are mixtures of end-member hydrothermal fluid with seawater (Mills *et al.* 1993*b*) demonstrating that seawater permeates into the mound and mixes with the circulating fluid before upwelling at the southern extremity of the mound. Again, these lower-temperature fluids are modified by reactions occurring within the mound (Mills *et al.* 1993*b*).

TAG mound formation

Black smoker chimneys

Models of chimney formation have been developed from petrological observations on East Pacific Rise chimneys (Haymon 1983; Goldfarb *et al.* 1983). Chimney growth is initiated by mixing of high-temperature fluid with seawater. A permeable anhydrite/sulphide mixture is precipitated rapidly (Haymon 1983). This feature gradually becomes less permeable as sulphide precipitation continues and clogs the pore spaces in the chimney walls (Haymon 1983). A chemical and mineral zonation is set up across the chimney wall, which develops over time. The fluid temperature increases as the chimney wall becomes less permeable and the minerals deposited become characteristic of higher temperatures (e.g. chalcopyrite and cubanite) (Tivey & McDuff 1990). Conversely, the chemistry of the chimney exterior is controlled by ambient sea water and fluid diffusing slowly through the chimney walls (Tivey & McDuff 1990).

The development of hydrothermal chimneys has also been studied by equilibrium path calculations (Janecky & Seyfried 1984). This approach allows a comparison of the calculated mineral assemblages with the observed petrographic distributions and therefore constraints can be made on fluid/seawater mixing, conductive heating and cooling. The effect of the physical environment on chemical interactions within the chimney walls has been modelled to assess the importance of the advection and diffusion of chemical species (Tivey & McDuff 1990).

Intact, active chimneys are relatively difficult to sample at TAG because of the dense black smoker shrouding the mound apex. Those chimneys that have been studied exhibit the characteristic zonation from chalcopyrite at the chimney–fluid interface, through marcasite and bornite to an anhydrite and sulphide mixture at the chimney exterior. TAG chimneys are often coated with a 1–3 mm layer of amorphous Fe oxide (Tivey *et al.* in press). Dendritic, amorphous Fe oxide also precipitates within the chimney walls and there is evidence that these Fe oxides are primary precipitates rather than the products of sulphide oxidation (Mills & Elderfield in press).

Equilibrium pathway and transport modelling of the observed petrographic relationships suggests that black smoker chimney formation at TAG is controlled by the mixing of seawater with high-temperature fluids and is modified by the diffusion of fluids both into and out of the chimney structure (Tivey *et al.* in press).

White smoker chimneys

White smoker chimneys can be more accurately described as a sulphide deposit with interconnected conduits that emits fluids at temperatures of 200–300°C. At the EPR, white smoker chimneys are covered with tube worms whose burrows provide pathways for seawater penetration of the chimney and for mixing of ambient seawater with the hydrothermal fluid (Haymon & Kastner 1981). White smoker chimneys are rich in sphalerite, amorphous silica and barite or anhydrite. The presence of sulphate minerals at the chimney interior and amorphous silica suggests extensive penetration of seawater and conductive cooling of the hydrothermal fluid (Haymon & Kastner 1981).

The TAG white smoker chimneys are bulbous in shape and shrimp feed on the crenulated surfaces (Tivey *et al.* in press). The white smoker fluid chemistry suggests that mixing with seawater occurs during circulation within the mound, along with the precipitation and dissolution of various sulphide and anhydrite phases (Tivey *et al.* in press). The mineral assemblages observed are dominated by sphalerite and other lower temperature sulphides and amorphous silica (Tivey *et al.* in press).

Mound deposits

Mound development is inferred to start with the clogging of chimneys and their weathering and collapse, and the diversion of active venting to new chimney structures (Haymon & Kastner 1981). As this process continues, the basal mound will become larger and the pore spaces will clog with alteration products and amorphous silica. Over time fluid circulation will be set up and sulphides will precipitate within the mound. It appears that the TAG mound has reached this mature stage as the basal mound is

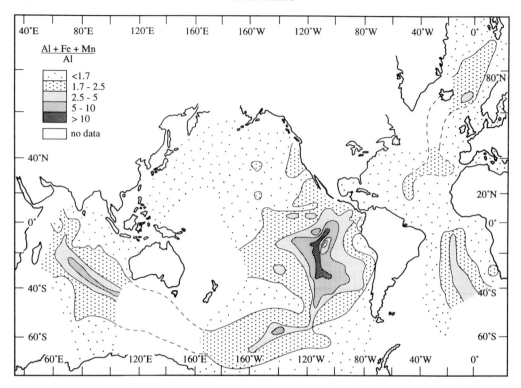

Fig. 5. Distribution of (Fe+Mn+Al)/Al from core tops (adapted from Boström *et al.* 1969). The high values delineate the mid-ocean ridge system.

large and there is evidence for fluid circulation within the mound (Tivey *et al.* in press).

The steep outer slopes of the active TAG mound are covered with sulphide talus which is undergoing weathering to a variety of secondary sulphate and oxide products. Atacamite (a secondary Cu chloride) is frequently associated with the surfaces of altered sulphides, suggesting local fluid transport followed by deposition at the mound/seawater interface (Hannington *et al.* 1988; Herzig *et al.* 1991; Hannington 1993). Gold enrichment has also been observed associated with the weathered sulphides (Hannington *et al.* 1988; Herzig *et al.* 1991). Deep red and yellow silica-rich ochres are present on the steep talus slopes of the inner mound and are sometimes associated with regions of upwelling of low-temperature fluids (Becker & Von Herzen 1993). These steep sloping oxide-rich portions of the mound are eventually transported to the surrounding sediments via mass wasting events and form one component within the TAG metalliferous sediments (Metz *et al.* 1988; German *et al.* 1993; Mills *et al.* 1993a). The distinction between mound and sediment is not well defined; the red/brown sulphide and oxide

material is gradually dominated by tan carbonate ooze away from the mound (Rona *et al.* 1993a). This sedimentary material is a thin blanket on the weathered basalt pillows which are typical of the Atlantic median valley floor.

History of TAG deposits

The large size of the active TAG mound and the presence of inactive deposits to the north imply that this site has been active for a long period. The range of deposits found at the active mound, especially the presence of aragonite (Thompson *et al.* 1988) suggest that hydrothermal activity has waxed and waned over time. Aragonite precipitation suggests near complete cooling of the mound at some point (Thompson *et al.* 1988).

TAG deposits have been dated using ^{210}Pb/Pb and ^{234}U/^{230}Th systematics of the sulphides (Lalou *et al.* 1990, 1993). This approach assumes that these deposits are closed systems with respect to U, Th and Pb and that the initial ^{210}Pb/Pb ratio has not varied over time, which may not be valid. Results suggest that chimney growth is rapid with all active chimneys giving

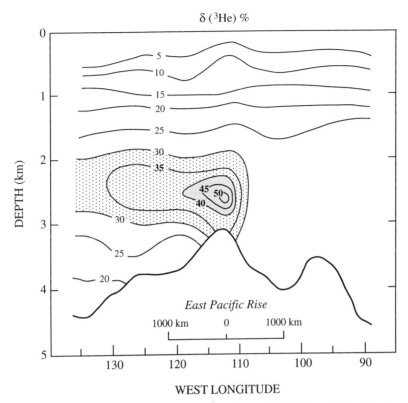

Fig. 6. Distribution of δ³He in the water column at 15°S on the East Pacific Rise (adapted from Lupton & Craig 1981).

ages that are less than 10 years (Lalou *et al.* 1993). Deposits from the active mound give ages up to 40 000 years and up to 100 000 years from the Mir zone (Lalou *et al.* 1993). Mn deposits from the eastern valley wall suggest an even longer history of up to 140 000 years (Lalou *et al.* 1993). The TAG site has a long and complex history of hydrothermal activity; this persistence makes it anomalous compared with other sites on fast-spreading ridges.

Metalliferous sediment formation

From the 1960s onwards, various workers recognized that metal-enriched sediments occur along the entire ocean ridge system (Arrhenius & Bonatti 1965; Skornyakova 1965; Bostrom & Peterson 1966, 1969; Bender *et al.* 1971; Dymond *et al.* 1973; Piper 1973; Bonatti 1975; Dymond & Veeh 1975). One way this can be demonstrated is by plotting the measured (Al+Fe+Mn)/Al ratio at the sediment–water interface. Metal enrichment at the MOR is characterized by high ratios. Figure 5 is adapted from an early compilation (Bostrom & Peterson

1969) and the (Al+Fe+Mn)/Al ratio effectively delineates the ocean ridge system. The pattern of metal enrichment could only be explained by a source of Fe and Mn at the ridge crest, which was then dispersed by mid-depth currents through the water column (Bostrom *et al.* 1969).

It was only after the direct observation of diffuse hydrothermal venting at Galapagos in 1977 (Corliss *et al.* 1979) and high-temperature venting at 21°N on the EPR (Edmond *et al.* 1979*a,b*) that it became apparent that all of these metal-enriched sediments represented different manifestations of the same phenomenon. Metal enrichment had also been widely reported from the sediment–basalt interface in Deep Sea Drilling Project (DSDP) holes (Cronan 1976; Leinen 1981). This observed basal metal enrichment could also be explained by the preservation of the ridge crest metalliferous sediment beneath pelagic sediment as the oceanic crust moved away from the ridge (Dymond *et al.* 1973).

Not only did the metal enrichment occur along the MOR crests, but it extended off-axis for large distances from the EPR. Concurrent

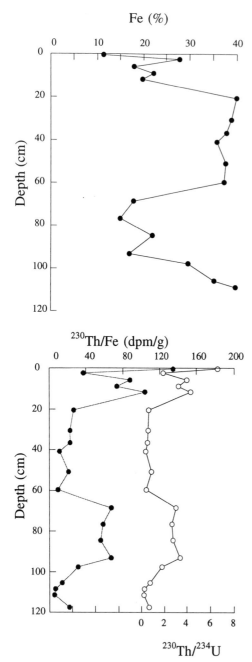

Fig. 7. Downcore variation of (**a**) Fe (%) and (**b**) ^{230}Th/Fe and ^{230}Th/^{234}U activity in a highly metalliferous core from the vicinity of the Alvin relict mound (data from Mills *et al.* 1993*a*). This core is plotted as an open circle in Fig. 3.

studies of the δ^3He anomaly in the water column (Lupton & Craig 1981) suggested that the hydrothermal effluent was dispersed from the

EPR by mid-depth water currents over several thousand kilometres (see Fig. 6). These currents also dispersed the fine-grained particulates derived from hydrothermal venting until they eventually settled out to form metal-enriched sediments beneath the overlying plume. The metal-rich lobe extending westwards from the EPR at 15°S (see Fig. 5) is attributed to dispersion of the neutrally buoyant plume at mid-depths as demonstrated by Fig. 6. By the early 1980s the general consensus was that metalliferous sediments were largely derived from the fallout of hydrothermal particles after dispersion in the water column.

TAG sediments

Metal-rich sediments have been recovered from the TAG area since the early 1970s (Scott *et al.* 1978; Shearme *et al.* 1983; Metz *et al.* 1988; Lisitsyn *et al.* 1989; German *et al.* 1993; Mills *et al.* 1993*a*). Within the median valley the sediment cover is generally patchy and thin with extensive exposure of pillow basalts separated by sediment ponds, making shipboard coring a difficult task. The formation of sediments at TAG is controlled by two factors: the first is mass wasting of mound material and the second is the fallout of particles from the neutrally buoyant plume (German *et al.* 1993; Mills *et al.* 1993*a*). These two types of input can be identified by the bulk geochemistry of the sediments.

Sulphide-derived sediments

Slumped mound material commonly consists of large angular clasts set in a fine sedimentary matrix. The bulk geochemistry is controlled by the mineral composition (Metz *et al.* 1988), which consists of sulphides and their alteration products. A systematic study of the rare earth element (REE) distributions in the active TAG mound and two metal-rich sediments has allowed identification of the types of mound material transported to the sediment (Mills & Elderfield in press). The sulphide and oxide material transported to the sediments is dominated by material from the steep talus slopes which is often coated with atacamite (Mills & Elderfield in press).

Sulphide-derived sediments are unstable in oxidizing seawater and may undergo oxidation after burial. Even after complete oxidation to Fe oxides, the bulk geochemical signature is distinct from plume-derived sediments. The uranium concentration in sulphide-derived sediments is high (*c.* 20 ppm) (Mills *et al.* 1994). This U is

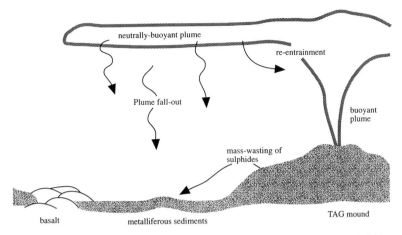

Fig. 8. Schematic diagram of metalliferous sediment formation at TAG, 26°N, Mid-Atlantic Ridge.

derived from seawater and the distribution is extremely heterogeneous within the sediment. High concentrations of U (up to 1000 ppm) are present within alteration rims on pyrite grains and a microbial mechanism of enrichment has been invoked to explain the observed distribution (Mills *et al.* 1994). The U is eventually released back to the seawater when the sediment is completely oxidized.

Plume-derived sediments

Particles from the neutrally buoyant plume above TAG have been sampled and are comparatively well characterized (German *et al.* 1990, 1991*a, b*; Feely *et al.* 1991). As the plume is dispersed through the water column the REEs and ^{230}Th are scavenged from seawater onto the Fe oxide surfaces (German *et al.* 1990). The REE/Fe and ^{230}Th/Fe ratios increase with distance from the source as scavenging from seawater continues (Olivarez & Owen 1989). These particles eventually form one component of the metalliferous sediments at TAG. Thorium-230 can be used to identify the source of metalliferous material to the sediment, as demonstrated in Fig. 7. Figure 7a is a plot of Fe down a highly metalliferous core retrieved from the vicinity of the relict Alvin mound (Metz *et al.* 1988). The Fe concentrations are high (11–40%) throughout the core. Figure 7b is a plot of ^{230}Th/Fe downcore compared with the values measured in the neutrally buoyant plume particles. The dual origin of these sediments is apparent from this plot; low ^{230}Th/Fe ratios characterize sulphidic, slumped material that has not been in contact with seawater, whereas high ^{230}Th/Fe ratios characterize particles that

have been transported through the water column and deposited from the neutrally buoyant plume. Also shown on this plot is the $^{230}Th/^{234}U$ ratio downcore. During periods of sulphide deposition $^{230}Th/^{234}U$ is <1, whereas excess ^{230}Th ($^{230}Th/^{234}U$>1) is present during periods of plume fallout.

A schematic diagram summarizing the controls on sediment formation at TAG is shown in Fig. 8. Sulphidic material slumps from both active and relict mound deposits in the TAG area. Particles settle from the neutrally buoyant plume to the underlying sediments contributing to the metal enrichment.

The Mn content of the near-field mound deposits and the neutrally buoyant plume particles is very low, yet Mn enrichment is commonly observed in metalliferous sediments from TAG (Scott *et al.* 1978; Shearme *et al.* 1983; Metz *et al.* 1988). Manganese deposits have been observed on the eastern valley wall of the TAG median valley (Scott *et al.* 1974; Thompson *et al.* 1985) and these could be transported to the sediments. Alternately, the Mn source may be from the oxidation and precipation of Mn in the distal hydrothermal plume.

Conclusions

The TAG hydrothermal site has been well characterized by shipboard and submersible studies over the last 22 years. A range of venting styles has been observed at TAG, associated with a variety of deposits from high-temperature active chimneys to inactive, relict mounds. Mound material eventually forms one component of metalliferous sediments within the median valley, the other component being particles from the neutrally buoyant plume dispersed away from the site of active venting. The proposed drilling of this site will give information on the subsurface structure and fluid flow that will enhance our understanding of how these deposits form.

Much of the research reviewed in this paper was carried out in the Department of Earth Sciences, Cambridge and more recently the Department of Oceanography, Southampton, funded by NERC. Particular thanks are extended to K. Saull who drafted the figures. The author thanks H. Elderfield, M. Greaves, M. Rudnicki, R. James, C. German, J. Thomson, M. Tivey and R. Nesbitt for support and encouragement during this period.

References

ALT, J. 1994. Subseafloor processes in mid-ocean ridge hydrothermal systems. *In*: HUMPHRIS, S. E. *ET AL.* (eds). *Physical, Chemical, Biological and Geological Interactivities within Hydrothermal Systems*. American Geophysical Union Monographs, in press.

ARRHENIUS, G. & BONATTI, E. 1965. Neptunism and vulcanism in the ocean. *In*: SEARS, M. (ed.) *Progress in Oceanography*. Pergamon Press, London, 7–22.

BECKER, K. & VON HERZEN, R. V. 1993. Conductive heat flow measurements using ALVIN at the TAG active hydrothermal mound, MAR at 26°N. *EOS, Transactions of the American Geophysical Union*. **74**, 99.

BENDER, M. L., BROECKER, W., GORNITZ, V., MIDDEL, V., KAY, R., SUN, S. S. & BISCAYE, P. 1971. Geochemistry of three cores from the East Pacific Rise. *Earth and Planetary Science Letters*, **12**, 425–433.

BONATTI, E. 1975. Metallogenesis at oceanic spreading centers. *Annual Reviews of Earth and Planetary Science*, **3**, 401–431.

BOSTROM, K. & PETERSON, M. N. A. 1966. Precipitates from the hydrothermal exhalations on the East Pacific Rise. *Economic Geology*, **61**, 1258–1265.

—— & —— 1969. The origin of aluminium-poor ferromanganoan sediments in areas of high heat flow on the East Pacific Rise. *Marine Geology*, **7**, 427–447.

——, ——, JOENSUU, O. & FISHER, D. E. 1969. Aluminium-poor ferromanganoan sediments on active oceanic ridges. *Journal of Geophysical Research*, **74**, 3261–3270.

CAMPBELL, A. C. 1991. Mineralogy and chemistry of marine particles by synchrotron x-ray spectroscopy, mossbauer spectroscopy and plasma-mass spectrometry. *In*: HURD, D. C. & SPENCER, D. W. (eds) *Marine Particles: Analysis and Characterization*. American Geophysical Union, Washington, 375–390.

CORLISS, J. B., DYMOND, J. & 9 others 1979. Submarine thermal springs on the Galapogos Rift. *Science*, **203**, 1073–1083.

CRONAN, D. S. 1976. Basal metalliferous sediments from the eastern Pacific. *Geological Society of America Bulletin*, **87**, 928–934.

DETRICK, R. S., HONNOREZ, J. & 12 others 1986. Drilling the Snake Pit hydrothermal sulfide deposit on the Mid-Atlantic Ridge, lat 23°22′N. *Geology*, **14**, 1004–1007.

DYMOND, J. & VEEH, H. H. 1975. Metal accumulation rates in the southeast Pacific and the origin of metalliferous sediments. *Earth and Planetary Science Letters*, **28**, 13–22.

——, CORLISS, J. B., HEATH, G. R., FIELD, C. W., DASCH, E. J. & VEEH, H. H. 1973. Origin of metalliferous sediments from the Pacific Ocean. *Geological Society of America Bulletin*, **84**, 3355–3372.

EDMOND, J. M., MEASURES, C. & 7 others 1979a. On the formation of metal-rich deposits on ridge crests. *Earth and Planetary Science Letters*, **46**, 19–30.

——, —— & 6 others 1979b. Ridge crest hydrothermal activity and the balances of the major and minor elements in the ocean: the Galapagos data. *Earth and Planetary Science Letters*, **46**, 1–18.

——, CAMPBELL, A. C., PALMER, M. R. & GERMAN, C. R. 1990. Geochemistry of hydrothermal fluids from the Mid-Atlantic Ridge: TAG and MARK 1990. *EOS, Transactions of the American Geophysical Union*, **71**, 1650–1651.

FEELY, R. A., TREFRY, J. H., MASSOTH, G. J. & METZ, S. 1991. A comparison of the scavenging of phosphorus and arsenic from seawater by hydrothermal iron oxyhydroxides in the Atlantic and Pacific oceans. *Deep-Sea Research*, **38**, 617–623.

GASS, I. G. 1968. Is the Troodos massif of Cyprus a fragment of Mesozoic ocean floor? *Nature*, **220**, 39–42.

GERMAN, C. R., BAKER, E. T. & KLINKHAMMER, G. 1995. The regional setting of hydrothermal activity. *This volume*.

——, CAMPBELL, A. C. & EDMOND, J. M. 1991a. Hydrothermal scavenging at the Mid-Atlantic Ridge: modification of trace element dissolved fluxes. *Earth and Planetary Science Letters*, **107**, 101–114.

——, FLEER, A. P., BACON, M. P. & EDMOND, J. M. 1991b. Hydrothermal scavenging at the Mid-Atlantic Ridge: radionuclide distributions. *Earth and Planetary Science Letters*, **105**, 170–181.

——, HIGGS, N. C. & 6 others 1993. A geochemical study of metalliferous sediment from the TAG hydrothermal mound, 26°08′N, MAR. *Journal of Geophysical Research*, **98**, 9683–9692.

——, KLINKHAMMER, G. P., EDMOND, J. M., MITRA, A. & ELDERFIELD, H. 1990. Hydrothermal scavenging of rare earth elements in the ocean. *Nature*, **316**, 516–518.

GOLDFARB, M. S., CONVERSE, D. R., HOLLAND, H. D. & EDMOND, J. M. 1983. The genesis of hot spring deposits on the East Pacific Rise, 21°N. *Economic Geology Monograph*, **5**, 184–197.

HANNINGTON, M. D. 1993. The formation of atacamite during weathering of sulfides on the modern seafloor. *Canadian Mineralogist*, **31**, 945–956.

——, THOMPSON, G., RONA, P. A. & SCOTT, S. D.

1988. Gold and native copper in supergene sulphides from the Mid-Atlantic Ridge. *Nature,* **333**, 64–66.

HAYMON, R. M. 1983. Growth history of hydrothermal black smoker chimneys. *Nature,* **301**, 695–698.

—— & KASTNER, M. 1981. Hot spring deposits on the East Pacific Rise at 21° N: preliminary description of mineralogy and genesis. *Earth and Planetary Science Letters,* **53**, 363–381.

HERZIG, P. M., HANNINGTON, M. D., SCOT, S. D., MAIOTIS, G., RONA, P. A. & THOMPSON, G. 1991. Gold-rich sea-floor gossans in the Troodos ophiolite and on the Mid-Atlantic Ridge. *Economic Geology,* **86**, 1747–1755.

HUMPHRIS, S. E., TIVEY, M. K. & FOUQUET, Y. 1993. Comparison of hydrothermal deposits at the Lucky Strike vent field with other mid-ocean ridge vent sites. *EOS, Transactions of the American Geophysical Union,* **74**, 100.

JANECKY, D. R. & SEYFRIED, W. E. 1984. Formation of massive sulfide deposits on oceanic ridge crests: incrememtal reaction models for mixing between hydrothermal solutions and sea water. *Geochimica et Cosmochimica Acta,* **48**, 2723–2738.

LALOU, C., REYSS, J.-L., BRICHET, E., ARNOLD, M., THOMPSON, G., FOUQUET, Y. & RONA, P. A. 1993. New age data for Mid-Atlantic Ridge hydrothermal sites: TAG and Snakepit chronology revisited. *Journal of Geophysical Research,* **98**, 9705–9713.

——, THOMPSON, G., ARNOLD, M., BRICHET, E., DRUFFEL, E. & RONA, P. A. 1990. Geochronology of TAG and Snakepit hydrothermal fields, Mid-Atlantic Ridge: witness to a long and complex hydrothermal history. *Earth and Planetary Science Letters,* **97**, 113–128.

LANGMUIR, C. H., FORNARI, D. & 17 others 1993. Geological setting and characteristics of the Lucky Strike vent field at 37°17'N on the Mid-Atlantic Ridge. *EOS, Transactions of the American Geophysical Union,* **74**, 99.

LEINEN, M. 1981. Metal-rich basal sediments from north-eastern Pacific Deep Sea Drilling Project sites, *In:* YEATS, R. S., HAQ, B. U. *et al.* (eds) *Initial Reports of the Deep Sea Drilling Program* **63** US Government Printing Office, Washington, 667–676.

LISITSYN, A. P., BOGDANOV, Y. A., ZONENSHAYN, L. P., KUZ'MIN, M. I. & SAGALEVICH, A. M. 1989. Hydrothermal phenomena in the Mid-Atlantic Ridge at Lat. 26°N (TAG hydrothermal field). *International Geology Review,* **31**, 1183–1198.

LUPTON, J. E. & CRAIG, H. 1981. A major helium-3 source at 15°S on the East Pacific Rise. *Science,* **214**, 13–18.

METZ, S., TREFRY, J. H. & NELSEN, T. 1988. History and geochemistry of a metalliferous sediment core from the Mid-Atlantic Ridge. *Geochimica et Cosmochimica Acta,* **48**, 47–62.

MILLS, R. A. & ELDERFIELD, H. 1992. Anhydrite precipitation at TAG – 26°N, Mid-Atlantic Ridge. *EOS, Transactions of the American Geophysical Union,* **73**, 585.

—— & —— Rare earth element geochemistry of

hydrothermal deposits from the active TAG mound, 26°N Mid-Atlantic Ridge. *Geochimica et Cosmochimica Acta,* in press.

——, —— & THOMSON, J. 1993a. A dual origin for the hydrothermal component in a metalliferous sediment core from the Mid-Atlantic Ridge. *Journal of Geophysical Research,* **98**, 9671–9681.

——, ——, ——, HINTON, R. W. & HYSLOP, E. 1994. Uranium enrichment in metalliferous sediments from the Mid-Atlantic Ridge. *Earth and Planetary Science Letters,* **124**, 35–47.

——, THOMSON, J., ELDERFIELD, H. & RONA, P. A. 1993b. Pore-water geochemistry of metalliferous sediments from the Mid-Atlantic Ridge: diagenesis and low-temperature fluxes. *EOS, Transactions of the American Geophysical Union,* **74**, 101.

MURTON, B. J., KLINKHAMMER, G. & 12 others 1993. Direct measurements of the distribution and occurrence of hydrothermal activity between 27°N and 30°N on the Mid-Atlantic Ridge. *EOS, Transactions of the American Geophysical Union,* **74**, 99.

OLIVAREZ, A. M. & OWEN, R. M. 1989. REE/Fe variations in hydrothermal sediments: implications for the REE content of seawater. *Geochimica et Cosmochimica Acta,* **53**, 757–762.

PIPER, D. Z. 1973. Origin of metalliferous sediments of the East Pacific Rise. *Earth and Planetary Science Letters,* **19**, 75–82.

ROBERTSON, A. & XENOPHONTOS, C. 1993. Development of concepts concerning the Troodos ophiolite and adjacent units in Cyprus. *In:* PRICHARD, H. M., ALABASTER, T., HARRIS, N. B. W. & NEARY, C. R. (eds) *Magmatic Processes and Plate Tectonics.* Geological Society, London, Special Publication, **76**, 85–119.

RONA, P. A. 1970. Comparison of continental margins of eastern north America at Cape Hatteras and northwestern Africa at Cape Blanc. *Bulletin of the American Association of Petroleum Geology,* **54**, 129–157.

—— 1980. TAG hydrothermal field: Mid-Atlantic Ridge crest at latitude 26°N. *Journal of the Geological Society of London,* **137**, 385–402.

—— 1988. Hydrothermal mineralization at oceanic ridges. *Canadian Mineralogist,* **26**, 431–465.

——, BOGDANOV, Y. A., GURVICH, E. G., RIMSKI-KORSAKOV, N. A., SAGALEVITCH, A. M., HANNINGTON, M. D. & THOMPSON, G. 1993a. Relict hydrothermal zones in the TAG hydrothermal field, Mid-Atlantic Ridge 26°N, 45°W. *Journal of Geophysical Research,* **98**, 9715–9730.

——, HANNINGTON, M. D., BECKER, K., VON HERZEN, R. P. V., NAKA, J., HERZIG, P. M. & THOMPSON, G. 1993b. Relict hydrothermal zones of TAG hydrothermal field, Mid-Atlantic Ridge, 26°N. *EOS, Transactions of the American Geophysical Union,* **74**, 98.

——, KLINKHAMMER, G., NELSEN, T. A., TREFRY, J. H. & ELDERFIELD, H. 1986. Black smokers, massive sulphides and vent biota at the Mid-Atlantic Ridge. *Nature,* **321**, 33–37.

SCOTT, M. R., SCOTT, R. B., MORSE, J. W., BETZER, P.

R., BUTLER, L. W. & RONA, P. A. 1978. Metal-enriched sediments from the TAG hydrothermal field. *Nature,* **276**, 811–813.

——, ——, RONA, P. A., BUTLER, L. W. & NALWALK, A. J. 1974. Rapidly accumulating manganese deposit from the median valley of the Mid-Atlantic Ridge. *Geophysical Research Letters,* **1**, 355–358.

SHEARME, S., CRONAN, D. S. & RONA, P. A. 1983. Geochemistry of sediments from the TAG hydrothermal field, MAR at latitude 26°N. *Marine Geology,* **51**, 269–291.

SKORNYAKOVA, I. S. 1965. Dispersed iron and manganese in Pacific ocean sediments. *International Geology Review,* **7**, 2161–2174.

SPEER, K. G. & RONA, P. A. 1989. A model of an Atlantic and Pacific hydrothermal plume. *Journal of Geophysical Research,* **94**, 6213–6220.

THOMPSON, G., HUMPHRIS, S. E., SCHROEDER, B., SULANOWSKA, M. & RONA, P. A. 1988. Active vents and massive sulfides at 26°N (TAG) and 23°N (Snakepit) on the Mid-Atlantic Ridge. *Canadian Mineralogist,* **26**, 697–711.

——, MOTTL, M. J. & RONA, P. A. 1985. Morphology, mineralogy and chemistry of hydrothermal deposits from the TAG area, 26°N Mid-Atlantic Ridge. *Chemical Geology,* **49**, 243–257.

TIVEY, M. K. & McDUFF, R. E. 1990. Mineral precipitation in the walls of black smoker chimneys: a quantitative model of transport and chemical reaction. *Journal of Geophysical Research,* **95**, 12 617–12 637.

——, HUMPHRIS, S. E., THOMPSON, G., HANNINGTON, M. D. & RONA, P. A. Deducing patterns of fluid flow and mixing within the active TAG mound using mineralogical and geochemical data. *Journal of Geophysical Research,* in press.

VON DAMM, K. L. 1990. Seafloor hydrothermal activity: black smoker chemistry and chimneys. *Annual Reviews of Earth and Planetary Sciences,* **18**, 173–204.

VON HERZEN, R., RONA, P. & 12 others 1993. Mid-Atlantic Ridge hydrothermal processes: prelude to drilling. *EOS, Transactions of the American Geophysical Union,* **74**, 382–383.

WOLERY, T. J. & SLEEP, N. H. 1976. Hydrothermal circulation and geochemical flux at mid-ocean ridges. *Journal of Geology,* **84**, 249–275.

Noble gas isotopes in 25 000 years of hydrothermal fluids from 13°N on the East Pacific Rise

F. M. STUART[1,2] P. J. HARROP[1], R. KNOTT[3], A. E. FALLICK[2], G. TURNER[1], Y. FOUQUET[4] & D. RICKARD[3]

[1]*Department of Geology, University of Manchester, Manchester M13 9PL, UK*
[2]*Isotope Geology Unit, SURRC, East Kilbride G75 0QU, UK*
[3]*Department of Geology, University of Wales, Cardiff CF1 3YE, UK*
[4]*IFREMER, BP70 29280 Plouzane, Brest Cedex, France*

Abstract: Noble gas isotopes have been measured in fluid inclusions in sulphides spanning 25 000 years of hydrothermal activity at 13°N on the East Pacific Rise. The ^{3}He/^{4}He ratios are typical of mid-ocean ridge hydrothermal fluids, albeit slightly higher than contemporary vent waters, and reveal no temporal variation or correlation with the δ^{34}S of the host sulphide. The absence of radiogenic He in fluids from the 25 000 year old mineralization on the SE Seamount suggests that the hydrothermal circulation occurred within an active magmatic system and not within the underlying 130 ka oceanic crust. This implies that seamount volcanism and hydrothermal activity occurred simultaneously off-ridge, and that magmatic activity shifted approximately 5 km off-ridge at this time. Helium concentrations in fluid inclusions from three samples are significantly greater than the end-member hydrothermal fluids at mid-ocean ridges. Small excesses of ^{40}Ar in the included fluids demonstrate that mantle-derived ^{40}Ar has been degassed along with primordial helium. Both are consistent with the direct addition of magmatic volatiles into the hydrothermal system at times during the history of hydrothermal activity at the site.

The morphological variability of mid-ocean ridges (MOR) (Gente *et al.* 1986; Haymon *et al.* 1991) and the small-sale chemical diversity of MOR basalts (e.g. Perfit *et al.* 1994) indicates that the magmatic and tectonic activity responsible for the generation of oceanic crust is episodic. Increasingly, it is becoming apparent that hydrothermal activity at MORs displays temporal episodicity on different time-scales, which probably reflects the control that magmatic and tectonic processes have on the thermal regime and the permeablity of the oceanic crust at ridge crests.

Active MOR vent fields display variations in the extent and type of venting (e.g. Haymon *et al.* 1993) and in the chemistry of vent fluids (e.g. Shanks *et al.* 1991) on the scale of days. Short time-scale changes in hydrothermal activity are closely linked in time and space with seismic and magmatic activity at MORs and are identified by dramatic changes in the relative abundances of volatiles and heat in vent waters (Lupton *et al.* 1989). For instance, the generation of a 'megaplume' over the Cleft segment on the southern Juan de Fuca Ridge in 1987 (Baker *et al.* 1987) was closely associated with the eruption of large volumes of basaltic lava (Embley *et al.* 1991). The 'megaplume' had a heat output comparable

with the annual output of the vent field (Lupton *et al.* 1989). At 9°N on the East Pacific Rise (EPR), a volcanic eruption in 1991 was closely followed by a dramatic increase in hydrothermal activity (Haymon *et al.* 1991). Short-lived, high-temperature (*c.* 400°C), Cl-depleted vent fluids at this site had He concentrations approximately an order of magnitude greater than the steady-state vent fluids (Lupton *et al.* 1991). The coincidence of three event plumes with extensive volcanic eruptions was recorded at Juan de Fuca Ridge during June and July, 1993 (Baker *et al.* 1995). Such variations are not restricted to MOR hydrothermal sites. Sedwick *et al.* (1994) recorded a 20-fold decrease in the He concentration and a 10-fold increase of CO_2/He ratios in vent fluids from the Loihi Seamount in the five years up to 1992.

Variations of hydrothermal activity on longer time-scales are more difficult to determine than that the lifetime of sulphide structures on the seafloor is short. U-series measurements of high and low temperature MOR hydrothermal deposits have been used to date the hydrothermal activity at both fast- and slow-spreading ridges. Variations occur on time-scales of the order of hundreds to thousands of years. For example, sulphides from the Trans-Atlantic Geotraverse

From PARSON, L. M., WALKER, C. L. & DIXON, D. R. (eds), 1995, *Hydrothermal Vents and Processes*, Geological Society Special Publication No. 87, 133–143.

133

hydrothermal site at 26°45'N on the Mid-Atlantic Ridge represent hydrothermal activity which occurred over the last 20 000 years. The distribution of ages suggests that high-temperature hydrothermal activity occurred intermittently, in pulses lasting between 4000 and 6000 years (Lalou *et al.* 1993). Off-ridge fossil sulphide deposits at 21°N on the EPR record hydrothermal activity which occurred up to 4000 years ago (Lalou & Brichet 1982). Low-temperature hydrothermal deposits at the Galapagos spreading centre may be up to 210 000 years old, while high-temperature activity occurred between 13 000 and 12 000 years BP with a renewal of activity approximately 8000 years BP (Lalou *et al.* 1989).

Between latitudes 12°40'N and 12°52'N on the EPR (13°N EPR), on- and off-axis sulphide deposits provide a record of intermittent hydrothermal activity over at least the last 25 000 years (Lalou *et al.* 1985). Deep-tow bathymetric profiles, SeaBeam surveys and submersible investigations have identified hydrothermal deposits at three sites in the area (Francheteau & Ballard 1983; Hekinian *et al.* 1983; Hekinian & Fouquet 1985). These deposits form a rarely observed profile of the temporal and spatial variability of hydrothermal activity at MORs. The youngest, active mineralization is located in the ridge axial graben and the deposits become older, more mature and ultimately inactive further from the axis (Fig. 1).

Helium isotopes as tracers of MOR vent fluids

$^3He/^4He$ values of MOR basalts (MORBs) have a characteristically narrow range, from seven to nine times the atmospheric ratio (R_A). High-temperature vent fluids have $^3He/^4He$ ratios that are indistinguishable from the basalts from which the He was extracted. Radiogenic He is generated from the decay of U and Th in MORBs with a $^3He/^4He$ ratio of approximately $0.01R_A$. The addition of radiogenic He lowers the $^3He/^4He$ ratio of oceanic basalts on 10^4–10^5 year time-scales (Graham *et al.* 1987). Such time-scales are similar to those for the transport and residence times of MOR magmas (Newman *et al.* 1982). In the case of seamount volcanism, the high radioelement content and the low He concentration of transitional and alkali basalts have conspired to lower $^3He/^4He$ ratios to values as low as $0.4R_A$ in basalts erupted on young seamounts from the East Pacific Rise (Graham *et al.* 1988).

Helium acquired during seawater circulation in the oceanic crust is isotopically indistinguish-

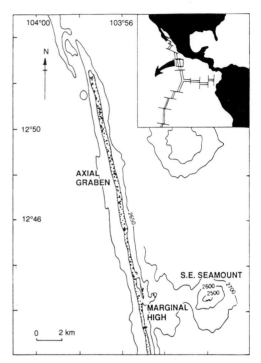

Fig. 1. Bathymetric map showing the topography of the East Pacific Rise between 12°42'N and 12°55'N and the location of the Marginal High and SE Seamount relative to the Axial Graben.

able from the basalt values. It has been demonstrated that the $^3He/^4He$ ratios of MOR hydrothermal fluids preserved as fluid inclusions in chimney sulphides are indistinguishable from the value of the contemporary vent fluids (Turner & Stuart 1992; Stuart *et al.* 1995). This advance allows $^3He/^4He$ measurements of fluid inclusions to trace past vent fluid compositions. To trace the evolution of the hydrothermal system at 13°N EPR, $^3He/^4He$ ratios have been measured in sulphide-hosted fluid inclusions from all three sites, spanning 25 000 years of hydrothermal activity.

Geological background and sample description

Hydrothermal activity at 13°N EPR is located on an axial high south of the Orozco fracture zone (Francheteau & Ballard 1983; Hekinian *et al.* 1983). A detailed discussion of the bathymetry of the site is given by Gente *et al.* (1986); a bathymetric map showing the topography of the 13°N EPR site is displayed in Fig. 1. Three hydrothermal fields have been identified and are outlined briefly.

Table 1. *Noble gas and sulphur isotope data from sulphides from 13°N on the East Pacific Rise*

Sample	^{4}He 10^{-8}cm^{3}STP/g	R/R_A	^{20}Ne 10^{-10}cm^{3}STP/g	^{36}Ar 10^{-10}cm^{3}STP/g	δ^{38}Ar(%)	δ^{40}Ar(%)	δ^{34}S(‰)
Axial Graben							
CY82-30-02	1.16 ± 0.01	8.4 ± 0.3	1.5 ± 0.1	11.1 ± 0.1	-0.17 ± 0.43	$+1.00 \pm 0.23$	$+2.9 \pm 0.2$
CY82-30-02*	0.21 ± 0.01	8.6 ± 0.5	3.2 ± 0.1	na	na	na	na
HR 1805	8.85 ± 0.01	8.8 ± 0.4	2.6 ± 0.1	6.6 ± 0.1	$+0.08 \pm 0.32$	$+1.65 \pm 0.23$	$+2.2 \pm 0.2$
Marginal High							
HR 2505	1.31 ± 0.01	8.1 ± 0.2	6.8 ± 0.2	5.0 ± 0.1	$+0.36 \pm 0.43$	$+2.64 \pm 0.23$	$+1.2 \pm 0.2$
HR 2504	1.65 ± 0.02	8.9 ± 0.4	7.8 ± 0.1	3.4 ± 0.1	-0.09 ± 0.31	$+0.68 \pm 0.23$	$+2.0 \pm 0.2$
HR 2502	0.46 ± 0.01	9.0 ± 0.6	2.9 ± 0.1	3.8 ± 0.1	-0.16 ± 0.35	$+0.92 \pm 0.23$	$+1.3 \pm 0.2$
SE Seamount							
CY82-14-04-1	7.65 ± 0.01	7.9 ± 0.2	8.2 ± 0.3	3.9 ± 0.1	-0.14 ± 0.42	$+1.78 \pm 0.23$	$+2.2 \pm 0.2$
CY82-14-04-2	0.94 ± 0.01	8.4 ± 0.4	bd	na	na	na	$+2.2 \pm 0.2$
CY82-14-4-G2*1	4.58 ± 0.02	7.4 ± 0.2	bd	19.4 ± 0.2	-0.23 ± 0.45	$+2.01 \pm 0.23$	$+2.6 \pm 0.2$
CY82-14-4-G2*2	2.41 ± 0.02	7.3 ± 0.2	bd	9.4 ± 0.2	-1.21 ± 0.62	$+2.21 \pm 0.23$	$+2.6 \pm 0.2$
CLDR-03-14	0.52 ± 0.03	8.3 ± 0.2	bd	1.4 ± 0.2	$+0.19 \pm 0.45$	$+0.41 \pm 0.23$	$+2.1 \pm 0.2$

* Anhydrite.
^{3}He/^{4}He ratios (R) are expressed relative to the air ratio (R_A) 1.4×10^{-6}.
na, not analysed; bd, below detection limit.

Axial Graben

Abundant active and inactive sulphide mounds are located in a 20 km long section of the ridge axial graben which is approximately 50 m deep and 0.5 km across. Focused, high temperature (280–340°C) fluids vent from chalcopyrite–anhydrite chimneys and diffuse lower temperature fluids (<280°C) vent from Fe–Zn sulphide mounds. The mound growth paragenesis is initiated by the growth of a Fe disulphide framework modified by infilling by granular sphalerite, pyrite and chalcopyrite. Hydrothermal activity in the axial graben is less than 170 years old.

Marginal High

Large Fe–Cu sulphide mounds and diffuse low temperature (<25°C) fluid flow are located between 500 to 1000 m east of the axial graben at 12°42′N. The deposits sit on top of a semi-circular volcanic structure which is approximately 3 km in radius and 40 m in height. In these mounds colloform pyrite is overgrown by crystalline pyrite and marcasite with late-stage replacement and void-filling by chalcopyrite. High-temperature hydrothermal activity has been restricted to between 1900 and 2100 years ago.

South Eastern (SE) Seamount

Approximately 5 km east of the Marginal High, inactive 8 m thick sulphide deposits cover an area of 0.5 km^2 on the summit and southern flank of an off-axis volcano. The seamount is approximately 350 m in height and has a basal diameter of 6 km. Magnetic anomaly data place the SE Seamount on oceanic crust that is between 110 000 and 130 000 years old. The hydrothermal deposits are large Fe(–Cu) mounds coated by Fe–Mn oxyhydroxides and are the surface expression of high-temperature hydrothermal activity that occurred between 12 000 and 25 000 years ago (Lalou *et al.* 1985). The link between hydrothermal activity and the seamount volcanism is unknown.

Noble gases have been analysed in fluid inclusions from minerals which were selected to span as much of the 25 000 years of hydrothermal activity as possible. Samples prefixed CY- in Table 1 were collected by the *Cyana* submersible in 1982 during the Cyatherm cruise (Hekinian *et al.* 1983) Sample CLDR03 was dredged from the SE Seamount during the same cruise. The samples prefixed HR- were collected as part of the Hero cruise in 1991.

CY82–30–02 (Axial Graben) is a massive chalcopyrite from the central conduit of a small high-temperature chalcopyrite–anhydrite 'black smoker' chimney. HR1805 (Axial Graben) is a coarse pyrite associated with high-temperature chalcopyrite-isocubanite intergrowths from a complex, zoned silicified Fe–Zn–Cu chimney. HR2502 and HR2505 (Marginal High) are pyrites from large, inactive Fe(–Cu) sulphide mound deposits. HR2502 is a fine granular sulphide and HR2505 is a massive pyrite aggregate. HR2504 (Marginal High) is a well-crystalline marcasite–chalcopyrite intergrowth representing the late-stage Fe(–Cu) sulphide mound mineralization. Sample CY82–14–4a (SE Seamount) is a pyrite similar to HR2502 and forms the main mound growth. CY82–14–4b (SE Seamount) is a massive marcasite aggregate from the thin outer zone of the mound. The main stage pyrite from CY82–14–4 has been dated at

24 600 ± 1600 years (Lalou et al. 1985). Sample CLDR03–04 (SE Seamount) is also marcasite-dominated and similar in form to HR2504. Samples from CLDR03 range in age from 12 100 ± 900 to 305 000 ± 2500 years old (Lalou et al. 1985).

Analytical procedure

Despite the inability to optically examine fluid inclusions in sulphides, it has been shown that sulphides trap and retain He from high-tempera-ture hydrothermal fluids better than sulphates in which fluid inclusion studies can be performed (Turner & Stuart 1992). To minimize cross-contamination from different generations of sulphide minerals, grains ranging from 2 to 10mm in diameter were hand-picked under a binocular microscope from coarsely crushed chips of chimney material. Separates were loaded into the ultra-high vacuum extraction system without prior ultrasonic cleaning in distilled water or acetone to avoid oxidizing the surface of the sulphides.

Fluid inclusion-hosted noble gases were liber-ated from 1–2 g separates by crushing in vacuum in a low volume crushing apparatus constructed from vacuum pipework. Crushing was performed by repeatedly dropping a 500 g weight onto the sample by manually agitating the apparatus. Movement was facilitated by a flexible pipe connecting the crusher to the purification line. This technique crushes the sample to a powder ($<100\,\mu m$). This releases the majority of the fluid inclusion-hosted noble gases, but retains post-precipitation radiogenic He which may have accumulated in the mineral lattice.

Noble gas measurements were made on an all-metal MAP 215 mass spectrometer using analytical procedures described previously (Stu-art et al. 1994b) with the exception that $^{3}He/^{4}He$ measurements were calibrated against $9.8 \times 10^{-7}\,cm^{3}$ STP aliquots of He which have a $^{3}He/^{4}He$ of 3.36×10^{-6} and He/Ne approxi-mately 13 000 times greater than the atmos-pheric value. The standard He was produced in an all-metal expansion line by mixing commer-cially available radiogenic He with an artificial mixture with a $^{3}He/^{4}He = 0.25$. The standard He has been calibrated against atmospheric He in both the Manchester and Zurich noble gas laboratories. Helium isotope measurements of the standard gas displays an external reproduci-bility of less than ±2%. Blank corrections never exceed 1% in all cases. Sulphur isotope measurements were made on the sulphide powders which remained after crush extraction of the noble gases. Details of the sample

preparation and measurement techniques have been reported elsewhere (Hall et al. 1991). He, Ne and Ar abundances and He, Ar and S isotope ratios of the sulphide and sulphate-hosted fluid inclusions from 13°N EPR are listed in Table 1. $^{3}He/^{4}He$ ratios (R) are reported relative to the atmospheric value of 1.4×10^{-6}.

Results

$^{3}He/^{4}He$ of the fluid inclusion-hosted solutions lie within a restricted range from 7.3 to 9.0 R_{A} (Fig. 2 and Table 1). These values are indis-tinguishable from vent fluids from sediment-free MOR hydrothermal systems, although they are slightly higher than the values of the hydro-thermal fluids currently venting from the axial region (Kim et al. 1984; Merlivat et al. 1987). $^{4}He/^{36}Ar$ ratios range from 10 to 196 and record a significant enrichment of He relative to seawater $(^{4}He/^{36}Ar = 0.037)$. The He enrichments are up to ten times higher than those measured in high-temperature hydrothermal fluids trapped in sulphides from bare-rock MOR sites on the EPR and Mid-Atlantic Ridge (Turner & Stuart 1992; Stuart et al. 1994a). Correcting the $^{3}He/^{4}He$ for atmosphere-derived He using the measured He/Ne ratio has no effect on the ratio because of the low abundance of He in seawater compared with the hydrothermal solutions.

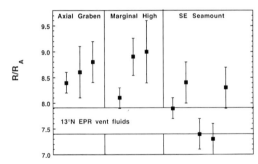

Fig. 2. The $^{3}He/^{4}He$ ratio (R), normalized to the air ratio (R_{A}), of trapped vent fluids from the three hydrothermal sites at 13°N on the East Pacific Rise. There is no measurable difference between the compositions of fluids over the 25 000 years of hydrothermal activity. The $^{3}He/^{4}He$ of the 25 000 year old SE Seamount hydrothermal fluids reveal no resolvable radiogenic component, which suggests that seamount volcanism was synchronous with hydrothermal activity. Vent fluid values are from Kim et al. (1984) and Merlivat et al. (1987).

Argon concentrations are similar to those recorded previously for fluids from MOR hydro-thermal sulphides (Turner & Stuart 1992). In contrast, the Ar isotopic composition of fluid inclusions from 13°N EPR reveal perceptible enrichments in ^{40}Ar in the trapped fluids (Table 1). The isotopic compositions are tabulated as ratio anomalies δ^xAr (%) calculated from

$$\delta^x\text{Ar} = \{[(^x\text{Ar}/^{36}\text{Ar})_{\text{measured}}/ \\ (^x\text{Ar}/^{36}\text{Ar})_{\text{air}}] - 1\} \times 100 \quad (1)$$

The δ^{40}Ar anomalies range up to 2.6% and cannot be the result of mass fractionation during fluid inclusion trapping or extraction (or as a result of phase separation) as δ^{38}Ar values are indistinguishable from air-saturated seawater. The ^{40}Ar excesses are derived from the oceanic crust along with juvenile He.

Discussion

SE Seamount: volcanism and hydrothermal activity

Extinct high-temperature hydrothermal sulphide mounds are located at two off-axial sites at 13°N EPR. The oldest site of hydrothermal activity is located on the SE Seamount and is 25 000 years old (Lalou et al. 1985). The seamount overlies oceanic crust that, from estimates of the current spreading rate, is up to 130 000 years old. The temporal relationship between the SE Seamount volcanism and the hydrothermal activity responsible for the sulphide mounds is unknown. The SE Seamount may have originated in one of two ways. (1) The seamount was erupted on-axis 130 000 years ago and the hydrothermal activity was independent of the magmatic activity responsible for the seamount formation. In this scenario the centre of hydrothermal activity shifted eastwards approximately 5 km from the axial graben 25 000 years ago and was decoupled from the active magma supply. (2) The seamount erupted off-axis 25 000 years ago and the hydrothermal activity was synchronous with the magmatism which produced the seamount. This requires that magmatic activity relocated 5 km off-axis 25 000 years ago.

The ingrowth of radiogenic He from the decay of U and Th in the oceanic crust on short time-scales is sufficient to generate significant ^{3}He/^{4}He reductions in the oceanic crust (Graham et al. 1987, 1988). In case 1, the hydrothermal fluids must have circulated within oceanic crust that was approximately 100 000 years old and which was capable of generating measurable quantities of radiogenic He (Graham et al. 1987). Subsequent seawater–crust interaction will generate hydrothermal fluids with ^{3}He/^{4}He ratios below typical MOR vent fluid values. In case 2, the hydrothermal fluids

circulated within recently formed oceanic crust and should have acquired normal MOR He ratios.

The effect of the accumulation of radiogenic He in the oceanic crust depends on the (U + Th)/He ratio. This ratio varies spatially due to the radically different chemistries of (U + Th) and He during melting of the mantle and the subsequent crystallization of the melts, and temporally, due to the removal of He by degassing and seawater–rock interaction. For instance, the strong partitioning of noble gases into the vapour phase results in the quantitative trapping of magmatic He in the vesicle phase of oceanic basalts (Kurz & Jenkins 1981). This leaves the quenched basaltic glasses with a low He abundance and consequently prone to the reduction of the ^{3}He/^{4}He ratio by the in situ decay of U and Th in the basalt glass (e.g. Graham et al. 1988). The chemistry, mineralogy and stable isotope composition of samples of ancient oceanic crust and the distribution and fluid inclusion content of hydrothermal veins in fossil hydrothermal systems preserved in ophiolites indicates that the sheeted dykes and gabbroic layers are the major chemical reaction zones within the oceanic crust (e.g. Gregory & Taylor 1981). The partitioning of inert gases into the melt and, ultimately, the vapour phase, leaves the crystalline units of the oceanic crust depleted in He. Consequently the (U + Th)/He ratios of the sheeted dykes and gabbros of the deep oceanic crust may be orders of magnitude higher than MOR basalts and thus susceptible to the ingrowth of radiogenic He.

The fractional change in ^{3}He/^{4}He ratios ($f_{3/4}$) as a result of radiogenic decay of U and Th is described by

$$f_{3/4} = 1 - [t\{1.207 + 0.278(^{232}\text{Th}/^{238}\text{U})\} \\ \times 10^{-13} (^{238}\text{U}/^4\text{He})] \quad (2)$$

where t is the age in years, U and Th are measured in ppm and ^{4}He is measured in cm^3 STP/g. Using U and Th concentrations of 0.1 and 0.3 ppm, respectively (Jochum et al. 1983), and He concentrations of 2.5×10^{-9} cm^3 STP/g (Staudacher & Allegre 1988), a 20% decrease in the ^{3}He/^{4}He ratio of the crystalline oceanic crust is generated in 100 000 years. This lowering of the ^{3}He/^{4}He ratio must be considered as a minimum value because EPR seamounts are commonly composed of alkali and transitional basalts, which have U contents of up to 1.5 ppm (Batiza & Vanko 1984). Therefore, the ingrowth of radiogenic He in the seamount basalts would significantly affect the ^{3}He/^{4}He ratio after 100 000 years. Helium extracted by hydrothermal seawater circulation in 100 000 year old

oceanic crust will display isotope ratios which
are significantly lower than those of fluids which
circulate in recently emplaced oceanic crust, e.g.
the Axial Graben.

The ^{3}He/^{4}He ratio of the Marginal High and
SE Seamount inclusion fluids are in the range
7.3–9.0R_A. This overlaps the range observed in
the fluid inclusions from sulphides from the
Axial Graben (Fig. 2) and clearly shows that the
fluids have not acquired measurable quantities
of radiogenic He from interaction with ancient
oceanic crust. This implies that the off-axial
hydrothermal circulation must have occurred
within recently erupted basalts rather than the
underlying 100 000 year old oceanic crust. This,
in turn, implies that the SE Seamount was
erupted approximately 25 000 years ago during a
period when magmatic activity occurred approximately 5 km east of the MOR axis. This is
consistent with the freshness of the seamount
basalts (observed during submersible dives) and
the lack of sediment cover on the sulphide
deposits themselves. The temporal connection
between volcanism and hydrothermal activity at
the SE Seamount is consistent with observations
from other MOR sites (e.g. Haymon *et al.* 1991).

Magmatic degassing?

The high He concentration of MOR basalts (up
to 10^{-4} cm^3 STP/g; Sarda & Graham 1990) and
its efficient extraction during seawater–basalt
interaction results in a narrow range of He
concentrations in end-member hydrothermal
fluids (3–6 × 10^{-5} cm^3 STP/cm^3). Such concentrations are orders of magnitude greater than
seawater (4 × 10^{-8} cm^3 STP/cm^3). Variations in
absolute He concentrations in ocean waters
result from mixing between the venting hydrothermal fluids and seawater (Jenkins *et al.* 1980).
The concentrations of ^{36}Ar in oceanic basalts are
low (*c.* 10^{-9} cm^3 STP/g) relative to seawater
(1.25 × 10^{-6} cm^3 STP/cm^3). Consequently, the
budget of ^{36}Ar in MOR hydrothermal fluids is
dominated by the seawater contribution, both
from recharge and from seawater entrainment
by vent fluids. As ^{36}Ar is always present in
seawater proportions, mixing between seawater
and hydrothermal fluid is reflected in variations
in the fluid He/^{36}Ar ratios. Helium enrichments
in the trapped hydrothermal fluids are displayed
in Fig. 3 using the *F*-notation, where

$$F(^4\text{He}) = (^4\text{He}/^{36}\text{Ar})_{\text{measured}}/(^4\text{He}/^{36}\text{Ar})_{\text{air}} \quad (3)$$

$F(^4\text{He})$ of the 13°N EPR fluid inclusions ranges
from 21 to 1180. This corresponds to hydrothermal fluid He concentrations in the range

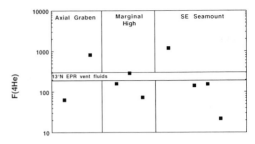

Fig. 3. Helium concentrations in the trapped
hydrothermal fluids relative to ^{36}Ar, where: $F(^4\text{He}) =$
$(^4\text{He}/^{36}\text{Ar})_{\text{measured}}/(^4\text{He}/^{36}\text{Ar})_{\text{air}}$. Three samples have
He concentrations greater than the vent fluid values
(Merlivat *et al.* 1987), which can be explained by the
direct addition of magmatic He to the hydrothermal
fluids.

5–246 × 10^{-6} cm^3 STP/cm^3, assuming the
measured ^{36}Ar is seawater-derived. In practice,
it is difficult to remove minor contamination by
atmospheric Ar (^{4}He/^{36}Ar = 0.167) during
sample analysis (Turner & Stuart 1992) so the
He concentrations calculated from the measured
He/^{36}Ar ratio must be treated as minima.

Helium concentrations reach 6 × 10^{-5} cm^3
STP/g in the 340°C fluids venting from sulphide
chimneys in the axial graben at 13°N EPR
(Merlivat *et al.* 1987). This concentration is
typical of the narrow range of He concentrations
in end-member hydrothermal solutions from the
global MOR system and probably reflects
equilibrium hydrothermal fluid circulation
within the oceanic crust. Clearly the $F(^4\text{He})$
values lower than those of vent fluid can result
from seawater dilution of high-temperature vent
fluids. However, three sulphide samples have
trapped fluid which have $F(^4\text{He})$ values, and
therefore absolute concentrations of He in the
fluids, which are in excess of the hydrothermal
fluid value (Fig. 3). The abnormally high He
concentrations in the fluid inclusions indicate
that the steady-state conditions existing now
have not persisted over the 25 000 years of
hydrothermal activity. High He concentrations
in the trapped fluids may be generated by two
processes, either (1) phase separation or (2) the
addition of magmatic volatiles into the hydrothermal system.

(1) Boiling of hydrothermal fluids during their
ascent, and mixing with pristine hydrothermal
fluids, can account for variations in the chemistry of vent fluids (e.g. Von Damm & Bischoff
1988). The strong tendency of the rare gases to
partition into the low-density vapour phase
means that boiling of MOR hydrothermal fluids

generates low He concentrations in the residual liquids (Kennedy, 1988) and extremely high He concentrations in the vapour fraction (Butterfield *et al.* 1990).

Contemporary high-temperature vent fluids sampled from the Axial Graben at 13°N EPR have Cl contents which are 30% higher than seawater (Michard *et al.* 1984). These fluids have Cl/Br ratios which are indistinguishable from seawater (Michard *et al.* 1984) and provide a strong indication that phase separation has affected the hydrothermal fluids. The range of He concentrations calculated for the 13°N EPR fluid inclusions from the measured $He/^{36}Ar$ ratios are consistent with the trapping of both vapour-rich and residual fluids. However, the differences in the solubilities of the noble gas species decrease with temperature (Crovetto *et al.* 1982). At 400°C, the temperatures of MOR hydrothermal systems, the difference between the solubilities of He and Ar in water/steam are predicted to be negligibly small and the separation of a vapour phase is not expected to significantly fractionate He and Ar in residual liquid (Crovetto *et al.* 1982). Kennedy (1988) demonstrated that residual waters sampled as vent fluids from the Juan de Fuca Ridge are depleted in all noble gases by over 30%, but show no measurable fractionation of the seawater-derived noble gas components. The He enrichments in the 13°N EPR hydrothermal fluids are not accompanied by similar enrichments of Ar, as expected if the noble gas content of the fluids was dominated by those of a vapour phase. Instead, $^{4}He/^{36}Ar$ ratios are up to five times those measured in the 13°N EPR vent fluids. High-temperature phase separation of hydrothermal fluids cannot account for the observed He/Ar ratios of the inclusion fluids from 13°N EPR.

(2) Short-lived vent fluids with abnormally high He concentrations have been observed as megaplumes above the Juan de Fuca Ridge (Baker *et al.* 1987) and as vent fluids at the eruption site on the EPR crest, 9°45′–52′N (Lupton *et al.* 1991). A likely explanation for the high volatile/heat ratio of megaplumes is the addition of magmatic volatiles into existing hydrothermal systems during volcanic eruption or dyke intrusion, either by direct degassing or by extraction during hydrothermal seawater interaction with volatile-rich basalts. The vesicularity of the most gas-rich MORBs is consistent with a significant pre-eruptive gas loss event (Sarda & Graham 1990). In this scenario, the degassing of recently intruded small volume melts is responsible for the generation of short-lived megaplumes. Indeed, pre-eruptive

magmatic degassing has been proposed to account for the 20-fold decrease in He concentrations in vent fluids from the Loihi Seamount over the last five years (Sedwick *et al.* 1994).

The presence of primordial He in the upper mantle means MOR magmas have $He/^{36}Ar$ ratios that are significantly higher than seawater; $F(He) = 2.7 \times 10^{6}$. The direct injection of magmatic volatiles into MOR hydrothermal fluids provides one way of achieving the high hydrothermal fluid $He/^{36}Ar$ ratios. The addition of magmatic volatiles to hydrothermal systems is closely linked to MOR volcanism. Where the aftermath of MOR volcanic eruptions have been observed, the superficial sulphide structures on the seafloor are largely destroyed (Haymon *et al.* 1993). The massive size and the form of the sulphide deposits, and the longevity of hydrothermal activity (Lalou *et al.* 1985), attest to the apparent long-term stability of the hydrothermal system at the SE Seamount site. The high He concentrations of trapped fluids in sulphides from the SE Seamount (and the Axial Graben) suggests that the steady-state venting has been periodically interrupted by magmatic events similar to those observed at 9°N EPR. From a detailed morphological study, Gente *et al.* (1986) have proposed a cyclic evolution of the MOR at 13°N EPR. A period of extension tectonics, forming an axial graben, was followed by local basaltic volcanism which eventually filled and overflowed the graben. Combined U-series dating and noble gas studies of sulphides may provide temporal constraints on the episodicity of tectono-volcanic activity and assess the control it has on hydrothermal activity.

Degassing of mantle Ar

The Ar isotopic composition of fluid inclusions from 13°N EPR reveal ^{40}Ar excesses of up to 2.6% in the trapped hydrothermal fluids (Fig. 4; Table 1). The volume of mantle-derived ^{40}Ar ($^{40}Ar_{mantle}$) in the hydrothermal fluids is calculated from

$$^{40}Ar_{mantle} = {}^{40}Ar_{assw} \times \delta^{40}Ar/100 \qquad (4)$$

where $^{40}Ar_{assw}$ is the ^{40}Ar concentration in air-saturated seawater (3.8×10^{-4} cm^3 STP/cm^3). The hydrothermal fluids have a maximum concentration of mantle-derived ^{40}Ar of 10^{-5} cm^3 STP/cm^3 in the hydrothermal fluids. The seawater ^{40}Ar concentration is high relative to that of normal MORBs ($c. 10^{-6}$ cm^3 STP/g) and consequently ^{40}Ar excesses have been largely undetected in MOR hydrothermal fluids. The mantle-derived ^{40}Ar in the 13°N EPR fluids may have been acquired from the mantle either

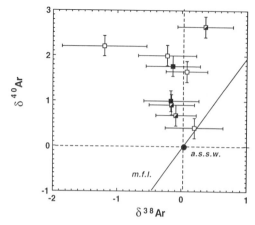

Fig. 4. Co-variation of Ar isotopes in fluid inclusions from hydrothermal sulphides from 13°N on the East Pacific Rise. Isotopic compositions are reported as ratio anomalies $\delta^x Ar$ (%). See text for method. ^{40}Ar enrichments are due to the presence of mantle-derived Ar in the hydrothermal fluids. Samples: closed squares, Axial Graben; open squares, SE Seamount; and half squares, Marginal High. mfl, Mass fractionation line; and assw, air-saturated seawater.

indirectly, by the extraction of ^{40}Ar from the oceanic crust by circulating hydrothermal fluids, or directly, as a result of the degassing of volatiles from recently injected MOR magmas (in a manner analogous to that proposed for He).

The presence of mantle-derived ^{40}Ar in the 13°N EPR hydrothermal solutions places severe constraints on the water–rock ratios of the hydrothermal system. ^{40}Ar concentrations range from 10^{-5} cm^3 STP/g in volatile-rich MORB [the so-called 'popping rocks' (Sarda & Graham 1990)] to 10^{-8} cm^3 STP/g in crystalline basalts (Staudacher & Allegre 1988), and probably lower in gabbros. Depending on the level of the oceanic crust from which the ^{40}Ar is acquired, effective water–rock ratios range from 1 ('popping rocks') to 0.001 (basalts). With the exception of the gas-rich 'popping rocks'', the water–rock ratios required to account for the ^{40}Ar excess are wholly inconsistent with those calculated from other chemical indices (e.g. Von Damm *et al.* 1985). The volatiles in 'popping rocks' are located in the vesicle phase; the high gas concentration is a consequence of the high degree of degassing the basalts have experienced and they are considered to be a product of the earliest degassing of MOR magmas (Sarda & Graham 1990). However, 'popping rocks' are a volumetrically insignificant volcanic product at

MORs and, to our knowledge, none has been dredged from the 13°N EPR site. Thus it is unlikely that the gases are extracted from 'popping rocks' themselves.

Normal MORBs display vesicle distributions and relative and absolute abundances of the noble gases that are consistent with the loss of volatiles in a degassing event before that which is preserved as the vesicle phase in MORBs (e.g. Jambon *et al.* 1985). It is likely that the early degassing event is responsible for the dramatic increases in the volatile contents of hydrothermal fluids which are asociated with ridge crest volcanic activity (Lupton *et al.* 1989, 1991). Small ^{40}Ar excesses have been detected in vent fluid samples from young 400°C fluids (Bender *et al.* 1991) which vent through 15–20 day old basalts at 9°N EPR (Haymon *et al.* , 1993). In contrast, vent fluids from mature hydrothermal systems at 9°N EPR (Bender *et al.* 1991) and hydrothermal fluids in fluid inclusions from 21°N EPR (Turner & Stuart 1992) appear to display no ^{40}Ar enrichments. Hydrothermal fluid ^{40}Ar excesses at 13°N EPR may be closely linked with the magmatism at the site, most likely as a result of the direct injection of mantle volatiles during magmatic degassing.

He–S isotope systematics

$\delta^{34}S$ of the sulphides display a range typical of MOR sulphides from +1.2 to +2.8‰ (Table 1). Such a variation results from the addition of seawater sulphate ($\delta^{34}S \approx 20$‰) to MORB-derived sulphur ($\delta^{34}S = -0.5 \pm 1$‰) in the high-temperature hydrothermal fluids. This may

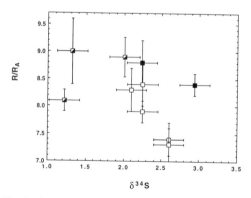

Fig. 5. Co-variation of He and S isotopes. The $^3He/^4He$ of trapped hydrothermal fluids is independent of $\delta^{34}S$. Symbols as in Fig. 4.

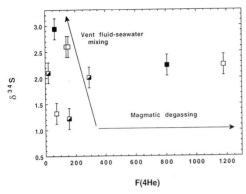

Fig. 6. Co-variation of the He/Ar ratio of fluid inclusions with the S isotopic composition of the host sulphide. Symbols as in Fig. 4. The addition of seawater to the hydrothermal fluid (represented by vent fluids) is reflected by the trend of decreasing $F(^4He)$ and increasing $\delta^{34}S$. This is masked in some samples by minor atmospheric contamination, the direct addition of magmatic He to the hydrothermal fluids or the derivation of isotopically heavy S by the dissolution of sulphate in the hydrothermal fluids.

occur either by the dissolution of anhydrite within the oceanic crust or chimney, or by the entrainment of seawater in the upflow zone. There is no correlation between the isotopic compositions of He and S in the sulphides (Fig. 5) suggesting that, although the (late) addition of seawater to hydrothermal fluids increases sulphide $\delta^{34}S$ values, it has had no measurable effect on the fluid $^3He/^4He$. This can be explained by the 10^4 difference in the He concentrations of the hydrothermal fluids and seawater.

The low He/Ar ratio of seawater means that the entrainment of seawater by the rising hydrothermal fluids results in a reduction of the fluid He/Ar along with increasing the sulphide $\delta^{34}S$ value. In Figure 6, fluid inclusion $F(He)$ values are plotted against $\delta^{34}S$ of the host sulphide. A trend of decreasing He/Ar with increasing $\delta^{34}S$ is observed in six samples which extends from notional vent fluid values ($\delta^{34}S >$ 1‰ (Woodruff & Shanks 1988), $F(He) \approx 250$ (Merlivat *et al.* 1987). This trend is qualitatively consistent with the control of vent fluid chemistries by hydrothermal fluid–seawater mixing in some cases. The lack of a strong correlation between $\delta^{34}S$ and He/Ar suggests that processes other than fluid mixing operate. The direct addition of magmatic He to the hydrothermal fluids increases He/Ar. Dissolution of sulphate S in anhydrite ($\delta^{34}S = 20‰$) by the hydrothermal fluids in the upflow zone may control vent fluid $\delta^{34}S$ but has no effect on the relative and absolute abundances of He and Ar in the hydrothermal fluids. Contamination by atmospheric noble gases is difficult to avoid (Turner & Stuart 1992) and it is apparent from Fig. 6 that small volumes of atmospheric noble gases released by crushing mask the Ar released from the inclusion fluids in some samples.

We are grateful to T. Dunai (ETH, Zurich) for cross-calibration of the He isotope standard, H.E. Hall (University of Manchester) for providing the high $^3He/^4He$ gas used for the preparation of the He standard, A. Boyce for S isotope analyses and D. Blagburn for freely imparting his technical expertise to F.S. over the past six years. Thoughtful reviews by D. Hilton and P. Jean-Baptiste clarified many of the ideas presented here.

References

BAKER, E. T., MASSOTH, G. J. & FEELY, R. A. 1987. Cataclysmic venting on the Juan de Fuca Ridge. *Nature*, **329**, 149–151.

——, ——, ——, EMBLEY, R. W., THOMSON, R. E. & BURD, B. J. 1995. Hydrothermal eventi plumes from the Co-Axial seafloor eruption site on the Juan de Fuca Ridge. *Geophysical Research Letters*, **22**, 147–150.

BATIZA, R. & VANKO, D. 1984. Petrology of young Pacific seamounts. *Journal of Geophysical Research*, **89**, 11235–11260.

BENDER, M., ORCHADO, J., LILLEY, M. D., LUPTON, J. E., VON DAMM, K. I. & EDMOND, J. M. 1991. ^{40}Ar outgassing from the mantle reflected in the $^{36}Ar/^{40}Ar$ ratio of hydrothermal vent fluids from the East Pacific Rise, 9°N. *EOS, Transactions of the American Geophysical Union*, **72**, 481.

BUTTERFIELD, D. A., MASSOTH, G. J., McDUFF, R. E., LUPTON, J. E. & LILLEY, M. D. 1990. Geochemistry of fluids from Axial Seamount Hydrothermal Emissions Study vent field, Juan de Fuca: subsea-floor boiling and subsequent fluid–rock interaction. *Journal of Geophysical Research*, **95**, 12 895–12 921.

CROVETTO, R., FERNANDEZ-PRINI, R. & JAPAS, M. L. 1982. Solubilities of inert gases and methane in H_2O and in D_2O in the temperature range 300 to 600 K. *Journal of Chemical Physics*, **76**, 1077–1086.

EMBLEY, R. W., CHADWICK, W. W. JR., PERFIT, M. R. & BAKER, E. T. 1991. Geology of the northern Cleft segment, Juan de Fuca Ridge: recent lava flows, seafloor spreading and the formation of megaplumes. *Geology*, **19**, 771–775.

FRANCHETEAU, J. & BALLARD, R. 1983. The East Pacific Rise near 21°N, 13°N and 20°S: inferences for along strike variability of axial processes of the mid-ocean ridge. *Earth and Planetary Science Letters*, **63**, 93–116.

GENTE, P., AUZENDE, J. M., RENARD, V., FOUQUET, Y. & BIDEAU, D. 1986. Detailed geological mapping by submersible of the East Pacific Rise axial graben near 13°N. *Earth and Planetary Science Letters,* **78,** 224–236.

GRAHAM, D. W., JENKINS, W. J., KURZ, M. D. & BATIZA, R. 1987. Helium isotope disequilibrium and geochronology of glassy submarine basalts. *Nature* **326,** 384–387.

——, ZINDLER, A., KURZ, M. D., JENKINS, W. J., BATIZA, R. & STAUDIGEL, H. 1988. He, Pb, Sr and Nd isotope constraints on magma genesis and mantle heterogeneity beneath young Pacific seamounts. *Contributions to Mineralogy and Petrology,* **99,** 446–463.

GREGORY, R. T. & TAYLOR, H. P. 1981. An oxygen isotope profile in a section of Cretaceous oceanic crust, Oman: evidence for $\delta^{18}O$ buffering of the oceans by deep (75 km) seawater–hydrothermal circulation at mid-ocean ridges. *Journal of Geophysical Research,* **86,** 2737–2755.

HALL, A. J., BOYCE, A. J., FALLICK, A. E. & HAMILTON, P. J. 1991. Isotopic evidence of the depositional environment of Late Proterozoic stratiform barite mineralisation, Aberfeldy, Scotland. *Chemical Geology,* **87,** 99–114.

HAYMON, R., FORNARI, D. J., EDWARDS, M., CARBOTTE, S. M., WRIGHT, D. & MacDONALD, K. C. 1991. Hydrothermal vent distribution along the East Pacific Rise crest (9°09′–54′N) and its relationship to magmatic and tectonic processes on fast spreading mid-ocean ridges. *Earth and Planetary Science Letters,* **104,** 513–534.

——, —— & 13 others 1993. Volcanic eruption of the mid-ocean ridge along the East Pacific Rise crest at 9°45′–52′N: direct submersible observation of seafloor phenomena associated with an eruption event in April, 1991. *Earth and Planetary Science Letters,* **119,** 85–101.

HEKINIAN, R. & FOUQUET, Y. 1985. Volcanism and metallogenesis of axial and off-axial structures on the East Pacific Rise near 13°N. *Economic Geology,* **80,** 1–29.

——, FEVRIER, M. & 9 others 1983. East Pacific Rise near 13°N: geology of new hydrothermal fields. *Science,* **219,** 1321–1324.

JAMBON, A., WEBER, H. W. & BEGEMANN, F. 1985. Helium and argon from an Atlantic MORB glass: concentration, distribution and isotopic composition. *Earth and Planetary Science Letters,* **73,** 255–268.

JOCHUM, K. P., HOFMANN, A. W., ITO, E. & WHITE, W. M. 1983. K, U and Th in mid-ocean ridge basalt glasses and heat production, K/U and K/Rb in the mantle. *Nature,* **306,** 431–436.

KENNEDY, B. M. 1988. Noble gases in vent water from the Juan de Fuca Ridge. *Geochimica et Cosmochimica Acta,* **52,** 1929–1935.

KIM, K. R., WELHAN, J. A. & CRAIG, H. 1984. The hydrothermal vent fields at 13°N and 11°N on the East Pacific Rise: Alvin results. *EOS, Transactions of the American Geophysical Union,* **65,** 973.

KURZ, M. D. & JENKINS, W. J. 1981. The distribution of helium in oceanic basalt glasses. *Earth and Planetary Science Letters,* **53,** 41–54.

LALOU, C. & BRICHET, E. 1982. Age and implication of East Pacific Rise sulphide deposits at 21°N. *Nature* **300,** 169–171.

——, BRICHET, E. & HEKINIAN, R. 1985. Age dating of sulfide deposits from axial and off-axial structures on the East Pacific Rise near 12°50′N. *Earth and Planetary Science Letters,* **75,** 59–71.

——, ——, & LANGE, J. 1989. Fossil hydrothermal sulfide deposits at the Galapagos spreading centre near 85°00′W: geological setting, mineralogy, and chronology. *Oceanologica Acta,* **12,** 1–18.

——, ——, ARNOLD, M., THOMPSON, G., FOUQUET, Y. & RONA, P. 1993. New age data for the Mid-Atlantic Ridge hydrothermal sites: TAG and Snakepit chronology revisited. *Journal of Geophysical Research,* **98,** 9705–9713.

LUPTON, J. E., BAKER, E. T. & MASSOTH, G. J. 1989. Variable ^{3}He/heat ratios in submarine hydrothermal systems: evidence from two plumes over the Juan de Fuca Ridge. *Nature,* **337,** 61–163.

——, LILLEY, M. D., OLSEN, E. & VON DAMM, K. L. 1991. Gas chemistry of vent fluids from 9–10°N on the East Pacific Rise. *EOS, Transactions of the American Geophysical Union,* **72,** 481.

MERLIVAT, L., PINEAU, F. & JAVOY, M. 1987. Hydrothermal vent waters at 13°N, East Pacific Rise: isotopic composition and gas concentration. *Earth and Planetary Science Letters,* **84,** 100–108.

MICHARD, G., ALBAREDE, F., MICHARD, A., MINSTER, J.-F., CHARLOU, J.-L. & TAN, N. 1984. Chemistry of solutions from the 13°N East Pacific Rise hydrothermal site. *Earth and Planetary Science Letters,* **67,** 297–307.

NEWMAN, S., FINKEL, R. & MacDOUGALL, J. D. 1982. ^{230}Th–^{238}U disequilibrium systematics in oceanic tholeiites from 21°N on the East Pacific Rise. *Earth and Planetary Science Letters,* **65,** 17–33.

PERFIT, M. R., FORNARI, D. J., SMITH, M. C., BENDER, J. F., LANGMUIR, C. H. & HAYMON, R. M. 1994. Small-scale spatial and temporal variations in mid-ocean ridge crest magmatic processes. *Geology,* **22,** 375–379.

SARDA, P. & GRAHAM, D. 1990. Mid-ocean ridge popping rocks: implications for degassing at ridge crests. *Earth and Planetary Science Letters,* **97,** 268–289.

SEDWICK, P. N., McMURTRY, G. N., HILTON, D. R. & GOFF, F. 1994. Carbon dioxide and helium in hydrothermal fluids from Loihi Seamount, Hawaii, USA: temporal variability and implications for the release of mantle volatiles. *Geochimica et Cosmochimica. Acta,* **58,** 1219–1227.

SHANKS, W. C. III, BOHLKE, J. K. & SEAL, R. R. II 1991. Stable isotope studies of vent fluids, 9–10°N East Pacific Rise: water-rock interaction and phase separation. *EOS, Transactions of the American Geophysical Union,* **72,** 481.

STAUDACHER, T. & ALLEGRE, C. J. 1988. Recycling of oceanic crust and sediments: noble gas subduction barrier. *Earth and Planetary Science Letters,* **89,** 173–183.

STUART, F. M., DUCKWORTH, R., TURNER, G. & SCHOFIELD, P. 1994*b*. Helium and sulfur isotopes in sulfides from the Middle Valley, northern Juan de Fuca Ridge. *In: Proceedings of the Ocean Drilling Program, Scientific Results*, **139**, 387–392.

——, TURNER, G., DUCKWORTH, R. C. & FALLICK, A. E. 1994*a*. Helium isotopes as tracers of trapped hydrothermal fluids in ocean-floor sulfides. *Geology*, **22**, 823–836.

TURNER, G. & STUART, F. M. 1992. He/heat ratios and the deposition temperatures of ocean-floor sulphides. *Nature*, **357**, 581–583.

VON DAMM, K. L. & BISCHOFF J. L. 1988. Chemistry of hydrothermal solutions from the southern Juan de Fuca Ridge. *Journal of Geophysical Research*, **92**, 11 334–11 346.

——, EDMOND, J. M., GRANT, B., MEASURES, C., WALDEN, B. & WEISS, R. 1985. Chemistry of submarine hydrothermal solutions at 21°N, East Pacific Rise. *Geochimica et Cosmochimica Acta*, **49**, 2197–2220.

WOODRUFF, L. G. & SHANKS, W. C. III 1988. Sulfur isotopes of chimney minerals and vent fluids at 21°N, East Pacific Rise: hydrothermal sulfur sources and disequilibrium sulfate reduction. *Journal of Geophysical Research*, **93**, 4562–4572.

Preliminary modelling of hydrothermal circulation within mid-ocean ridge sulphide structures

PENNY DICKSON, ADAM SCHULTZ & ANDREW WOODS

University of Cambridge, Institute of Theoretical Geophysics, Department of Earth Sciences, Downing Street, Cambridge CB2 3EQ, UK

Abstract: The spatial distribution of venting sites at mid-ocean ridge crest hydrothermal fields, in combination with direct observations of the temperature and velocity of (primarily diffuse) effluent, provides constraints on the circulation of fluids both within exposed hydrothermal edifices and in the seafloor below. Before any constraints can be placed on either the interior structure of these bodies or the associated heat and mass transfer between them, their role in controlling the flow needs to be investigated. A simplified finite difference forward model of an idealized sulphide edifice allows the calculation of the interior velocity and temperature distributions and, more importantly, observation of the effect that the flow-governing components have on the circulation. Changes in permeability structure and in basement heat and mass fluxes have an important effect on the flow.

Large-scale, diffuse venting recently observed at many vent fields (Schultz *et al.* 1992) suggests that a large mass of relatively cool water (3–40°C) is involved in the hydrothermal circulation as well as the very hot, white and black smoker fluid (200–380°C). Field measurements of the temperature and heat flux of this cooler, diffusely venting fluid put new constraints on the nature of the hydrothermal circulation and the structure of the crust and exposed, polymetallic sulphide mounds. To begin to gain a better understanding of such circulation and the importance of these constraints, we developed a numerical model of hydrothermal circulation within sulphide structures on the seafloor. The purpose of this paper is to introduce the model and describe some simple solutions which elucidate the nature of the velocity and thermal fields in hydrothermal circulation systems.

In the hydrothermally deposited sulphide structures, the internal temperature and velocity fields of fluid circulation are largely governed by the internal permeability structure and the distribution of heat and fluid volume flux from below. We examine the pattern of flow as a function of the Rayleigh number and the permeability using a two-dimensional finite difference solution. We also consider the importance of a source of hydrothermal fluid through the structure supplied from within the ocean crust.

The model

The focused upwelling associated with deeper crustal convection governs the spatial distri-
bution of zones of hydrothermal discharge at the seafloor. In turn, this hot fluid emerging from the upper surface of the crust is supplied to the base of the exposed hydrothermal sulphide structures within the vent fields. The pattern of fluid circulation within these structures is largely determined by the bottom boundary conditions, as well as the nature of heat and mass flux exchange.

For simplicity, we have developed a steady-state model of the fluid flow within a two-dimensional porous block heated from below. This enables us to examine the pattern of the flow and its dependence on the permeability structure and the strength of the heat transfer. By applying appropriate boundary conditions and suitable scalings, such a model can be used to investigate aspects of the interior flow pattern of an exposed sulphide structure and gives some idea of the nature of the flows within such structures. The two-dimensional flow is a simplification of the actual three-dimensional convecting system. However, in a narrow elongate sulphide structure, the flow may indeed be dominantly two dimensional. Even for more complicated flow geometries, much of the essential physics can be captured through consideration of the two-dimensional case. We also note that if the flow is dominated by abrupt changes in permeability, as would be expected of highly fractured surface rock, our model is not appropriate. In such a situation it might be more reasonable to model the temperature field as that due to focused high-temperature flow in narrow zones separated by small-scale impermeable blocks, inside which the temperature is determined by thermal conduction. However, such models are beyond the scope of the present study.

From the assumed nature of formation of the sulphide edifices, we allow the structure to have different horizontal and vertical permeabilities but, for simplicity, we assume that the permeability is homogeneous throughout the domain of the structure. We also assume that the individual flow paths in the system are sufficiently small that the flow is, on the pore scale, of low Reynolds number (Phillips 1991) and therefore governed by Darcy's law

$$\underline{u} = -\frac{k}{\mu}(\nabla P + \Delta\rho g\hat{z}) \tag{1}$$

where $\underline{u} = (\hat{x}u + \hat{z}w)$ is the transport velocity, k is the permeability of the medium, μ is the dynamic viscosity of fluids within the structure, P is the pressure, $\Delta\rho$ is the density difference determined by the temperature distribution, g is gravitational acceleration and $\hat{x}, \hat{z}$ are, respectively, the horizontal and vertical Cartesian unit vectors. We follow the development of Phillips (1991) and write the velocity in terms of a stream function ψ, where $u = \partial\psi/\partial z$ and $w = -\partial\psi/\partial x$. We shall also non-dimensionalize the stream function and temperature so that our resultant models can later be applied to a particular case. The dimensionless temperature and stream function (θ, ϕ) are related to the corresponding dimensional quantities (T, ψ) by

$$T = T_0\theta \tag{2}$$

$$\psi = \phi Ra\frac{l}{h}\kappa \tag{3}$$

where Ra is the Rayleigh number

$$Ra = \frac{\alpha g T_0 k_v h}{\nu\kappa} \tag{4}$$

h and l are height and quarter length of the structure, respectively, α is the thermal expansion coefficient, T_0 is the maximum basement temperature above ambient, ν is the kinematic viscosity and κ is the thermal diffusivity.

The vorticity equation follows from Darcy's law (Phillips 1991) and may be written

$$\frac{\partial^2\phi}{\partial X^2} + S^{-1}\frac{\partial^2\phi}{\partial Z^2} = -\frac{\partial\theta}{\partial X} \tag{5}$$

where $X = xl$ and $Z = zh$ are the non-dimensional horizontal and vertical coordinates, and S is a geometrical scaling factor

$$S = \frac{k_h}{k_v}\left(\frac{h}{l}\right)^2 \tag{6}$$

and k_h and k_v are the horizontal and vertical permeabilities, respectively. In the steady state, the conservation of thermal energy equation has the form (Phillips 1991)

$$Ra\left(\frac{\partial\phi\partial\theta}{\partial Z\partial X} - \frac{\partial\phi\partial\theta}{\partial X\partial Z}\right) = \frac{h}{l}\frac{\partial^2\theta}{\partial X^2} + \frac{\partial^2\theta}{\partial Z^2} \tag{7}$$

It is crucial to set appropriate boundary conditions so that the velocity and temperature distributions may be determined at zones of outflow. These are described in the following sections. The above equations, together with the boundary conditions, were solved using an iterative scheme. An initial temperature distribution ($\theta = 0$ inside the domain of the model, a 41 node by 21 node finite difference grid) is assumed. This is used to solve equation (5) for the stream function everywhere within the domain of the medium. The solution for the stream function is inserted into equation (7) to obtain a new temperature distribution. The process iterates until it has converged. The convergence criterion we use is that the $L\infty$ norm of the difference between both the temperature and stream function fields at iteration i and at iteration $i - 1$ must be less than a critical value, ϵ

$$\epsilon_T \geq \left\|\{|T_{kl}^i - T_{kl}^{i-1}|\}\right\|_\infty,$$

$$(k = 1, \ldots, N; l = 1, \ldots, M)$$

$$\epsilon_\psi \geq \left\|\{|\Psi_{kl}^i - \Psi_{kl}^{i-1}|\}\right\|_\infty, \tag{8}$$

$$(k = 1, \ldots, N; l = 1, \ldots, M)$$

where N and M are the number of finite difference nodes in the x and y directions, respectively, within the domain of the porous medium; the superscripts refer to iteration i and $i - 1$; and (for all results shown here) $\epsilon = 10^{-6}$. During a typical run for low values of Ra, convergence is reached within less than 20 iterations. However, for geometries and conditions leading to the development of many small-scale convection cells, more than 100 iterations may be necessary.

Boundary conditions for exposed polymetallic sulphide structures

The two-dimensional model of the sulphide structures which develop on the seafloor requires boundary conditions for both the flow and the temperature (Dirichlet condition) or heat flux (Neumann condition) on each boundary.

Impermeable basement

We consider two sets of bottom boundary conditions. In the first, the porous medium representing the sulphide structure sits on top of an impermeable seafloor. Fluid is allowed to enter and leave through the side walls and upper surface of the structure, but is totally inhibited from entering from below, so that the only fluid in the interior domain is advected seawater. The second set of bottom boundary conditions, to be considered later, allows for a mass source of hydrothermal fluid from below. For the first case the temperature along the basement decays exponentially from the centre.

Side boundaries

Flow is allowed through the side walls. The boundary condition we have applied is that the pressure on the side boundaries equals the hydrostatic pressure in the ambient seawater (Phillips 1991). The temperature along the vertical sides of the sulphide mound is held constant and equal to that of the ambient seawater. This condition follows by noting that fluid is entrained into the side walls and that the ambient turbulence and mixing in the surrounding ocean is likely to be much stronger than any convective transport of heat to these boundaries by the flow in the porous layer.

Upper surface

Flow is allowed through the upper surface and, again, the boundary condition we have applied is that the pressure equals the hydrostatic pressure in the ambient seawater. The thermal boundary condition at the upper surface is more difficult to define and depends on the Rayleigh number of the flow in the porous layer. In the case of low Rayleigh number, the convective heat flux is comparable with the diffusive heat flux. These are therefore characterized by speeds of order κ/h, where as before κ is the thermal diffusivity and h is the height of the structure. This is approximately of order 10^{-7}–10^{-8} m s^{-1} (Tivey & McDuff 1990; Schultz et al. 1992). In contrast, the mean currents in the ocean are much stronger (of the order of 10^{-2} m s^{-1} or larger) and so we expect that the upper boundary temperature will be very close to ambient, even though there is upflow through the structure. Phillips (1991) considered an isothermal ($\theta = 0$) boundary condition on the sides and top surface of two-dimensional porous blocks. At higher Rayleigh numbers the convective upflow may be much stronger and therefore the temperature of the upper boundary may exceed that of the ambient seawater. Seafloor observations (Schultz et al. 1992) show the temperature of the top surface of hydrothermal mounds, and of the diffuse component of the effluent as it flows from such structures, may be as much as 10–50°C greater than that of ambient seawater. To properly establish the temperature at the top of the structure, the details of the flow above the upper boundary should be determined. In the limiting case of negligible background flow, there is little mixing of the venting fluid with the overlying ambient fluid just above the structure and so the vertical temperature gradient just above the structure is nearly zero. However, with a strong current in the overlying water there will be a greater degree of mixing and the upper boundary temperature will decrease towards zero, as for low Rayleigh number convection within the structure.

More detailed boundary conditions which take account of turbulent convection in the water column above the sulphide structures are possible (e.g. Woods & Delaney 1992). For low Rayleigh number convection within the structure, such conditions, however, lead to the prediction that the upper surface temperature is approximately zero. In the absence of very strong ocean currents and at higher Rayleigh number, the two-dimensional structure of the temperature distribution on the upper surface, and the dominance of convective transport at the hotter regions, requires consideration of the details of plume formation at and above the boundary between the structure and the overlying water column. We are presently developing a more detailed model for this intermediate situation based on the mixing which occurs just above the surface.

For simplicity, in the present model, we examine the two limiting cases, $\theta = 0$ and $\theta_z = 0$, which are strictly only valid at low and high Rayleigh numbers, respectively. The calculations should be interpreted as denoting the trend in the flow pattern as the Rayleigh number decreases for the Dirichlet condition $\theta = 0$, or as the Rayleigh number increases for the Neumann condition $\theta_z = 0$, rather than denoting simulations of the actual situation.

We consider here the nature of hydrothermal structures to which high and low Ra flow would relate. If we assume that the coefficient of thermal expansion and the viscosity take on the values for water, i.e. $\alpha \approx 10^{-4}$ K^{-1}, $\upsilon \approx 10^{-6}$ m^2 s^{-1}, $g \approx 10$ m s^{-2} and $\kappa \approx 10^{-6}$ m^2 s^{-1}, then from equation (4) $Ra \approx 10^9 k_\upsilon h\Delta T$. From seafloor observations we know that $\Delta T \approx 10^2$ K

Ra=30, kh/kv=10, r=0.25

Temperature Fields

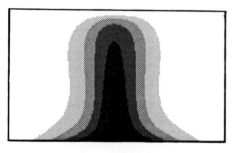
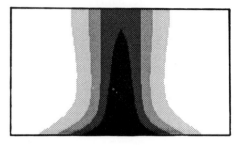

Isothermal ($\theta = 0$)   ### Constant heat flux ($\theta z = 0$)

0   **1**

Fig. 1. Variation of internal temperature field with two end-member top boundary conditions. (left) $\theta = 0$ on top boundary, (right) $\theta_z = 0$ on top boundary. A Rayleigh number of 30 was used for these examples. The horizontal permeability is a factor of 10 greater than the vertical permeability. The vertical to horizontal aspect ratio of the structure is 1:4.

and $h \approx 10\,\mathrm{m}$, which gives a value of $Ra \approx 10^{12}k_v$. Therefore, for the internal flow to be of low Rayleigh number ($\ll 1$), the suphide mound must have a vertical permeability of $k_v \ll 10^{-12}\,\mathrm{m}^2$. Repeating this analysis for the high Rayleigh number regime results in a vertical permeability of the order of $10^{-10}\,\mathrm{m}^2$ for $Ra = 100$.

Model predictions

Figure 1 depicts the temperature profiles for cross-sections through a suphide mound for each of the top surface thermal boundary conditions. As can be seen, the upper surface boundary condition has an important effect on the pattern of flows particularly in the domain in the vicinity of the upper surface. Initially, we investigate the effect of individually varying the basement temperature and permeability distributions.

We illustrate how the flow patterns change with Rayleigh number in Fig. 2. At low *Ra* values the temperature profile is that of purely conductive heat flow, but as *Ra* is increased advection gradually takes over and a plume-like pattern evolves in the centre of the domain above the hottest region on the base. This plume-like pattern consists of a thin thermal boundary layer along the base of the domain and relatively elevated temperatures along its central axis owing to the upward advection of heat. The rest of the model domain is at, or is very close to, ambient seawater temperature.

In the case $\theta_z = 0$, as *Ra* increases, the strength of the convection increases and the upper surface temperature at the centre of the site of outflow also increases due to the formation of the plume. At higher *Ra*, when the plume has formed, an increased volume flux of fluid is drawn in through the sides. This means that, in a relatively short but wide structure, the cooling of the side walls by the entrained seawater is more effective and there is only a thin thermal boundary layer on the base of the structure. Therefore, although the temperature of the basement and upper surface increases, the sides are actually becoming cooler (Schultz *et al.* 1992).

kh/kv=10, r=0.25

Temperature Fields

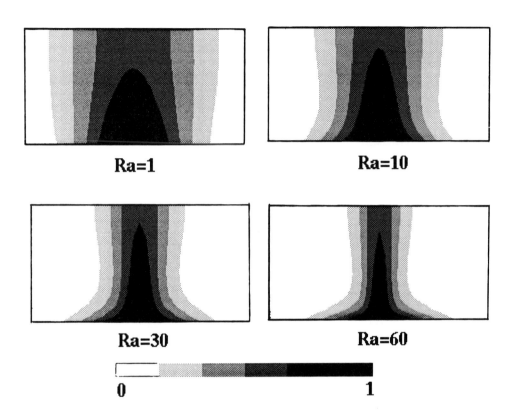

Fig. 2. Variation of the internal temperature field with different Rayleigh numbers, for $\theta_z = 0$ top boundary condition, aspect ratio 1:4, and with horizontal permeability a factor of 10 higher than vertical permeability. This top boundary condition is used for all subsequent examples. Increasing the Rayleigh number is equivalent to increasing the temperatures along the base of the structure. The transition from a conduction-dominated temperature field at lower base temperatures ($Ra = 1$) to an advection-dominated field at higher base temperatures ($Ra = 60$) is evident from the inward compression of the isotherms.

Figure 3 shows the corresponding stream function pattern as a function of Ra. In two-dimensional space, the gradient of the stream function in a given direction equals the velocity along the streamline (i.e. the velocity in the orthogonal direction). Therefore regions of intense grey-scale change denote high velocity flow, whereas regions of constant grey-scale correspond to relatively slow flow. In these figures the fluid is entering the model via the sides and/or edges of the upper surface and is leaving the model through the upper surface near the central axis. Despite the conductive

temperature profile for $Ra = 1$, the stream function plot shows that there is actually a simultaneous weak flow, i.e. forced convection, throughout the structure which is driven by the horizontal temperature gradient. As Ra is increased the strength of this flow increases and this is particularly highlighted in the stream function plots at the outflow zone. Gradual increases in the intensity of the grey-scale change across the outflow region indicate a continual increase in the velocity of the fluid leaving the structure, whereas the decrease in the width of the zone of outflow can be inferred

kh/kv=10, r=0.25

Stream Functions

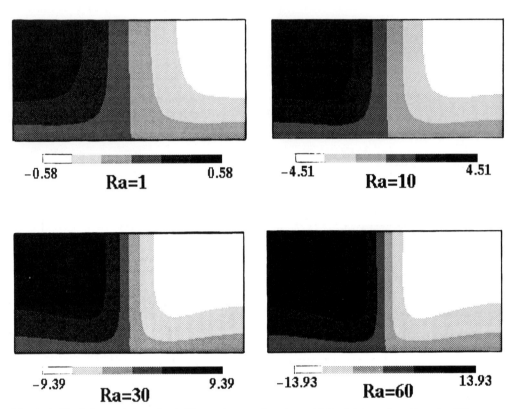

Fig. 3. As Fig. 2, but variation of scaled internal stream functions with Rayleigh number. Note the difference in stream function magnitudes for the four cases as denoted by the different colour scales. The stream functions shown here have been scaled according to $1/\kappa$, i.e. the value of the true stream function is κ times that shown in these figures. This scaling is used in all subsequent stream function plots. The spatial gradients of the stream functions are considerably greater at higher *Ra*, hence the corresponding velocities in the advected plume show a commensurate increase.

from the decreasing extent over which the change takes place. This is in accord with the appearance of the plume-like pattern emerging in the corresponding temperature field plots (Fig. 2). Also, as *Ra* increases more fluid is drawn in from the periphery of the upper surface, as larger pressure gradients are established across the domain as a result of the rapid buoyancy-driven flow at the centre.

It is interesting to note that these simple solutions identify regions within the sulphide structures where secondary mineralization reactions may occur. If we assume that such

reactions result from changes in the solubility of various minerals as a function of the temperature, then as fluid is advected through the domain and its temperature changes, it will cease to be in chemical equilibrium with the sulphide structure (Tivey & McDuff 1990; Phillips 1991). Phillips (1991) has shown that such gradient-driven reactions occur at a rate which is limited by the advective supply of fluid along the temperature gradient. This may be expressed in terms of the rock alteration index (RAI)

$$RAI = \underline{u} \cdot \nabla T \tag{9}$$

kh/kv=10, r=0.25

Rock Alteration Indices

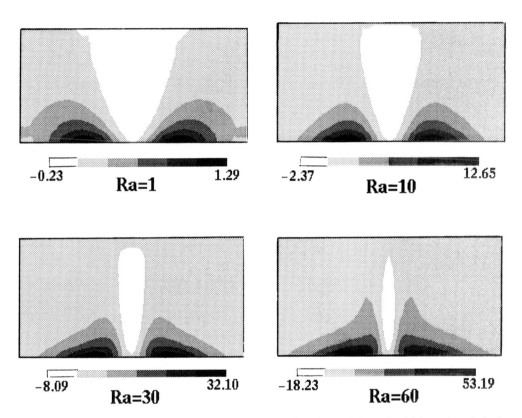

Fig. 4. As Fig. 2, but variation of scaled rock alteration index (RAI) with different Rayleigh number. As in the previous figure, note that different grey-scales hold for each of the four cases shown. The RAIs shown here have been scaled according to $h^2/(\kappa T_0)$, i.e. the value of the true RAI is $\kappa T_0/h^2$ times that shown in these figures. This scaling is used in all subsequent RAI plots. Note the thin region of most intense alteration at high Ra in the centre of the advected plume and along the thin thermal boundary layer at the base of the model.

Regions where |RAI| are large denote regions in which considerable mineral alteration may occur. Whether the specific alteration is mineral dissolution or deposition depends on the sign of the RAI and the relationship between the saturation concentration of the particular mineral and its temperature. Figure 4 shows that at low Ra the zone of mineral alteration is broad and weak, whereas at higher Ra alteration is relatively intense near the base of the domain where there is a large temperature gradient and near the central axis where there is a large advective flux. In the case of a fixed upper temperature of $\theta = 0$ (not shown here), there is a significant thermal

boundary layer near the upper surface where further mineralization may occur.

Figures 5, 6 and 7 show the effect of varying the ratio of horizontal to vertical permeability while Ra and r, the aspect ratio, are kept constant. With a small horizontal to vertical permeability ratio, the pattern of the temperature field is similar to that of thermal conduction, even at relatively high Ra values. Low relative horizontal permeability inhibits the horizontal advection of entrained seawater into the structure and limits the fluid source for the upward plume. However as the horizontal permeability increases the advection-dominated plume-like

Ra=30, r=0.25

Temperature Fields

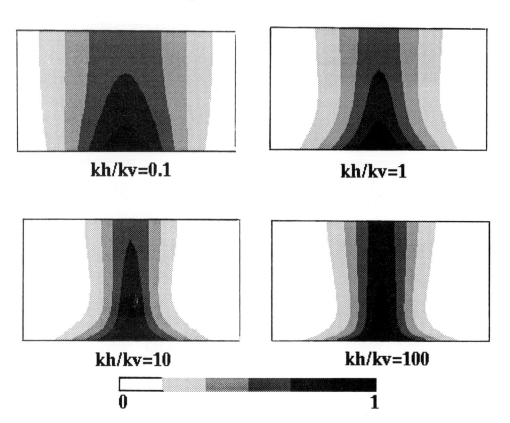

Fig. 5. Variation of internal temperature field with different permeability ratios, for a Rayleigh number of 30, and vertical to horizontal aspect ratio of 1:4. The ratio of horizontal to vertical permeability is 1:10 (top left), 1:1 (top right), 10:1 (bottom left) and 100:1 (bottom right). Formation of an advected plume is inhibited at low relative horizontal permeabilities and enhanced at high relative horizontal permeabilities.

pattern can develop and large |RAI| values appear at the base and along the central axis. Also, as the permeability ratio increases, the discharge velocity increases due to the formation of the plume (Fig. 6), as for the increasing *Ra* case. However, the width of the outflow zone increases due to the increasing ease with which the fluid can travel horizontally relative to flow in the vertical direction.

Mass source from below

We have applied a second set of conditions to the bottom boundary of the sulphide structure to investigate the effect of hot source fluid entering the structure through a crack in the impermeable base (Fig. 8). This source fluid was added through a narrow gap at the centre of the domain with non-dimensional temperature, $\theta = 1$. Along the rest of the base, the temperature is assumed to decay uniformly such that the temperature is zero at the edge of the structure. In Fig. 9, a series of images displays the effect of increasing the liquid flux as measured by the dimensionless Peclet number

$$Pe = \left.\frac{wh}{\kappa}\right|_{z=0} \qquad (10)$$

Ra=30, r=0.25

Stream Functions

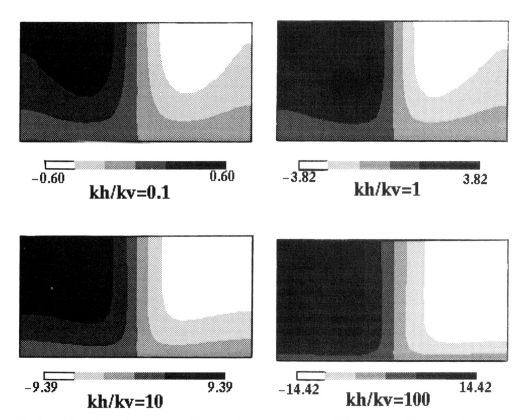

-0.60 0.60 -3.82 3.82

kh/kv=0.1   **kh/kv=1**

-9.39 9.39 -14.42 14.42

kh/kv=10   **kh/kv=100**

Fig. 6. As Fig. 5, but variation of scaled internal stream functions with different permeability ratios. Note the formation of the strong central plume at high relative horizontal permeabilities. At low k_h/k_v, weak advection of ambient fluid occurs throughout the sides and upper periphery of the structure, whereas at high k_h/k_v, intense advection into the structure occurs, but is concentrated on the sides near the base of the structure.

where w is the vertical velocity of the source fluid entering through the basement crack. Increasing Pe is equivalent to increasing the velocity of the hydrothermal fluid entering the structure from below. The Rayleigh number, permeability ratio, aspect ratio and the crack width as a fraction of the total length of the structure, are all kept constant.

As the upwardly advected flux increases, the plume becomes stronger owing to the greater total heat flux, and this can be seen in Fig. 9 where each isotherm in the successive images is translated to a greater height, resulting in increases in the temperature at outflow on the upper surface. The corresponding stream func-

tion patterns are shown in Fig. 10. As the Peclet number increases, the fraction of the surface area through which outflow emerges increases as the inflow of fluid through the side walls becomes progressively localized towards the base of the structure. However, the volume flux of the entrained fluid does not decrease so dramatically, rather it becomes an increasingly concentrated flow. Furthermore, the mass fraction of the effluent discharging from the surface of the structure which originates from entrainment through the sides of the structure decreases with Peclet number, so that, at high enough Pe, the venting fluid is composed mainly of hydrothermal fluid supplied from depth.

Ra=30, r=0.25

Rock Alteration Indices

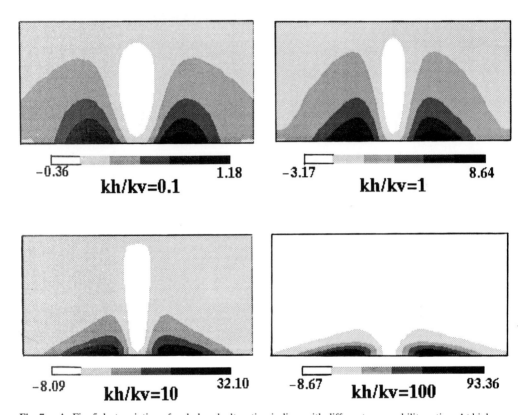

kh/kv=0.1   −0.36 ... 1.18

kh/kv=1   −3.17 ... 8.64

kh/kv=10   −8.09 ... 32.10

kh/kv=100   −8.67 ... 93.36

Fig. 7. As Fig. 5, but variation of scaled rock alteration indices with different permeability ratios. At high k_h/k_v ratios, intense alteration occurs over much of the interior domain of the model, with particularly high values along the bottom boundary. At low k_h/k_v ratios, there are broad zones on the sides of the structure where little alteration takes place.

A numerical study has shown that, for the $Ra = 30$ case depicted in Fig. 9, the mass fraction of the effluent derived from fluid entrained through the side walls of the structure falls below 10% when $Pe = 173$ and below 1% when $Pe = 484$. These values, however, also depend on Rayleigh number. We find that as the Rayleigh number increases from 1 to 60, the Peclet number at which the mass fraction of entrained fluid in the effluent falls below 10% (1%) increases from $Pe = 5$ (10) to $Pe = 280$ (1025).

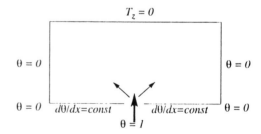

Fig. 8. Lower boundary condition modified to permit flux from below of mass of set temperature distribution and velocity through a crack of defined width.

Summary

This paper describes a technique which will enable us to make deductions from seafloor

Ra=30,kh/kv=1,r=0.25,width=0.1

Temperature Fields

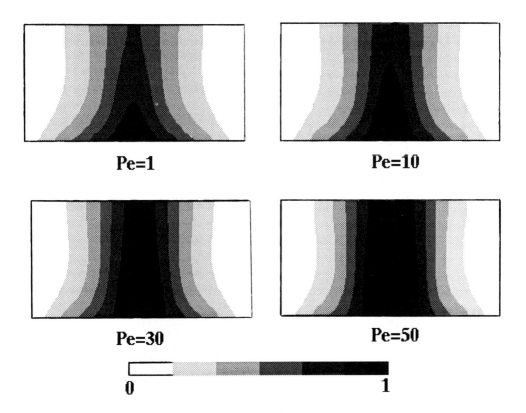

Fig. 9. Variation of internal temperature field with different Peclet number, *Pe*. The crack width is 10% the length of the structure. Fluid entering through the crack is isothermal at the base of the structure, with $\theta = 1$. Increasing the Peclet number is equivalent to increasing the velocity of the fluid entering the structure from below and has the effect of advecting a given isotherm to a higher position within the structure.

observation about circulation within exposed polymetallic sulphide structures. From the spatial distribution of venting sites over a somewhat larger scale, i.e. that comparable with an entire ridge crest vent field, it is possible in principle to infer features of fluid circulation within the crust, including the heat and mass flux of the fluid leaving the seafloor. This information, along with seafloor measurements of velocity and temperature of diffuse hydrothermal effluent (e.g. Schultz *et al.* 1992), will ultimately allow us to place better constraints on the interior structure, such as permeability distribution, of the hydrothermal mounds. Independent information on the permeability structure returned

from geophysical means and other sources such as Ocean Drilling Program cores, will allow us to constrain more tightly such models.

The results presented herein show that the permeability structure and the presence of a fluid volume source from below can have a fundamental effect on the mass of entrained fluid, the scale of fluid velocities in the medium and the regions of significant chemical alteration. It is particularly important to understand the effect of these factors on the discharge velocities and temperatures of the effluent under the experimental ranges through which we have exposed our model. This is of more than theoretical interest as one of the authors (A.S.) has obtained

Ra=30,kh/kv=1,r=0.25,width=0.1

Stream Functions

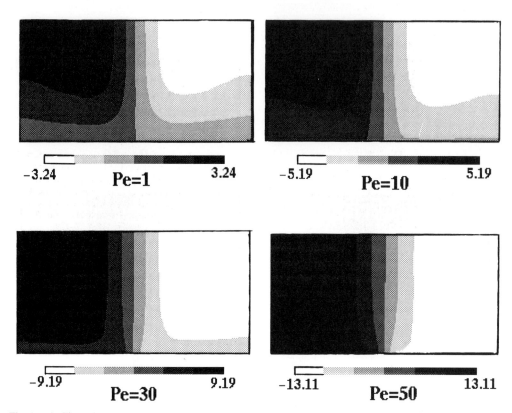

Fig. 10. As Fig. 9, but variation of scaled internal stream functions with different Peclet number.

direct measurements of diffuse hydrothermal effluent temperatures and velocities at the Endeavour vent field, Juan de Fuca Ridge, and the Snake Pit, TAG and Broken Spur vent fields on the Mid-Atlantic Ridge. The work presented here is a preliminary step toward the quantitative analysis of data of this sort.

In the low Rayleigh number regime, the temperature profile is conductive and hence the discharge temperature is not much greater than ambient. Above a critical Rayleigh number, the formation of a plume, whose definition increases with Rayleigh number, replaces the conductive temperature profile, gradually increasing the surface temperature. The presence of this plume-like profile depends also on the permeability ratio of the structure, as conductive

features will dominate if k_h/k_v is low (i.e. <1) even if the Rayleigh number is fairly high ($\gg$1). As this permeability ratio is increased, however, convection will eventually replace conduction. The temperature profile is also influenced by the presence and speed of fluid entering through the base of the structure.

As *Ra* is increased, the width of the discharge zone decreases while the speed of the emerging fluid at outflow increases. However, as k_h/k_v increases, the speed of the effluent again increases, but the width of the discharge zone for this case, also increases. The presence of fluid being vertically ejected into the base of the structure also has an effect on the velocity and width of the outflow; as its velocity increases the width of outflow increases.

Although a steady-state two-dimensional model with homogenous permeability is inadequate to describe fully the details of hydrothermal circulation at ridge crests, some seafloor observations suggest that elements of these models do reflect gross features of the circulation (e.g. the anti-correlation observed between the temperature and velocity of diffuse effluent at the Juan de Fuca Ridge; Schultz *et al.* 1992). Recent observations of the morphology of the TAG mound made by the authors (A.S. and P.D.) during the BRAVEX/94 expedition to the Mid-Atlantic Ridge suggest that the advection of cool seawater, locally, into a steep slope on the west of the TAG mound bordering the main high-temperature black smoker complex might be associated with a region of anomalously low conductive heatflow detected by K. Becker (pers. comm.). The importance of the local lateral advection of seawater into hydrothermal structures, forced by the pressure drop due to the nearby buoyant plume, is a significant feature of the models in the present work.

The work reported here was carried out, in part, under the support of NERC grant ESD/08/13/04/01. P.D. was supported by a NERC studentship.

References

HOLMAN, J. P. (1986). *Heat Transfer*. McGraw-Hill, New York.

PHILLIPS, O. M. 1991. *Flow and Reactions in Permeable Rocks*. Cambridge University Press, Cambridge.

SCHULTZ, A., DELANEY, J. R. & McDUFF, R. E. 1992. On the partitioning of heat flux between diffuse and point source seafloor venting. *Journal of Geophysical Research*, **97**(12), 299–314.

TIVEY, M. K. & McDUFF, R. E. 1990. Mineral precipitation in the walls of black smoker chimneys: A quantitative model of transport and chemical reaction. *Journal of Geophysical Research*, **95**(B8), 12 617–12 637.

WOODING, R. A. 1963. Convection in a saturated porous medium at large Rayleigh number or Peclet number. *Journal of Fluid Mechanics*, **15**, 527–544.

WOODS, A. W. & DELANEY, J. R. 1992. The heat and fluid transfer associated with the flanges on hydrothermal venting structures. *Earth and Planetary Science Letters*, **112**, 117–129.

Modelling diffuse hydrothermal flow in black smoker vent fields

A. RACHEL PASCOE & JOHNSON R. CANN

Department of Earth Sciences, University of Leeds, Leeds LS2 9JT, UK

Abstract: An important end-member of hydrothermal circulation at mid-ocean ridges is the low-temperature diffuse flow found at all known black smoker vent fields. Diffuse flow is the result of the subsurface mixing of cold seawater with hot black smoker fluid. The conditions that allow this mixing to take place are poorly understood. To gain further understanding of diffuse flow a simple pipe model has been used to explore the relationship between the sub-seafloor permeability structure and the temperature of the exiting fluid. Calculations with this simple model show a wide and continuously varying exit temperature from black smoker temperatures to diffuse flow temperatures as the permeability structure is changed. The results show that for low temperatures to be observed at the surface, the upper part of the circulation for that part of the discharge has to be relatively permeable, and that the deeper sections of the upflow must be much more impermeable. The model suggests that the early stages of evolution of a hydrothermal system are characterized by widespread diffuse flow and that black smokers develop as subsurface precipitation reduces the permeability of the upper section of the crust.

Seafloor hydrothermal systems have complex structures. A typical hydrothermal vent field is 50–200 m across and contains a wide variety of vents of many types. In some places high-temperature fluid emerges, usually through discrete chimneys, and mixes with the overlying seawater to form the characteristic turbulent, buoyant, sulphide-charged black smoker plumes. In other places lower temperature fluid emerges from less well-defined chimneys to mix with seawater to produce grey or white smoker plumes. Over large areas of most vent fields low-temperature fluids emerge as diffuse flow from cracks in the seafloor, producing a shimmering structure in the water, but no visible precipitates. Chemical analyses of water from a single vent field suggest that typically all of these fluids are derived from a small range of primary high-temperature fluids by dilution with cold seawater percolating down into the seafloor within and around the vent field. As this subsurface dilution takes place, sulphide and quartz are precipitated in a stockwork of cracks below the seafloor. Such stockworks, with veins filled by pyrite and quartz, are seen to underlie sulphide deposits in ophiolites and these can give evidence on the geometry and relationships in the sub-seafloor parts of active systems. Little is known about the reasons for the emission of such a variety of fluids from a single vent field. Here we seek to construct a simple model that can throw some light on this question.

Until recently much of the modelling work carried out on seafloor hydrothermal systems has concentrated on the high-temperature black smoker vents (e.g. Speer & Rona 1989; Tivey & McDuff 1990; Rudnicki & Elderfield 1992). This is particularly true when considering the sub-seafloor system (e.g. Lister 1974; Strens & Cann 1982, 1986; Lowell & Rona 1985; Cann *et al.* 1985; Cann & Strens 1989; Lowell & Burnell 1991). Yet the diffuse flow, characterized by temperatures up to about 100°C, is much more widespread in most vent fields, even if less spectacular. Diffuse flow is important for two reasons: (1) it is the most significant source of nutrients for the biological communities around vents, because of its low temperature and apparently lower toxicity and (2) it contributes a large part, perhaps on average well over half, of the thermal energy emitted by a vent field. Schultz *et al.* (1992) collected measurements of diffuse hydrothermal flow velocities and temperatures from a source within a vent field at the Endeavour Segment of the Juan de Fuca Ridge. He used these data to calculate the heat flux density from the top of the sulphide body due to the diffuse flow, concluding that the heat flow out of the system due to diffuse flow may exceed that due to high-temperature venting by a factor of five. Trivett (1994) developed a model for the plume generated by the diffuse flow and used this to guide the collection of measurements on the southern Juan de Fuca Ridge and in the calculation of the resulting heat fluxes, (Trivett & Williams 1994). His results show that the diffuse flow does contribute a significant amount to the total heat flux from a vent field. Other workers support these views, including Elderfield *et al.* (1993), who suggest that diffuse flow

From Parson, L. M., Walker, C. L. & Dixon, D. R. (eds), 1995, *Hydrothermal Vents and Processes*, Geological Society Special Publication No. 87, 159–173.

appears to account for about three-quarters to six-sevenths of the total heat flux for the TAG vent field on the Mid-Atlantic Ridge, and Baker et al. (1993) who suggest a figure of c. 70% for the north Cleft vent field on the Juan de Fuca Ridge.

Diffuse flow may represent hot hydrothermal fluid which has been cooled conductively or may be produced when cold seawater mixes with hot black smoker fluid below the seafloor. This question can be addressed by looking at the chemistry of the end-member fluid. Evidence that diffuse flow may be the result of cold water mixing with hotter fluid comes from the Galapagos Spreading Centre at 86°W in the East Pacific. Here there are several vent fields spread out along the mid-ocean ridge axis, but all flow is of the diffuse low-temperature type. The vent fields range in diameter from 30 to 100 m and the diffuse flow can be seen emanating from many widespread sources, such as from between pillow basalts and in one case from between talus blocks (Corliss et al. 1979). Chemical analyses carried out on these fluids suggest that they are the result of cold seawater mixing with 300°C hydrothermal fluids below the seafloor, (Edmond et al. 1979).

Why are the Galapagos vent fields uniformly discharging diffuse flow? Does this represent a waning stage of flow as the system cools after high-temperature venting? Or is it an early stage in the history of hydrothermal circulation, with black smoker flow still to develop? Why do vent fields from other areas show such a variety of exit temperatures in a small area? It is clear that permeability plays an important part in controlling the hydrothermal circulation (Lowell 1991), but to what extent can it affect the type of venting observed, whether low temperature diffuse flow or high-temperature black smokers?

As an attempt to answer some of these questions we set up a simple model to investigate the genesis of diffuse flow.

Pipe models of convecting systems

Convecting hydrothermal systems can be modelled in a number of ways, and the whole field is well discussed in Lowell (1991). We choose the type of model known as a pipe model. In this type of model, the downflow (recharge), cross-flow (heater) and upflow (discharge) limbs of the system are modelled as pipes filled with porous material through which water flows. A good physical parallel to such a model is a gravity-fed domestic hot water system (a system with no pump) in which hot water convects by rising in one pipe from the furnace, losing heat in the radiators, and sinking in another pipe to return to the furnace. Another way of viewing pipe models is as a special case of porous medium flow, in which an extremely heterogeneous distribution of permeability restricts flow to pipe-like zones (Lowell 1991).

A pipe model does have limitations. In particular, it does not allow for any exchange of fluid between the discharge and recharge limbs except where specified by the pipes, whereas such exchange can happen in natural systems at any point where it is driven by local pressure gradients. However, it does have the advantage that it is simpler and more transparent in use than, say, large two-dimensional numerical models.

Experience in their use (Bodvarsson 1950; Lowell & Burnell 1991; Cann et al. 1985; Cann & Strens 1989) shows that, though pipe models cannot show the full complexity of natural systems and cannot be used to describe the flow in a given system, they are, however, very useful in showing the general and fundamental features of the dynamics of hydrothermal systems and it is in this sense that we use one here. For complex systems, a given run of the model may apply to only part of the system, perhaps to one flow path through it, while other flow paths may require adjustment of model parameters.

In a pipe model the driving force to the circulation is the difference in pressure between the base of the cold downflow pipe and the base of the hot upflow pipe. This pressure difference is called the buoyancy pressure. At steady-state flow, the buoyancy pressure is balanced by the flow resistance pressure generated by the flow through the pipe network. The directions and magnitudes of flow in a system of pipes is governed by the pressure gradients and permeability structure (Cann & Strens 1989). For horizontal pipes the driving pressure gradient is simply the difference in pressures at the ends divided by the length of the pipe. However, for vertical pipes the hydrostatic pressure has to be considered as well. Darcy's law can be used to express the non-hydrostatic, or resistance, pressures (Phillips 1991). This states that the volume flow rate

$$Q = \frac{kA}{\mu} \frac{dP}{dz} \qquad (1)$$

where k is the permeability and A the cross-sectional area of the pipe, μ is the viscosity of the fluid. These three terms can be combined as a resistance term $r = \mu/kA$.

Consider a vertical pipe with pressures P_1 at the bottom and P_0 at the top. If the fluid is rising

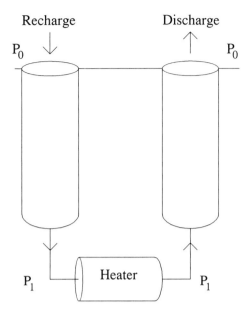

Recharge　　　　Discharge

$P_0 \downarrow$　　　　$\uparrow$　P_0

P_1　　Heater　　P_1

Fig. 1. The simplest convecting pipe model consists of two vertical pipes connected by a non-resistive heater. The system is recharged from, and discharges into, a large well-mixed cold ocean above. P_1 and P_0 are the pressures at the bottom and top of the pipes, respectively. Flow in the model is controlled by the resistance to flow in both limbs and the nature of the heater. At the limit, two contrasting states are possible: (**a**) the resistance in the discharge is much greater than in the recharge and pressure in the system is everywhere cold hydrostatic (discharge-dominated); or (**b**) the resistance in the recharge is much greater than in the discharge and pressure in the system is everywhere hot hydrostatic (recharge-dominated). See text for detailed discussion.

the pressure gradient becomes the sum of the hydrostatic and buoyancy forces

$$(P_1 - P_0) = \rho_{up} g z + \frac{Q\mu_{up}z}{k_{up}A_{up}} = \rho_{up}gz + Qr_{up}z \quad (2)$$

and similarly, is the difference of these for a downflow pipe

$$(P_1 - P_0) = \rho_{down}gz - Qr_{down}z \quad (3)$$

If these pipes are then connected by a non-resistive heater unit, Fig. 1, so that cold dense water descends one pipe, is heated and hot less dense water rises in the other one, the unknown pressures P_1 and P_0 can be eliminated so that

$$(\rho_{down} - \rho_{up})gz = Q(r_{up} + r_{down})z \quad (4)$$

This equation can be simplified to one of two end-members in the followingway. If the resistance to flow in the upflow pipe is much greater

than that in the downflow (discharge-dominated system), then the equation becomes

$$(\rho_{down} - \rho_{up})gz = Qr_{up}z \quad (5)$$

Substituting this back into the pressure equation for the upflow (2) gives

$$(P_1 - P_0) = \rho_{down}gz \quad (6)$$

In such a discharge-dominated system, the pressure everywhere is close to cold (recharge) hydrostatic pressure, even in the hot discharge limb.

Similarly, if all the resistance to flow is in the downflow, or recharge, then Equation (4) becomes

$$(\rho_{down} - \rho_{up})gz = Qr_{down}z \quad (7)$$

and substituting this into the pressure equation for the downflow (3) gives

$$(P_1 - P_0) = \rho_{up}gz \quad (8)$$

Hence the pressure gradient in a recharge-dominated system is approximately hot hydrostatic everywhere, even in the cold recharge.

It is counter-intuitive that, in a discharge-dominated system, the pressure in a pipe of rising hot fluid is cold hydrostatic, yet this is probably the normal state of black smoker systems. The fact that black smokers emit hot undiluted hydrothermal fluids implies that cold seawater is often not able to enter the conduits, and hence that the pressure in the hot fluid must be close to cold hydrostatic and the system a discharge-dominated one. Our model is based on this conclusion.

The model

The simple pipe model used is shown in Fig. 2. The hydrothermal circulation is divided into five zones or limbs which are each represented by different pipes. The resistances of each of these pipes can be defined separately, where the resistance is a function of the area and permeability of the pipe. In the model all the limbs are assumed to be isothermal and there is assumed to be no resistance to flow in either the heater limb or the connections between pipes. If these features were included, they would complicate the algebra without changing the essential features of the results. Cold, ambient seawater is assumed to enter pipe a (the recharge zone) and pipe d (which represents cold seawater percolating down to mix with the hot hydrothermal fluids without being heated by the magmatic heat source). Pipes b and c can be thought of as the discharge zone, with pipe b carrying the hot undiluted hydrothermal fluid and pipe c carrying the cooler fluid resulting

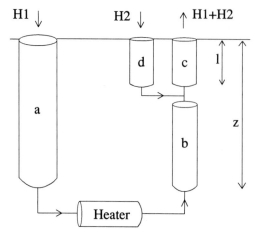

H1 ↓   H2 ↓   ↑ H1+H2

Fig. 2. Convecting pipe model used in this paper, which allows subsurface mixing of cold seawater with hot upflow water. Limb a is the main recharge to the heater, which we consider as having low or, eventually, negligible resistance to flow compared with the total resistance to flow in the other limbs combined. The heater is taken to have no resistance, including resistance to flow here complicates the algebra without changing the physics. H_1 and H_2 are flows in the two loops that compose the model: H_1 is positive in all calculations in the direction shown, but H_2 may reverse and become negative so that hot water flows up the d limb. Depths below the seafloor are l and z. See text for detailed discussion.

from the mixing of the fluid in b and the cold water entering from d.

Such a simple model clearly cannot represent the full complexity of natural hydrothermal systems containing many vents emitting fluid of different temperatures. Instead, we envisage it as representing one of the many flow paths within a complex system. Each vent emitting water at a given temperature is linked with a common source of hot fluid and with its own source of cold, diluting water. Each vent would thus share a common a limb (recharge limb) and b limb (lower discharge limb) with the others, but would have its own c limb (upper discharge limb) and d limb (source of cold water), with a particular resistance related to each. The different combinations of c and d limbs would account for the variety of fluid vent temperatures distributed across a single vent field.

Equations for the flow in each of the pipes can be written down as discussed in the previous section. By visualizing the flow as occurring in two loops through the model, the unknown pressures at the ends of the pipes can be eliminated, giving the two following flow equations

$$\rho_a g z - \rho_c g l - \rho_b g (z - l)$$
$$= \frac{H_1 z r_a}{\rho_a} + \frac{(H_1 + H_2) l r_c}{\rho_c} + \frac{H_1 (z - l) r_b}{\rho_b} \quad (9)$$

$$\rho_d g l - \rho_c g l = \frac{H_2 l r_d}{\rho_d} + \frac{(H_1 + H_2) l r_c}{\rho_c} \quad (10)$$

where the resistance, $r_n = \mu / k_n a_n$, assuming μ, the viscosity, is the same in all limbs. As cold water is assumed to enter pipe d then $\rho_d = \rho_a$ and equation (10) simplifies slightly.

A mixing equation can be derived relating the flows and temperatures at the junction of pipes b, c and d using conservation of energy above 0°C.

$$H_1 T_b = (H_1 + H_2) T_c \quad (11)$$

This equation assumes that the specific heat capacity integrated over the range $0–T_b(°C)$ is independent of temperature for $T_b < 350°C$ and the calculations of Cann et al. (1985) show that this is a reasonable assumption.

The equations for this model were developed assuming a particular flow regime, i.e. that water enters pipes a and d and comes out of pipe c. Referring to Fig. 2 this is the case with both H_1 and H_2, the mass flow rates, positive. These are the directions of flow that are observed to be happening in nature. The flow equations remain the same regardless of the flow directions, but the mixing equation would be affected. This is discussed in a later section.

In addition to the flow equations there is an equation for the heat exchange taking place in the heater limb. The heater limb is assumed to be of length l_h with constant cross-sectional area a_h and porosity ϕ. A general advection–diffusion equation for the temperature field in a fluid can be written as

$$\rho c_p \left\{ \frac{\partial T}{\partial t} + \boldsymbol{u} \cdot \nabla T \right\} = k \nabla^2 T + J \quad (12)$$

where k is the thermal conductivity of the fluid, c_p its specific heat capacity and J is the rate of heat generation per unit volume (Tritton 1990).

The relative importance of the advection of heat to the conduction (diffusion) of heat is given by the Péclet number

$$Pe = \frac{UL}{\kappa} = \frac{UL \rho c_p}{k} \quad (13)$$

κ being the thermal diffusivity).

In this instance the Péclet number is very high ($\approx 10^{10}$) and the diffusion term in Equation (12) can be neglected. For the one-dimensional case, the equation becomes

$$\rho c_p \left\{ \frac{\partial T}{\partial t} + u \frac{\partial T}{\partial x} \right\} = J \quad (14)$$

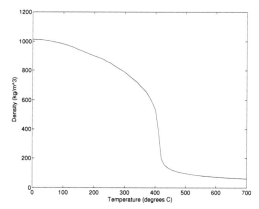

Fig. 3. Variation of fluid density with temperature used to construct a look-up table for the model. This curve is taken from Cann *et al.* (1985) and was constructed for seafloor pressures by applying corresponding state theory to values experimentally derived at other pressures.

The power supplied by the heat source can be written as

$$W = \frac{w_0(T_m - T)}{T_m} \qquad (15)$$

which is Newtonian heating with w_0 being a constant of proportionality. This form of W ensures that the temperature of the fluid cannot exceed that of the heat source, T_m.

So the power per unit volume is

$$J = \frac{w_0(T_m - T)/T_m}{\phi l_h a_h} \qquad (16)$$

Also $u = H_1/(\phi a_h \rho)$ and Equation (14) becomes

$$\frac{\partial T}{\partial t} + \frac{H_1}{\phi a_h \rho} \frac{\partial T}{\partial x} = \frac{w_0(T_m - T)/T_m}{\phi l_h a_h \rho c_p} \qquad (17)$$

Putting $T_b = T(l_h, t)$ gives

$$\frac{dT_b}{dt} = \frac{w_0(T_m - T_b)/T_m - H_1 l_h c_p \frac{\partial T}{\partial x}\big|_{x = l_h}}{\phi l_h a_h \rho c_p} \qquad (18)$$

Assuming the gradient at $x = l_h$ is equal to the average gradient gives

$$\frac{dT_b}{dt} = \frac{w_0(1 - T_b/T_m) - H_1 c_p T_b}{\phi l_h a_h \rho c_p} \qquad (19)$$

which is the heater equation used in this model. It has the properties that for slow flow rates the temperature of the fluid becomes comparable with that of the heat source and at high flow rates very little heat is added to the fluid.

Writing $\lambda = l/z$, $\rho_a = \rho_0$ and applying the Boussinesq approximation [where variations in all fluid properties other than density are

ignored completely and variations in the density are only considered where they give rise to a gravitational force (Tritton 1990)], Equations (9) and (10) become

$$g\rho_0^2\left[1 - \frac{\rho_c}{\rho_0}\lambda - \frac{\rho_b}{\rho_0}(1 - \lambda)\right]$$
$$= H_1 r_a + (H_1 + H_2)\lambda r_c + H_1(1 - \lambda)r_b \qquad (20)$$

$$g\rho_0^2\left(1 - \frac{\rho_c}{\rho_0}\right) = H_2 r_d + (H_1 + H_2)r_c \qquad (21)$$

To simplify these equations they can be non-dimensionalized using the following relationships, where the prime refers to the dimensionless quantities

$$T_n' = \frac{T_n}{T_m} \qquad H_n' = \frac{r_a H_n}{g\rho_0^2}$$

$$\delta_n' = \frac{\rho_n}{\rho_0} \qquad t' = \frac{w_0 t}{T_m l_h a_h c_p \rho_0 \phi}$$

The flow, heater and mixing equations become

$$1 - \delta_c'\lambda - \delta_b'(1 - \lambda)$$
$$= H_1'[1 + \lambda R_c + (1 - \lambda)R_b] + \lambda R_c H_2' \qquad (22)$$

$$1 - \delta_c' = H_2'(R_d + R_c) + R_c H_1' \qquad (23)$$

$$\frac{dT_b'}{dt'} = 1 - T_b' - \gamma H_1' T_b' \qquad (24)$$

$$H_1' T_b' = (H_1' + H_2')T_c' \qquad (25)$$

where

$$\gamma = \frac{g\rho_0^2 c_p T_m}{r_a w_0}$$

and $R_n = r_n/r_a$.

In addition to Equations (22)–(25) a look-up table is used which provides a non-linear relationship between density and temperature, (Cann *et al.* 1985) and is shown graphically in Fig. 3. By specifying γ, R_b, R_c and R_d the system can be solved for the exit temperature, T_c, and mass flow rates H_1 and H_2. In particular, here we are interested in the exit temperature, T_c, and how this varies with changing the resistances in the pipes. The results are investigated over a range of γ as this controls the amount of heat allowed into the system.

The equations are solved using SIMULINK, which is a dynamic system simulation package, part of the MATLAB system. For all the parameter ranges tried, the solutions converge to a stable equilibrium and hence transient behaviour is not discussed.

Estimation of model parameters

Estimates of the cross-sectional areas of these limbs and the permeability of the recharge can be made from studies of both modern and fossil hydrothermal systems. A good idea of the lateral extent of these zones can be gained by assuming that convection cells are equidimensional. This would imply that the lateral scale is twice the depth of circulation. The depth of circulation can be estimated from both geochemical calculations (e.g. Von Damm & Bischoff 1987) and from field observations in ophiolites (Richardson et al. 1987), both of which concur at depths of about 2 km below the sea floor. Seismic profiling methods give good estimates of the depth to the magma chamber, which then give a lower bound for the depth of circulation, if it is assumed that water cannot penetrate deeper than the magma. On the East Pacific Rise, seismic experiments have constrained the depth to the top of the magma chamber to between 1.2 and 2.4 km below the seafloor, with an average depth of about 1.5 km (Detrick et al. 1987). The depth has been observed to vary systematically with spreading rate, with, for example, the depth being about 3 km at the Valu Fa Ridge (Collier & Sinha 1990).

The discharge zone is assumed to be confined to narrow conduits so that little heat is lost by conduction to the overlying rocks during ascent, contrasting with the large recharge zone (Bischoff 1980). There is little direct evidence as to where the recharge actually occurs (Strens & Cann 1982), but it is thought to be channelled by areas of high permeability such as in faults (Strens & Cann 1986). The regions of discharge, however, can be mapped in ophiolite complexes such as Troodos, Cyprus and estimates gained of their dimensions. The sulphide deposits in Troodos are underlain by mineralized stockwork zones and then by vertical alteration pipes (Constantinou & Govett 1973; Richards et al 1989). These represent the channels of ascending hydrothermal solutions, the discharge limb in the model. These alteration pipes, such as at Pitharokhoma, show concentric patterns of mineral zonation and are 50–200 m in diameter (Richards 1987).

The hardest variable to constrain is the permeability of the recharge zone. Estimates for this can be made from measurements taken in, for example, Hole 395A and Hole 504B of the Ocean Drilling Program. The first of these gives an estimate of 10^{-13} m^2 (Becker 1990). In Hole 504B in situ permeabilities measured within the upper kilometre are of the order 10^{-13}–10^{-14} m^2, (Anderson et al. 1985). Below this, bulk permeabilities are much lower, of the order of 10^{-17} m^2. These measurements are all in crust that is older than that at the spreading axis where vigorous hydrothermal circulation is taking place. Anderson et al. (1985) suggest that at 504B, to have achieved the observed mineral alterations, higher permeabilities must have been required in the past. Hence an estimate of the permeability for the recharge zone may be an order of magnitude or so higher than the 10^{-13} m^2 measured in these holes. These values are for the bulk permeability and do not take into account heterogeneities in the various limbs.

A range of values for gamma can be estimated by returning to the heater equation (19) which at equilibrium gives

$$w_0(-T_b/T_m) = H_1 c_p T_b \qquad (26)$$

If we assume that a black smoker system is at steady state and can be modelled using Equation (19) for the heat source, then values of H_1 and T_b can be taken from observations on the seafloor, which in turn gives an estimate for w_0, and hence γ.

So if $H_1 \approx 100$–500 kg s^{-1} and $T_b \approx 350°C$ for $T_m = 500°C$ (e.g. MacDonald et al. 1980; Converse et al. 1984), $c_p = 4000$ J kg^{-1} then $w_0 \approx 4 \times 10^8$–$4 \times 10^9$ kg m^2 s^{-3}.

Hence if the upper discharge permeability is taken as 10^{-9} m^2, its area as 10^3–10^4 m^2, then a range of reasonable values for γ is approximately 5–500.

Results of the full model

It is interesting to examine the system qualitatively to begin with. If the system is assumed to be discharge-dominated, where all the resistance is in pipes b and c, then for cold seawater to enter pipe d the pressure at the top of b has to be less than cold hydrostatic. One way of achieving this is by having a high resistance to flow in b. The resistance is inversely proportional to the permeability of the pipe times the cross-sectional area. Hence a high resistance to flow can be achieved by either having a narrow pipe or a low permeability, or both. Hence if there is a high resistance in b, cold seawater may be entrained in the hot hydrothermal fluid and cool temperatures are observed at the surface. Alternatively, if the resistance to flow was low throughout the system, flow rates would increase, less heat would be input to the system and lower temperatures observed. However, the observed exit velocities for diffuse flow tend to be very low (0.002–0.1 m s^{-1}), favouring the first explanation.

If the permeability in b was suddenly increased by, for example, fracturing then the resistance to flow would be decreased, less water would enter d and hotter temperatures would be

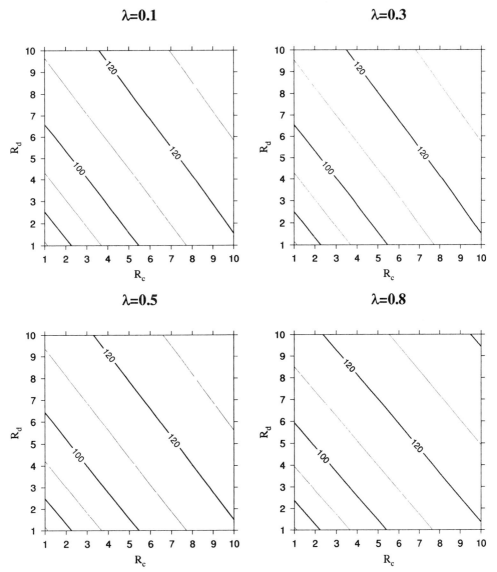

Fig. 4. Contours of T_c, the discharge temperature from limb c, in degrees Celsius, for four different values of λ, the ratio of the depth of the mixing point to the depth to the heater. Here $\gamma = 30$ and $R_b = 1000$. Even a large variation in λ has little effect on the discharge temperature.

observed. In the upper limbs of the discharge, c and d, greater flow resistances favour higher temperatures.

This qualitative argument suggests that it is the resistances in the pipes, the permeability structure and/or areas, which control the surface observations.

Apart from the ratios of resistances, the only dimensionless parameters for this model are λ and γ. In all of the results shown later in this

paper, λ, the ratio of the depth of the mixing point to the depth of total circulation, has been set to 0.1. So if the total depth of circulation is taken as 1000 m, cold seawater is mixing with the hot hydrothermal fluids at about 100 m below the seafloor. Varying λ does not have a large effect on the results; this is shown in Fig. 4. The graphs show how the discharge temperature T_c, in degrees Centigrade, varies with the resistances in b, c and d (R_b, R_c and R_d respectively). The

Table 1. *Effects of* γ, R_b, R_c *and* R_d *on the discharge temperature* T_c *(see text for explanation)*

				T_c (°C)	
R_b	R_c	R_d	γ = 400	γ = 70	γ = 30
1	1–10	1–10	80–140	130–260	170–330
10	1–10	1–10	80–140	130–240	160–290
100	1–10	1–10	60–140	130–200	130–250
1000	1–10	1–10	40–120	70–135	70–140
10,000	1–10	1–10		45–65	45–65
100–1000	1	1–20	60–140	70–195	60–225
1–1000	1	1–20	60–130		70–240
100–1000	10	1–20	100–160	120–225	120–265
100–1000	1–20	1	60–150	70–225	70–280
100–1000	1–20	10	100–160	110–245	120–290

resistances are in their non-dimensional form as the ratio of the resistance to the recharge resistance.

The parameter γ has a larger effect on the results as it controls the amount of heat input into the system. Table 1 shows the effect of changing γ for a range of permeability structures for the model. Decreasing γ results in much higher temperatures in the system as γ is inversely proportional to w_0, and the higher w_0 the more heat is supplied to the fluid.

R_b has been shown in the qualitative discussion above to be potentially important in controlling the temperature of the discharge. Figure 5 shows the effect of changing R_b at γ = 30 for a range of values of R_c and R_d. This highlights the importance of R_b in the model, with a large value for R_b bringing the discharge temperature well below 100°C.

To achieve temperatures lower than 100°C R_c and R_d both have to be low (less than 10). Increasing either R_c or R_d will increase the exit temperature T_c, but increasing both simultaneously is more effective. The value of R_b for which discharge temperatures are less than 100°C depends on γ, but in general raising R_b decreases the temperature. So, for example, if γ = 400, R_b has to be greater than 100 to ensure low temperatures.

The model values can be converted into the properties of a natural system. If the permeability of the recharge is taken as 10^{-12} m^2 and the cross-sectional area as 4×10^6 m^2 then $r_a = 250$ kg m^{-5} s^{-1}, assuming μ $= 10^{-3}$ kg m^{-1} s^{-1}. This implies that for $R_b > 100$, the product $k_b a_b$ has to be $< 4 \times 10^{-8}$ m^4. So if the cross-sectional area for the lower discharge zone is taken as 1000 m^2 the permeability has to be smaller than 4×10^{-11} m^2 for low temperatures in the discharge.

Similarly, if R_c and R_d have to be less than 10, this means that the product of their areas and permeabilities has to be greater than 4×10^{-7} m^4. Thus if the area of c, the upper discharge, is 1500 m^2 then its permeability has to be greater than 2.67×10^{-10} m^2; and if the area of d is 4×10^4 m^2 its permeability has to be greater than 10^{-11} m^2. Such permeabilities are higher than those measured within the ocean crust to date but, as indicated above, measurements have all been made so far in crust that has become partly sealed by earlier circulation.

Negligible resistance in the recharge limb

If a discharge-dominated system is being considered, a simplification can be made by assuming that the resistance in the recharge limb, r_a, is negligible compared with the other resistances. This leaves only two parameters in the problem if r_b and r_d are now ratioed to r_c, so that R_{BC} and R_{DC}, respectively, are r_b/r_c and r_d/r_c. The resistance r_a only appears in the first flow equation (9) in which it can be neglected. The equations can then be non-dimensionalized using $H'_n = r_c H_n / g \rho_0^2$ giving

$$1 - \delta'_c \lambda - \delta'_b (1 - \lambda)$$
$$= H'_1 [\lambda + (1 - \lambda) R_{BC}] + H'_2 \lambda \quad (27)$$

$$1 - \delta'_c = H'_1 + H'_2 (1 + R_{DC}) \quad (28)$$

$$\frac{dT'_b}{dt'} = 1 - T'_b - \gamma H'_1 T'_b \quad (29)$$

where

$$\gamma = \frac{g \rho_0^2 c_p T_m}{r_c w_0}$$

These equations, together with the mixing equation and the non-linear density-temperature

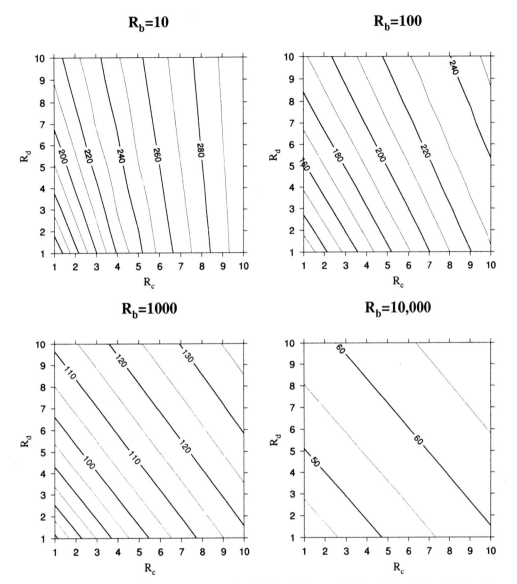

Fig. 5. Contours of T_c, the discharge temperature from limb c, in degrees Celsius, for four different values of R_b, the ratio of resistance in the lower part of the discharge to the resistance in the recharge limb. Here $\gamma = 30$ and $\lambda = 0.1$. High values of R_b, coupled with low values of R_c and R_d, result in low temperature discharge.

relationship, can be used to solve for the exit temperature, T'_c, and the mass flow rates, H'_1, H'_2, as before.

Figure 6 shows a contour map of the discharge temperature T_c against R_{BC} and R_{DC} in the ranges 1–10 000 and 0.1–10 000, respectively. For $R_{DC} \leqslant 10$ and $R_{BC} > 10$, increasing R_{BC} lowers the temperature. Increasing R_{DC} increases the temperature regardless of R_{BC}. For

$R_{DC} > 100$ increasing R_{BC} initially raises the temperature, but for $R_{BC} > 500$ the temperature decreases with further increases of R_{BC}. To accompany these temperature results, Fig. 7 shows a contour map of H_2/H_1, the ratio of the mass flow rate of the cold seawater to that of the circulating hydrothermal fluid. Together these figures show that when the H_2 is high the temperatures are lower, as expected. Figure 7

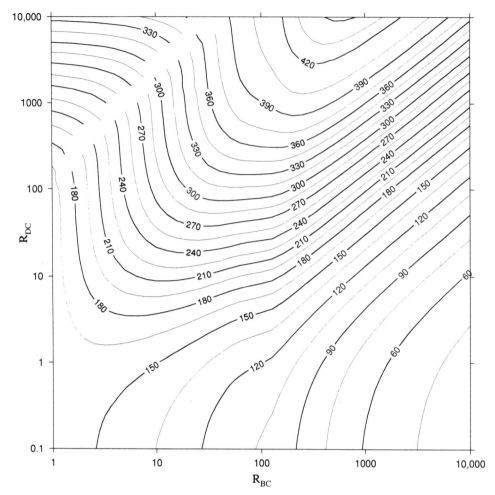

Fig. 6. Contours of T_c, the discharge temperature, for the version of the model with negligible resistance to flow in the recharge limb. R_{BC} and R_{DC} are ratios of resistance to flow in limbs b and d to that in limb c, respectively. The system parameter γ is set at 30.

also shows where the flow in limb d becomes negligible with small R_{BC} and high R_{DC}, i.e. a high resistance in limbs c and d.

For comparison, Figs 8 and 9 show T_c and H_2/H_1 calculated for $\gamma = 400$. As the previous results showed, the effect of increasing γ is clearly shown, with high γ leading to lower temperatures overall.

These results show how the resistances in limbs d and b affect the flow when they are considered as ratios of the resistance in limb c. If the resistance in d is fairly low, less than ten times the resistance in c, then the lower discharge, b, resistance controls the resulting temperatures. If this resistance is increased, the pressure is reduced and allows more cold water to enter d resulting in cooler discharge temperatures. However, if the resistance in d is more

than a hundred times that in limb c, there has to be at least fifty times more resistance in the lower discharge before significant cold water can enter d. This means that until the lower discharge resistance is high enough, increasing it only results in slower flow rates and consequently hotter temperatures in limbs b and c. Once the resistance in b is high enough, enough cold water is once again able to enter limb d and further increases in b reduce the temperature as before.

Other flow patterns

The directions of the flows through the system of pipes depends on the resistances in the pipes and for most cases discussed so far the flows H_1 and H_2 are both positive. This is where cold water enters limb d and exits limb c. However, if the

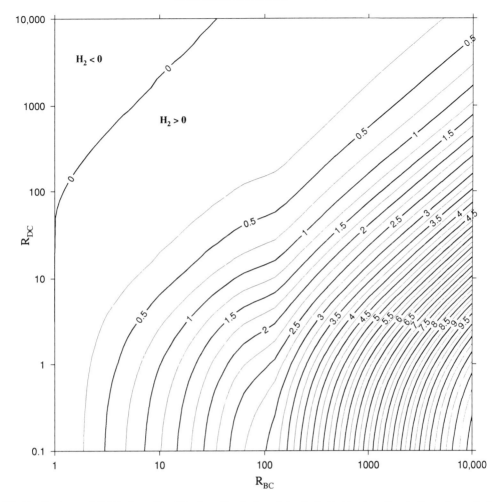

Fig. 7. Contours of H_2/H_1, the ratio of the downflow in the mixing limb d to the downflow in the recharge limb a, for the same conditions as for Fig. 6: the resistance of the recharge limb is negligible and $\gamma = 30$.

resistance in c becomes significantly larger than the resistance in b then the flow H_2 will reverse. An example of where there is flow reversal is illustrated in Fig. 7, which shows that H_2 is zero for high R_{DC} (>100) and lower R_{BC} (<100). Beyond that line H_2 is negative: flow has reversed in the mixing limb. So if the resistance in limb d is high and the resistance in limb b is low, then no cold water penetrates down. Beyond that point, hot water will flow up limb d.

At low resistances, flow reversal does not seem to depend too strongly on the resistance in d. However, once the flow H_2 has reversed, the resistance in d will affect the stability of this reversal.

Once the flow has reversed then heated water will rise through limbs c and d. By analogy with

electric circuits this could be thought of as having two resistors in parallel. Hence the resistance to flow through c and d could be replaced by a single limb, q say, which has a resistance, r_q where

$$\frac{1}{r_q} = \frac{1}{r_c} + \frac{1}{r_d} \qquad (30)$$

By considering the resistance as such it is possible to explore how R_d affects the stability of the reversed flow. If $R_d < R_b$ then $1/r_d > 1/r_b$ and $r_q < r_b$. Having a lower resistance in the upper part of the system implies that the pressure at the top of b would allow cold water to enter pipe c and the original pattern of flow would be obtained with limbs c and d taking opposite roles.

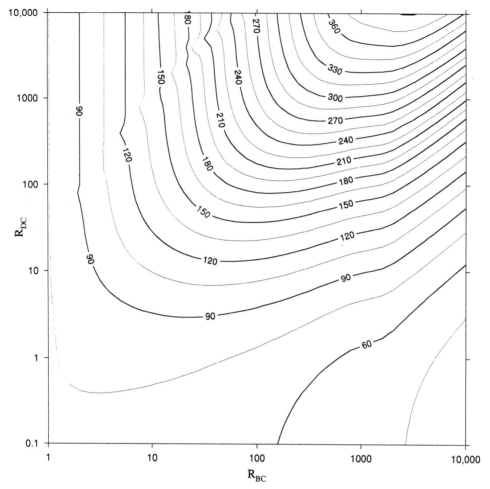

Fig. 8. Contours of T_c, the discharge temperature, for the version of the model with negligible resistance to flow in the recharge limb. Conditions are the same as for Fig. 6 except that $\gamma = 400$. Note that temperatures are lower than in Fig. 6.

If $R_d > R_b$ then the 'stability' of the reversed flow depends on the relative sizes of R_b, R_c and R_d.

If $R_b = n$ and $R_c = m$, where $m > n$ then

$$R_d < \frac{nm}{(m-n)} \Rightarrow r_q < r_b$$

$$R_d > \frac{nm}{(m-n)} \Rightarrow r_q > r_b$$

The case with $r_q < r_b$ is the same as above with $R_d < R_b$. However, if $r_q > r_b$ then fluid will continue to flow up both limbs c and d, and the flow reversal is 'stable'. This emerging fluid is therefore undiluted with cold seawater and emerges at higher temperatures.

The flow rate H_1 does not reverse for any of the resistances discussed here.

Discussion

Although pipe models are very simplified they are powerful tools in understanding complex systems. By using them it is possible to see how physical changes in the system affect the flow rates and temperatures. Observations on the seafloor can be interpreted in the light of these models to lead to insights about the sub-seafloor characteristics of the hydrothermal system.

Low discharge temperatures arise in the model in two distinct parts of Figs 6 and 8. The low temperatures in the lower right-hand part of the figures corresponds to high values of the flow ratio H_2/H_1. Hot fluid is being diluted by copious flow of entrained cold water. This happens when $R_{BC} \gg R_{DC}$, or when the resistance in the b limb

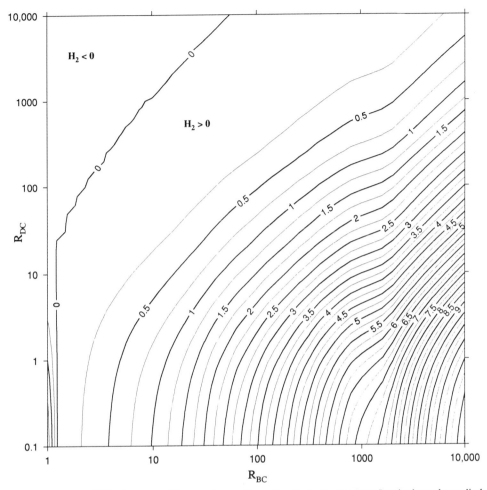

Fig. 9. Contours of H_2/H_1, the ratio of the downflow in the mixing limb d to the downflow in the recharge limb a, for the same conditions as for Fig. 8: the resistance of the recharge limb is negligible and $\gamma = 400$.

is much greater than the resistance in the d limb, as both are ratioed to the resistance in the c limb. Under these conditions, as predicted qualitatively above, the high resistance to flow in the lower part of the discharge reduces the pressure at the point where cold water can enter and allows it to dilute the rising hot water extensively. Such a distribution of resistance, with a low permeability layer underlying a high permeability layer in the discharge, is typical of normal oceanic crust, with more permeable lavas over less permeable sheeted dykes. It seems likely then that low temperature discharge at the seafloor will be the result of early stages in the evolution of a hydrothermal system, before precipitation of hydrothermal minerals can have altered the original permeability structure. Low temperature discharge

would also be expected for the same reasons at the margins of older active systems as the upflow expands outwards into the unaltered rocks around.

The other area of low temperatures in Figs 6 and 8 is along the left-hand edge, where R_{BC} is low. This is a region where H_2/H_1 is also low, so that the low temperatures here reflect the low temperature of fluid in the b limb. These low temperatures arise from the overall low resistance to flow in the system and the rapid flow of water through the heater. At such low resistances the pattern of flow is unstable and flow reversal can occur, as discussed above. The low temperatures in this region of Figs 6 and 8 are a property of the whole hydrothermal system and in systems of this type high-temperature flow cannot occur. Natural systems with evidence of high-temperature flow cannot lie in this part of parameter space.

High temperatures in the model are confined to regions where R_{DC} is high and r_d about ten times greater than r_b. Both of these factors act to exclude cold water from the hot flow. One trivial example that illustrates this is black smoker flow through a pipe, the walls of which are coated with an impermeable layer of, say, quartz and sulphides. The impermeable layer forms part of the d limb in the formal model, giving a high r_d and thus allowing the hot water to remain undiluted. On a larger scale, precipitation of hydrothermal minerals in cracks in the upper parts of the flow increases r_d and therefore increases the temperature of the fluid emerging.

Although these are only the steady-state solutions, this model can be used to investigate the evolution of a system as permeability changes, provided changes happen slowly enough to allow the system to be considered as a series of successive steady states. Permeability may decrease with time because of the precipitation of hydrothermal minerals, especially in limbs d and c where mixing of fluids takes place. A converse increase in permeability may come from a fissuring event or from a hydrothermal brecciation event [as suggested by Cann & Strens (1989) in connection with megaplume genesis]. The system parameter γ would decrease if a new influx of magma to the heater region led to an increase in w_0 or if r_c increased because of mineral precipitation. All of these changes can lead to changing styles of venting.

So, for example, if a heat source is introduced into an area of no venting, with rubbly lavas overlying less permeable lower lavas and sheeted dykes, then once the hydrothermal circulation had established itself, the model would predict that the venting would be dominated by diffuse flow. As the upper layer became clogged by the precipitation of sulphides, its resistance would increase, leading to an increase in exit temperatures. Similarly, if the lower discharge region became fissured, this would lead to an increase in the permeability and a decrease in the resistance and, once more, higher temperatures would be observed. Once higher temperature circulation had established itself there is the possibility of hydrothermal brecciation, leading to a decrease in the upper layer resistance and a corresponding decrease in temperature, re-establishing diffuse flow conditions.

We have shown that a simple pipe model can be used to explain the occurrence of diffuse hydrothermal flow and give some insights into the evolution of a vent field.

We acknowledge the RIDGE Theoretical Institute, 1993, for the excellent talks and discussions which were the inspiration for this work. A.R.P. was supported by Natural Environment Research Council studentship ES-/91/PES/4.

References

ANDERSON, R. N., ZOBACK, M. D., HICKMAN, S. H. & NEWMARK, R. L. 1985. Permeability versus depth in the upper oceanic crust: in situ measurements in DSDP Hole 504B, eastern Equatorial Pacific. *Journal of Geophysical Research*, **90**, 3659–3669.

BAKER, E. T., MASSOTH, G. M., WALKER, S. L. & EMBLEY, R. W. 1993. A method for quantitatively estimating diffuse and discrete hydrothermal discharge. *Earth and Planetary Science Letters*, **118**, 235–249.

BECKER, K. 1990. Measurements of the permeability of the upper oceanic crust at Hole 395A, ODP Leg 109. *In*: DETRICK, R., HONNOREZ, J., BRYAN, W. B., JUTEAU, T., *et al.*, *Proceedings of the ODP, Scientific Results*, **106/109**, 213–222.

BISCHOFF, J. L. 1980. Geothermal system at 21°N, East Pacific Rise: physical limits on geothermal fluid and role of adiabatic expansion. *Science*, **207**, 1465–1469.

BODVARSSON, G. 1950. Geophysical methods in prospecting for hot water in Iceland. *Timarit Verkfraedingafelags Islands*, **35**, 49–59. [Translated from the Danish by Mr Grainger, Dept. of Scientific and Industrial Research, Wellington, New Zealand].

CANN, J. R. & STRENS, M. R. 1989. Modeling periodic megaplume emission by black smoker systems. *Journal of Geophysical Research*, **94**, 12227–12237.

——, —— & RICE, A. 1985. A simple magma-driven thermal balance model for the formation of volcanogenic massive sulphides. *Earth and Planetary Science Letters*, **76**, 123–134.

COLLIER, J. & SINHA, M. 1990. Seismic images of a magma chamber beneath the Lau Basin back-arc spreading centre. *Nature*, **346**, 646–648.

CONSTANTINOU, G. & GOVETT, G. J. S. 1973. Geology, geochemistry, and genesis of Cyprus sulfide deposits. *Economic Geology*, **68**, 843–858.

CONVERSE, D. R., HOLLAND, H. D. & EDMOND, J. M. 1984. Flow rates in the axial hot springs of the East Pacific Rise (21°N): implications for the heat budget and the formation of massive sulphide deposits. *Earth and Planetary Science Letters*, **69**, 159–175.

CORLISS, J. B., DYMOND, J. & 9 others. 1979. Submarine thermal springs on the Galapagos rift. *Science*, **203**, 1073–1083.

DETRICK, R. S., BUHL, P., VERA, E., MUTTER, J., ORCUTT, J., MADSEN, J. & BROCKER, T. 1987. Multi-channel seismic imaging of a crustal magma chamber along the East Pacific Rise. *Nature*, **326**, 35–41.

EDMOND, J., MEASURES, C., & 6 others. 1979. Ridge crest hydrothermal activity and the balances of the major and minor elements in the ocean: the

Galapagos data. *Earth and Planetary Science Letters*, **46**, 1–18.

ELDERFIELD, H., MILLS, R. A. & RUDNICKI, M. D. 1993. Geochemical and thermal fluxes, high-temperature venting and diffuse flow from mid-ocean ridge hydrothermal systems: the TAG hydrothermal field, Mid-Atlantic Ridge 26°N. *In*: PRICHARD, H. M., ALABASTER, T., HARRIS, N. B. W. & NEARY, C. R. (eds) *Magmatic Processes and Plate Tectonics*. Geological Society, London, Special Publication, **76**, 295–307.

LISTER, C. R. B. 1974. On the penetration of water into hot rock. *The Geophysical Journal of the Royal Astronomical Society*, **26**, 515–535.

LOWELL, R. P. 1991. Modelling continental and submarine hydrothermal systems. *Reviews of Geophysics*, **29**, 457–476.

——— & BURNELL, D. K. 1991. Mathematical modelling of conductive heat transfer from a freezing, convecting magma chamber to a single-pass hydrothermal system: implications for seafloor black smokers. *Earth and Planetary Science Letters*, **104**, 59–69.

——— & RONA, P. A. 1985. Hydrothermal models for the generation of massive sulfide ore deposits. *Journal of Geophysical Research*, **90**, 8769–8783.

MACDONALD, K. C., BECKER, K., SPIESS, F. N. & BALLARD, R. D. 1980. Hydrothermal heat flux of the 'black smoker' vents on the East Pacific Rise. *Earth and Planetary Science Letters*, **48**, 1–7.

PHILLIPS, O. M. 1991. *Flow and Reactions in Permeable Rocks*. Cambridge University Press, Cambridge, 25–29.

RICHARDS, H. G. 1987. *Petrology and geochemistry of hydrothermal alteration pipes in the Troodos ophiolite, Cyprus*. PhD Thesis, University of Newcastle-upon-Tyne.

———, CANN, J. R. & JENSENIUS, J. 1989. Mineralogical zonation and metasomatism of the alteration pipes of Cyprus sulfide deposits. *Economic Geology*, **84**, 91–115.

RICHARDSON, C. J., CANN, J. R., RICHARDS, H. G. & COWAN, J. G. 1987. Metal-depleted root zones of the Troodos ore-forming hydrothermal systems, Cyprus. *Earth and Planetary Science Letters*, **84**, 243–253.

RUDNICKI, M. D. & ELDERFIELD, H. 1992. Theory applied to the Mid-Atlantic Ridge hydrothermal plumes: the finite difference approach. *Journal of Volcanology and Geothermal Research*, **50**, 161–172.

SCHULTZ, A., DELANEY, J. R. & McDUFF, R. E. 1992. On the partitioning of heat flux between diffuse and point source seafloor venting. *Journal of Geophysical Research*, **97**, 12 299–12 314.

SPEER, K. G. & RONA, P. 1989. A model of an Atlantic and Pacific hydrothermal plume. *Journal of Geophysical Research*, **94**, 6123–6220.

STRENS, M. R. & CANN, J. R. 1982. A model of hydrothermal circulation in fault zones at mid-ocean ridge crests. *Geophysical Journal of the Royal Astronomical Society*, **71**, 225–240.

——— & ——— 1986. A fracture loop thermal balance model of black smoker circulation. *Tectonophysics*, **122**, 307–324.

TIVEY, M. K. & McDUFF, R. E. 1990. Mineral precipitation in the walls of black smoker chimneys: a quantitative model of transport and chemical reaction. *Journal of Geophysical Research*, **95**, 12 617–12 637.

TRITTON, D. J. 1990. *Physical Fluid Dynamics*. Oxford University Press, Oxford, 162–196.

TRIVETT, D. A. 1994. Effluent from diffuse hydrothermal venting, part I: a simple model of plumes from diffuse hydrothermal sources. *Journal of Geophysical Research*, **99**, 18403–18415.

——— & WILLIAMS III, A. J. 1994. Effluent from diffuse hydrothermal venting, part II: measurement of plumes from diffuse hydrothermal vents at the southern Juan de Fuca Ridge. *Journal of Geophysical Research*, **99**, 18417–18432.

VON DAMM, K. L. & BISCHOFF, J. L. 1987. Chemistry of hydrothermal solutions from the southern Juan de Fuca Ridge. *Journal of Geophysical Research*, **92**, 11 334–114346.

Mineralogy and sulphur isotope geochemistry of the Broken Spur sulphides, 29°N, Mid-Atlantic Ridge

ROWENA C. DUCKWORTH[1,5], RICHARD KNOTT[1],
ANTHONY E. FALLICK[2], DAVID RICKARD[1], BRAMLEY J. MURTON[3],
& CINDY VAN DOVER[4]

[1]*Department of Earth Sciences, PO. Box 914, University of Wales, Cardiff CF1 3YE, UK*
[2]*Isotope Geoscience Unit, Scottish Universities Research and Reactor Centre,
East Kilbride, Glasgow G75 0QU, UK*
[3]*Institute of Oceanographic Sciences, Deacon Laboratory, Wormley, Surrey GU8 5UB,
UK*
[4]*Woods Hole Oceanographic Institution, Woods Hole, MA 02543, USA*
[5]*Present address: National Key Centre in Economic Geology, Department of Earth
Sciences, James Cook University, Townsville 4811, Queensland, Australia*

Abstract: Massive sulphide–sulphate–oxides from the newly discovered Broken Spur hydrothermal field at 29°10′N on the slow-spreading Mid-Atlantic Ridge have been examined to determine their mineralogical, chemical and sulphur isotope characteristics. The chimney and spire samples are dominated by chalcopyrite–isocubanite–pyrrhotite–anhydrite, with the uppermost active areas composed of anhydrite-rich 'beehive' structures which act as diffusers for the hydrothermal fluids and scaffolding for the chimney construction. Delicate 'organ-pipe' chimney tips grow on the beehive structures. Samples from mounds and the base of active edifices are more enriched in Fe disulphides and sphalerite, with significant amounts of Fe oxyhydroxides and amorphous silica. Sulphur isotope values of sulphides range from −0.8 to 2.4‰, similar to those reported from the Snakepit hydrothermal vent field. This range is significantly lower than the range reported from the fast-spreading East Pacific Rise, and this may be a function of fluid flow paths in differing hydrothermal settings.

The Broken Spur hydrothermal field was discovered on 4 March 1993 by the RRS *Charles Darwin* cruise 76 (Murton *et al.* 1993). It is located at 29°10′N on the slow-spreading Mid-Atlantic Ridge (Fig. 1). The area was further investigated during RRS *Charles Darwin* cruise 77 (March–April 1993) (Elderfield *et al.* 1993). *Alvin* dives 2624 and 2625 in June 1993 (Murton & Van Dover 1993) sampled the massive sulphides and hydrothermal sediments described here, as well as vent fluids and biota.

The Broken Spur hydrothermal field is located in the axial graben of the neovolcanic ridge of the Mid-Atlantic Ridge and extends over an area of 150 m by 60 m. This volcanic graben structure is approximately 100 m wide and 30 m deep, and active vent sites lie along an east–west traverse crossing the northeast–southwest trending graben at 3050–3091 m water depth. The detailed geology of the area is described in Murton *et al.* (this volume). During the two *Alvin* dives, five sites were investigated (Fig. 2). At Sites 1 and 2, low temperature diffuse

venting is indicated by the presence of fauna and video images of shimmering water. Site 3 contains 'The Spire', a spectacular 18 m high, and 1 m diameter black smoker edifice (before sampling), with a 'beehive' structure on the top, vigorously belching black smoke from a central orifice. At Site 4, 'Saracen's Head' is a large sulphide mound approximately 37 m high and 20 m in diameter, with active chimney and beehive structures growing from the top. The 'Wasp's Nest', at Site 5 is another active vent, 15 m tall and 10 m in diameter, with chimneys and large conical beehive structures. Measurements taken with a temperature probe at active vents indicate fluid temperatures between 357 and 366°. Active and inactive sulphide structures are surrounded by sulphide rubble and oxide sediments.

Analytical procedures

The Broken Spur samples were studied using transmitted and reflected light microscopy,

From PARSON, L. M., WALKER, C. L. & DIXON, D. R. (eds), 1995, *Hydrothermal Vents and Processes*,
Geological Society Special Publication No. 87, 175–189.

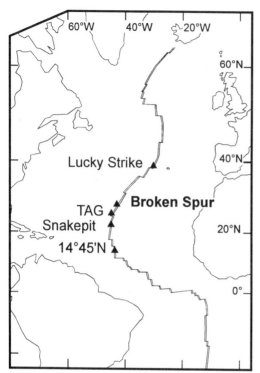

Fig. 1. Location of the Broken Spur hydrothermal field (29°10′N) and other sites described on the Mid-Atlantic Ridge. Lucky Strike (37°18′N; Langmuir *et al.* 1993); TAG (26°08′N; Rona *et al.* 1993); Snakepit (23°22′N; Fouquet *et al.* 1993); 14°45′N (Batuyev *et al.* 1994).

scanning electron microscopy (SEM), electron microprobe (EDS), X-ray diffractometry (XRD) and sulphur isotope analysis. Electron probe microanalysis was made by analytical SEM with a Link AN10000 EDS system, at 20 kV and 1.0 nA probe current. The XRD analyses were performed on freeze-dried, randomly oriented powders using a Phillips PW1820 powder diffractometer and CoKα radiation.

For the sulphur isotopic study, concentrated sulphide and sulphate samples were prepared by precision drilling and hand-picking under a binocular microscope. Pyrite and chalcopyrite samples were further purified using a mixture of removing acid-soluble phases using the method of Hall *et al.* (1988). Samples were thermally decomposed with a mixture of Cu_2O at 1076°C (Robinson & Kusakabe 1975) and the extracted SO_2 was cryogenically purified and analysed on a SIRA II dual-inlet mass spectrometer. The results are given in standard $\delta^{34}S$ notation relative to the Cañon Diablo troilite (CDT) standard. The precision of the data is ±0.2‰ (1σ).

Mineralogy of the massive sulphide samples

Overall, the mineralogy of the Broken Spur samples is marcasite, pyrrhotite, pyrite, chalcopyrite, sphalerite, wurtzite, isocubanite, anhydrite, Fe oxyhydroxides, amorphous silica, minor covellite, bornite, chalcocite, digenite, magnetite, haematite, aragonite and trace barite, galena and jordanite.

The sulphide–sulphate–oxide samples have been classified into three different types based on hydrothermal setting, morphology and mineralogy (Table 1). Type I samples are hydrothermally immature, anhydrite-rich, 'beehive' diffusers (type Ia) and delicate chimney tips or 'organ-pipes', (type Ib). Type II samples are hydrothermally mature, chalcopyrite-rich chimney samples. Type III samples are dominated by iron disulphides and are from the base of chimneys or from the outer crust of sulphide mounds. Additionally, unconsolidated hydrothermal sediment was collected. The mineralogy and mineral abundances of each of the sample types are shown in Fig. 3. The samples are described in the following sections on a site by site basis.

Table 2 summarizes the mineralogical types (I–III) observed and sampled at each site. No samples were taken from Site 2, which is a site of low temperature diffuse venting similar to Site 1.

Site 1, extinct

Alvin video images of Site 1, which is located within the axial graben at 3075 m water depth (Fig. 2), show a 5 m tall, 3 m wide mound surrounded by a 20 m wide apron of hydrothermal sediments. No focused discharge was seen at this site, which is approximately 150 m north of the high-temperature vents at Sites 3, 4 and 5. Hydrothermal activity is confined to shimmering water discharging from cracks and flange structures.

Samples from this site consist of red–brown Fe oxide sediment and a large, irregular, zoned sample of massive sulphide which is partly oxidized. This coherent type III sample consists mainly of Fe and Zn sulphides [ca. 43% pyrite, 20% marcasite, 30% sphalerite (with associated galena grains), 5% chalcopyrite–isocubanite and 2% pyrrhotite]. The sample is zoned with a Zn sulphide-rich outer zone and an inner zone with abundant Fe sulphides. Texturally, it appears that sphalerite is replacing the chalcopyrite in this sample. Coarser sphalerite grains contain chalcopyrite–isocubanite inclusions.

Sphalerite appears to have precipitated around dendritic pyrite–marcasite grains ('bird's

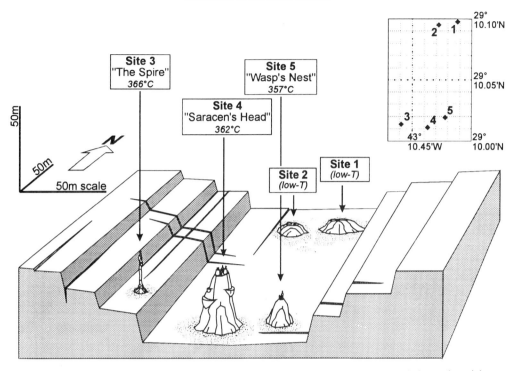

Fig. 2. Schematic drawing of the setting and morphology of the hydrothermal deposits relative to the axial graben. The axial graben is situated on the neovolcanic ridge, within the median valley of the Mid-Atlantic Ridge. Inset shows the location of the five deposits discovered during *Alvin* dives 2624 and 2625. Modified from Murton *et al.* (this volume).

Table 1. *Summary of the different types of sulphides sampled from the Broken Spur area*

	Type I: chimney (immature)	Type II: modified chimney (mature)	Type III: crust/carapace	Hydrothermal sediments
Morphology	Beehive diffusers; organ pipe chimney tips; pieces in oxide sediment	Spire-type thick chimney walls	Zone crusts; pieces in oxide sediment	Coating on chimney-mound structures and forming an apron around mounds
Activity	Focused hydrothermal black smoker. Max. $T = 366°C$	Focused hydrothermal black smoker	Diffuse hydrothermal. $T \ll 366°C$	Associated with type III mineralization and diffuse flow
Composition	Major constructional anhydrite–chalcopyrite	Major constructional chalcopyrite. Outer zone oxidized to secondary Cu sulphides	Fibrous Fe disulphides. Later void filling Zn- and Cu–Fe sulphides	Fe-oxyhydroxide Sulphides present due to mass wasting of types I, II and III mineralization
Notes	Chimney growth zonation: e.g. Haymon (1983)	Multiple-walled concentric, radial growth	A special type includes anhydrite crusts	Form by sulphide oxidation and/or low-T precipitation

	Type Ia Beehive	Type Ib Organ-pipe	Type II Chimney	Type III Crust	Sulphide sediment	Oxide sediment
Pyrrhotite	major	minor		minor		
Pyrite	minor	trace	trace	major	trace	
Marcasite		trace		abundant		
Sphalerite	minor	trace	trace	abundant	trace	
Wurtzite	trace	trace		trace		
Chalcopyrite	abundant	abundant	major	minor	minor	
Isocubanite	abundant	minor	trace	minor		
Bornite		minor	abundant			
Digenite			minor	trace		
Covellite			trace	trace		
Galena				trace		
Jordanite				trace		
Magnetite	minor	trace				
Haematite	minor		trace			
Goethite				minor	minor	major
Lepidocrocite	abundant		trace		major	minor
Akaganeite						abundant
FeOOH (amorph.)	trace	trace	trace	trace	abundant	abundant
Silica (amorph.)	trace	trace	trace	abundant		abundant
Anhydrite	major	major			abundant	
Barite				trace		
Aragonite		trace	trace			
Talc	abundant					

KEY: ■ major (>50%); ■ abundant (<50%); ▮ minor (<10%); ▪ trace (<2%).

Fig. 3. Summary of mineralogy and relative abundance for each of the sulphide–sulphate sample types.

foot' texture, Fig. 4a), which grades into colloform pyrite with subhedral crystalline pyrite overgrowths towards the centre of the sample. Late amorphous silica occurs in voids and around many of these dendritic grains. The colloform pyrite is also overgrown and partially replaced by sphalerite which contains inclusions of isocubanite–chalcopyrite, pyrrhotite and galena rimmed by jordanite (Fig. 4b). The sphalerite is, in turn, overgrown by an intergrowth of chalcopyrite and isocubanite. The clusters of sphalerite grains often have a core of an 'intermediate Fe sulphide product' mixed with some relict grains of pyrrhotite. In addition, the sphalerite clusters contain radiating spheroids of marcasite which are ca. 0.25 mm wide. This marcasite is also partly replaced by an intermediate Fe sulphide product. Unaltered spheroidal structures are repeatedly zoned with alternating layers of marcasite and pyrite, and this zonation is a function of changing physicochemical conditions of the hydrothermal fluids. Some pyrite grains have a porous overgrowth of second generation pyrite; this rim may be a result of partial alteration from a precursor Fe sulphide phase. Where the entire pyrite grain is porous, the alteration is complete. Some fine-grained interstitial pyrrhotite laths appear to precipitate last in the paragenetic sequence (Fig. 4c). The paragenesis of this silicified type III sample is

$$\text{marcasite} + \text{pyrite} \rightarrow \text{isocubanite} + \text{chalcopyrite} \rightarrow \text{sphalerite} \rightarrow \text{isocubanite} + \text{chalcopyrite} \rightarrow \text{pyrrhotite} \rightarrow \text{amorphous silica}$$

This is illustrated in Fig. 4d, which shows gradational zoning from sphalerite to isocubanite (cross-cut by pyrrhotite lath) to pyrrhotite.

Table 2. *Classification and location of the Broken Spur* Alvin *samples*

	Site 1	Site 2	Site 3 'The Spire'	Site 4 'Saracen's Head'	Site 5 'Wasp's Nest'
Location:	29°10.098'N, 43°10.412'W	29°10.095'N, 43°10.428'W	29°10.014'N, 43°10.459'W	29°10.011'N, 43°10.437'W	29°10.019'N, 43°10.422'W
Depth (m):	3075	3070	3050	3057	3091
Activity:	Low T diffuse flow	Low T diffuse flow	366°C black smokers; diffuse flow	362°C black smokers; diffuse flow	357°C black smokers; diffuse flow
Type Ia: beehive chimney			2624-3-3 2624-3-4c	2625-4-3	2625-5-1c
Type Ib: organ pipe chimney			2624-3-1 2624-3-2 2625-3-1	2625-4-1 2625-4-2	Observed*
Type II: mature chimney			2625-3-2 2625-3-3 2625-3-4 2625-3-5		
Type III: crust (sulphide)	2624-1-2	Observed*	2624-3-4b	2625-4-4a 2625-4-5b 2625-4-6	2625-5-1b
Type III: crust (anhydrite)	Observed*		Observed*	2625-4-7	Observed*
Hydrothermal sediments	2624-1-1	Observed*	2624-3-4a	2625-4-4a 2625-4-5a	2625-5-1a

* Indicates mineral types were seen on *Alvin* video and still camera, but not sampled.

Site 3, 'The Spire'

The Spire is situated on the top of the eastern graben wall, approximately 20 m above the other sites which lie within the axial graben (Fig. 2). The single, contorted, 1–2 m diameter chimney was 18 m tall before sampling in June 1993, when approximately 4 m was knocked off. At the top of the chimney, before sampling, were multiple small chimney structures reminiscent of organ pipes, which often coalesce to form 'pan-pipe' arrays. Below these was a large bulbous beehive-shaped structure. At the base of the chimney is a 10 m wide, 4 m high mound of sulphide rubble and sediments. *Alvin* temperature probe measurements show that fluid was venting at 366°C from the top of this chimney. The water depth at this site is 3050 m. Samples from this site consist of *in situ* fragments from the top of this spectacularly tall active chimney structure, and weathered sulphides from the base.

There are three types of active chimney samples. Firstly, there are the 'diffuser' beehive samples (type Ia). *Alvin* video images of this active beehive-shaped structure show it to be composed of stacked horizontal, concentric, growth zones of anhydrite, with shimmering water diffusing from the basal zones. This growth pattern gives it a characteristic ribbed appearance. Temperature measurements made at the surface of the active diffuser indicate 200°C. Similar beehive structures have been described by Fouquet *et al.* (1993) from the Snakepit site, Mid-Atlantic Ridge, and Koski *et al.* (1994) from the Monolith Vent on the Cleft segment of the southern Juan de Fuca Ridge. The beehives at Broken Spur are also mineralogically zoned around the central chalcopyrite conduit, with an inner anhydrite + chalcopyrite zone, a middle anhydrite + pyrrhotite zone and an oxidized outer pyrite + magnetite + magnesium silicate zone.

Secondly, there are delicate chimney tips that form 'organ-pipe' structures that grow on top of the beehive (type Ib). The largest of these is 3–4 cm in diameter, with conduit walls 3–4 mm thick. Others are smaller, only 1–2 cm in diameter. Mineralogically, they are similar to the beehive structures, but with thinner and more sharply defined concentric growth zones.

Thirdly, there are concentrically zoned chalcopyrite conduit wall samples (type II) from the main supporting vertical chimney below the beehive structure. These chimney wall samples are 2–4 cm thick.

The weathered sulphides from the base of the Spire are an example of sample type III.

Diffuser 'beehive' structures. These samples are dominated by anhydrite, chalcopyrite and

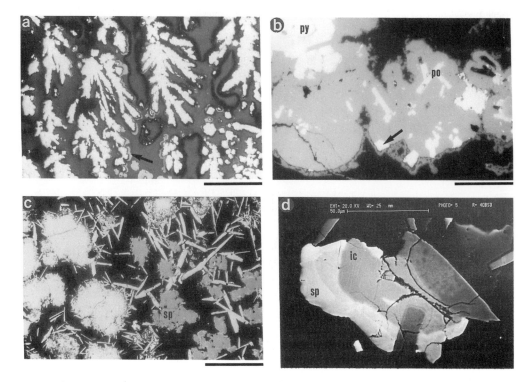

Fig. 4. Mineral textures typical of type III mineralization. (**a**) Reflected light photomicrograph showing dendritic marcasite ('bird's foot' texture) with overgrowths of sphalerite (arrow). Scale bar 400 µm. (**b**) Reflected light photomicrograph showing pyrite (py) overgrown by sphalerite with inclusions of galena (arrow) and pyrrhotite (po). Scale bar 50 µm. (**c**) Reflected light photomicrograph showing pyrrhotite aggregates and laths (light grey) co-precipitated with sphalerite grains (sp). Scale bar 150 µm. (**d**) Backscattered SEM image showing sphalerite (sp) and isocubanite (ic) overgrowth on pyrrhotite (dark grey).

pyrrhotite laths. The pyrrhotite laths cross-cut and replace anhedral to subhedral sphalerite and euhedral wurtzite. Sparse anhedral isocubanite with characteristic wedge-shaped chalcopyrite lamellae has co-precipitated with the wurtzite and sphalerite. Chalcopyrite grains, up to 75 µm, appear to have co-precipitated with pyrrhotite. The chalcopyrite forms complex intergrowths with sphalerite, and is also overgrown by sphalerite and pyrrhotite. The pyrrhotite laths, which are up to 75 µm long, have radial growth habits indicative of open space precipitation. They are commonly embedded in a matrix of coarse-grained anhydrite and extremely fine-grained magnetite and pyrite (Fig. 5a). Fine-grained magnetite and pyrite also rim the pyrrhotite laths (Fig. 5b) and partly replace the pyrrhotite.

In samples from the outer edge of the beehive, bladed magnetite grains pseudomorph primary pyrrhotite laths and are partially altered to haematite (Fig. 5c). Scanning electron microscope EDS analysis also reveals magnesium silicates in the matrix of these samples as void

infillings; optical and XRD methods determine this to be fibrous talc (Fig. 5d). The paragenesis of these type Ia diffuser samples is

anhydrite → isocubanite + wurtzite + sphalerite → chalcopyrite + pyrrhotite → pyrite + magnetite → haematite + serpentine + talc

'Organ pipe' structures. These delicate chimney tips have a similar mineralogy and paragenesis to the large, bulbous beehive structures, and are therefore classified as type Ib samples. The thin-walled (3–4 mm) chimney tips have an inner zone of chalcopyrite and isocubanite which partially replaces anhydrite. The Cu sulphides become less abundant away from the central conduit and pyrrhotite laths become increasingly common. These appear to have precipitated with anhydrite. An outer veneer of Fe oxides and amorphous silica is present on all these samples. The mineralogical zones are sharper than in the beehive samples and this may reflect steeper physicochemical gradients.

Fig. 5. Mineral textures typical of type Ia beehive mineralization. (**a**) Reflected light photomicrograph showing corroded anhydrite grain (top) and boxwork pyrrhotite laths. Scale bar 400 μm. (**b**) Backscattered SEM image showing rims of magnetite and pyrite (dark grey) around fine-grained pyrrhotite laths (light grey). (**c**) Reflected light photomicrograph showing magnetite (centre) pseudomorphing pyrrhotite laths. The magnetite has subsequently been partially altered to haematite around the rim (arrows). Scale bar 150 μm. (**d**) Transmitted light photomicrograph showing fibrous talc infilling voids between fine-grained sulphides. Scale bar 150 μm.

The paragenesis of these type Ib samples is

anhydrite → chalcopyrite + isocubanite → pyrrhotite → amorphous silica and Fe oxyhydroxides

Chalcopyrite chimneys. These chimney pieces are composed primarily of chalcopyrite with interstitial anhydrite. Chalcopyrite replaces the coarse-grained anhydrite matrix and, towards the outer edge of these conduit wall samples, bornite replaces chalcopyrite, digenite replace bornite, and covellite replaces digenite.

The inner part of the chimneys are fine-grained with concentric growth zones. However, towards the outer edge of the chimney the grains are coarser and grow radially outwards (Fig. 6). In some thick-walled chimney samples this outward growth pattern is repeated several times, with late-stage pyrite at the outer edge of the chimney. Paragenetically late anhydrite infills radial fractures in the chimneys.

Late-stage aragonite, also shown by XRD, is

present in these samples as overgrowths at the outer edge of the chimneys.

The paragenesis of these type II samples is

anhydrite → chalcopyrite → bornite → digenite → covellite → pyrite → anhydrite → aragonite

Weathered samples from base of The Spire. These samples consist mainly of Fe oxides and oxyhydroxides with fine-grained type III Fe–Zn sulphides. Fragments of type Ia beehive mineralization are composed of sulphides in a matrix of anhydrite, Fe oxyhydroxides and amorphous silica. Voids are lined with grey colloform Fe oxyhydroxide precipitates. Pyrrhotite laths are common, as are subhedral to anhedral pyrite grains (up to 100 μm). Some isocubanite atolls are infilled with sphalerite grains and resemble microconduit structures. These are an example of sample type III.

Site 4, 'Saracen's Head'

Site 4 is a large sulphide mound, 37 m high and

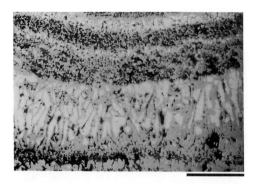

Fig. 6. Reflected light photomicrograph showing bornite and digenite (dark grey) alteration of radially growing chalcopyrite in the outer part of a type II chimney wall. Scale bar 400 μm.

20 m in diameter, which is located within the central part of the axial graben, east of Site 3 (Fig. 2). The top of this mound is approximately level with the top of the enclosing graben, and the water depth at the base of this site is 3057 m. Active beehive and chimney structures at the top of the mound were discharging black particle-rich fluid at 362°C. Samples from this site consist of pieces of weathered sulphide and sulphide sediment from the mound (type III).

Under a 5–6 m wide ledge on the side of the mound there was a mass of filamentous bacteria forming a white cloud which was associated with shimmering water diffusing from the underside of the ledge (similar to a flange structure). A texturally distinctive type III massive sulphide, associated with these bacterial mats, was sampled from this flange-like structure near the base of the mound. The outer part of this sample is composed of dendritic and skeletal marcasite, whereas the inner part is composed dominantly of pyrite. In the central part of the sample porous marcasite is overgrown by clean euhedral pyrite (Fig. 7a), which is locally overgrown by isocubanite, and then by euhedral wurtzite and subhedral sphalerite. This central zone contains marcasite–pyrite clusters with a preferred orientation (Fig. 7b). Iron disulphides are dominant, with lesser sphalerite and amorphous silica, which finely coats the Fe sulphides cementing the rock and preventing oxidation.

The paragenesis of this type III sample is

marcasite → pyrite → isocubanite → wurtzite
→ sphalerite → amorphous silica

Other samples from this site consist of sulphide pieces in sediment. In these samples radiating blades of marcasite form spherical and hemispherical colloform structures which appear to be growing out from Fe oxyhydroxides and amorphous silica (Fig. 7c). Some covellite is seen in amorphous silica in these colloform zones (Fig. 7d). In one of these samples, one end of a thin section is composed of colloform marcasite, the other of sphalerite growing out from the marcasite. The sphalerite is red and translucent, with coarse-grained 'pyrite disease' (Fig. 7e) as well as chalcopyrite disease (Barton and Bethke 1987). The triangular pyrite inclusions appear to grow out from relatively Fe-rich zones at the edges of the crystal or from cleavage planes.

The textural change from the outer to inner parts of these samples – that is, from colloform to feathery to euhedral – is indicative of increasing crystal maturity.

The mineral paragenesis of these type III samples is

iron oxyhydroxides → amorphous silica →
marcasite → sphalerite → pyrite → chalcopyrite

Site 5, 'Wasp's Nest'

This site is approximately 20 m southeast of Site 4, within the axial graben (Fig. 2). The sulphide mound at site 5 is smaller than at Site 4, but with similar actively venting chimneys, beehives and flange-like structures. The temperature of the exiting fluids was 357° and the water depth is 3091 m. Samples from this site are delicate anhydrite-dominated diffuser fragments from active beehive structures (type Ia samples).

In all samples anhydrite is more abundant than total sulphides. The anhydrite forms coarse, euhedral grains ca. 0.5–4 mm long, and is more common at the inner edge of conduit walls. The anhydrite is overgrown and partly replaced by chalcopyrite (Fig. 8) and both phases are overgrown by amorphous silica. Fine-grained aragonite also occurs in the matrix.

Chalcopyrite, the dominant phase lining conduits in these bulbous samples, occurs as subhedral grains ca. 0.1–0.2 mm long, with an equidimensional morphology. This chalcopyrite becomes finer grained away from the conduits, and decreases in abundance, whereas pyrite becomes more abundant. Some sphalerite co-precipitated with chalcopyrite and isocubanite. Minor bornite and digenite replace chalcopyrite and pyrite.

Away from the conduit, the chimney contains very fine-grained (0.01 mm) boxwork pyrrhotite with beaded rims composed of magnetite. The 10–500 μm long pyrrhotite laths form an interesting dendritic, 'bird's foot' texture, with

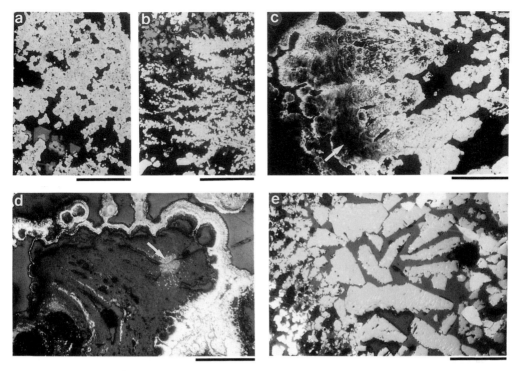

Fig. 7. Further textures of type III massive sulphides. (**a**) Reflected light photomicrograph showing porous marcasite overgrown by fine-grained, inclusion-free euhedral pyrite. Scale bar 400 μm. (**b**) Reflected light photomicrograph showing strongly oriented growth fabric of Fe disulphides. Scale bar 400 μm. (**c**) Reflected light photomicrograph showing colloform Fe disulphides apparently growing from an outer colloform crust of Fe oxyhydroxide (arrow). Scale bar 400 μm. (**d**) Reflected light photomicrograph showing covellite (arrow) in dark grey amorphous silica, overgrown and partly replaced by Fe disulphides (white). Scale bar 400 μm. (**e**) Reflected light photomicrograph showing sphalerite grains with coarse-grained chalcopyrite inclusions. Scale bar 150 μm.

branches of pyrrhotite overgrown by sphalerite. This is similar to a texture typically found in type III samples (Fig. 4a). Some of the Zn sulphide grains are hexagonal wurtzite and are heavily diseased with chalcopyrite, pyrite and pyrrhotite.

The paragenesis of these type Ia samples is

anhydrite → aragonite → wurtzite →
chalcopyrite → isocubanite → pyrrhotite →
sphalerite → pyrite → bornite → digenite →
amorphous silica

Hydrothermal sediments

Sulphide sediments from all the sampled sites (Table 2) are dark brown in colour and contain sulphide–sulphate debris up to 5 cm in size. Chalcopyrite, isocubanite, pyrrhotite, anhydrite and pyrite are major components. The oxide component consists of variable proportions of amorphous Fe-oxyhydroxide and lepidocrocite,

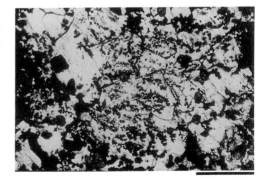

Fig. 8. Transmitted light photomicrograph showing type Ib texture of fine-grained chalcopyrite (black) replacing coarse-grained anhydrite along cleavage planes. Scale bar 400 μm.

with minor goethite. These phases appear to be derived from *in-situ* seawater oxidation of sulphide debris.

Table 3. *Iron, Cd, and Cu content of sphalerite and wurtzite, Wt.% Zn and S have been omitted, but the total is presented*

Sample	Type and paragenesis	Fe (wt.%) mean ($\pm 1\sigma$)	Cd (wt.%) (range)	Cu (wt.%) (range)	Total (mean) (wt.%)
2624-3-3	Ia: Assoc. cpy	13.0 ($\pm$ 1.2)	<0.1–0.3	0.9–1.4	95.5 ($N = 3$)
2625-3-4	II: Zoned; assoc. po	24.2 ($\pm$ 7.6)	<0.1–0.4	<0.1–3.5	96.0 ($N = 5$)
2625-4-5b	III: Early, assoc. FeS$_2$	10.3 ($\pm$ 2.8)	<0.1	<0.1–2.0	93.0 ($N = 3$)
2624-1-2	III: Main, assoc. cpy	4.3 ($\pm$ 1.4)	<0.1–0.5	<0.1–0.7	74.9 ($N = 3$)
2625-4-5b	III: Coarse bladed (e.g. Fig. 7e)	7.1 ($\pm$ 0.6)	<0.1–0.3	<0.1–0.4	92.4 ($N = 6$)
2624-1-2	III: Late, assoc. galena	17.0 ($\pm$ 0.1)	<0.1	0.9–1.3	87.7 ($N = 2$)

N = Number of individual grains analysed to determine mean.
Mineral abbreviations; cpy, chalcopyrite; po, pyrrhotite.

Oxide sediments are red–brown in colour. They are composed of fine-grained oxides with irregular pieces of silicified material. The presence of amorphous silica is indicated by a broad, low intensity peak at around 3.9 Å (Jones & Segnit 1971). In some cases, XRD spectra gave only very low intensity peaks for crystalline Fe-oxyhydroxide phases, although the sediments were obviously red and oxidic. In these instances, the presence of amorphous Fe-oxyhydroxide is assumed. Goethite, akaganeite [β-FeO(OH,Cl)], and minor lepidocrocite were positively identified by XRD. The absence of any sulphides and association with amorphous silica indicate that this sediment may be a direct hydrothermal precipitate from diffuse low temperature hydrothermal fluids, but alternatively it may represent the complete oxidation of sulphides. However, the sulphide and oxide sediments were sampled together and it is seems unlikely that only localized patches of sulphide material would completely oxidize in this way.

The seawater oxidation of sulphide results in an assemblage dominated by amorphous Fe-oxyhydroxide and lepidocrocite. This is observed in samples of the sulphide rubble on and around mound structures and in the outer carapace of a beehive structure (sample 2625–4–3). Low temperature hydrothermal precipitates take the form of amorphous Fe-oxyhydroxide, akaganeite, and goethite, associated with amorphous silica. The goethite–silica mineralization forms accumulations on and around the deposits, and as an outer carapace on porous massive Fe–(Zn) sulphide mound samples.

Microprobe analysis

Selected phases were quantitatively analysed using EDS electron probe microanalysis. The results of a systematic study of sphalerite and wurtzite are summarized in Table 3.

Zn-sulphides

Iron, Cd and Cu were all detected in analyses of Zn sulphides. Low totals (75–96 wt.%) may indicate the presence of hydrated or thiosulphate phases within the Zn sulphide structure (Kucha & Stumpfl 1992). Considerable variation in Fe content was detected, both within grains and between different paragenetic associations.

Type III Fe–Zn sulphides. Sphalerite from all paragenetic stages of these Fe–Zn sulphide mound samples was analysed. Variation within individual grains (maximum 6.5 wt.% Fe) is insignificant when compared with the variation between different stages of the paragenesis (range 2.7–17.1 wt.% Fe). The Fe content decreases through the main paragenetic stage from 10.3 ($\pm$2.8) wt.% for the sphalerite associated with FeS$_2$, to 4.3 ($\pm$1.4) wt.% for grains associated with chalcopyrite. However, late sphalerite, associated with galena, contains on average 17 wt.% Fe.

Types I and II Cu-rich chimney samples. In contrast with the analyses from type III sulphides, sphalerite and wurtzite from chimney and beehive structures have, on average, higher Fe contents (range 12.0–36.4 wt.%), and are zoned with respect to both Fe and Cd. Wurtzite from the outer parts of a beehive structure (sample 2624-3-3) contains 13.0 ($\pm$1.2) wt.% Fe. Sphalerite from the outer part of the chalcopyrite chimney is associated with pyrrhotite and has the highest Fe content, mean 24.2 ($\pm$7.6) wt.%. Individual grains are zoned, with homogenous Fe-rich (maximum 36.4 wt.%)

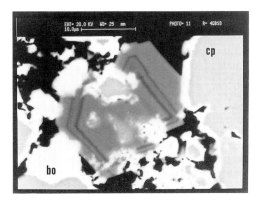

Fig. 9. Backscattered SEM image of growth-zoned magnetite crystal associated with chalcopyrite (cp) and bornite (bo) from a type II massive chalcopyrite chimney sample.

Table 4. *Sulphur isotopic data of the Broken Spur samples*

Type	Sample		Mineral	$\delta^{34}S$ (‰)
Ia	2625-4-3	(inner)	Cpy	1.8
Ia	2624-3-3	(outer)	Po	0.4
Ia	2624-3-3	(inner)	Anhy	19.3
1b	2624-3-2		Cpy	2.0
II	2625-3-4	(inner)	Cpy	2.0
II	2625-3-4	(massive)	Cpy	1.8
II	2625-3-4	(outer)	Cpy-bo	2.2
II	2625-3-4	(outer)	Cpy-dig	2.4
III	2624-1-2	(outer)	Py	−0.8
III	2624-1-2	(inner)	Py	1.3
III	2625-4-6	(outer)	Py-(ZnS)	0.9
III	2625-4-6	(inner)	Py-(cpy)	1.4
III	2625-4-5b		Py-cpy	0.8
III	2625-5-1b		Py	0.9

Mineral abbreviations: anhy, anhydrite; bo, bornite; dig, digenite; cpy, chalcopyrite; and py, pyrite.

cores, and delicately zoned, relatively Fe-poor rims (16.3–22.0 wt.%).

Zoned magnetite from chimney sample

Euhedral magnetite grains are associated with chalcopyrite–bornite aggregates in type II chimney samples. These grains exhibit multiple thin growth zones less than 1 μm thick (Fig. 9). These zones are enriched in Mg (at least 1.6 wt.% Mg; beam resolution unable to sample pure dark zone). The cores of these grains also contain up to 1.1 wt.% Mg. The magnetite contains 0.5–0.9 wt.% Cu and it is proposed that as chalcopyrite is oxidized by seawater to bornite in the outer chimney wall, the liberated Fe subsequently precipitates as magnetite, with the incorporation of seawater-derived magnesium. This process has also been described by Oudin (1983) and Woodruff & Shanks (1988).

Sulphur isotopic analysis

The results of a preliminary sulphur isotope study of the Broken Spur sulphide and sulphates are shown in Table 4 and Fig. 10. The $\delta^{34}S$ range for the sulphides is from −0.8 to 2.4‰, which is similar to the range reported for the Snakepit sulphides (Kase *et al.* 1990).

The mean $\delta^{34}S$ value is 1.3‰ for the Broken Spur sulphide minerals and the median is 1.55‰; however, there is a definite difference between the $\delta^{34}S$ measured from Fe–Zn sulphide and Cu–Fe sulphide samples. The types I and II Cu–Fe sulphide samples are systematically isotopically heavier than the Fe sulphides (Fig. 10) and this is obviously not due to equilibrium precipitation. There is also a similar range in the

spread of data (*c.* 2‰) for both the Fe sulphides and the Cu sulphides. The data cannot be subdivided on the basis of sample site (i.e., hydrothermally active or extinct sites; Table 4). Anhydrite has a $\delta^{34}S$ value of 19.3‰ which is similar to that of modern day seawater ($\delta^{34}S = 21$‰, Rees *et al.* 1978), and supports the idea that the anhydrite precipitates from conductively heated seawater around the black smoker edifices (Haymon 1983).

Discussion

On the basis of mineralogy and micro-textures it is possible to subdivide the Broken Spur samples into four main types which reflect their paragenesis and hydrothermal maturity.

Type I

Type Ia samples are the 'beehive' diffuser type immature chimney samples. These are mineralogically dominated by anhydrite and pyrrhotite. It appears as though these structures act as heat exchangers between the hydrothermal fluids and cold seawater. The *Alvin* video images indicate that cold seawater is entrained up through the base of these structures and the resulting mixed fluid diffuses outwards through the ribbed shell and is seen as shimmering water around the beehives. Oxidation of pyrrhotite to magnetite and haematite, and the precipitation of Mg-bearing silicates in voids in the outer parts of these structures, suggest cooling of the hydrothermal fluids by mixing with seawater. Similar

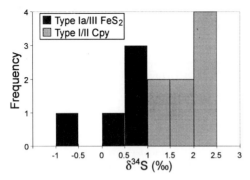

Fig. 10. Frequency histogram of $\delta^{34}S$ data from the Broken Spur sulphide samples.

seawater oxidation of sulphides has been documented in an extinct sulphide mound at Middle Valley on the northern Juan de Fuca Ridge (Duckworth *et al.* 1994), but in that case it appears to be a result of low-temperature seawater-dominated fluids reacting with warm sulphides.

Vigorous discharge of black smoke is also occurring from the tops of these beehives, suggesting that this fluid mixing is localized and does not affect the main jet of discharging hot fluid. Therefore, the beehives appear to act as a scaffolding during the construction of the inner chalcopyrite-rich 'backbone' of the type II chimney or spire by insulating the precipitating chalcopyrite wall from the ambient seawater. As the chimney grows, these beehive structures break off or dissolve due to the retrograde solubility of anhydrite (Blount & Dickson 1969). Rickard *et al.* (in press) discuss the reactions that occur during the formation of these structures. Similarly, the formation of beehive structures from the Juan de Fuca Ridge is described by Koski *et al.* (1994), and Fouquet *et al.* (1993) discuss those from the Snakepit hydrothermal field on the Mid-Atlantic Ridge.

Type Ib samples are the delicate organ pipe chimneys that grow from the beehive structures at the top of chimneys. These structures often coalesce to form 'pan-pipe' arrays. They have a similar growth history to the 'classic' black smoker model as proposed by Haymon (1983), and represent the youngest part of the active chimney.

Type II

Type II samples are characterized by thick chimney walls entirely composed of chalcopyrite, which have been oxidized around their outer edges to bornite, digenite and covellite. These samples show radial and concentric growth textures with the chalcopyrite crystals growing outwards from the fluid conduit. Some samples are multi-walled with repeated growth patterns. These samples represent the major, high-temperature, constructional part of chimneys. The paragenetic history (anhydrite replaced by Cu sulphides) is similar to that proposed by Haymon (1983) for black smoker chimneys from 21°N on the East Pacific Rise, but taken to an extreme not seen at the surface of fast-spreading ridge sites.

Type III

Type III samples are dominated by iron sulphide minerals and are also texturally different from the immature (type I) and mature (type II) chimney samples. These samples are from sulphide mounds and usually have a strongly oriented growth fabric. The outer wall of these samples is dominated by colloform Fe oxyhydroxides that appear to be a primary precipitate. Iron sulphides, usually marcasite, grow inward from this crust, indicating that the sulphides are forming by sulphidation of the Fe oxides. Iron-poor sphalerite is replacing and growing outward from the marcasite and pyrite and is, therefore, paragenetically later. The low Fe sphalerite may result either from variations in sulphur fugacity in the hydrothermal fluids, or precipitation of abundant Fe sulphides and oxides in these mounds. Samples from Site 1 (extinct) and Site 4 are characteristic of this type.

Type III samples from Site 4 are associated with bacterial mats and are texturally distinctive. However, the paragenesis is similar to other type III samples, except for the absence of early Fe oxyhydroxides and the presence of an amorphous silica cement.

Hydrothermal sediments

The XRD data from sulphide sediments indicates that the sulphide material in these sediments is derived from the surface weathering of the adjacent sulphide mounds. Anhydrite has a short residence time in the sediments because of its solubility in ambient seawater (Blount & Dickson 1969) and therefore the presence of anhydrite indicates that recently formed black smoker beehives and chimneys are the major contributor to this sulphide sediment. Some oxide-rich material may be a primary precipitate from the low temperature diffuse hydrothermal fluids (shimmering water).

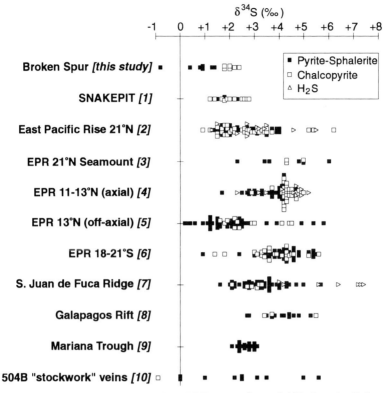

Fig. 11. Diagram showing sulphur isotopic variation of different seafloor sulphide deposits. References: [1] Kase *et al*. (1990); [2] Woodruff & Shanks (1988); [3] Alt (1988); (4) Bluth & Ohmoto (1988); [5] Knott (pers. comm.); [6] Marchig *et al*. (1990); [7] Shanks & Seyfried (1987); [8] Knott *et al*. (this volume); [9] Kusakabe *et al*. (1990); and [10] Alt *et al*. (1989).

Sulphur isotope geochemistry

The sulphur isotopic ratios for the sulphides are low compared with data from other unsedimented ridge settings in the Pacific, such as axial East Pacific Rise deposits and southern Juan de Fuca Ridge sulphides (Fig. 11), but similar to that reported from the Snakepit sulphide deposit (Kase *et al*. 1990), which is south of Broken Spur (Fig. 1). The off-axial deposits at 13°N, East Pacific Rise also have low $\delta^{34}S$ values, but the range of data is greater than at Broken Spur. Negative $\delta^{34}S$ values have only previously been reported from sulphides in veins from ODP Hole 504B (Alt *et al*. 1989). The Cu sulphides at Broken Spur are isotopically heavier than the Fe sulphides, which is the reverse of that expected in equilibrium precipitation conditions. The isotopically lighter Fe sulphides from the mounds are probably older than the chalcopyrite samples from the currently active chimneys, and temporal variations in the $\delta^{34}S$ H₂S in the fluids may have occurred. One cause of such variations

may be mixing of the hydrothermal fluids with seawater in the subsurface stockwork zone (Janecky & Shanks 1988). It is also probable that the hydrothermal regime that forms the mounds is different to that which forms the chimneys. This idea is supported by the different mineralogy and textures of the type III mound samples, compared with the types I and II chimney samples.

Summary

The samples collected from the Broken Spur site show a range in textures and mineralogy that can be explained by different precipitation mechanisms and differences in hydrothermal maturity. The mound samples (type III) are Fe disulphide-rich compared with the anhydrite–chalcopyrite–pyrrhotite-dominated chimney samples (types I and II). This difference in mineralogy and paragenesis suggests that the mounds form independently of the chimneys. Interestingly, it appears as though the growth of mound samples starts from a primary hydrothermal precipitate of Fe oxyhydroxides, and the growth of chimney structures is aided by the construction of anhydrite-rich beehive 'diffusers'.

Copper sulphides are isotopically heavier than Fe sulphides, and this may be a result of temporal variations in the $\delta^{34}S$ (H_2S) of the hydrothermal fluid. The sulphur isotope data are low compared with bare-ridge seafloor sulphide deposits from the faster spreading southern Juan de Fuca Ridge and East Pacific Rise.

Kase *et al.* (1990) attribute the low $\delta^{34}S$ values from the Snakepit sulphides as reflecting a smaller contribution of seawater sulphate to the deposit than in the Pacific deposits; this suggests low fluid to rock ratios. It may be that the slow spreading of the Mid-Atlantic Ridge, and the resulting tectonic structures, restrict the fluid paths through the basalts, allowing less fluid–rock interaction and resulting in more focused discharge forming deposits with isotopically lighter sulphur. This enhanced tectonic stability also facilitates the growth of spectacularly tall chimneys, although the contorted form of The Spire may reflect occasional, tectonically induced collapse followed by regrowth.

This work at Cardiff was funded by NERC grant GR9/603 to David Rickard, a NERC Research Fellowship to Rowena Duckworth and a NERC studentship to Richard Knott. The Scottish Universities Research and Reactor Centre (SURRC) is supported by NERC and a consortium of Scottish universities. We would like to thank the technicians at SURRC for the speedy sulphur isotope analyses and also Pete Fisher at Cardiff for help with the SEM analyses. Many thanks to Kevin Blake for his essential trans-world communication contribution. The comments of two anonymous reviewers are also acknowledged.

References

ALT, J. C. 1988. The chemistry and sulphur isotope composition of massive sulphide and associated deposits on Green Seamount, Eastern Pacific. *Economic Geology*, **83**, 1026–1033.

——, ANDERSON, T. F. BONELL, L. 1989. The geochemistry of sulphur in a 1.3 km section of hydrothermally altered oceanic crust, DSDP Hole 504B. *Geochimica et Cosmochimica Acta*, **53**, 1011–1023.

BARTON, P. B. & BETHKE, P. M. 1987. Chalcopyrite disease in sphalerite: pathology and epidemiology. *American Mineralogist*, **72**, 451–467.

BATUYEV, B. N., KROTOV, A. G., MARKOV, V. F., CHERKASHEV, G. A., KRASNOV, S. G. & LISITSYN, YE. D. 1994. Massive sulphide deposits discovered and sampled at 14°45′N, Mid-Atlantic Ridge. *BRIDGE Newsletter*, **6**, 6–10.

BLOUNT, C. W. & DICKSON, F. W. 1969. The solubility of anhydrite ($CaSO_4$) in $NaCl–H_2O$ from 100 to 450°C and 1 to 1000 bars. *Geochimica et Cosmochimica Acta*, **33**, 227–245.

BLUTH, G. J. & OHMOTO, H. 1988. Sulphide-sulphate chimneys on the East Pacific Rise, 11 and 13°N latitude. Part II: sulphur isotopes. *Canadian Mineralogist*, **26**, 505–515.

DUCKWORTH, R. C., FALLICK, A. E. & RICKARD, D. Mineralogy and sulphur isotopic composition of the Middle Valley massive sulphide deposit, northern Juan de Fuca Ridge. *In*: MOTTL, M. J, DAVIS, E. E., *et al.* (eds) *Proceedings of the Ocean Drilling Program, Scientific Results*, **139**, 373–385.

ELDERFIELD, H., GERMAN, C. R. & PALMER, M. R. 1993. Hydrothermal activity on the Mid-Atlantic Ridge at 29°N: results of RRS *Charles Darwin* Cruise 77 (BRIDGE Cruise no. 8). *BRIDGE Newsletter*, **5**, 7–10.

FOUQUET, Y., WAFIK, A., CAMBON, P., MEVEL, C., MEYER, G. & GENTE, P. 1993. Tectonic setting and mineralogical and geochemical zonation in the Snake Pit sulphide deposit (Mid-Atlantic Ridge at 23°N). *Economic Geology*, **88**, 2018–2036.

HALL, G. E. M., PELCHAT, J. C. & LOOP, J. 1988. Separation and recovery of various sulphur species in sedimentary rocks for stable isotopic determination. *Chemical Geology*, **67**, 35–45.

HAYMON, R. 1983. Growth history of hydrothermal black smoker chimneys. *Nature*, **301**, 695–698.

JANECKY, D. R. & SHANKS, W. C. III. 1988. Computational modelling of chemical and sulphur isotopic reaction processes in seafloor hydrothermal systems: chimneys, massive sulphides, and subjacent alteration zones. *Canadian Mineralogist*, **26**, 805–825.

JONES, J. B. & SEGNIT, F. R. 1971. The nature of opal, I. Nomenclature and consistent phases. *Journal Geological Society Australia*, **18**, 57–68.

KASE, K., YAMAMOTO, M. & SHIBATA, T. 1990. Copper-rich sulphide deposits near 23°N, Mid-Atlantic Ridge: chemical composition, mineral chemistry, and sulphur isotopes. *In*: DETRICK R., HONNOREZ, J., *et al.* (eds) *Proceedings of the Ocean Drilling Program, Scientific Results*, **106/109**, 163–172.

KNOTT, R., FALLICK, A. E., RICKARD, D. & BÄCKER, H. Mineralogy and sulphur isotopic characteristics of a massive sulphide boulder, Galapagos Rift 85°55′W, this volume.

KOSKI, R. A., JONASSON, I. R., KADKO, D. C., SMITH, V. K. & WONG, F.L. 1994. Compositions, growth mechanisms, and temporal relations of hydrothermal sulphide–sulphate–silica chimneys at the northern Cleft segment, Juan de Fuca Ridge. *Journal of Geophysical Research*, **99**, 4813–4832.

KUCHA, H. & STUMPFL, E. F. 1992. Thiosulphates as precursors of banded sphalerite and pyrite at Bleiberg, Austria. *Mineralogical Magazine*, **56**, 165–172.

KUSAKABE, M., MAYEDA, S. & NAKAMURA, E. 1990. S, O and Sr isotope systematics of active vent materials from the Mariana backarc basin spreading axis at 18°N. *Earth and Planetry Science Letters*, **100**, 275–282.

LANGMUIR, C. H., FORNARI, D. & 17 others. 1993. Geologic setting and characteristics of the Lucky Strike vent field at 37°17'N on the Mid-Atlantic Ridge. *EOS, Transactions, American Geophysical Union*, **74**, 99.

MARCHIG, V., PUCHELT, H., ROSCH, H. & BLUM, N. 1990. Massive sulphides from ultra-fast spreading Ridge, East Pacific Rise at 18–21°S: a geochemical stock report. *Marine Mining*, **9**, 459–493.

MURTON, B. J. & VAN DOVER, C. 1993. ALVIN dives on the Broken Spur hydrothermal field at 29°10'N on the Mid-Atlantic Ridge. *BRIDGE Newsletter*, **5**, 11–14.

——, Becker, K., Briais, A., Edge, D., Hayward, N., Klinkhammer, G., Millard, N., Mitchell, I., Rouse, I., Rudnicki, M., Sayanagi, K. & SLOAN, H. 1993. Results of a systematic approach to searching for hydrothermal activity on the Mid-Atlantic Ridge: the discovery of the Broken Spur vent site. *BRIDGE Newsletter*, **4**, 3–6.

——, VAN DOVER, & SOUTHWOOD, E. The geological setting and ecology of the Broken Spur hydrothermal vent field: 29°10'N on the Mid-Atlantic Ridge, this volume.

OUDIN, E. 1983. Hydrothermal sulphide deposits of the East Pacific Rise (21°N). Part 1: descriptive mineralogy. *Marine Mining*, **4**, 39–72.

REES, C. E., JENKINS, W. J. & MONSTER, J. 1978. The sulphur isotopic composition of oceanic water sulphate. *Geochimica et Cosmochimica Acta*, **42**, 377–382.

RICKARD, D., KNOTT, R., DUCKWORTH, R. C. & MURTON, B. J. Organ pipes, beehive diffusers and chimneys at the Broken Spur hydrothermal sulphide deposits, 29°N, MAR. *Mineralogical Magazine*, in press.

ROBINSON, B. W. & KUSAKABE, M. 1975. Quantitative preparation of SO_2 for $^{34}S/^{32}S$ analysis from sulphides by combustion with cuprous oxide. *Analytical Chemistry*, **47**, 1179–1181.

RONA, P. A., HANNINGTON, M. D., RAMAN, C. V., THOMPSON, G., TIVEY, M. K., HUMPHRIS, S. E., LALOU, C. & PETERSON, S. 1993. Active and relict sea-floor hydrothermal mineralization at the TAG hydrothermal field, Mid-Atlantic Ridge. *Economic Geology*, **88**, 1989–2017.

SHANKS, W. C. III. & SEYFRIED, W. E. JR 1987. Stable isotope studies of vent fluids and chimney minerals, southern Juan de Fuca Ridge: sodium metasomatism and seawater sulphate reduction. *Journal of Geophysical Research*, **92** (B11), 11 387–11 399.

WOODRUFF, L. G. & SHANKS, W. C. III. 1988. Sulphur isotope study of chimney minerals and hydrothermal fluids from 21°N, East Pacific Rise: hydrothermal sulphur sources and disequilibrium sulphate reduction. *Journal of Geophysical Research*, **93** (B5), 4562–4572.

Hydrothermal processes and contrasting styles of mineralization in the western Woodlark and eastern Manus basins of the western Pacific

STEVEN D. SCOTT[1] & RAYMOND A. BINNS[2]

[1]*Marine Geology Research Laboratory, Department of Geology, University of Toronto, Toronto, Ontario, Canada M5S 3B1*

[2]*CSIRO Division of Exploration and Mining, PO Box 136, North Ryde, New South Wales 2113, Australia*

Abstract: The western Woodlark Basin (initial rifting of continental crust) and eastern Manus Basin (rifted arc crust) offshore eastern Papua New Guinea display contrasting styles of hydrothermal activity and mineralization. In the eastern Manus basin, en echelon felsic and mafic volcanic ridges have formed in a pull-apart basin of the rifted New Britain arc terrane. Here, the PACMANUS Cu–Zn–Pb–Ag–Au sulphide deposit is forming within an area of about $800 \times 350\,m$ on the flank of a dacite lava dome atop a prominent 20 km long and 250–350 m high volcanic ridge. The ridge is andesitic in its lower reaches, dacitic to rhyolitic on top and is adjacent to an extensive field of basalt. At Woodlark, submarine rhyolite domes are devoid of hydrothermal products, but extensive Fe–Si–Mn oxyhydroxide deposits are forming from low temperature fluids on Franklin Seamount, an axial basaltic andesite volcano near the tip of the oceanic propagator. Protruding through and perhaps underlying these oxyhydroxides are inactive, higher temperature, precious metal-rich (Ag to 545 ppm, Au to 21 ppm), barite–silica spires. The Franklin Seamount deposits are thought to cap a disseminated sulphide deposit within the volcano and represent a failed massive sulphide system. Both Franklin Seamount and PACMANUS provide models for ancient ores on land.

Base and precious metal sulphide ores of volcanic association, both modern and ancient, are known from a variety of tectonic settings (continental margin, mid-ocean ridge, back-arc, within-arc, forearc and intraplate seamounts) and in rocks ranging in composition from basalt to rhyolite (Sawkins 1984; Scott 1992; Rona & Scott 1993). In the ancient geological record, the largest and richest volcanogenic polymetallic ore districts are found where felsic volcanic rocks are particularly abundant. Where tectonic settings have been deciphered, they are commonly island arcs. Prominent ancient examples include the Hokuroku district of Miocene age in northern Japan (Ohmoto and Skinner 1983), the Bathurst district of Ordovician age in New Brunswick (van Staal *et al.* 1992), the Mount Reid Volcanics of Cambrian age in western Tasmania (Large 1992) and the Skellefte district of Proterozoic age in Sweden (Rickard & Zweifel 1975). Even older and of uncertain tectonic setting are the Archean ore districts such as those of the Canadian Shield, as exemplified by Noranda in Quebec (Rive *et al.* 1990), which have associated volcanic rocks not unlike those of a modern island arc.

It is well known that some arc terranes (or suspected arc terranes such as Noranda) are very important for ore genesis. There is an almost universal association of polymetallic ores with abundant felsic volcanism even though mafic volcanic rocks typically dominate regionally, and such ore districts are commonly large and rich. The current model for massive sulphide ore formation (e.g. Sangster & Scott 1976; Lydon 1988; Scott 1992; Cathles 1993) does not explain these observations. The model simply requires a subcrustal heat source of any composition in a submarine setting that is responsible for high-temperature reactions between seawater and rock and discharge of the resulting metal-enriched hydrothermal fluid onto the seafloor where metallic sulphides are precipitated.

Of the 139 seafloor sites tabulated by Rona & Scott (1993), three-quarters are in the Pacific and about one-third of these are in island arc or other terranes (Fig. 1) where felsic volcanism might be expected. To try to understand the significance of felsic volcanism in ore generation, we have examined two modern seafloor sites in the western Pacific, the western Woodlark Basin and eastern Manus Basin,

From PARSON, L. M., WALKER, C. L. & DIXON, D. R. (eds), 1995, *Hydrothermal Vents and Processes*, Geological Society Special Publication No. 87, 191–205.

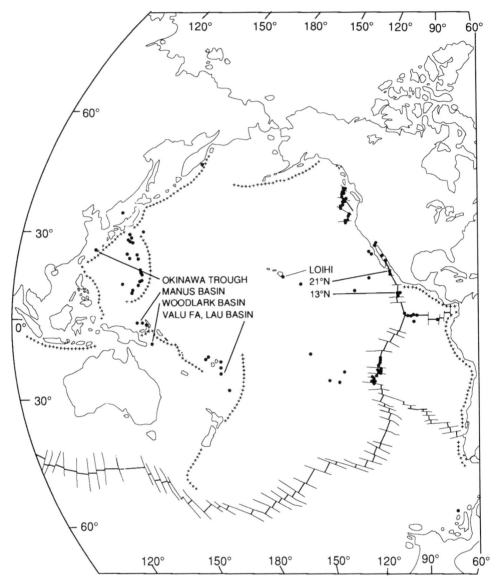

Fig. 1. Active and fossil hydrothermal vent sites in the Pacific Ocean. Locations are from Rona and Scott (1993).

where felsic volcanic rocks are known and hydrothermal activity might be anticipated. The main objective has been to study the origins and geological environments of modern seafloor hydrothermal deposits, preferentially in felsic volcanic rocks, as a means of improving exploration models for ancient volcanogenic massive sulphide and related ore types on land. The locations and regional tectonic settings of these two sites are shown in Fig. 2. Seagoing activities during six of our expeditions have included

detailed bathymetric mapping, magnetic surveys, deep-tow camera/video traverses, water column measurements and sampling, dredging, sediment coring and, in western Woodlark, a series of Mir submersible dives. Other work has included the collection and processing of side-scan SONAR imagery and seismic data (Taylor *et al.* 1991, 1993; Mutter *et al.* 1992; Goodliffe *et al.* 1993). The simplicity of Fig. 2 belies the tectonic complexity of the region undergoing transpressional deformation and rotation of at

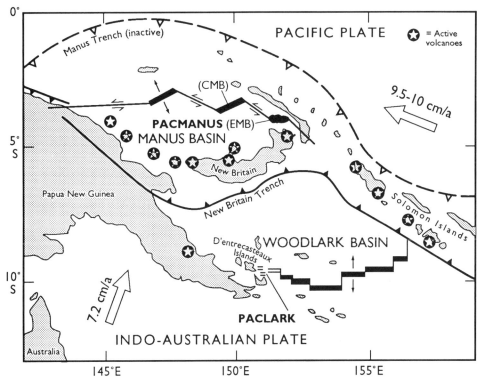

Fig. 2. Tectonic settings of the Manus and Woodlark Basins with the PACLARK and PACMANUS sites located. The complex tectonics of the region is a consequence of the oblique convergence of the Indo-Australian and Pacific Plates (large open arrows). Woodlark is a marginal basin that is propagating into the continental margin of the Indo-Australian plate in the vicinity of the D'Entrecasteaux Islands. Manus is the back-arc basin of the volcanically active New Britain island arc. The PACMANUS site is in the Eastern Manus pull-apart basin (EMB). Sulphide deposits are also known in the Central Manus back-arc basin (Both *et al.* 1986; Tufar 1990).

least three microplates caused by the oblique convergence of the Indo-Australian and Pacific plates (Weissel *et al.* 1982; Taylor *et al.* 1991; Martinez & Taylor 1993; Benes *et al.* 1994).

The work in the eastern Manus Basin has led to the discovery of an actively forming polymetallic sulphide deposit with many characteristics of ancient volcanogenic massive sulphide ores. On the other hand, polymetallic sulphides do not occur in the western Woodlark Basin despite the presence of young felsic volcanic rocks. Instead there are large Fe–Si–Mn oxyhydroxide deposits that are perhaps equivalent to ancient Fe formations, and what appears to be a little recognized type of Au mineralization associated with barite, silica and minor sulphides. As will be shown, these deposits are probably surface manifestations of a failed massive sulphide system.

Eastern Manus Basin

The eastern Manus Basin (Fig. 2), part of the Manus back-arc system, is a pull-apart structure in extended island arc crust. Extension is occurring between two transform faults in a 60 km wide zone within which are exposed subparallel volcanic ridges with intervening sediments (Taylor *et al.* 1991). The presumed arc basement has not been sampled despite several attempts to dredge it. The apparent neovolcanic zone is a 20 km long, high-standing, NE–SW-striking, Y-shaped edifice, informally named Pual Ridge ('Pual' means 'fork' in a local Papua New Guinea dialect). The bathymetry in the vicinity of Pual Ridge is shown in Fig. 3. Pual Ridge itself consists predominantly of highly vesicular dacite, rhyodacite and rhyolite on its upper flanks with andesite at greater depths. The

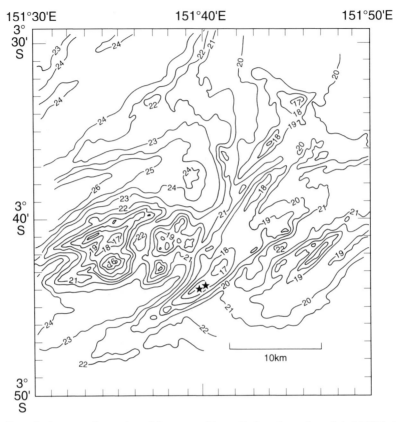

Fig. 3. SeaBeam bathymetry of a portion of the eastern Manus Basin redrawn from Sakai (1991). Contours are in hundreds of metres. The PACMANUS deposit (stars) lies within a zone of hydrothermal activity extending for 3.5 km along the top of Pual Ridge. Pual Ridge is the Y-shaped edifice outlined by the 1900 m contour.

ridge displays two prominent, 40 m high dacite lava domes near the bathymetric minimum at 1630 m water depth. A smaller felsic volcanic ridge lies just to the east of Pual and the hummocky terrain to the west is basalt. The area was deemed to be prospective for polymetallic sulphides because of the presence of pronounced CH_4 and Mn anomalies in the water column and a small active vent site discovered in a basalt cauldron 30 km to the east of Pual Ridge on an earlier expedition (Sakai 1991).

An actively forming polymetallic sulphide deposit, named PACMANUS (stars in Fig. 3), lies along the flank of the southwestern dacite lava dome on Pual Ridge at 1650–1675 m water depth (Binns and Scott 1993). The size of the deposit is known only from camera/video tows and appears to occur intermittently within an area of 800 × 350 m. Another actively forming deposit of unknown size and hosted by altered dacite lies 7 km to the northeast and there are several other inactive hydrothermal deposits

scattered along the ridge. The total strike length of discontinuous hydrothermal deposits along the crest of Pual Ridge is about 8 km. Videos of the PACMANUS deposit show large spires and chimneys up to 4 m high and mounds up to 100 m across. Fauna (crabs, clams, mussels, snails, shrimp, small fish and tubeworms) are abundant in the hydrothermal areas.

Dredging the PACMANUS deposit returned large pieces, both fresh and oxidized, of mound and chimney material. The oxidized samples have pores plugged with Fe oxide and are heavily coated with Mn oxide. The mineralogy consists of high temperature assemblages of chalcopyrite, bornite, pyrite and anhydrite and lower temperature assemblages of sphalerite or wurtzite, galena, tennantite, silica and barite. The bornite and tennantite are Ag-rich. Chemical analyses are given in Table 1 and are compared with other back-arc sites in Table 2. The PACMANUS samples are rich in Cu, Zn, Pb and Ba, which is typical of massive sulphides,

Table 1. *Chemical analyses of samples dredged from the PACMANUS deposit (Analysts: L. Dotter, CSIRO and R. Moss, Toronto)*

	Range	Mean	No. of analyses
Wt.%			
Cu	Tr.–35.2*	10.9	26
Zn	Tr.–55.7	26.9	26
Pb	Tr.–2.6	1.7	26
Fe	1.7–32.8	14.9	26
SiO₂	0.1–2.8	0.8	6
Ca	0.1–0.7	0.3	8
Ba	0.2–13.5	7.3	26
S(total)	28.5–30.1	29.5	6
As	Tr.–2.3	1.1	26
ppm			
Cd	992–1330	1155	6
Bi	3–15	8	4
Hg	15–21	17	8
Sb	146–2659	1130	8
Co	1–8	3	7
Mo	7–11	9	4
Ag	19–400	230	26
Au	1–54	15	26

* Tr. = trace amount.

both modern and ancient, in arc environments (Franklin *et al.* 1981; Scott 1992; Fouquet *et al.* 1993). The Au values of PACMANUS samples in Table 1 are abnormally high and, at the present state of knowledge, are several times greater than those found in samples from other back-arcs and from all mid-ocean spreading ridges.

The average Cu–Zn–Pb ratio of the PAC-MANUS samples is compared with deposits from different settings in Fig. 4, which clearly shows the influence of tectonic and volcanic settings on composition (Fouquet *et al.* 1993). The PACMANUS samples (EMB in Fig. 4) plot in the middle of the field for Valu Fa Ridge of the Lau back-arc basin, which also has felsic volcanic rocks. Both are more Pb-rich than either mid-ocean ridges or a mature back-arc such as the North Fiji Basin, where only basalts are present, but are less Pb-rich than back-arc deposits such as the modern Jade in Okinawa Trough or ancient Kuroko of Japan where there is continental crust. Within the Manus tectonic system itself, samples from the PACMANUS deposit in the eastern Manus Basin contain considerably more Pb than those from the Vienna Woods site (Tufar 1990) in the mature basaltic central Manus back-arc basin (CMB in Fig. 4).

Western Woodlark Basin

Woodlark is a young oceanic basin (Fig. 2) that for the past 5 Ma has been propagating westward at an average rate of 12 cm/a into the Cretaceous–Tertiary continental crust of Papua New Guinea; spreading rates range from 6.0 cm/a at its eastern end to 2.7 cm/a at its western end (Weissel *et al.* 1982; Benes *et al.* 1994). The bathymetry near the tip of the propagator (Fig. 5) is complex, with the relatively smooth region of seafloor spreading in the east giving way westward to rotated blocks of metamorphic continental crust producing en echelon depressions such as South Valley and North Valley and high-standing horsts such as Moresby 'Seamount' (*sic*). The transition from seafloor spreading to continental rifting is accomplished by means of an accommodation zone of transfer faults (Benes *et al.* 1994) such as can be seen from the steep, linear bathymetric contours immediately to the east of Normanby Island in Fig. 5. Volcanism in the accommodation zone consists of basalts and Fe–Ti basaltic andesites of mid-ocean ridge affinity on the floor of East Basin and evolved andesite on the neighbouring Cheshire Seamount. Ahead of the seafloor spreading in the vicinity of Dobu Seamount, submarine volcanism of high Mg

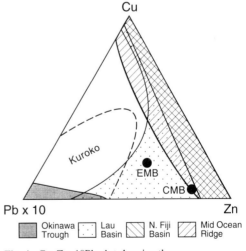

Fig. 4. Cu–Zn–10Pb plot showing the mean composition of the PACMANUS (EMB) samples in Table 1 and central Manus Basin (CMB) samples from Tufar (1990). These are superimposed on fields from Fouquet *et al.* (1993) for sulphide deposits in different settings and, for comparison, the Miocene age Kuroko ores of Japan which formed in an island arc with a thick basement of continental crust. See Fouquet *et al.* (1993) for details.

Table 2. *Average bulk compositions of samples from seafloor sulphide deposits in back-arc basins*

	PACMANUS East Manus Basin	North Fiji Basin	Mariana Trough at 18°N	Valu Fa, Lau Basin	Jade, Okinawa Trough
No. of samples	26	24	11	47	17
Host*	D	B	A	B,A,D	D,R
Source†	1	2	3	4	3,5
Wt.%					
Cu	10.9	7.5	1.2	4.6	3.1
Zn	26.9	6.6	10.0	16.1	24.5
Pb	1.7	0.06	7.4	0.3	12.1
Fe	14.9	30.1	2.4	17.4	4.8
SiO_2	0.8	16.2	1.2	12.5	10.2
Ca	0.3	0.2	3.7	0.6	na
Ba	7.3	0.8	33.3	11.6	3.4
ppm					
Cd	1155	260	465	482	620
Hg	17	na	22	>1	na
As	11 000	na	126	2213	31 000
Sb	1130	na	190	51	na
Ag	230	151	184	256	1160
Au	15	1.0	0.8	1.4	3.3

* Host: A, andesite; B, basalt; D, dacite; and R, rhyolite.
na, Not analysed.
† Sources: 1, Table 1; 2, Bendel *et al.* (1993); 3, Hannington *et al.* (1990); 4, Fouquet *et al.* (1993); and, 5, K. Marumo (pers. comm. 1990).

potassic andesites, basaltic andesites and mildly peralkaline sodic rhyolites is emplaced on thinned continental crust (Binns and Whitford 1987). Similar rhyolites, indicative of continental rifting, are found with subaerial active hot springs on the nearby islands (Smith 1976).

It was knowledge of this rhyolite and associated hot springs that suggested an actively thinning continental margin as a possible site for the formation of felsic volcanic-hosted massive sulphide mineralization that was not related to subduction. This notion was reinforced by the discovery of particulate plumes in the bottom waters. Although the plume signals were weak, filtered samples of seawater recovered particles of Mg silicate and Cu–Fe minerals that were clearly of hydrothermal origin. Mir submersible dives on Dobu Seamount at the far western end of South Valley and East Basin (Binns *et al.*

1990; Lisitzin *et al.* 1991) did not find black smokers or sulphide deposits, however, so the venting there is probably very diffuse. Hydrothermal activity of a different sort was encountered on Franklin Seamount, a small axial volcano near the western tip of the seafloor spreading regime (Fig. 5), where Fe–Si–Mn oxyhydroxides had been dredged on earlier cruises.

Franklin (Fig. 6) is a young submarine basaltic andesite volcano with a pronounced collapsed summit caldera open to the west and with rare occurrences of sodic rhyolite (seen on scree slopes). The top of the volcano is at a water depth of 2138 m and sits 250 m above a prominent volcanic ridge, which itself stands 600 m above the surrounding seafloor. Complexly shaped hydrothermal spires and mounds of Fe–Si–Mn oxyhydroxide up to several meters

Fig. 5. Bathymetry of the western Woodlark Basin where seafloor spreading is propagating into continental crust. Steep slopes to the south and west of Franklin Seamount mark the continental margin. Seafloor spreading extends eastward to Franklin Seamount and then jumps to East Basin where the most recent basalt is seen. Basins to the north, south and west of Moresby 'Seamount' (sic), itself a block of continental crust, are in extended continental crust. Submarine rhyolites occur northeast of Dobu Seamount and subaerial rhyolites with hot springs have been described from nearby islands by Smith (1976). Revised from Benes *et al.* (1994). Place names are informal.

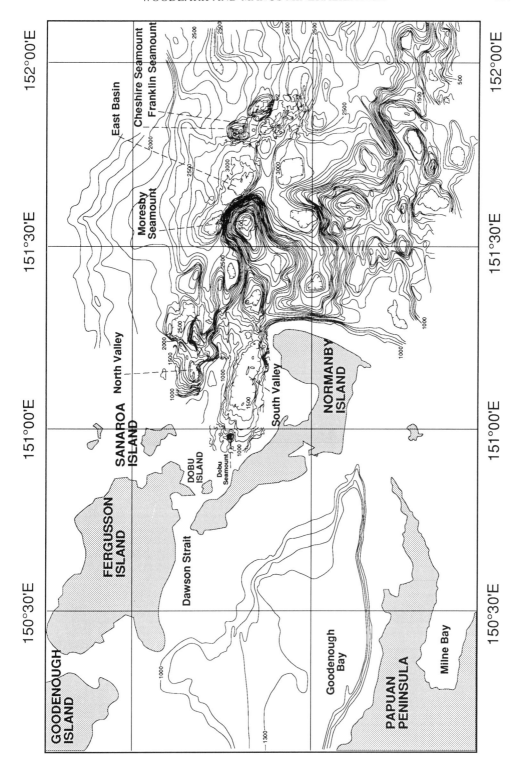

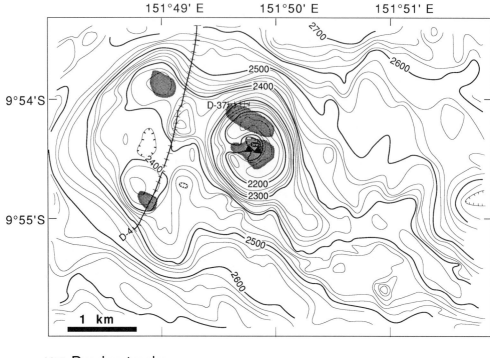

H+H **Dredge tracks**

▲ **Barite spires**   ▦ **Fe-Si-Mn oxyhydroxide field**

Fig. 6. SeaBeam bathymetry of Franklin Seamount with Fe–Si–Mn oxyhydroxides and barite spires located. Location and size of deposit dredged by D–4 is very approximate. Revised from Boyd *et al.* (1993).

thick and 100–200 m in extent are widely distributed at 2143–2366 m depth on and near the seamount (Binns *et al.* 1993). The mounds and spires look very much like those formed by black smokers but in the case of Franklin Seamount there are no sulphides, just oxyhydroxides. The yellow–orange to red–brown deposits consist mainly of amorphous silica and Fe oxyhydroxide, much of it in the form of filaments of probable microbial origin (Fig. 7). Bright green nontronite is crystallized in internal patches and veinlets. Rare earth element patterns and Sr isotope ratios (Binns *et al.* 1993) confirm the importance of seawater components in their formation. The range and average composition of the Fe–Si–Mn oxyhydroxides are in Table 3.

Some spires are venting a 20–30°C fluid at about $5 \, \mathrm{cm \, s^{-1}}$ from irregularly shaped orifices, overhanging ledges, cracks and crevices. A 450-ml sample containing about 98% entrained sea water was taken with a Walden-type titanium syringe. The estimated composition of

the end-member hydrothermal fluid, calculated by extrapolation of the measured values to 0% Mg (Edmond *et al.* 1982), is given in Table 4 together with representative end-members from some other sites. The 270–350°C temperature of the end-member fluid was estimated from the dissolved SiO_2 content of the ambient fluid. The pH of the fluid immediately on being taken out of the sampling syringe was 6.1, but slowly drifted upward to 6.3, probably due to loss of CO_2, although the CO_2 content was not measured. The ambient fluid from Franklin Seamount, and more so its calculated end-member, is mildly acidic and enriched in most elements relative to seawater. This is also the case for fluids from other representative Fe–Si–Mn oxyhydroxide deposits (Galapagos and Loihi) in Table 4. The compositions of the vent fluids from Fe–Si–Mn oxyhydroxide deposits are remarkably similar to those from sulphide-producing vents in Table 4 at mid-ocean spreading ridges (e.g. 21°N and 13°N East Pacific Rise) and back-arc basins (e.g. Valu Fa Ridge in Lau

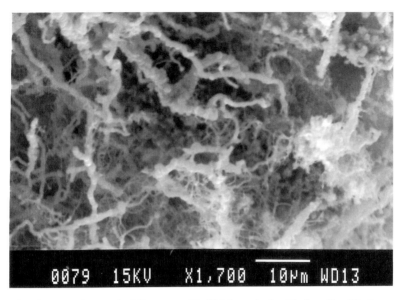

0079 15KV   X1,700   10μm WD13

Fig. 7. Scanning electron micrograph of filamentous Fe–Si–Mn oxyhydroxide from Franklin Seamount.

Basin and the Jade site in the Okinawa Trough). The only significant difference is their H_2S contents. Compared with sulphide-producing vent fluids from black smokers, the H_2S in vent fluids from Fe–Si–Mn oxyhydroxides is considerably less and is zero at Franklin Seamount. This, of course, explains the lack (or paucity at Galapagos) of sulphide minerals in the Fe–Si–Mn oxyhydroxide deposits, but also has further implications which are discussed below.

Spires up to 1 m in height of barite, silica and minor sulphides were found during Mir dives at two localities within the caldera of Franklin Seamount. They were typically coated with Mn oxides and in one case the spire appeared to be poking through Fe–Si–Mn oxyhydroxides. A core sample from the bottom of an oxyhydroxide mound had a high Ba content, which suggests that barite–silica material may underlie the mounds, at least in part. The spires are exceedingly porous and are composed largely of frondescent barite and botryoidal silica with about 2% of sulphides (pyrite and sphalerite with lesser amounts of galena and chalcopyrite). Some spires have a central orifice lined with euhedral barite crystals. Liquid + vapour fluid inclusions in these gave pressure corrected filling temperatures of 184–244°C (average 224°C, $n = 14$) and salinities of 3.4–5.8 equivalent wt.% NaCl (average 4.7 wt.%, $n = 10$). The Franklin Seamount spires have morphology, mineralogy and fluid inclusions that are very similar to the much larger spires described by Hannington and

Scott (1988) from the caldera of Axial Seamount on the Juan de Fuca Ridge.

Chemical analyses of the barite–silica spires are given in Table 5. Particularly notable are the very high values for Ag (to 550 ppm) and Au (to 21 ppm) despite the low sulphide content. Most of the Au is 'invisible' – that is, it cannot be resolved as a separate phase at the highest optical magnification. The strong positive correlation of Au ($R^2 = 0.924$) and Ag ($R^2 = 0.907$) with As + Sb in Fig. 8 suggests that both precious metals are in or closely associated with unidentified sulphosalt minerals. Sulphosalt inclusions of submicron size in pyrite, analysed by electron microprobe, do contain Ag but Au was not detected. These relationships are also similar to those found by Hannington and Scott (1988) at Axial Seamount.

Discussion

PACMANUS, in the eastern Manus back-arc basin, is a classic felsic volcanic-hosted massive sulphide deposit of the type recognized in the ancient geological record. Such deposits are usually enriched in Pb (except for Archean examples), Ag and Au, compared with those in mafic volcanic rocks. The reason for PAC-MANUS being so exceedingly rich in precious metals is unknown, although this is not likely to be an artifact of sampling because all analyses are high in Au and Ag. Whether it is simply a consequence of source rocks that are enriched in

Table 3. *Analyses of Fe–Si–Mn oxyhydroxides from Franklin Seamount by instrumental neutron activation and X-ray fluorescence (T. Boyd, analyst)*

	Range*	No. of analyses	Average
Wt.%			
SiO_2	9.3–64.7	32	26.1
TiO_2	bd–0.6	19	0.1
Fe_2O_3 (total)	10.1–52.9	19	35.2
Al_2O_3	0.04–6.5	32	1.4
MnO	0.08–25.7	19	4.9
MgO	0.5–2.7	32	1.4
CaO	0.4–19.7	32	3.2
K_2O	0.09–1.5	19	0.9
Na_2O	0.8–2.8	19	1.9
P_2O_5	0.5–5.3	19	1.9
S	0.09–0.5	23	0.2
LOI	17.5–42.8	19	23.4
Total			100.6
ppm			
Zn	bd–309	46	93
Pb	bd–26	24	bd
Cu	bd–308	24	64
Mo	2–2020	46	270
Ni	bd–540	46	120
Co	bd–269	46	65
As	13–4020	46	935
Sb	1–31	46	11
Hg	bd–4.8	46	0.7
Ag	bd–4	46	bd
Au (ppb)	10–181	46	32

* bd = Below detection.

precious metals, or perhaps a consequence of direct input of the elements from a magmatic fluid (cf. Urabe 1987; Urabe and Marumo 1991) will have to await the sampling and analysis of the relevant basement rocks and vent fluids. However, both the eastern Manus Basin and western Woodlark Basin are within an Au-rich metallogenic province extending from Kyushu in southern Japan through the western Pacific arcs to North Island of New Zealand (Herzig *et al.* 1993).

The tectonic settings of Precambrian massive sulphides are obscure, but they are also typically closely spatially associated with felsic volcanic rocks. A good example is the Noranda district of Quebec in Canada, where some massive sulphides (e.g. Millenbach mine; Knuckey *et al.* 1982) formed on ridges of coalesced rhyolite lava domes that are underlain and surrounded by basalt and andesite. This volcanology is practically identical to that of the Pual Ridge and vicinity. As such and given its apparent large size, the PACMANUS site represents the best

modern analogue yet found of ancient felsic volcanic-hosted massive sulphide ores.

The mere presence of felsic volcanic rocks does not by itself mean that conditions were correct for massive sulphides to form, as our results in the western Woodlark Basin demonstrate. Although weak hydrothermal venting was detected, this was from basaltic areas and not in the vicinity of the submarine rhyolites where there is no evidence of present or former hydrothermalism. Subduction is not occurring here so Woodlark is not a back-arc basin. Neither is seafloor spreading occurring in the vicinity of the rhyolites so, by both accounts, there is not likely to be a near-surface magma chamber such as exists in back-arc basin and mid-ocean spreading centres and which is necessary for sustained hydrothermal circulation. The neovolcanic rocks in the continental crustal portion of the western Woodlark west of East Basin (Fig. 5) have trace element and isotopic signatures of a mantle origin without contamination by continental crust (R.A. Binns and G.E. Wheller, unpublished data), which further suggests that the area lacks near-surface magma reservoirs.

Hydrothermal deposits do occur in the western Woodlark Basin, on Franklin Seamount, but these are not massive sulphides. The barite–silica spires and the Fe–Si–Mn oxyhydroxide deposits and vent fluids here are conspicuous by their paucity to total lack of sulphide. This is puzzling because the reaction between hot basalt and seawater is expected to produce abundant H_2S as it does at mid-ocean spreading centres and back-arc basins. In all other regards, the composition of the end-member fluid from Franklin Seamount has the appearance of a black smoker from a mid-ocean spreading ridge or back-arc (Table 4). It is an inescapable conclusion that the reduced sulphur was lost beneath the seafloor within the volcano, most likely by precipitation of metallic sulphides in a stockwork or as disseminations. The presence of nontronite (an Fe^{2+} mineral) in the vent areas of the oxyhydroxide deposits precludes the loss of reduced sulphur by oxidation of the vent fluid through mixing with fresh ambient seawater beneath the seafloor, as does the presence of the barite–silica spires. Barite is extremely insoluble in a sulphate-containing fluid and Ba would have been totally titrated from the fluid before it reached the seafloor. The barite–silica and Fe–Si–Mn oxyhydroxide mineralization must therefore be the manifestation of a failed massive sulphide system in which the metals were deposited beneath the sea floor.

Ferruginous cherts, into which the Fe–Si–Mn

Table 4. *Composition of vent fluid from Franklin Seamount compared with other Fe–Si–Mn oxyhydroxide deposits, sulphide-forming black smokers and seawater. Data from Von Damm (1990) and Scott (in press) unless indicated otherwise*

	Franklin Seamount (ambient)*	Franklin Seamount (end-member)*	Galapagos Mounds (end-member)	Loihi Seamount (end-member)	21°N EPR (end-member)	13°N EPR (end-member)	Valu Fa Ridge, Lau (end-member)	Okinawa Trough (end-member)	Seawater
Temperature (°C)	20–30(est)	270–350	<13	na	350	380?	334	320	2
pH (measured)	6.1	na	na	na	3.5	3.2	2	4.7	7.8
Cl(ppt)	18.89	24.46	16.44	6.13	18.26	25.96	27.25	19.50	19.18
Cl(mmol kg⁻¹)	533	690	464	173	515	733	769	550	541
mmol kg⁻¹									
H_2S	0	0	~3	>0.002	7.5	5.2	na	12.4	0
CO_2	na‡	na	10.3	6474	5.7	13.8	na	198	2.3
SiO_2	0.33	13.5	21.9	24.9	17.5	20.3	14.2	12.9	0.16
Ca	10.5	48.5	33.4	31.9	16.2	52	39.4	22.3	10.2
NH_4	7.5	na	na	na	<0.01	na	na	5.0	<0.01
Li	0.04	0.7	1.0	0.3	1.0	0.6	0.6	2.5	0.026
µmol kg⁻¹									
Mn	10	630	593	459	885	1914	6800	110	<0.001
Fe	38	2400	~600	21956	1429	6640	1710	2.8	<0.001
Zn	0.4	25	na	na	85	36	2400	7.6	0.01
Cu	0.8	48	0	na	22	36	24	0.003	0.007
nmol kg⁻¹									
Pb	1.2	50	na	na	261	59	4900	36	0.01
Ag	0.23	7	na	na	26	24	na	na	0.02

End-member calculated for Mg = 0. Seawater also contains 27.9 mmol kg⁻¹ SO₄ and 52.7 mmol kg⁻¹ Mg.
na, Not analysed or not calculated.
* Lisitzin *et al.* (1991) and Binns *et al.* (1993). Average of two analyses of ambient bottom water at Franklin Seamount used to calculate end-member. End-member temperature calculated from silica geothermometer by Binns *et al.* (1993).
† Inferred end-member *T* = 350°C for Galapagos.
‡ CO_2 detected by upward drift of pH but not measured.

Table 5. *Composition of barite–silica spires from Franklin Seamount (n = 6). Data from Binns et al. (1993)*

	Minimum	Maximum
Wt.%		
$BaSo_4$	52	88
SiO_2	2	40
FeS_2	1	2
ppm		
Cu	80	350
Zn	220	6200
Pb	170	630
As	190	370
Sb	90	590
Hg	60	470
Ag	130	550
Au	4	21

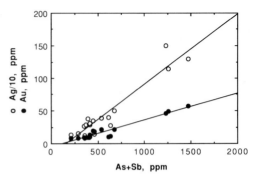

Fig. 8. Correlation of Ag and Au with As + Sb content for barite spires from Franklin Seamount. The three samples with the highest values for As + Sb are from sulphide concentrates obtained by heavy liquid extraction. Analysts: T. Boyd, Toronto and L. Dotter, CSIRO.

oxyhydroxide deposits will be eventually transformed through lithification, are fairly common in the ancient geological record. In Table 6, the Franklin Fe–Si–Mn oxyhydroxides have been recalculated to an anhydrous equivalent and compared with the mean compositions of two types of ancient ferruginous cherts. The Japanese Kuroko and Noranda massive sulphide ore bodies have in their hanging walls and along-strike from the ore distinctive sediments, 'tuffaceous exhalites' (Kalogeropoulos and Scott 1983, 1989), which consist of recognizable tuffaceous and hydrothermal constituents. The latter consists mainly of Fe oxide (hematite or magnetite), metallic sulphides and silica (chert). These rocks are remarkably similar from the two localities despite the vast differences in their ages (Miocene in Japan and Archean at Noranda). As Table 6 shows, the Fe–Si–Mn oxyhydroxide deposits of Franklin Seamount are very different from the chemical tuffaceous exhalites, having lower contents of Al_2O_3 and sulphur, a higher content of Mn and a higher Fe_2O_3/SiO_2 ratio. The Franklin Seamount deposits are, in fact, not dissimilar to average Precambrian oxide facies iron formations (column 5 in Table 6) and add credence to a hydrothermal origin for these enigmatic ancient chemical sediments.

Conclusions

The eastern Manus Basin and nearby western Woodlark Basin offer contrasting types of present day hydrothermal activity and deposits, both of which are relevant to interpreting the ancient geological record. The PACMANUS

site in eastern Manus Basin is a classic felsic volcanic hosted massive sulphide in a back-arc setting and offers, together with similar occurrences such as Jade in the Okinawa Trough (Halbach *et al.* 1993) and Valu Fa Ridge in the Lau Basin (Fouquet *et al.* 1993), a distinctive and relevant contrast with better known basalt-hosted sites on mid-ocean ridges. Further study could, for example, clarify the relative roles of magmatic- and seawater-derived hydrothermal fluids in a felsic back-arc setting, which may differ significantly from mid-ocean ridges. The answer has two vital implications for land-based mineral exploration concepts: whether or not fractionated volcanic rocks are more prospective in general, and what is the role of magmatic fluids in generating large, high-grade 'world class' ore bodies. The Franklin Seamount occurrences confirm that hydrothermal activity may be associated with submarine volcanism where seafloor spreading propagates into a continental margin. This setting may have numerous ancient analogues, but conditions appear not to be correct for spawning massive sulphides despite the presence of felsic volcanic rocks. The Fe–Si–Mn oxyhydroxide deposits at Franklin Seamount support an exhalative origin for iron formations associated with ancient volcanic sequences and the associated ore-grade Au and Ag in barite–silica chimneys suggest targets for land-based exploration. Both are the probable surface manifestation of extensive sub-seafloor sulphide mineralization of a failed massive sulphide system. Further study may find evidence from fluid and rock chemistry for why the hydrothermal system failed and how mineralization of this type may be distinguished from real stockworks beneath hidden massive sulphide

Table 6. *Comparison of analyses of Fe–Si–Mn oxyhydroxides from Franklin Seamount (recalculated to LOI = 0) with chemical constituent of ferrugenous chert (tetsusekiei) from a Japanese Kuroko massive sulphide deposit (Kalogeropoulos and Scott 1983), a similar material (chemical constituent of Main Contact Tuff) from a Noranda massive sulphide deposit (Kalogeropoulos and Scott 1989) and average Precambrian Lake Superior-type oxide facies iron formation (Gross 1988)*

	Franklin Seamount (LOI = 0)	Kuroko tetsusekiei	Noranda Main Contact Tuff	Average Precambrian oxide facies iron formation
Wt.%				
SiO_2	33.8	47.1	53.0	47.7
TiO_2	0.1	0.2	0.7	0.0
Fe_2O_3 (total)	45.6	25.3	20.6	44.3
Al_2O_3	1.8	5.5	10.2	1.3
MnO	6.3	0.1	0.2	0.6
MgO	1.8	2.9	2.1	1.2
CaO	4.1	0.7	1.0	1.6
K_2O	1.2	0.8	2.1	0.2
Na_2O	2.5	na	0.6	0.1
P_2O_5	2.5	0.4	0.1	0.1
S	0.3	14.5	8.3	0.0
LOI	0	0.7	0.8	3.9
Total	100.0	98.1	99.7	101.0
Fe_2O_3/SiO_2	1.3	0.5	0.4	0.9

ores in ancient terranes. These issues cannot be effectively addressed using samples or geophysical data collected at the seafloor alone, nor do studies of ancient mineralized environments, affected by metamorphism and deformation, allow conclusive answers. The essential requirement for these studies is pristine samples from the interior of hydrothermally active volcanic edifices such as Pual Ridge and Franklin Seamount, the best representatives currently known for the two 'end-members' of the hydrothermal deposits spectrum.

This paper is a product of our ongoing research offshore eastern Papua New Guinea. We thank our many colleagues and ship's officers and crew who have participated on the PACLARK, SUPACLARK and PACMANUS series of oceanographic expeditions, L. Dotter (CSIRO), T. Boyd (Toronto) and R. Moss (Toronto) for providing various chemical analyses, and H. Sakai (Yamagata) and K. Tamaki (Tokyo) for making their bathymetric map of the eastern Manus Basin available to us prior to its publication. Scott's research is funded by the Natural Sciences and Engineering Research Council of Canada and the Bank of Nova Scotia. Binns' research is funded by CSIRO. This paper was written while Scott was on leave at the Hawaii Undersea Research Laboratory, University of Hawaii. We thank I. Wright, D. Rickard and L. Parson for thoughtful reviews.

References

BENDEL, V., FOUQUET, Y., AUZENDE, J.-M., LAGABRIELLE, Y., GRIMAUD, D. & URABE, T.

1993. The White Lady hydrothermal field, North Fiji back-arc basin, southwest Pacific. *Economic Geology*, **88**, 2237–2249.

BENES, V., SCOTT, S. D. & BINNS, R. A. 1994. Tectonics of rift propagation into a continental margin: Western Woodlark Basin, Papua New Guinea. *Journal of Geophysical Research*, **99**, 4439–4455.

BINNS, R. A. & SCOTT, S. D. 1993. Actively forming polymetallic sulfide deposits associated with felsic volcanic rocks in the eastern Manus back-arc basin, Papua New Guinea. *Economic Geology*, **88**, 2226–2236.

—— & WHITFORD, D. J. 1987. Volcanic rocks from the western Woodlark basin, Papua New Guinea. *Pacific Rim Congress '87 Proceedings*, **1**, 531–535.

——, Scott, S. D. & 9 others. 1993. Hydrothermal oxide and gold-rich sulfate deposits of Franklin Seamount, western Woodlark Basin, Papua New Guinea. *Economic Geology*, **88**, 2122–2153.

——, ——, WHELLER, G. E. & BENES, V. 1990. Report on the SUPACLARK cruise, Woodlark Basin, Papua New Guinea, April 8–28, 1990, RV Akademik Mstislav Keldysh. *CSIRO Division of Exploration Geoscience, Restricted Report 176R*, 1–55.

BOTH, R., CROOK, K. & 7 others. 1986. Hydrothermal chimneys and associated fauna in the Manus back-arc basin, Papua New Guinea [abstract]. *Eos, Transactions of the American Geophysical Union*, **67**, 489–491.

BOYD, T., SCOTT, S. D. & HEKINIAN, R. 1993. Trace element patterns in Fe–Si–Mn oxyhydroxides at three hydrothermally active sea-floor regions. *Journal of the Society of Resource Geology, Special Issue*, **17**, 83–95.

CATHLES, L. M. 1993. A capless 350°C flow zone model to explain megaplumes, salinity variations, and high temperature veins in ridge axis hydrothermal systems. *Economic Geology*, **88**, 1977–1988.

EDMOND, J., VON DAMM, K. L., McDUFF, R. E. & MEASURES, C.I. 1982. Chemistry of hot springs on the East Pacific Rise and their effluent dispersal. *Nature*, **297**, 187–191.

FOUQUET, Y., VON STACKELBERG, U., CHARLOU, J. L., ERZINGER, J., HERZIG, P.M., MÜHE, R. & WIEDICKE, M. 1993. Metallogenesis in back-arc environments: the Lau Basin example. *Economic Geology*, **88**, 2154–2181.

FRANKLIN, J. M., LYDON, J. W. & SANGSTER, D. F. 1981. Volcanic-associated massive sulfide deposits. *Economic Geology 75th Anniversary Volume*, 485–627.

GOODLIFFE, A., TAYLOR, B., HEY, R. & MARTINEZ, F. 1993. Seismic images of continental breakup in the Woodlark Basin, Papua New Guinea [abstract]. *Eos, Transactions of the American Geophysical Union, Supplement*, **74**, 606.

GROSS, G. 1988. *Gold Content and the Geochemistry of Iron Formation in Canada*. Geological Survey of Canada Paper 86–19.

HALBACH, P., PRACEJUS, B. & MÄRTEN, A. 1993. Geology and mineralogy of massive sulfide ores from the central Okinawa Trough, Japan. *Economic Geology*, **88**, 2210–2225.

HANNINGTON, M. D. & SCOTT, S. D. 1988. Mineralogy and geochemistry of a silica–sulfide–sulfate spire in the caldera of Axial Seamount, Juan de Fuca Ridge. *Canadian Mineralogist*, **26**, 603–625.

——, HERZIG, P. M. & SCOTT, S. D. 1990. Auriferous hydrothermal precipitates on the modern sea-floor. *In*: FOSTER, R. P. (ed.) *Gold Metallogeny and Exploration*, Blackie, Glasgow, 249–282.

HERZIG, P. M., HANNINGTON, M. D., FOUQUET, Y., VON STACKELBERG, U. & PETERSEN, S. 1993. Gold-rich polymetallic sulfides from the Lau back arc and implications for the geochemistry of gold in sea-floor hydrothermal systems of the southwest Pacific. *Economic Geology*, **88**, 2182–2209.

KALOGEROPOULOS, S. I. & SCOTT, S. D. 1983. Mineralogy and Geochemistry of Tuffaceous Exhalites (Tetsusekiei) of the Fukazawa Mine, Hokuroku District, Japan. *Society of Economic Geologists, Economic Geology Monograph*, **5**, 412–432.

—— & —— 1989. Mineralogy and geochemistry of an Archean tuffaceous exhalite: the Main Contact Tuff, Millenbach Mine area, Noranda, Quebec. *Canadian Journal of Earth Science*, **26**, 88–105.

KNUCKEY, M. J., COMBA, C. D. A. & RIVERIN, G. 1982. Structure, metal zoning and alteration at the Millenbach deposit, Noranda, Quebec. *Precambrian Sulphide Deposits*. Geological Association of Canada, Special Paper, **25**, 255–295.

LARGE, R. R. 1992. Australian volcanic-hosted massive sulfide deposits: features, styles, and genetic models. *Economic Geology*, **87**, 471–510.

LISITZIN, A. P., BINNS, R. A. & 7 others. 1991. [Present-day hydrothermal activity of Franklin Seamount and the western part of the Woodlark Sea, Papua New Guinea]. *Isvestiya, Russian Academy of Science, Geological Series*, **8**, 125–140 [in Russian]; English translation *International Geology Review*, **33**, 914–929.

LYDON, J. L. 1988. Volcanogenic massive sulphide deposits. Part 2: genetic models. *Geoscience Canada*, **15**, 43–65.

MARTINEZ, F. & TAYLOR, B. 1993. Manus Basin, Bismarck Sea: an epitome of microplate deformation[abstract]. *Eos, Transactions of the American Geophysical Union, Supplement*, **74**, 605.

MUTTER, J. C., DIEBOLD, J. B., MUTTER, C. Z., ABERS, G., SCOTT, S., BENES, V., LISTER, G. & PAHL, A.-K. 1992. Continental breakup by rift propagation in the Woodlark Basin/D'Entrecasteaux Islands mimics oceanic propagation [abstract]. *Eos, Transactions of the American Geophysical Union, Supplement*, **73**, 536.

OHMOTO, H. & SKINNER, B.J. (eds) 1983. *The Kuroko and Related Volcanogenic Massive Sulfide Deposits*. Society of Economic Geologists, Economic Geology Monograph, **5**.

RICKARD, D. T. & ZWEIFEL, H. 1975. Genesis of Precambrian sulfide ores, Skellefte district, Sweden. *Economic Geology*, **70**, 255–274.

RIVE, M. *et al.* (eds) 1990. *The Northwestern Quebec Polymetallic Belt*. Canadian Institute of Mining and Metallurgy, Special Volume, **43**.

RONA, P. A. & SCOTT, S. D. 1993. A special issue on sea-floor hydrothermal mineralization: New perspectives – preface. *Economic Geology*, **88**, 1933–1976.

SAKAI, H. 1991. *Expedition East Manus Basin Hydrothermal Field, Hakuro-Maru Cruise KH90-3, Leg 2. A Brief Summary Report for SOPAC*. South Pacific Applied Geoscience Commission, SOPAC Cruise Report, **138**.

SANGSTER, D. F. & SCOTT, S. D. 1976. Precambrian, stratabound massive Cu–Zn–Pb sulfide ores in North America. *In*: WOLF, K. H. (ed.) *Handbook of Stratabound and Stratiform Ore Deposits*, Vol. 7, Elsevier, Amstgerdam, 129–222.

SAWKINS, F. J. 1984. *Metal Deposits in Relation to Plate Tectonics*. Springer-Verlag, Berlin.

SCOTT, S. D. 1992. Polymetallic sulfide riches from the deep: fact or fallacy? *In*: HSU, K. J. & THIEDE, J. (eds) *Use and Misuse of the Seafloor*. Wiley, Chichester, 87–114.

——. Submarine hydrothermal systems and deposits. *In*: BARNES, H. L. (ed.) *Geochemistry of Hydrothermal Ore Deposits*, 3rd Edn, Wiley, Chichester, in press.

SMITH, I. E. M. 1976. Peralkaline rhyolites from the D'Entrecasteaux Islands, Papua New Guinea. *In*: JOHNSON, R. W. (ed.) *Volcanism in Australia*. Elsevier, Amsterdam, 275–285.

TAYLOR, B. J., CROOK, K. A. W., SINTON, J. L. & PETERSEN, L. 1991. *Manus Basin, Papua New Guinea*. Hawaii Institute of Geophysics, Pacific Sea Floor Atlas, Sheets 1–7.

——, MARTINEZ, F., HEY, R. & GOODLIFFE, A. 1993. A new view of continental rifting and initial sea-floor spreading: the Woodlark Basin, PNG [abstract]. *Eos, Transactions of the American Geophysical Union, Supplement*, **74**, 606.

——, SINTON, J., CROOK, K. A. W. & SHIPBOARD PARTY 1986. Extensional transform zone, sulfide chimneys and gastropod vent fauna in the Manus Back-arc Basin [abstract]. *Eos, Transactions of the American Geophysical Union, Supplement*, **67**, 377.

TUFAR, W. 1990. Modern hydrothermal activity, formation of complex massive sulfide deposits and associated vent communities in the Manus back-arc basin (Bismarck Sea, Papua New Guinea). *Österreichishe Geologische Gesellschaft Mitteilungen*, **82**, 183–210.

URABE, T. 1987. [Kuroko deposit models based on magmatic hydrothermal theory]. *Mining Geology*, **37**, 159–176 [in Japanese].

—— & MARUMO, K. 1991. A new model for kuroko-type deposits of Japan. *Episodes*, **14**, 246–251.

VAN STAAL, C. R., FYFE, L. R., LANGTON, J. P. & McCUTCHEON, S. R. 1992. The Ordovician Tetagouche Group, Bathurst camp, northern New Brunswick, Canada: history, tectonic setting, and distribution of massive sulfide deposits. *Exploration and Mining Geology*, **1**, 93–103.

VON DAMM, K. L. 1990. Seafloor hydrothermal activity: black smoker chemistry and chimneys. *Annual Review of Earth and Planetary Sciences*, **18**, 173–204.

WEISSEL, J. K., TAYLOR, B. & KARNER, G.D. 1982. The opening of the Woodlark Basin, subduction of the Woodlark spreading system, and the evolution of northern Melanesia since mid-Pliocene time. *Tectonophysics*, **87**, 253–277.

Mineralogy and sulphur isotope characteristics of a massive sulphide boulder, Galapagos Rift, 85°55′W

RICHARD KNOTT[1], ANTHONY E. FALLICK[2], DAVID RICKARD[1]
& HARALD BÄCKER[3]

[1]*Department of Earth Sciences, University of Wales, Cardiff CF1 3YE, UK*
[2]*Isotope Geosciences Unit, SURRC, East Kilbride, Glasgow G75 0QU, UK*
[3]*GEOMAR Technologie GmbH, Wischofstrasse 1–3, Geb. 11 24148 Kiel, Germany*

Abstract: The submarine hydrothermal sulphide deposits of the Galapagos Rift are characterized by being exceptionally Cu-rich. Fe- and Cu–Fe sulphides are extremely heterogeneous, display granular textures and evidence of open space growth. Core was examined from a 1.5 m diameter massive sulphide boulder collected by TV grab on the GARIMAS II cruise. The main growth stage is dominated by pyrite and chalcopyrite, forming granular aggregates from <1 mm to >5 cm. These sulphides have $\delta^{34}S$ from +2.7 to +5.5‰. The Fe–Cu sulphides are modified by minor seawater oxidation and a late stage marcasite–sphalerite–galena–barite assemblage. $\delta^{34}S$ varies along the core in a systematic manner and there appears to be an isotopically heavy deviation from the modal +3.9‰. This is related to particular zones and suggests a second, isotopically distinct S source, which is interpreted to be a hydrothermal diagenetic solution derived from a shallow seawater–hydrothermal circulation system.

The Galapagos Rift between 85°50.5′W and 85°58.5′W hosts large, faulted, Cu-rich massive sulphide deposits (Malahoff *et al.* 1983). Present hydrothermal activity is entirely low temperature, forming Fe and Mn oxide and silica deposits (Corliss *et al.* 1979). The GARIMAS II (SO39) cruise by the RV Sonne in 1985 sampled 19.9 tonnes of massive sulphide, using a TV-controlled electro-hydraulic grab (GARIMAS II Cruise Report 1986). This paper describes the detailed study of a core taken through the centre of the largest sample recovered (3.3 tonnes and 1.8 by 1.4 by 1.2 m in dimension).

Recently formed submarine hydrothermal deposits (i.e. those less than a few thousand years old) show considerable divergence in textural, mineralogical and geochemical characteristics, both between each other (e.g. Graham *et al.* 1988; Paradis *et al.* 1988) and with ancient volcanic-hosted massive sulphide (VMS) deposits (e.g. Oudin *et al.* 1981; Halbach et al. 1989). Many of the petrological characteristics of ancient deposits are derived from later, sub-seafloor hydrothermal diagenetic processes superimposed on the primary hydrothermal precipitates. The Galapagos deposits are interesting because they are large, Fe–Cu sulphide dominated, and among the oldest of the recent seafloor sulphide deposits (8000 to 12 000 years, Lalou *et al.* 1989). Other deposits of similar character have been described by Fouquet *et al.*

(1988) at off-axis sites on the East Pacific Rise at 13°N (2000 to 20 000 years, Lalou *et al.* 1985); by Alt (1988) at the Green Seamount (70 000 to 140 000 years, Alt *et al.* 1987); and by Scott *et al.* (1991) from the Southern Explorer Ridge (undated). This study is aimed at investigating in detail the Galapagos mineralization to examine the contribution of post-depositional processes to the evolution of the deposit.

Regional geological setting

The Galapagos Rift between longitude 86°14′W and 85°48′W is spreading at a medium rate (3.0 cm a⁻¹). It has been described by Embley *et al.* (1988) as a structurally complex double rift, separated by a narrow fault-bounded ridge (Fig. 1). The central ridge has been interpreted as a tectonically uplifted block (Embley *et al.* 1988), although Ridley *et al.* (1994) reinterpret the structure as a relict part of an axial summit caldera. Inactive, disrupted massive sulphide deposits occur around this central ridge and are spatially related to axis-parallel fault structures. Mounds, chimneys and crusts of Fe–Si–Mn oxyhydroxide are ubiquitous in the rift, but active deposits are confined to a neovolcanic zone situated to the north of the central ridge (Fig. 1). Venting of low temperature (<17°C) fluids is precipitating Fe–Si–Mn oxyhydroxide

From PARSON, L. M., WALKER, C. L. & DIXON, D. R. (eds), 1995, *Hydrothermal Vents and Processes*, Geological Society Special Publication No. 87, 207–222.

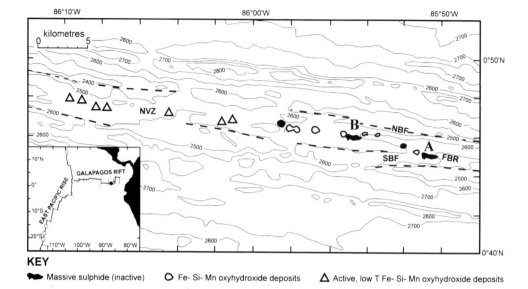

KEY

🦴 Massive sulphide (inactive)   ⭕ Fe- Si- Mn oxyhydroxide deposits   △ Active, low T Fe- Si- Mn oxyhydroxide deposits

Fig. 1. Bathymetric map of the Galapagos Rift between 86°12′W and 85°50′W, showing location of hydrothermal deposits. The Northern Boundary Fault (NBF), Southern Boundary Fault (SBF), neovolcanic zone (NVZ) and central fault-bounded ridge (FBR) are also indicated. The sample under investigation comes from site B. Modified from GARIMAS II Cruise Report (1986).

and is attracting abundant biological activity (Corliss *et al.* 1979).

Massive sulphides

Inactive hydrothermal sulphide deposits (Fig. 1) have been investigated at 85°52′W (site A) and at 85°54.75′W (site B). The deposits have been mapped using submersible, TV grab and deep-tow camera sled, and cover areas of up to 650 m by 100 m (GARIMAS II Cruise Report 1986). At site A, faulting has exposed the core of a sulphide mound and the related subsurface stockwork and alteration zones (Embley *et al.* 1988; Ridley *et al.* 1994). It is estimated that the deposit at site A amounts to 1.6×10^6 tonnes. The deposit at site B is in a broadly similar structural setting and could be of comparable size.

Massive sulphides occur on the seafloor as upstanding edifices up to 3 m high, projecting through a cover of gossanous sediments consisting of Fe oxyhydroxides, sulphides and silica (Marchig *et al.* 1987). Chimney structures (Ballard *et al.* 1982) and large blocks of fault talus (Embley *et al.* 1988) have been observed. It is one such structure, sampled from site B, that forms the basis of this investigation. It is not certain if this structure is a disaggregated block

or *in situ*, perhaps linked to more extensive subsurface sulphide mineralization.

The age of the massive sulphides from sites A and B has been determined by Lalou *et al.* (1989) to be in two groups, at 8000 and 12 000 years. Sulphides of both ages are present at site B, with the older samples confined to the southern periphery of the deposit. The sample under investigation is from a group of structures dated 8900 ± 700 years to 9000 ± 1000 years (Lalou *et al.* 1989).

Methods

The large massive sulphide structure (SO39–126 GTVC) was recovered from site B (0°46.00′N 5°54.75′W) in a water depth of 2579 m, and was sampled by a continuous 5.5 cm diameter, 1 m long core drilled subvertically through the centre. Macro-scale observation and porosimetry measurements were made on the archived half of the core. Position is assigned as distance along the core from 0 to 100 cm. Mineralogy was studied by optical microscopy, scanning electron microscopy (SEM) and X-ray diffraction (XRD). Bulk chemical analyses were by X-ray fluorescence (XRF) and trace element analysis of individual mineral grains was by electron

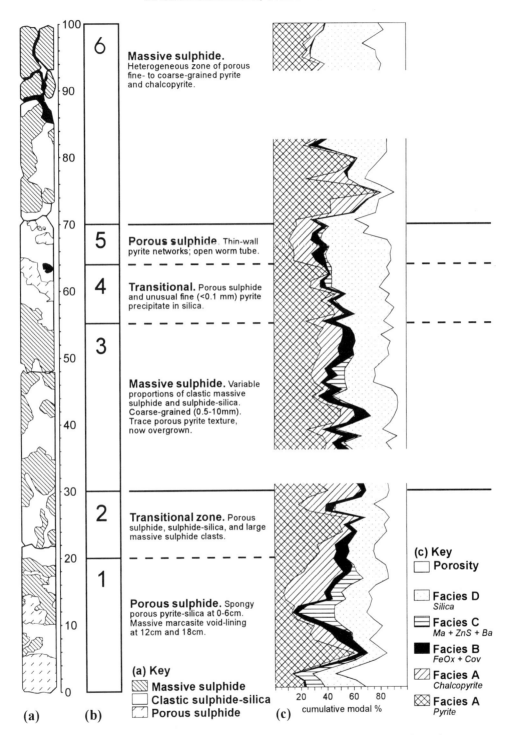

Fig. 2. Massive sulphide core description. (**a**) Map of the core section showing the textural zonation; (**b**) summary log showing the zone definitions, based on mineralogy and texture (solid and broken lines indicate sharp and gradational zone boundaries, respectively; (**c**) mineralogical variation based on point-count data for 1 cm intervals from polished sections, showing how the mineral facies vary systematically despite the textural heterogeneity.

probe microanalysis (EPMA). Sulphur isotope analysis was made on sulphide mineral separates, prepared by precision drilling and hand picking, and checked for purity by powder camera XRD. Samples were roasted with Cu_2O following the method of Robinson & Kusakabe (1975) and the resulting SO_2 was purified by vacuum distillation. Isotopic analyses were carried out at the Scottish Universities Research and Reactor Centre (SURRC) on a SIRA II dual-inlet mass spectrometer. $\delta^{34}S$ is reported relative to Cañon Diablo troilite (CDT). Analytical uncertainty is $\pm 0.2‰$ (1σ).

Sample description

The sample investigated is representative of those massive sulphides that form as relief features on the seafloor. Initial inspection revealed an extreme heterogeneity of both texture and mineralogy, with zones of massive sulphide, massive silica–sulphide and porous sulphide (Fig. 2a). The boulder is dominantly composed of granular pyrite and chalcopyrite which varies from fine-grained (<0.5 mm) individual grains and aggregates to large, massive composite bodies greater than 5 cm in size. The sulphide component is cemented by varying amounts of silica. The core also has porous zones that are texturally and mineralogically distinct.

Three styles of mineralization are defined. Macro- and micro-scale features of these textural types are illustrated in Fig. 3a–d.

1. Granular massive sulphide. Massive sulphides are defined as total sulphide (pyrite and chalcopyrite) exceeding 60%. The sulphides commonly occur as massive areas with sharply defined outlines, but may also grade to finer, disrupted silica–sulphide. Some massive Fe disulphides display remnants of primary growth textures, preserving the outlines of fossilized worm tubes with thin walls of pyrite.

2. Granular sulphide–silica. Sulphide–silica mineralization is composed of >40% amorphous silica. Sulphides occur as angular grains and aggregates typically <2 mm in size, enclosed in silica cement.

3. Porous sulphide. Porous zones of Fe disulphides and minor silica contain large (>5 mm) cavities. These zones exhibit three-dimensional networks of thin-walled sulphide, which may be fossilized worm tubes. A distinct subtype of this mineralization occurs as mineral overgrowths (dominantly marcasite) lining large (>2 cm) void spaces.

It is possible to subdivide the core into six zones on the basis of the styles of mineralization defined above. The zones are numbered 1 to 6, and summarized in Fig. 2b. Zones 1 (0–20 cm) and 5 (64–70 cm) are dominantly composed of porous sulphide. Zone 3 (30–55 cm) is composed of variable proportions of granular massive sulphide and sulphide–silica. Zones 2 (20–30 cm) and 4 (55–64 cm) are transitional, with some porous sulphide present. Zone 6 (70–100 cm) is predominantly massive sulphide, but distinctly finer-grained and less siliceous than zone 3 massive sulphide.

Mineralogical variations

The boulder is dominantly composed of pyrite and amorphous silica, with abundant chalcopyrite and minor marcasite. Covellite, sphalerite, barite and amorphous Fe oxyhydroxide are present as minor to trace phases, with rare galena, bornite and idaïte. Four mineral facies can be recognized in the sample. These are, in order of formation, (A) main stage dominated by Fe–Cu sulphide, (B) covellite and Fe oxyhydroxide assemblage, (C) later Fe–Zn–Pb–Ba mineralization and (D) amorphous silica precipitation. The mineralogy and textures which characterize these facies are illustrated in Fig. 3e–h.

Fig. 3. Textural and mineralogical features of the Galapagos massive sulphides. (**a**) Detail of two parts of the sulphide core between 0 and 18 cm and between 38 and 56 cm. Typical porous sulphide at 3 cm; granular sulphide–silica between 6 and 10 cm; marcasite-lined void at 12 cm; massive pyrite–chalcopyrite between 38 and 56 cm. (**b**) Porous and massive sulphide. Typical textures of porous thin-wall colloform pyrite (PyI) and massive polycrystalline pyrite (PyII), fractured and cemented by amorphous silica (arrow). Scale bar 400 μm. (**c**) Granular sulphide–silica. Diverse colloform pyrite (PyI) and euhedral pyrite (PyIII; indicated by arrows) aggregates and corroded chalcopyrite (labelled cp) enclosed in massive amorphous silica. Scale bar 400 μm. (**d**) Primitive pyrite texture. Very fine-grained crystallites of FeS_2 forming spheroidal aggregates in amorphous silica. Scale bar 20 μm. (**e**) Corroded chalcopyrite (labelled cp1), surrounded by fibrous covellite (labelled cv) and idaïte (arrow) aggregates in amorphous silica. Scale bar 100 μm. (**f**) Bladed chalcopyrite showing covellite alteration and overgrown by later marcasite (Mall; indicated by arrows) and amorphous silica. Scale bar 400 μm. (**g**) Tabular barite partly overgrown by globular amorphous silica. Scale bar 50 μm. (**h**) Sulphides (black) overgrown by amorphous silica with dusting of Fe oxide. Transmitted light; scale bar 100 μm.

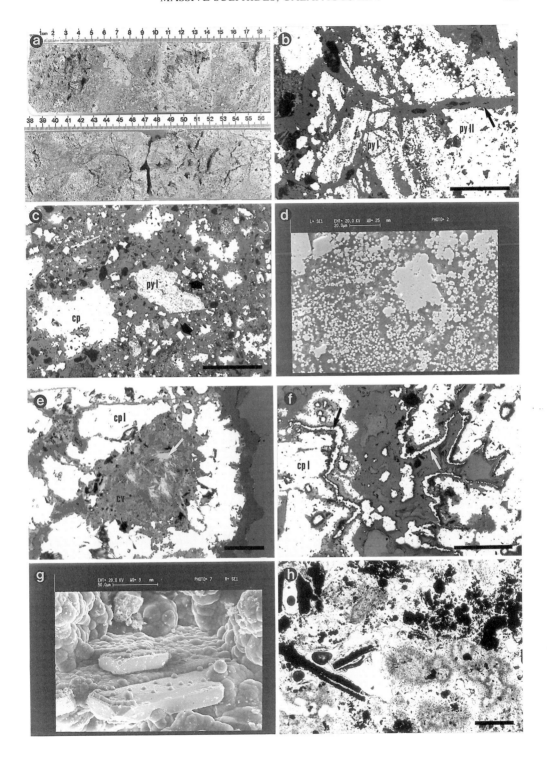

Table 1. *Summary of mineralogical types and related microprobe analyses*

Mineral	Facies	Description	Associations	Trace elements
Pyrite I	A	Microcrystalline, colloform, spheroidal habits. Forms thin-wall texture. Gradational to PyII	Apparently builds on silica foundations. Localized MaI	800–1050 ppm Co; <90–350 ppm Se; <170–270 ppmAs ($N = 26$)
Pyrite II	A	Polycrystalline, microporous or massive. Gradational to PyIII	Chalcopyrite and sphalerite inclusions	<160–1150 ppm Co; <90–940 ppm Se; <170–770– ppmAs ($N = 10$)
Pyrite III	A	Euhedral 50 μm to 2 mm. Commonly included or zoned	CpyI; chalcopyrite and sphalerite inclusions	260–2160 ppm Co; <90–1150 ppm Se; <170–530 ppm As ($N = 7$)
Chalcopyrite I	A	Bladed 200 μm to 2 mm grains. Open-space growth	PyIII; covellite-silica	<500–1010 ppm Zn; <170–640 ppm Co; <80–360 ppm Se ($N = 14$)
Chalcopyrite II	A	Massive, fracture- or porosity-filling	Rare replacement of PyI and PyII	na
Covellite	B	Fibrous, bladed, microcrystalline	Idaïte-bornite. Replaces chalcopyrite and sphalerite	na
Marcasite I	A	Microporous, tabular or fibrous	Pseudomorph of pyrrhotite	na
Marcasite II	C	Clean, euhedral grains	(CpyI); sphalerite– galena–barite	<150–960 ppm Co; no Se; <170–280 ppm, As; Cu max. 0.25 wt.% ($N = 7$)
Sphalerite	C	Small <50 μm euhedral crystals	MaII–galena–barite	2.92–4.54 wt.% Fe; 0.45–0.69 wt.% Cu ($N = 5$)

na indicates not analysed

Facies A: Fe–Cu sulphides. Pyrite and chalcopyrite are major constructional minerals. Pyrite forms an initially porous foundation which has been modified by extensive massive void-filling and overgrowth of chalcopyrite and pyrite. Three texturally distinct types of pyrite are recognized (Table 1) and can be related to the growth history of the sample. Primitive colloform, spheroidal or microcrystalline pyrite (PyI) forms at the initial stages of FeS$_2$ precipitation. This is overgrown by polycrystalline pyrite (PyII), which is the main constituent of the massive pyrite areas. Ghost textures of colloform PyI indicate that polycrystalline pyrite also forms by recrystallization of primitive pyrite. Polycrystalline PyII coarsens outwards to, and is overgrown by, euhedral PyIII. Chalcopyrite and sphalerite inclusions are present in coarse polycrystalline and euhedral

pyrite, and marcasite inclusions are seen in primitive and fine polycrystalline pyrite. Euhedral pyrite crystals commonly display concentric growth zones, picked out by trails of inclusions or micropores.

Chalcopyrite is the second most abundant sulphide. It is invariably coarse-grained or massive, and is associated with PyIII of main stage facies A. Different textural types of chalcopyrite are recognized (Table 1). The dominant form is coarse, euhedral, bladed crystals (CpyI) indicative of growth into open space. Anhedral chalcopyrite (CpyII) forms due to replacement of pyrite (rare), or where growth has been impeded in fractures and interstitial to porous pyrite aggregates.

Pyrite generations from all parts of the core ($N = 43$) were analysed for Cu, Zn, Se, Co and As to study the variation within and between generations. The results are summarized in Table 1. Cu and Zn contents of up to 0.2 wt% are due to microscopic and submicroscopic inclusions of chalcopyrite and sphalerite. Co was detected in 98% of analyses, ranging from <160 to 2160 ppm. Se ranges from <90 to 1150 ppm, detected in 35% of the analyses. In 28% of analyses, As was detected ranging from <170 to 770 ppm. The highest concentrations of Co, Se and As are in euhedral pyrite (PyIII).

Chalcopyrite was analysed for Co, Se and Zn ($N = 14$). Grains contain <160 to 640 ppm Co (detected in 10 analyses), <80 to 360 ppm Se (detected in six analyses), and <290 to 1010 ppm Zn (detected in all analyses).

Facies B: covellite – Fe oxyhydroxide assemblage. Evidence of oxidation is pervasive throughout the core in the form of amorphous Fe oxyhydroxides and secondary Cu sulphides. Oxidation reactions occur at a stage between the deposition of main-stage facies A and the late silicification (facies D) and result in corrosion and alteration of the sulphides. Chalcopyrite is altered on grain edges and along fractures to covellite, with trace amounts of bornite and idaïte. The covellite forms fine polycrystalline, bladed aggregates. Amorphous Fe oxyhydroxide and hydrous Fe sulphates (melanterite and rozenite) precipitate around the sulphides and are preserved as a dusting within silica.

Facies C: late Fe–Zn–Pb–Ba mineralization. Local textural modification of porous primary Fe–Cu sulphides (facies A) takes the form of overgrowths of marcasite, sphalerite, barite and rare galena. This assemblage occurs in void spaces, and is enclosed by silica (facies D). Marcasite forms coarse, clean polycrystalline overgrowth, is free of inclusions and has a low trace element content. Maximum values of 960 ppm Co and 280 ppm As were determined and Se was not detected. Some marcasite is considerably enriched in Cu (up to 0.25 wt%), although inclusions are not visible. Coarse, fibrous, bladed marcasite aggregates with a high microporosity occur locally, indicating possible replacement and pseudomorphing of pyrrhotite (Murowchick 1992).

Although sphalerite is a minor/trace sulphide in the sample under investigation, other mineralogical studies indicate that sphalerite is locally a major constituent of the massive sulphides (Tufar *et al.* 1986). Regardless of abundance it is always associated with marcasite, barite and amorphous silica. Grains are small (<50 μm), euhedral, enclosed by amorphous silica and associated with covellite

and galena. The sphalerite is pale yellow in transmitted light, indicating low Fe and Cd contents and the absence of chalcopyrite inclusions. Microprobe analysis gives 2.9 to 4.5 wt% Fe, 0.4 to 0.7 wt% Cu and Cd below detection levels (<0.1 wt%). Galena is a trace mineral which occurs as 10 μm euhedral grains and overgrowths on sphalerite. Barite is a minor mineral in voids, associated with marcasite, sphalerite and amorphous silica. Crystals are generally tabular and fine-grained, or form bow-tie aggregates. Barite crystals were analysed by EDS microprobe and contain 0.32 to 5.5 wt% Sr ($N = 32$; mean 1.75 wt%); the most Sr-enriched barite crystals also contain up to 0.43 wt% Ca.

Facies D: amorphous silica. Silica is an abundant and locally dominant phase. It is ubiquitous in the sample as late 20 μm coatings of globular habit, lining all cavities. Elsewhere, silica may completely cement aggregates of granular Fe–Cu sulphide. Filamentous silica in zones of high porosity is similar in morphology to that described by Juniper & Fouquet (1988) as being coatings on filamentous bacteria.

The granular sulphide and sulphide–silica zones of the boulder are composed dominantly of pyrite and chalcopyrite (facies A). Oxidation was pervasive before the precipitation of amorphous silica (facies D). There is evidence of multiple facies A–B–D mineralization in the form of chalcopyrite overgrowths on silica-covellite aggregates. Facies C Fe–Zn–Pb–Ba mineralization appears more closely related to the silica precipitation and is especially favoured in porous, primitive pyrite-dominated zones.

The degree to which the facies vary along the core was quantitatively assessed by measuring modal mineral abundance on oriented polished thin sections (Fig. 2c). In zone 1, facies A Fe–Cu mineralization (the sum of pyrite and chalcopyrite in Fig. 2c) is low (20% total), but increases sharply where massive sulphide bodies are intersected. There is a general increase in facies A through the transitional zone 2, to a maximum of 60% between 30 and 40 cm. Thereafter there is a gradual decrease through zones 3 and 4 to 30% total in porous zone 5. At 70 cm there is a sharp discontinuity where the core intersects a distinctive massive sulphide zone (zone 6). Chalcopyrite is highly variable within this facies A mineralization, but locally constitutes over 50% total sulphide.

Facies B (dominantly covellite plus Fe oxyhydroxide) occurs in the core between 0 and 70 cm. Facies C (marcasite, with minor barite and sphalerite) is most abundant in porous sulphide

Table 2. *Bulk geochemical data. Summarized as mean value for each zone and including only those elements detected by XRF*

Zone	S (wt.%)	Fe (wt.%)	Cu (wt.%)	SiO_2 (wt.%)	TiO_2 (ppm)	Al_2O_3 (ppm)	MnO (ppm)	CaO (ppm)	K_2O (ppm)	P_2O_5 (ppm)
1	41.66	33.32	4.55	21.34	108	863	116	76	59	98
2	36.45	31.94	13.25	17.73	65	500	33	41	61	173
3	34.33	31.14	5.61	27.09	87	2620	58	52	132	106
4&5	26.53	24.07	6.90	44.37	99	3555	68	49	51	113
6	37.94	34.73	10.85	16.51	36	3110	36	<9	81	149

Zone	Co (ppm)	Se (ppm)	As (ppm)	Mo (ppm)	Te (ppm)	Ga (ppm)	Zn (ppm)	Cd (ppm)	Pb (ppm)	Sb (ppm)
1	109	43	46	53	12	6	400	11	14	2
2	90	74	38	27	17	5	198	10	<7	<2
3	114	35	90	44	14	6	279	9	21	2
4&5	133	38	52	30	12	5	316	10	22	<2
6	26	35	74	43	16	6	395	12	10	2

Zone	Ba (ppm)	Sr (ppm)	Sr/Ba (atomic)	Cl (ppm)	Ni (ppm)	V (ppm)	Zr (ppm)	Nb (ppm)	U (ppm)
1	4087	95	0.0363	236	9	<6	10	<2	<4
2	1194	28	0.0370	273	11	3	7	<2	<4
3	2012	31	0.0241	274	10	8	9	2	<4
4&5	6581	148	0.0350	280	11	29	12	<2	<4
6	189	10	0.0809	209	10	17	6	<2	<4

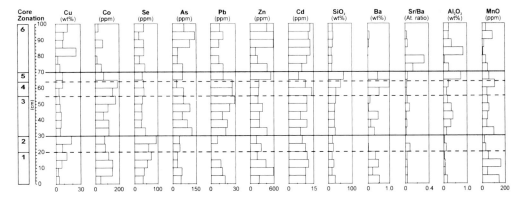

Fig. 4. Variation of selected major and trace elements along the core showing sharp changes at 30 and 70 cm and a generally symmetrical distribution of the sulphide-related trace elements through zones 2 to 5. Results are derived from analysis of 5 cm core intervals and are also summarized in Table 2.

zones (especially in zone 1), although it may be localized elsewhere (e.g. around 47 cm). Zone 6 differs in both its low degree of oxidation and the absence of late Fe–Zn–Ba mineralization.

Amorphous silica is ubiquitous, serving to reduce microscale porosity to less than 30%. There was, however, a variation in the porosity of the sample before the pervasive, late silicification. This minus-cement porosity is measured as combined porosity plus silica in Fig. 2c. The lowest minus-cement (around 40%) porosity is in zone 3. This increases in zones 4 and 5 to 60%.

Geochemistry

Bulk major element composition for the whole core (weighted mean) is 7.9 wt% Cu, 31.4 wt% Fe, 35.2 wt% S, 26.6 wt% SiO_2, 0.23 wt% Ba and 0.22 wt% Al_2O_3. Other elements are present at <400 ppm concentration. The results are summarized in Table 2.

Statistically significant positive correlation (99% confidence interval) occurs between the elements P_2O_5–Cu, Al_2O_3–V, Ba–(Sr–SiO_2–Ca–MnO), Sr–MgO and Zn–As. Negative correlation occurs between Cu–Pb and Te–(Mn, Ba, Sr, Co). Two trace element associations are related to sulphide phases. Se and Te are associated with main-stage Fe–Cu sulphides (facies A), and Mo–As–Pb is associated with the late-stage Fe–Zn–Ba assemblage (facies C). Co is the most abundant trace element in sulphides, but does not correlate with either group, although there is a significant negative correlation with Te.

The variation in selected major and trace elements along the core is illustrated in Fig. 4. Discontinuities in the general trend are seen for

all elements at 30 and 70 cm, corresponding to zone transitions. Cu and SiO_2 are both major elements. Cu increases through zones 1 and 2 (4.6 to 13.3 wt.%); it is low in Zones 3 to 5 (mean 6.3 wt.%) and it is erratic and locally high in Zone 6 (maximum 24.8 wt.%). Silica shows a gradual increase through zones 1 to 5 and is low thereafter. The Ba distribution is erratic in the core, related to variable barite content. Sr correlates with Ba due to substitution in the barite lattice, but a plot of the Sr/Ba atomic ratio reveals zones of Sr excess. Microprobe analyses show that the rare, small barite crystals from the high Sr/Ba part of zone 6 (between 75 and 80 cm) are consistently Sr-enriched (mean Sr/Ba atomic ratio 0.062 ± 0.017; $N = 8$) relative to the more abundant barite in zone 3 at 40 cm (Sr/Ba = 0.031 ± 0.022; $N = 8$). Hannington & Scott (1988) proposed that the variation of Sr/Ba ratios in barite from a sulphate-silica spire was due to the increased availability of Sr during seawater mixing. The presence of Sr-rich zones are consistent with localized seawater entry into the sulphide structure.

Sulphide-related trace elements (Co, Se, As, Pb, Zn and Cd) show an approximately symmetrical distribution, especially between the major zone transitions at 30 and 70 cm. Se, Co and As have been shown by EPMA to concentrate in particular generations of pyrite and chalcopyrite. The high Se in zone 2 corresponds to a high Cu zone and positive correlation between Cu and Se has also been demonstrated in other deposits (Auclair et al. 1987). The concentration of Co is moderate to high through to Zone 4, and low thereafter. The opposite pattern of distribution is shown by As, with highs at >90 cm. Pb, Zn and Cd variation records the

Table 3. *Sulphur isotope data for bulk sulphide analyses*

Position (cm)	Sulphide (modal %)	Mineral	$\delta^{34}S$ (‰)	Description
3	35.4	Pyrite	4.1	Coarse PyIII
6	64.0	Pyrite	4.4	Massive PyII
20	53.1	Pyrite	4.8	Clastic pyrite
24	57.8	Pyrite	5.3	Fine PyIII
35	51.2	Pyrite	3.9	Clastic pyrite
40	60.6	Pyrite	4.4	Massive PyII
47	49.6	Pyrite	4.0	Clastic pyrite
54	51.0	Pyrite	4.6	Massive PyII
60	42.0	Pyrite	4.2	Fine Py in silica
73	75.9	Pyrite	3.7	Massive PyII
76	55.0	Pyrite	3.5	Porous PyI
90	33.7	Pyrite	3.2	Massive Py
97	39.2	Pyrite	3.8	Massive Py
11	35.8	Marcasite	2.8	Massive MaI aggregate
13	43.6	Marcasite	2.7	Massive MaI aggregate
13	43.6	Chalcopyrite	3.1	Massive CpyI
31	67.1	Chalcopyrite	5.5	Massive CpyI
40	60.6	Chalcopyrite	4.1	Massive CpyI
66	33.4	Chalcopyrite	3.2	Massive CpyI
83	37.2	Chalcopyrite	3.7	Massive CpyI
95	28.7	Chalcopyrite	3.4	Massive CpyI
83	37.2	$FeSo_4 \cdot nH_2O$	3.3	Associated with cpy

distribution of facies C sphalerite–galena. Other trace elements were detected (Table 2) but they show no systematic variation.

Sulphur isotopes

Mineral separates of pyrite ($N = 13$), marcasite ($N = 2$) and chalcopyrite ($N = 6$) sampled at approximately 5 cm intervals along the core were analysed for S isotopes (Table 3). $\delta^{34}S$ is between +2.7 and +5.5‰, with a modal value of +4.0‰. No systematic isotopic difference is seen between spatially coexisting FeS_2 and chalcopyrite. A single analysis of primary hydrous Fe sulphate gave a $\delta^{34}S$ value that is 0.4‰ lower than the associated sulphide. Measured $\delta^{34}S$ lies within the range reported for submarine massive sulphides and fossil VMS deposits (Bluth & Ohmoto 1988). Analyses reported here are distinctly lower than the range (+5.4 to +6.3‰) reported by Skirrow & Coleman (1982) from sulphides at site A, 7 km to the east.

Systematic differences are seen between zones and facies when $\delta^{34}S$ is plotted against distance along sample (Fig. 5). Main-stage Fe–Cu sulphides (facies A) are isotopically heavy in zones between 20 and 30 cm and between 54 and 60 cm (+4.2 to +5.5‰), and light between 35 and 50 cm (+3.2 to +4.4‰). Fibrous aggregates of marcasite at 12 cm are the lightest sulphides analysed (+2.7 to +3.1‰). From 70 to 100 cm (zone 6), $\delta^{34}S$ is uniform, between +3.2 and +3.7‰.

Discussion

Sample growth history

It is possible to develop a generalized growth history for the whole core where the textural and mineralogical variation within the sample is a function of the degree to which a particular diagenetic stage has developed. This growth history is summarized in Fig. 6, and subdivided into three stages.

Growth

Growth starts with the construction of an initial foundation at the seawater interface, which is subsequently modified. The initial framework is a porous network of 'thin-walled' pyrite and/or marcasite, which develops by overgrowth and replacement of amorphous silica. In most parts of the sample the growth stage is obscured by the later mineralization. However, the presence of fossilized worm tubes of thin-walled pyrite

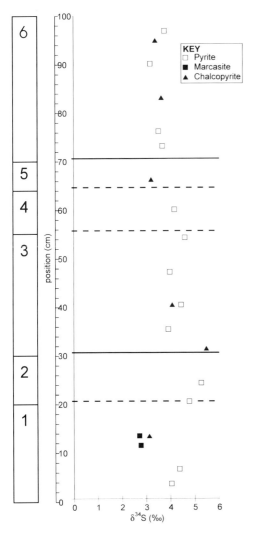

Fig. 5. Sulphur-isotope variation along the core showing systematically heavy isotope zones between 20 and 30 cm and between 54 and 60 cm.

Two substages are recognized, a 'main' or prograde hydrothermal stage and a 'late", or retrograde stage. The main stage is a record of intensifying hydrothermal conditions and is responsible for the main Fe–Cu mineralization. Within this stage, the evolution of pyrite is seen from early 'primitive' colloform textures to later euhedral pyrite. There is also an increase in the trace element loading. Peak conditions result in euhedral pyrite associated with chalcopyrite. The pyrite is zoned and contains highly variable trace element concentrations, indicating that the peak hydrothermal conditions fluctuated greatly. Locally high Se concentrations attest to this; Se is chemically mobile and therefore incorporated into the sulphide lattice only when there is no mixing of oxidizing seawater (Auclair *et al.* 1987).

The late diagenesis is a retrograde stage, occurring as hydrothermal conditions declined. The presence of marcasite as the stable Fe disulphide and the occurrence of sphalerite, galena and barite all indicate very different conditions from the Fe–Cu dominated deposition. Marcasite is formed more rapidly at low pH (Murowchick & Barnes 1986); the low pH is probably maintained by minor oxidation of sulphide by mixed seawater–hydrothermal solutions within the deposit. In these low pH conditions below 250°C, Cu is increasingly soluble as Cu^{2+} (no sulphide in solution) or as bisulphide complexes (Crerar & Barnes 1976). The mobility of Cu may explain the particularly high Cu content of marcasite, and the development of supergene Cu sulphides such as covellite. Incremental reaction models (Janecky & Shanks 1988) have shown that chalcopyrite–pyrite mineralization occurs in response to near-adiabatic mixing, but sphalerite–barite–silica deposition requires extensive conductive cooling. This and the presence of barite indicates that seawater sulphate is present in the diagenetic system.

Oxidation is important in the late diagenetic stage as the relative proportion of ambient temperature seawater increases. The cessation of hydrothermal flow is not sudden and the late diagenetic stage records a gradual transition from high-temperature hydrothermal fluid-dominated reactions, through mixed hydrothermal/seawater-dominated reactions, to ambient temperature seawater-dominated reactions. The low Mn concentrations in the sample (Table 2) demonstrate that the Fe oxyhydroxides are related to sulphide oxidation and not to the low-temperature Fe–Si–Mn oxyhydroxide mineralization observed in the surrounding area.

suggests that the main structural growth occurred on a biogenic foundation, in the manner described by Juniper *et al.* (1992), rather than in an anhydrite scaffolding (e.g. the classic chimney growth model of Haymon 1983).

Hydrothermal diagenesis

This stage represents the main mineralizing processes. It takes place in the subsurface, within a porous structure, effectively insulated from exterior seawater at ambient temperature.

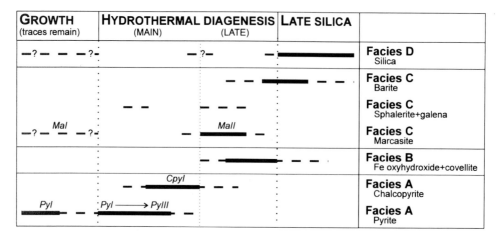

Fig. 6. Summary of the deposit growth history subdivided into early growth main diagenesis and late modification stages.

Late silica

Amorphous silica associated with barite gives oxygen isotope equilibration temperatures of 60 to 170°C (Skirrow & Coleman 1982). In contrast, silica associated with Fe–Mn oxides indicates formation temperatures of 32 to 42°C (Herzig *et al.* 1988). Simple mixing of end-member hydrothermal fluids with seawater will not precipitate silica because the mixing line does not intersect the amorphous silica saturation curve (Janecky & Seyfried 1984) and so conductive cooling of the hydrothermal fluids is required. The presence of amorphous silica coatings is important for two reasons. Silica protects the sulphides from oxidation by seawater and the silica is also an important cement, holding together fractured and disrupted sulphide structures. Both properties have contributed to the preservation of these massive sulphides for over 9000 years.

The transition from prograde Fe–Cu sulphide deposition to retrograde Fe–Zn–Pb–Ba–Si mineralization reflects increasing conductive cooling and seawater mixing. Minor oxidation of sulphide by seawater during the retrograde stage is important for marcasite formation and metal remobilization. A permeability change in the hydrothermal upflow zone or in the deposit itself would enhance conductive cooling and mixing, resulting in sulphide precipitation shifting more deeply subsurface and silica precipitating near-surface. The scenario is consistent with metal-enriched silica–barite associations described here and for the Mariana (Stüben *et al.* 1994) and Woodlark (Binns *et al.* 1993) deposits. It is likely that silica precipitation was triggered by the event that caused extensive fracturing and disruption of the sulphides.

Sulphur isotope variations

The variation of $\delta^{34}S$ along the core from this sample is not random. There are systematic variations with respect to the zonation defined on the basis of mineralogical and textural features. The isotope profile (Fig. 5) and that of percent total sulphide (Fig. 2c) appear to be generally similar in shape. We speculate that a relationship exists between $\delta^{34}S$ and either total sulphide or the minus-cement porosity (i.e. silica plus porosity). A general positive correlation between $\delta^{34}S$ and total sulphide (i.e. a negative correlation with minus-cement porosity) for all samples is seen in Fig. 7. There is no *a priori* reason why $\delta^{34}S$ should vary with either porosity or total sulphide unless this is a record of changing $\delta^{34}S$ with increasing hydrothermal diagenesis. The trend to increasing $\delta^{34}S$ with increasing sulphide in the structure must be recording sequential overgrowths of isotopically heavier sulphide. This hypothesis could be tested by laser probe analysis of the sulphide overgrowths.

The isotopic variation in submarine sulphides is due to mixing of S from two end-member sources, MORB-derived sulphide ($\delta^{34}S$ of $+0.1 \pm 0.5\permil$; Sakai *et al.* 1984) and reduced seawater sulphate ($\delta^{34}S$ of *c.* $+21\permil$; Rees *et al.* 1978). This may occur locally, within the sulphide structure, or deeper in the hydrothermal reaction system. Previous studies (Bluth & Ohmoto 1988; Woodruff & Shanks 1988; Crowe & Valley 1992) have described systematic

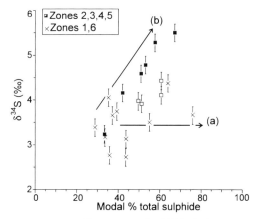

Fig. 7. Plot of $\delta^{34}S$ versus total sulphide (point-count data) showing a general positive correlation. Two mixing trends are defined, corresponding to (a) constant end-member hydrothermal $\delta^{34}S$ and variable total sulphide and (b) mixing of this constant $\delta^{34}S$ with a second, isotopically heavy S source. Filled symbols are from isotopically heavy analyses through zones 3 and 4. Linear regression through these gives $y = 0.068x + 1.16$; $R^2 = 0.957$.

$\delta^{34}S$ variation in some (not all) chimney structures, which has been related to mixing sulphur from end-member H_2S in deeply circulated hydrothermal fluid ($\delta^{34}S$ of c. +1.0 to +1.5‰; Shanks & Seyfried 1987) and reduced seawater sulphate, either within the hydrothermal structure or in a shallow subsurface reaction zone. In general, $\delta^{34}S$ of the sulphides increases in the outer parts of the chimney, consistent with the increasing incorporation of reduced seawater sulphate. However, the limited reducing potential of the hydrothermal fluid can only account for $\delta^{34}S$ variations of +1.5 to +4.5‰, assuming adiabatic or conductive cooling and sulphide-sulphate disequilibrium (Janecky & Shanks 1988). To explain measured $\delta^{34}S$ in excess of +4.5‰ requires additional sulphate reduction within the sulphide structure or deeper in the hydrothermal reaction system. In contrast to the observations from chimney structures, many ancient VMS deposits show a poorly defined trend of decreasing $\delta^{34}S$ upward through the ore lens (Franklin *et al.* 1981). This has been explained by the limited sulphate reducing potential of the mineralizing system and progressive exclusion of seawater from the deposit as it grows (Janecky & Shanks 1988).

Two models are proposed to explain the isotopic variation observed in the core, involving either *in situ* sulphate reduction within the deposit, or mixing of two hydrothermal solutions, one having a reduced sulphate signature.

The simplest explanation for the measured range of $\delta^{34}S$ and the variation through the sample is local seawater mixing and sulphate reduction within the deposit, which modifies a hydrothermal solution of constant $\delta^{34}S$. The three heaviest measured $\delta^{34}S$ all come from zone 2, which may represent a zone of seawater access and major sulphate reduction. Sulphate reduction within chimney sulphide deposits is not normally considered significant because of slow reaction kinetics (Woodruff & Shanks 1988). However, there is greater potential (from both kinetic and chemical considerations) for sulphate reduction in a large deposit undergoing long-lived hydrothermal diagenesis. Hydrothermal diagenesis would thus result in sulphides with increasingly heavier, sea-water derived $\delta^{34}S$. However, appealing to local sulphate reduction enhanced by hydrothermal diagenesis does not adequately explain the observed $\delta^{34}S$ variation in sulphides from all parts of the sample growth history and the mixing trends observed in Fig. 7.

In addition to the general positive trend in Fig. 7, analyses from the zones of transition above and below the massive sulphide of zone 3 display a statistically significant linear relation with total sulphide. The trends in Fig. 7 are therefore interpreted to indicate (a) variable proportion of total sulphide but constant $\delta^{34}S$, (i.e. an end-member hydrothermal source of c. +3.5‰) and (b) mixing two end-members of constant isotopic composition. Mixing between these two trends then fills the triangle. We can regard the end-member mixing (b) as involving two components with isotopic compositions δ_1 and δ_2, and atomic abundance S_1 and S_2. By mass balance, the total number of atoms of the element S_T is

$$S_T = S_1 + S_2 \tag{1}$$

and the isotopic composition of the bulk S (the measured value) is δ_T, where

$$S_T \delta_T = S_1 \delta_1 + S_2 \delta_2 \tag{2}$$

and using Equation (1)

$$S_T \delta_T = S_1 \delta_1 + (S_T - S_1)\delta_2$$

$$\therefore \delta_T = \frac{S_1}{S_T}\delta_1 + \delta_2 - \frac{S_1}{S_T}\delta_2$$

$$\therefore \delta_T = \delta_2 + \frac{S_1}{S_T}(\delta_1 - \delta_2) \tag{3}$$

For the case when the two components comprise one of constant abundance (i.e. S_1 is constant) but the second is variable (i.e. S_2 and hence S_T are variable), then Equation (3) is a straight line

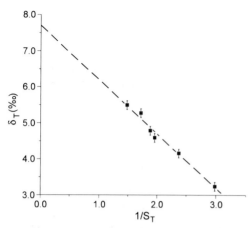

Fig. 8. Plot of δ_T (i.e. $\delta^{34}S$) versus $1/S_T$ (i.e. 100/ modal % total sulphide) for the mixing trend (b) defined in Fig. 7 by the isotopically heavy analyses from zones 3 and 4. Linear regression gives $\delta_T = 7.72 - 1.51/S_T$; $R^2 = 0.988$. The intersection with the δ_T axis defines the end-member isotopic composition δ_2 of 7.7‰.

The mineralizing hydrothermal fluids still ultimately owe their isotopic character to mixing of MORB and seawater-derived sulphur, controlled by the plumbing and sulphate-reducing potential of deep and shallow hydrothermal systems. However, the isotopic variation within this massive sulphide structure is caused by the mixing of two isotopically distinct hydrothermal solutions. This has implications for the processes of hydrothermal diagenesis. It appears that the evolution of the deposit (or at least the isotopic characteristics) is controlled by the combination of deep-circulated end-member hydrothermal solutions (high temperature, metal-rich) with shallow-circulated, conductively heated, modified seawater diagenetic solutions. This mixing occurs at the hydrothermal deposit scale; within an individual structure the degree to which these solutions modify mineralogy, texture and chemistry is controlled by fluid flow pathways and the localized entry of oxidising ambient temperature seawater. In a large, mature deposit such as that sampled here these variables may homogenize with time to give the observed trends.

in a plot of δ_T versus $1/S_T$ and the intercept is δ_2. This can be understood when, as $S_T \rightarrow \infty$, the relatively small and constant amount of component S_1 becomes negligible, and so $\delta_T \rightarrow \delta_2$.

In the seafloor hydrothermal deposit situation, it is assumed that the two components are magmatic sulphur and seawater sulphur. If the former is delivered at a constant rate and the latter is variable, then a plot of measured $\delta^{34}S$ ($=\delta_T$) versus 100/modal total sulphide ($=1/S_T$) should be a straight line which intercepts the y-axis at the $\delta^{34}S$ of the second end-member (δ_2). The analyses lying on the trend in Fig. 7 fit a straight line ($R^2 = 0.98$) which intercepts the y-axis at +7.7‰ (Fig. 8). This result is significant in two ways. Firstly, it demonstrates that the isotope variation in the sulphide structure is due to the mixing of two isotopically distinct hydrothermal solutions and is not due to *in situ* mixing of end-member hydrothermal and seawater solutions. Secondly, it appears to distinguish the isotopic character of the second end-member solution, whose $\delta^{34}S$ of 7.7‰ is indicative of a component of seawater-derived S, which may originate in the shallow subsurface hydrothermal reaction zone. This inferred end-member solution $\delta^{34}S$ is slightly higher than the maximum $\delta^{34}S$ measured from Galapagos sulphides (Skirrow & Coleman 1982), a relationship also commonly observed in active submarine hydrothermal fields (e.g. Shanks & Seyfried 1987).

Conclusions

Detailed textural analysis has demonstrated that massive sulphides from the Galapagos Rift have formed by the diagenetic modification of a porous sulphide structure. Sulphur-isotope data indicate an initial stage of sulphide mineralization dominated by fluids with an end-member, deep-circulated hydrothermal isotopic signature ($\delta^{34}S$ of *c.* +4‰). Superimposed on this is a long-lived hydrothermal diagenesis stage, characterized by increased mixing of hydrothermal solutions with a seawater component. This is recorded in sulphides as a gradual shift to increasing $\delta^{34}S$ (up to +5.5‰). The hydrothermal diagenetic fluids originate in a shallow sub-seafloor hydrothermal circulation system. The diagenetic growth mechanism has the potential to form large, Cu-rich (*c.* 8 wt% Cu) deposits during long-lived hydrothermal activity and as such is very different in style to the ephemeral black smoker type activity that is more familiar in surface expression.

Work at Cardiff was supported by NERC grant GR9/603 to D.R. and NERC studentship GR4/91/GS/ 127 to R.K. We thank GEOMAR for providing the samples, and background information on the geological setting, and the technical staff at SURRC and Cardiff for their assistance in isotope and geochemical analyses. The manuscript was greatly improved by the comments of S. D. Scott and an anonymous reviewer.

References

ALT, J. C. 1988. The chemistry and sulphur isotope composition of massive sulphide and associated deposits on Green Seamount, Eastern Pacific. *Economic Geology*, **83**, 1026–1033.

——, LONSDALE, P., HAYMON, R. & MUEHLENBACHS K. 1987. Hydrothermal sulphide and oxide deposits on seamounts near 21°N, East Pacific Rise. *Geological Society of America Bulletin*, **98**, 157–168.

AUCLAIR, G., FOUQUET, Y. & BOHN, M. 1987. Distribution of selenium in high-temperature hydrothermal sulphide deposits at 13°N, East Pacific Rise. *Canadian Mineralogist*, **25**, 577–588.

BALLARD, R. D., VAN ANDEL, T. H. & HOLCOMB, R.T. 1982. The Galapagos rift at 86°W, 5. Variations in volcanism, structure, and hydrothermal activity along a 30-km segment of the rift valley. *Journal of Geophysical Research*, **87**, 1149–1161.

BINNS, R. A., SCOTT, S. D. & 9 others. 1993. Hydrothermal oxide and gold-rich sulphate deposits of Franklin Seamount, western Woodlark basin, Papua New Guinea. *Economic Geology*, **88**, 2122–2153.

BLUTH, G. J. & OHMOTO, H. 1988. Sulphide-sulphate chimneys on the East Pacific Rise, 11° and 13°N latitude. Part II: sulphur isotopes. *Canadian Mineralogist*, **26**, 505–515.

CORLISS, J. B., DYMOND, J. & 9 others 1979. Submarine thermal springs on the Galapagos Rift. *Science*, **203**, 1073–1083.

CRERAR, D. A. & BARNES, H. L. 1976. Ore solution chemistry – V. Solubilities of chalcopyrite and chalcocite assemblages in hydrothermal solutions at 200° to 300°. *Economic Geology*, **71**, 772–794.

CROWE, D. E. & VALLEY, J. W. 1992. Laser microprobe study of sulphur isotope variation in a sea-floor hydrothermal spire, Axial Seamount, Juan de Fuca Ridge, eastern Pacific. *Chemical Geology*, **101**, 63–70.

EMBLEY, R. W., JONASSON, I. R. & 5 others 1988. Submersible investigation of an extinct hydrothermal system on the Galapagos Ridge: sulphide mounds, stockwork zone, and differentiated lavas. *Canadian Mineralogist*, **26**, 517–539.

FOUQUET, Y., AUCLAIR, G., CAMBON, P. & ETOUBLEAU, J. 1988. Geological setting and mineralogical and geochemical investigations on sulphide deposits near 13°N on the East Pacific Rise. *Marine Geology*, **84**, 145–178.

FRANKLIN, J. M., LYDON, J. W. & SANGSTER, D. F. 1981. Volcanic-associated massive sulphide deposits. *Economic Geology 75th Anniversary Volume*, 485–627.

GARIMAS II CRUISE REPORT 1986. *Galapagos Rift massive sulphides (SONNE-39)*. Preussag Marine Technology, Hannover, 201 pp.

GRAHAM, U. M., BLUTH, G. J. & OHMOTO H. 1988. Sulphide-sulphate chimneys on the East Pacific Rise, 11° 13°N latitudes. Part I: mineralogy and paragenesis. *Canadian Mineralogist*, **26**, 487–504.

HALBACH, P., NAKAMURA, K. & 16 others 1989. Probable modern analogue of Kuroko-type massive sulphide deposits in the Okinawa Trough back-arc basin. *Nature*, **338**, 496–499.

HANNINGTON, M. D. & SCOTT, S. D. 1988. Mineralogy and geochemistry of a hydrothermal silica–sulphide–sulphate spire in the caldera of Axial Seamount, Juan de Fuca ridge. *Canadian Mineralogist*, **26**, 603–625.

HAYMON, R. M. 1983. Growth history of hydrothermal black smoker chimneys. *Nature*, **301**, 695–698.

HERZIG, P. M., BECKER, K. P., STOFFERS, P., BÄCKER, H. & BLUM, N. 1988. Hydrothermal silica chimney fields in the Galapagos Spreading Centre at 86°W. *Earth and Planetary Science Letters*, **89**, 261–272.

JANECKY, D. R. & SEYFRIED, W. E. JR 1984. Formation of massive sulphide deposits on oceanic ridge crests: Incremental reaction models for mixing between hydrothermal solutions and seawater. *Geochimica et Cosmochimica Acta*, **48**, 2723–2738.

——, & SHANKS, W. C. III. 1988. Computational modelling of chemical and isotopic reaction processes in seafloor hydrothermal systems: chimneys, massive sulphides, and subjacent alteration zones. *Canadian Mineralogist*, **26**, 805–825.

JUNIPER, S. K. & FOUQUET, Y. 1988. Filamentous iron–silica deposits from modern and ancient hydrothermal sites. *Canadian Mineralogist*, **26**, 859–869.

——, JONASSON, I. R., TUNNICLIFFE, V. & SOUTHWARD, A. J. 1992. Influence of a tube-building polychaete on hydrothermal chimney mineralization. *Geology*, **20**, 895–898.

LALOU, C., BRICHET, E. & HEKINIAN, R. 1985. Age dating of sulphide deposits from axial and off-axial structures of the East Pacific Rise near 12°50′N. *Earth and Planetary Science Letters*, **75**, 59–71.

——, BRICHET, E. & LANGE, J. 1989. Fossil hydrothermal deposits at the Galapagos spreading centre near 85°00 West; geological setting, mineralogy and chronology. *Oceanologica Acta*, **12**, 1–18.

MALAHOFF, A., EMBLEY, R., CRONAN, D. S. & SKIRROW, R. 1983. The geological setting and chemistry of hydrothermal sulphides and associated deposits from the Galapagos Rift at 86°W. *Marine Mining*, **4**, 123–137.

MARCHIG, V., ERZINGER, J. & RÖSCH, H. 1987. Sediments from a hydrothermal field in the central valley of the Galapagos Rift Spreading Center. *Marine Geology*, **76**, 243–251.

MUROWCHICK, J. B. 1992. Marcasite inversion and the petrographic determination of pyrite ancestry. *Economic Geology*, **87**, 1141–1152.

——, & BARNES, H. L. 1986. Marcasite precipitation from hydrothermal solutions. *Geochimica et Cosmochimica Acta*, **50**, 2615–2629.

OUDIN, E., PICOT, P. & POUIT, G. 1981. Comparison of sulphide deposits from the East Pacific Rise and Cyprus. *Nature*, **291**, 404–407.

PARADIS, S., JONASSON, I. R., LE CHEMINANT, G. M. & WATKINSON, D. H. 1988. Two zinc-rich chimneys from the Plume Site, southern Juan de Fuca Ridge. *Canadian Mineralogist*, **26**, 637–654.

REES, C. E., JENKINS, W. J. & MONSTER, J. 1978. The sulphur isotopic composition of oceanic water sulphate. *Geochimica et Cosmochimica Acta*, **42**, 377–382.

RIDLEY, W. I., PERFIT, M. R., JONASSON, I. R. & SMITH, M. F. 1994. Hydrothermal alteration in ocean ridge volcanics: a detailed study at the Galapagos Fossil Hydrothermal Field. *Geochimica et Cosmochimica Acta*, **58**, 2477–2494.

ROBINSON, B. W. & KUSAKABE, M. 1975. Quantitative preparation of SO_2 for $^{34}S/^{32}S$ analysis from sulphides by combustion with cuprous oxide. *Analytical Chemistry*, **47**, 1179–1181.

SAKAI, H., DES MARAIS, D. J., UEDA, A. & MOORE, J. G. 1984. Concentrations and isotope ratios of C, N, and S in ocean floor basalts. *Geochimica et Cosmochimica Acta*, **48**, 2433–2441.

SCOTT, S. D., CHASE, R. L., HANNINGTON, M. D., MICHAEL, P. J., McCONACHY, T. F. & SHEA, G. T. 1990. Sulphide deposits, tectonics, and petrogenesis of southern Explorer Ridge, northeast Pacific Ocean. *In*: MALPAS, J. *ET AL.* (eds) *Ophiolites: Oceanic Crustal Analogues: Symposium 'Troodos 1987'*. Geological Survey Department, Nicosia, Cyprus, 719–733.

SHANKS, W. C. III. & SEYFRIED, W. E. JR 1987. Stable isotope studies of vent fluids and chimney minerals, southern Juan de Fuca Ridge. Sodium metasomatism and seawater sulphate reduction. *Journal of Geophysical Research*, **92**, 11 387–11 399.

SKIRROW, R. & COLEMAN, M. L. 1982. Origin of sulphur and geothermometry of hydrothermal sulphides from the Galapagos Rift, 86°N. *Nature*, **299**, 142–144.

STÜBEN, D., TAIBI, N. E., McMURTRY, G. M., SCHOLTEN, J., STOFFERS, P. & ZHANG, D. 1994. Growth history of a hydrothermal silica chimney from the Mariana backarc spreading center (southwest Pacific, 18°13′N). *Chemical Geology*, **113**, 273–296.

TUFAR, W., TUFAR, E. & LANGE, J. 1986. Ore paragenesis of recent hydrothermal deposits at the Cocos–Nazca plate boundary (Galapagos Rift) at 85°51′ and 85°55′W: complex massive sulphide mineralizations, non sulphide mineralizations and mineralized basalts. *Geologische Rundschau*, **75**, 829–861.

WOODRUFF, L. G. & SHANKS, W.C. III. 1988. Sulphur isotope study of chimney minerals and hydrothermal fluids from 21°N, East Pacific Rise: hydrothermal sulphur sources and disequilibrium sulphate reduction. *Journal of Geophysical Research*, **93**, 4562–4572.

Hydrothermal input into sediments of the Mid-Atlantic Ridge

G. A. CHERKASHEV

Institute of Geology and Mineral Resources of the Ocean, 1 Angliysky Avenue,
190121 St Petersburg, Russia

Abstract: The analysis of geochemical profiles in metalliferous sediment cores obtained from the Mid-Atlantic Ridge rift valley at the TAG, MARK and 15°N hydrothermal vent fields and near 27°10′N allowed a study of the history of hydrothermal processes. Of the three hydrothermal zones of the TAG field, the North hydrothermal zone reached its maximum activity between 6000 and 25 000 years ago, the MIR Mound 10 000 years ago and the Active mound 15 000 years ago. Evidence of intensive hydrothermal sedimentation is not seen near the MARK field. The MARK sulphide mounds were probably formed in the course of a single hydrothermal episode that began 4000 years ago. Most of the sulphides of the newly found hydrothermal field near 15°N were formed in the course of an event traced by a Fe-enriched layer 70–90 cm below the seafloor. At present, this field is only slightly hydrothermally active. Three Fe-enriched layers correspond to the possible stages of ore formation at a site of presumed hydrothermal activity near 27°10′N. Their relative intensity is decreasing with time.

During detailed investigations of the Mid-Atlantic Ridge (MAR) hydrothermal zones in research cruises of the Sevmorgeologiya Association (Krasnov *et al.* this volume), metalliferous sediments of the TAG, MARK and the newly discovered 14°45′N hydrothermal fields were studied. The geochemical profiles in sediment cores studied from these fields and from a site of presumed hydrothermal activity at 27°10′N showed evidence of the variation of hydrothermal input into sediments at the different segments of the MAR with time.

Studies of 'proximal' metalliferous sediments sampled at distances of up to several hundred metres from hydrothermal vents showed some geochemical features different from those of usual 'distal' metalliferous sediments more distant from the vents. The most significant features were the extremely high concentrations of Fe (up to 40%) and Cu and Zn (up to several per cent). The 'proximal' metalliferous sediments also have specific rare earth element patterns characterized by the presence of a positive Eu anomaly and the absence of the negative Ce anomaly typical of 'distal' metalliferous sediments. Bands of these highly ferrigenous sediments were distinguished by their bright red colour. All of these specific features were typical of metalliferous sediments enriched in sulphides and their oxidation products (Cherkashev 1990; Mills *et al.* 1993; German *et al.* 1993). There are three possible mechanisms (processes) for the transfer of sulphides and the products of their oxidation into the sediments: (1) fallout from hydrothermal plumes; (2) gravity mass wasting; and (3) transport by near-bottom streams from the hydrothermal mounds. Processes (2) and (3) may be responsible for the transport of hydrothermal material only over short distances. For example, a mass flow was observed across the slope below the black smoker complex during the crossing of the northwest part of the Active Mound of the TAG field in a MIR submersible dive (cruise 15 R/V *Akademik M. Keldysh* 1988). The material was transported for a distance of no more than a few metres. We presume that fallout from the hydrothermal plume is the main mechanism by which hydrothermal matter is transferred into sediments (especially for distances of more than 100 m from the vents). This conclusion is confirmed by the synchronous appearance of Fe-enriched layers in the four cores around the MIR Mound, TAG field (Fig. 1, stations 6–9). The downcore variations in metal concentrations mostly reflect differences in the intensity of high-temperature ore-forming hydrothermal activity at all the vent sites.

TAG field

The zone of surface metalliferous sediments of the TAG area surrounding the three sulphide deposits [Active Mound, MIR Mound and the relict North Zone (Rona *et al.* 1993)] occupy an area of 4×6 km. It is the largest of all hydrothermal fields in the Atlantic. According

From PARSON, L. M., WALKER, C. L. & DIXON, D. R. (eds), 1995, *Hydrothermal Vents and Processes*, 223
Geological Society Special Publication No. 87, 223–229.

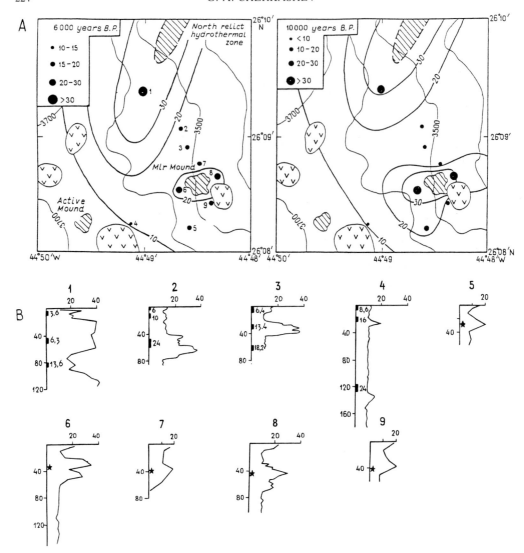

Fig. 1. (**A**)Position of sediment cores (closed circles) from the TAG hydrothermal field relative to volcanic mounds (shown as v-hatching), sulphide mounds and the hydrothermal North Zone of the TAG Field (shown in diagonal hatching after Rona *et al.* 1993). Core 1 is from Metz *et al.* (1988); cores 2 to 4 are from Lisitsyn *et al.* (1989); core 5 is from Shearme *et al.* (1983); and the rest of the cores are from this study. Circle sizes and heavy lines show Fe wt.% concentrations in sediments, recalculated on a carbonate-free basis; thin lines are isobaths from Rona *et al.* (1993). Left-hand panel, 6000 years BP; right-hand panel 10 000 years BP.
(**B**) Sediment cores and vertical distribution of Fe in sediments (as a percentage; for stations 2–9, recalculated on a carbonate-free basis). Figures near vertical bars are ^{14}C dates for sediments. Stars indicate red bands in sediments.

to dating of the sulphides from the Active Mound (Lalou *et al.* 1990, 1993) the present activity began only 50 years ago. This is a reason for the weak influence of the Active Mound on metal concentrations in the surface sediments compared with those of the deeper layers (Fig. 1).

Four sediment cores sampled at distances of up to 150 m from the MIR Mound have similar geochemical profiles, which are characterized by a large peak of Fe, Cu and Zn enrichments at 20–50 cm and a smaller peak in the subsurface sediments (Fig. 1). The analytical data for all four cores are given by Krasnov *et al.*, this

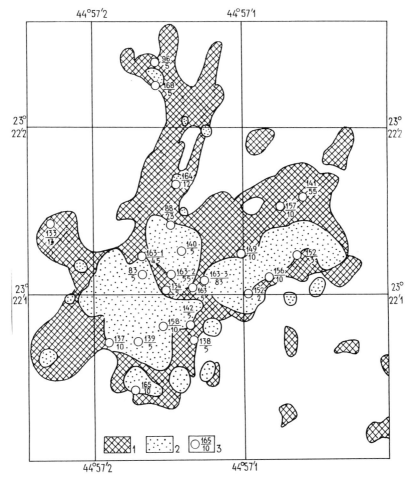

Fig. 2. Distribution of metalliferous sediments in the central part of the MARK hydrothermal field (data from cruise 3 of R/V *Professor Logachev* 1992). 1, Areas covered by metalliferous sediments; 2, sulphide deposits; and 3, station/established thickness of the sediment (cm).

volume, their table 5. Age data on sulphides of the MIR Mound (Rona *et al*. 1993) suggest that the last event, which led to the formation of the most metal-enriched layer in this sediment, occurred 10 000 years ago.

Based on these dates and those on five additional cores (stations 1–5) we reconstructed the distribution of Fe in the TAG sediments for the periods 6000 and 10 000 year ago (Fig. 1). The dates of the sediments and sulphides (Lalou *et al*. 1990, 1993; Rona *et al*. 1993) suggest asynchrony for the high-temperature events at North Zone. Active and Mir Mounds. The North Zone and Mir Mound were active during the hiatus period of the Active Mound between 10 000 and 4000 years BP (Lalou *et al*. 1990). The earlier events, which took place 15 000 and 25 000 years ago were also more intensive in the

North Zone and MIR Mound than in the Active Mound.

The cores did not reach the sedimentary layers corresponding to the events dated at 50 000 and 100 000 years in sulphides of the MIR Mound (Lalou *et al*. 1993). The comparison of the total intensity of hydrothermal activity in each of the three zones measured by Fe concentrations in sediments shows that the North hydrothermal zone reached its maximum activity 6000 and 25 000 years ago, the MIR Mound 10 000 and the Active Mound 15 000 years ago.

MARK field

Metalliferous sediments mostly of the 'proximal' type were sampled to a depth of 83 cm at 22 sites in the vicinity of the sulphide mounds (Fig. 2).

Table 1. *Estimated resources of Fe, Cu, Zn, Au and Ag in the MARK hydrothermal field*

| | Weight (tonnes) | | | | | |
	Total	Fe	Cu	Zn	Au	Ag
Massive sulphides	999 000	377 000	40 000	28 000	1.6	35
Metalliferous sediments	47 000	18 000	3100	600	0.03	0.2

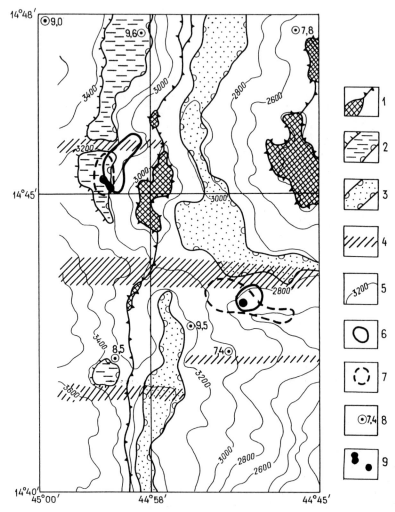

Fig. 3. Fe distribution in surface sediments in hydrothermal field at 14°45′N on the Mid-Atlantic Ridge. 1, Crest and crestal surfaces of linear highs; 2, bathymetric steps on the rift valley slope; 3, deeps between linear highs; 4, transverse tectonic dislocations; 5, isobaths (m); 6, contours of sulphide activity anomalies; 7, contours of electrical field anomalies; 8, Fe concentrations (as a percentage recalculated on a carbonate-free basis) in surficial sediments of grab samples; and 9, hydrothermal deposits.

This thin sedimentary cover overlying basalts and some sulphide mounds reflects the relatively young age of this field (up to 4000 years (Lalou *et al.* 1990)). The vertical distribution of iron in the cores ranges from 22 to 47%, without any of the distinct peaks in the profiles that would indicate events of hydrothermal activity. In some of the cores fresh sulphides alter towards the surface

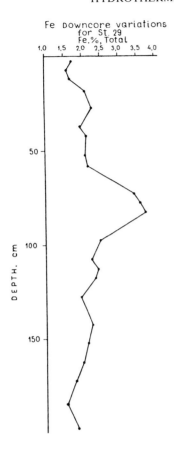

Fe Downcore variations
for St. 29
Fe,%, Total

Fig. 4. Downcore Fe variations for station 29 and its position relative to sulphide deposits at 14°45'N on the Mid-Atlantic Ridge. 1, Rift valley floor; 2, neovolcanic zones; 3, crests of linear highs flanking the rift valley; 4, crests of linear highs on rift valley slopes; and 5, transverse dislocations. Star indicates the newly discovered sulphide deposits (Krasnov *et al.*, this volume). The box shows the position of the area in Fig. 3.

into oxidized products. Sometimes the upcore decrease of S and Fe concentrations is accompanied in these cores by Au enrichment and a decrease in the Eu/Sm ratio.

The total Fe, Cu, Zn, Au and Ag resources in metalliferous sediments were evaluated with respect to those in the massive sulphides (Table 1).

14°45'N field

Red-coloured 'proximal' metalliferous sediments were sampled by TV-equipped grab sampler, together with and close to massive sulphides at the newly found active hydrothermal field near 14°45'N MAR (Krasnov *et al.*, this volume). The surface sediments taken by the grab sampler several kilometres from the sulphide deposits were slightly enriched in Fe (Fig. 3). The maximum Fe concentrations in a core taken 10 km from the hydrothermal field are 70–90 cm below the seafloor (Fig. 4). Many pyrite and Fe-oxyhydroxides were found in the same layer, which probably corresponds to the event of high-temperature activity which formed the sulphide deposit.

27°10'N

A core 130 cm long (Fig. 5) was taken near 27°10'N at a site of presumed hydrothermal activity (our data; Murton *et al.* 1993). Three hydrothermal events are seen in the core, reflected in Fe enrichment above the low background typical of carbonate-rich sediments. These events took place at equal temporal intervals (assuming a constant sedimentation rate). Their relative intensity decreased with time.

Future dating of massive sulphides and metalliferous sediments of hydrothermal fields will allow a detailed reconstruction of the history of MAR hydrothermal sedimentation.

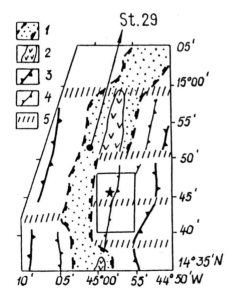

St.29

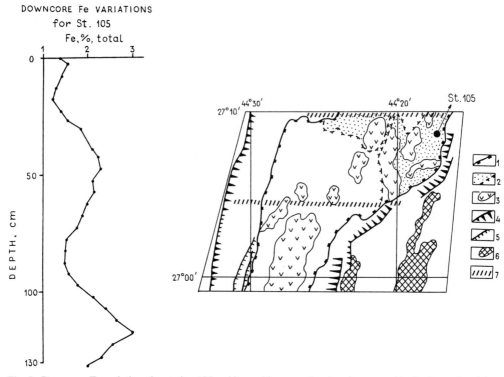

DOWNCORE Fe VARIATIONS
for St. 105
Fe,%, total

Fig. 5. Downcore Fe variations for station 105 and its position near the site of presumed hydrothermal activity at 27°10′N on the Mid-Atlantic Ridge. 1, Contours of the rift valley floor; 2, bathymetric steps; 3, volcanic highs and separate volcanic edifices; 4, major fault scarps; 5, minor fault scarps; 6, crestal surfaces of rift mountains; and 7, zone of transverse dislocations.

References

CHERKASHEV, G. A. 1990. [*Metalliferous sediments from the sites of oceanic sulphide formation (example of the northern East Pacific Rise)*]. PhD Thesis. Leningrad [in Russian].

GERMAN, C. R., HIGGS, N. S., THOMSON, J., MILLS, R., ELDERFIELD, H., BLUSZTAIN, J., FLEER, A. P. & BACON, M. P. 1993. A geochemical study of metalliferous sediment from the TAG hydrothermal mound, 26°08′N, Mid-Atlantic Ridge. *Journal of Geophysical Research*, **98**, 9683–9692.

KRASNOV, S. G., & 10 others. 1995. Detailed geological studies of hydrothermal fields in the North Atlantic. *This volume*.

LALOU, C., REYSS, J.-L., BRICHET, E., ARNOLD, M., THOMPSON, G., FOUQUET, V. & RONA, P. A. 1993. New age data for Mid-Atlantic Ridge hydrothermal sites. TAG and Snakepit chronology revisited. *Journal of Geophysical Research*, **98**, 9705–9713.

——, THOMPSON, G., BRICHET, E., DRUFFEL, E. & RONA, P. A. 1990. Geochronology of TAG and Snakepit hydrothermal fields, Mid-Atlantic Ridge: witness of a long and complex hydrothermal history. *Earth and Planetary Science Letters*, **97**, 113–128.

LISITSYN, A. P., BOGDANOV, Y. A., ZONENSHAIN, L. P., KUZMIN, M. I. & SAGALEVICH, A. M. 1989. Hydrothermal phenomena in the Mid-Atlantic Ridge at latitude 26°N (TAG hydrothermal field). *International Geology Review*, **31**, 1183–1198.

METZ, S., TREFRY, J. H. & NELSEN, T. 1988. History and geochemistry of a metalliferous sediment core from the Mid-Atlantic Ridge at 26°N. *Geochimica et Cosmochimica Acta*, **52**, 2369–2378.

MILLS, R., ELDERFIELD, H. & THOMSON, J. 1993. A dual origin for the hydrothermal component in a metalliferous sediment core from the Mid-Atlantic Ridge. *Journal of Geophysical Research*, **98**, 9671–9681.

MURTON, B. J., BECKER, K., BRIAIS, A., EDGE, D., HAYWARD, N., KLINKHAMMER, G., MILLARD, N., MITCHELL, I., ROUSE, I., RUDNICKI, M., SAYANAGI, K. & SLOAN, H. 1993. Results of a systematic approach to searching for hydrothermal activity on the Mid-Atlantic Ridge: the discovery of the 'Broken Spur' vent site. *BRIDGE Newsletter*, **N4**, 3–6.

RONA, P. A., BOGDANOV, Y. A., GURVICH, E. G., RIMSKI-KORSAKOV, N. A., SAGALEVICH, A. M., HANNINGTON, M. D. & THOMPSON, G. 1993. Relict hydrothermal zones in the TAG hydrothermal Field, Mid-Atlantic Ridge 26°N, 45°W. *Journal of Geophysical Research*, **98**, 9715–9730.

SHEARME, S., CRONAN, D. S. & RONA, P. A. 1983. Geochemistry of sediments from the TAG hydrothermal field, MAR at latitude 26°N. *Marine Geology*, **51**, 269–291.

Hydrothermal sedimentation at ODP Sites 834 and 835 in relation to crustal evolution of the Lau Backarc Basin

R. A. HODKINSON & D. S. CRONAN

Department of Geology, Imperial College of Science, Technology and Medicine, London SW7 2BP, UK

Abstract: Hydrothermally metal-enriched sediments from ODP Leg 135 Sites 834 and 835 in the Lau Backarc Basin have been studied with respect to basin evolution. Multivariate statistics (factor analysis) indicate element associations in the non-detrital, carbonate-free sediment fraction characteristic of hydrothermal oxide plume fallout. Selective chemical leach, X-ray diffraction and lithological studies confirm that the hydrothermal enrichment in the sediments at both sites results from distal oxide plume fallout. No evidence of a sulphide/weathered sulphide component is seen, suggesting that the sites have never been proximal to a hydrothermal discharge source. Factor scores for the hydrothermal oxide phase of the sediments reflect the hydrothermal activity associated with the propagation of the Eastern Lau Spreading Centre (ELSC) into the basin. The subsequent propagation of the Central Lau Spreading Centre (CLSC) into the basin is not clearly recorded. Non-detrital element accumulation rates for Mn and Fe confirm this. However, the results suggest a 'time lag' between the closest passage of the ELSC past the latitude of Sites 834 and 835 and maximum hydrothermal plume fallout flux to the sediment section. This may be due to a delay between ridge formation and either the initiation of hydrothermal systems on the ridge and/or their full development. In addition, however, plume dispersion patterns may have been affected by palaeo-current directions and basin topography. Element accumulation rate values show that the hydrothermal flux from the ELSC was of much greater intensity than that from the CLSC. The latter was comparable with values in presently accumulating sediments proximal to the CLSC and within the range typical of East Pacific Rise (EPR) values. The ELSC accumulation rates are comparable with, or higher than, the highest reported values from the EPR. Compared with EPR sediments, the Lau Basin hydrothermal plume fallout is rich in Mn relative to Fe. In particular, plume fallout associated with the ELSC exhibits a two- to four-fold enrichment of Mn relative to Fe compared with the EPR.

Metalliferous seafloor sediments are those that exhibit metal enrichment from hydrothermal fluxes. The metals most commonly enriched in such deposits include Fe, Mn, Cu and Zn. Because of hydrothermal contributions, metalliferous sediments are found either in or adjacent to hydrothermally active submarine volcanic areas, most commonly mid-ocean ridges. However, in recent years such deposits have also been found in island arcs and in backarc basins. It is the last of these settings which characterize most of the metalliferous sediments in the southwest Pacific.

The origin of most metalliferous sediments is related to the precipitation of phases from hydrothermal plumes. The process starts with the diffusion of seawater into newly formed volcanic rocks below the seafloor in submarine volcanic areas. It is heated and reacts with the rocks, becoming enriched in metals, before being discharged back into seawater. Much of the dissolved material carried in the hydrothermal solutions precipitates in the immediate vicinity of the discharge vents, constructing chimneys and mounds of sulphide and associated minerals, but some fraction of the vent fluids also escapes from the seafloor in a hydrothermal plume which disperses away from the site of discharge, partly under its own dynamics and partly under the influence of ocean currents. Such currents can carry dissolved and particulate materials of hydrothermal origin for considerable distances from the vent site, as far as several hundred kilometres on the East Pacific Rise (EPR), for example (Edmond *et al.* 1982). Gradually these materials precipitate from seawater to give rise to metalliferous sediments. It is evident that this process will result in a localized area of highly concentrated hydrothermal precipitates in the vicinity of the vents and that these will be surrounded by, or be adjacent to (depending on

From PARSON, L. M., WALKER, C. L. & DIXON, D. R. (eds), 1995, *Hydrothermal Vents and Processes*, Geological Society Special Publication No. 87, 231–248.

the direction of plume dispersal), a much larger area of hydrothermally enriched metalliferous sediments.

Basal metalliferous sediments and those within the sediment column are thought to be the ancient analogues of modern metalliferous sediments which have moved to their present positions as a result of seafloor spreading. Like surface metalliferous sediments, they contain high concentrations of several metals such as Fe, Mn, Cu and Zn compared with normal pelagic clays.

Studies of surface deposits and short cores in the central Lau Basin have shown the modern sediments to consist of mixtures of both detrital and non-detrital components, principally biogenic carbonate and volcaniclastic detritus, with a superimposed hydrothermal metalliferous component. The hydrothermal component has been demonstrated to be widespread in the central Lau Basin and to result from hydrothermal discharge at the two active spreading ridges, the Eastern Lau Spreading Centre (ELSC) and the Central Lau Spreading Centre (CLSC) (Cronan 1983; Cronan et al. 1984, 1986; Hodkinson et al. 1986; Riech 1990; Riech et al. 1990; Walter et al. 1990; Hodkinson & Cronan 1991). Some off-axis hydrothermalism has also been postulated (Cronan et al. 1986; Hodkinson & Cronan 1991; Parson et al. 1994b). Surface sediments proximal to the ELSC and CLSC have been shown to contain sulphide/weathered sulphide hydrothermal components, whereas the hydrothermal component of the more distal sediments consists solely of oxide plume fallout (Hodkinson & Cronan 1991).

These studies have focused on modern sediments and have not addressed the question of the long-term temporal variability in the flux of the hydrothermal component of the sediments during the evolution of the basin, particularly with respect to ELSC and CLSC propagation. One of the aims of ODP Leg 135 was to recover complete sedimentary sections down to basement at different localities in the basin to assess this variability.

Results of ODP Leg 135 drilling are given in Parson et al. (1992a) and Parson et al. (1994a). The hydrothermal inputs to all backarc sites (834 to 839) have been assessed using multivariate statistical analysis (factor analysis) of geochemical data for the HCl-soluble, carbonate-free sediment phase (Hodkinson & Cronan 1994). In this paper we present more detailed results of studies on the hydrothermal phase of sediments from Sites 834 and 835 alone, including multivariate statistical analysis, selective chemical leach data and metal accumulation rates, with

particular reference to ELSC and CLSC propagation into the basin.

Geological setting

Morphotectonic setting

The Lau Basin, part of the Lau Basin, Havre Trough, Tonga–Kermadec Ridge arc–backarc complex, is a small, shallow (2–3 km deep) marginal basin located between the islands and atoll reefs of the Lau Ridge in the west and the volcanic islands, atoll reefs and uplifted platform carbonates of the Tonga Ridge in the east (Fig. 1). Backarc extension has resulted in the separation of the Lau and Tonga Ridges and the formation of the Lau Basin.

Studies in recent years have shown the Lau Basin to have a complex origin. Models of tectonic evolution of the basin are discussed in Parson et al. (1990, 1992a, 1992b), Hawkins et al. (1994) and Parson & Hawkins (1994) (and references cited therein). The model of tectonic evolution of the Lau Basin adopted here is that described by Parson & Hawkins (1994), which is based on a compilation of bathymetric, magnetic and sidescan sonar data and the results of drilling during ODP Leg 135 (Fig. 1).

The western part of the central Lau Basin comprises a region of attenuated, heterogeneous crust that probably represents a mixture of relict arc fragments, ephemeral rift grabens and volcanic constructs derived from easterly migrating arc magmatism. The area exhibits irregular topography with a complex pattern of discontinuous ridges and basins, locally showing a north–south trend. Basins are partly sediment-filled and are separated by uplifted, elongate basement ridges and highs which have only a thin sediment cover.

The central part of the Lau Basin contains two major morphotectonic features: the ELSC and the CLSC (Fig. 1). The ELSC, lying close to the Tofua Arc, extends for over 180 km from 19°20′S to around 21°S, where it merges with the Valu Fa Ridge. The Valu Fa Ridge extends southwards to around 23°S where its propagating rift tip is thought to be located. Parson & Hawkins (1994) proposed that the ELSC propagated southwards at an estimated average rate of 120 mm/a, from the southeast end of a structure that became the Peggy Ridge, commencing at around 5.5 Ma. It passed through the horst and graben dominated extensional terrain derived from the fragmented Lau/Tonga composite arc. Propagation is thought to have taken place in an intermittent manner, probably controlled by the pre-existing rifted fabric of the western Lau

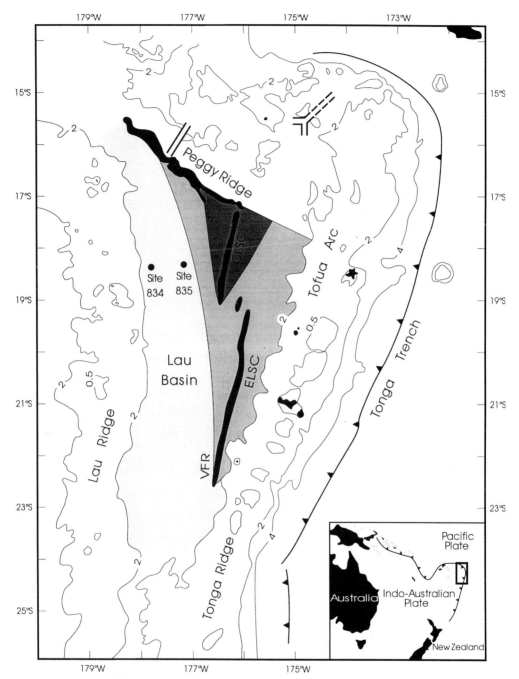

Fig. 1. Location, simplified bathymetry, simplified morphotectonic domains and major tectonic features of the central Lau Basin (after Parson & Hawkins 1994). CLSC = Central Lau Spreading Centre, ELSC = Eastern Lau Spreading Centre, VFR = Valu Fa Ridge. Light shading = pre-spreading crust, medium shading = ELSC generated crust and dark shading = CLSC generated crust. Isobaths in kilometres.

Basin. The estimated age of the passage of the southwardly propagating ELSC past the latitude of Sites 834 and 835 is 3.6 Ma. The ELSC has generated the majority of crust seen in the easternmost central Lau Basin.

The CLSC is a gently curving axial ridge, concave to the east, extending from around 17°10′S to around 19°20′S. Parson & Hawkins (1994) proposed that the CLSC commenced southward propagation from the southeastern limit of the Peggy Ridge at between 1.2 and 1.5 Ma, with a similar propagation rate to that of the ELSC (120 mm/a). Continued propagation, at the expense of the ELSC, into ELSC generated crust, has formed a wedge-shaped area of crust (Fig. 1). The estimated age of the passage of the southwardly propagating CLSC past the latitude of Sites 834 and 835 is 0.5 to 0.6 Ma (Parson & Hawkins 1994).

ODP Sites 834 and 835 are located in two north–south trending adjacent sub-basins in the western part of the central Lau Basin, at about the 18°30′S latitude, approximately 150 and 60 km, respectively, west of the CLSC (Fig. 1). The geometries of the two basins are very different. The Site 834 basin has relatively gently sloping margins (3° to 4°, rising to 1600 m above the basin floor in the west and 1100 m in the east), is about 8 km wide at the latitude of Site 834 and widens southwards. The Site 835 basin has very steep sloping margins (4° to 30°, rising 1400 m above the basin floor in the west and 1100 m in the east) and is about 9 km wide at Site 835 (Rothwell et al. 1994b).

Sediment sequences at Sites 834 and 835

Sediment sequences (continuously cored) at Sites 834 and 835 are 112.5 and 155 m thick, respectively. Sediments at both sites are described in Parson et al. (1992a) whereas the sedimentary history and sedimentary processes are discussed in Rothwell et al. (1994a, b) and Parson et al. (1994b). Figure 2 shows the generalized lithological columns for both sites. They show a similar sequence of clayey nannofossil oozes and mixed sediments (hemipelagic and redeposited) with interbedded epiclastic vitric sands and silts (both graded and massive) and occasional pyroclastic fallout ash layers. The vitric sands and silts are mainly restricted to the lower part of the sediment section and generally show an increase in thickness and frequency downhole. The clayey nannofossil ooze and mixed sediment sections are pervasively coloured by Fe and Mn oxyhydroxides. At Site 834 several periods of enhanced sediment redeposition have been identified (Parson et al.

1994b), notably an increase in redeposited volcaniclastic sediments between 3.6 and 4.4 Ma. Enhanced volcaniclastic inputs also occur between 2.4 and 2.9 Ma, whereas predominantly calcareous turbidites are more frequent in sediments younger than 0.2 Ma. At Site 835 the clayey nannofossil ooze sediment section is anomalously thick compared with that at Site 834. It contains several medium to thick beds of matrix-supported mudclast conglomerate and large amounts of clayey nannofossil ooze turbidites and coherent rafted blocks of older hemipelagic sediment (Rothwell et al. 1994b). Based on the presence of these redeposited intervals, Parson et al. (1994b) have proposed that periods of instability of the area around Site 835 were pronounced between 0 and 0.4, 0.9 and 1.0, 1.4 and 1.7 and 2.1 and 2.9 Ma.

A significant proportion of the sedimentary sections at all backarc sites sampled during ODP Leg 135 contain both redeposited clayey nannofossil ooze intervals and/or intervals of epiclastic volcanic ash deposition (see above). To assess the downhole variability in hydrothermal inputs to the hemipelagic, non-redeposited sediment intervals alone the thickness of the sediment section at each site has been recalculated by subtracting the thicknesses of redeposited intervals. This enables a more accurate assessment of the temporal variability in hydrothermal input at each site to be made. Detailed work by Rothwell et al. (1994b) has distinguished the redeposited intervals on sedimentological criteria. These include variation in colour hue and chroma, the presence or absence of bioturbation, the presence or absence of scattered foraminifera, grain size characteristics, variability in calcium carbonate content, the presence or absence of pumice clasts and their micropalaeontology. Removal of the redeposited sediment intervals would reduce the thickness of the sediment section from 112.5 to 74 m at site 834 (66% of the original sediment thickness) and from 155 to 52 m at Site 835 (34% of the original sediment thickness). From this, the depth below seafloor that individual samples would have been in the absence of redeposition can be recalculated. Having done this, all depths referred to in the subsequent descriptions of downhole variability are quoted as 'recalculated depths below seafloor' (mbsf*).

To compare variations in hydrothermal input between the sites, the ages of prominent peaks in hydrothermal input have been determined based on a combination of shipboard and post-cruise data. At both sites, Rothwell et al. (1994b) have estimated sediment ages using shipboard palaeomagnetic data and additional

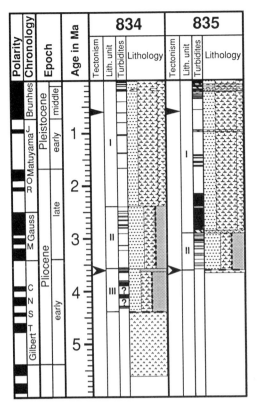

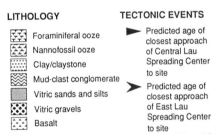

LITHOLOGY

Foraminiferal ooze
Nannofossil ooze
Clay/claystone
Mud-clast conglomerate
Vitric sands and silts
Vitric gravels
Basalt

TECTONIC EVENTS

► Predicted age of
closest approach
of Central Lau
Spreading Center
to site

► Predicted age of
closest approach
of East Lau
Spreading Center
to site

Fig. 2. Simplified lithological columns for Sites 834 and 835 with age (Ma) as a vertical scale (after Rothwell *et al.* 1994*a*). Allochthonous beds (determined by Rothwell *et al.* 1994*b*) are shown in the 'Turbidites' column. Thick broken lines indicate that the lithologies shown are interbedded (see Parson *et al.* 1992*a*). Predicted ages of the closest approach of the CLSC (0.5 to 0.6 Ma) and ELSC (3.6 Ma) are shown (see Parson & Hawkins 1994). Recovered sediment sequences are 112.5 m at Site 834 and 155 m at Site 835. Recalculation to subtract the redeposited intervals would reduce these thicknesses to 74 m at Site 834 and 52 m at Site 835 (see text).

biostratigraphic data obtained post-cruise for the hemipelagic intervals in Unit I sediments (see Fig. 2). Below Unit I, sediment ages have

been established from shipboard palaeomagnetic data applied to the recalculated hemipelagic sedimentary intervals identified by Rothwell *et al.* (1994*b*).

Analytical methods and data processing

Before chemical analysis, sediment samples ($n = 467$) were air-dried at room temperature over silica gel and finely ground using an agate mortar and pestle. Bulk chemical analyses were performed by inductively coupled plasma atomic emission spectrometry (ICP-AES) after total digestion with a mixture of nitric, perchloric and hydrofluoric acids and subsequent leaching with 1 M HCl (Thompson & Walsh 1989). Accuracy was checked with the use of certified international reference materials, along with in-house reference materials. No systematic errors were observed and accuracy was better than ±5% relative for the elements studied here. Analytical precision was checked with the use of duplicate samples and was found to be better than ±3% relative.

To assess the geochemical partitioning of elements between the constituent phases of the sediments, a series of selective leaches were undertaken on some samples ($n = 85$ for leaches 1 and 2, $n = 467$ for leach 3). These leaches are based on methods described by Chester & Hughes (1967) and subsequently modified by Cronan (1976*a*). The leaches used provide a means of assessing which phase of the sediments the hydrothermal enrichments are associated with. These phases are operationally defined as follows: (1) carbonates, loosely adsorbed phases and interstitial water evaporates (leached using 25% acetic acid); (2) as (1), plus reducible ferromanganese oxides and associated elements (leached using a combined acid-reducing agent of 1 M hydroxylammonium chloride in 25% acetic acid); (3) as (2), plus more crystalline Fe oxides, clays, altered detrital phases and partial sulphide dissolution (leached using hot 50% HCl); and (4) resistant detrital phases [by subtraction of (3) from bulk concentrations].

After filtration through 0.45 μm cellulose nitrate filter papers, decomposition of the leachates using HNO_3 and leaching of the residues in 1 M HCl, the leachate compositions were determined by ICP-AES techniques. Analytical precision was found to be better than ±4% relative for the elements studied here. The weight per cent of individual samples soluble in each leach was also determined by weighing the dried filter residues.

To assess inter-element associations in the non-detrital, carbonate-free, fraction of the sediments, factor analysis was performed on geochemical data for the HCl-soluble, carbonate-free sediment phase. The factor analysis technique enables a larger number of variables (in this case, chemical elements) to be grouped into a lesser number of factors, each of which is a linear function (transformation) of the element concentrations and is based on the element associations inherent in a suitably conditioned correlation matrix (Howarth 1983). The relative influence of each factor (in this case the hydrothermal oxide factor) on individual samples is then determined by factor scores. These scores enable the relative intensity of hydrothermal oxide plume fall-out throughout the sediment section at each site be assessed.

By applying this technique to geochemical data for the HCl-soluble, carbonate-free sediment phase, the effects of two major sources of variance in the bulk uncorrected data – biogenic carbonate and volcaniclastic detritus – have been removed. The effect of these two dominant and antipathetically varying sediment components is to mask variations in the hydrothermal component of the sediments. Their removal by selective leach and statistical methods thus allows a closer assessment of the variability in the hydrothermal sediment phase to be made (Hodkinson & Cronan 1991).

Non-detrital metal accumulation rates for Mn and Fe have been calculated using the equation

$$\text{Accumulation rate } (\mu g\,cm^{-2}\,ka^{-1}) = M \cdot \rho \cdot S$$

where M = element concentration (ppm) in the HCl-soluble sediment phase; ρ = dry, in situ sediment density $(g\,cm^{-3})$; and S = sedimentation rate $(cm\,ka^{-1})$.

Dry, in situ density values were calculated from shipboard physical properties measurements for both sites (Parson et al. 1992a). Sedimentation rates are based on the recalculated hemipelagic sediment thicknesses for each site (see above, Rothwell et al. 1994b).

X-ray diffraction (XRD) measurements on air-dried mounts of finely ground sample powders ($n = 15$) were made using a Phillips PW1820 goniometer/PW1830 generator XRD system with automated divergence slit geometry. X-ray patterns were recorded using Cu Kα radiation with a monochromator, a step size of 0.02°, counting for six seconds per step with generator settings of 40 kV and 50 mA over a peak angle range of 5° to 90°. Data were processed using Phillips APD 1700 software.

Results

Shipboard smear slide analysis of clayey nannofossil ooze sediment sections at both sites

(Parson et al. 1992a) show that Fe and Mn oxyhydroxides are abundant in the sediments and occur as both amorphous aggregates and surface coatings on sediment grains. Also, XRD analysis carried out on selected samples from both sites shows that a significant X-ray amorphous ferromanganese oxide component occurs in the sediments. No crystalline Fe or Mn oxide, Fe silicate or sulphide mineral phases could be detected.

The bulk, HCl-soluble and carbonate-free HCl-soluble geochemistry of the sediments studied in this work has been described by Hodkinson & Cronan (1994). In summary, the bulk sediment composition reflects its two major components, biogenic carbonate and volcaniclastic detritus, with a superimposed metal-rich hydrothermal phase. For the principal elements of the metal-rich hydrothermal phase, Mn and Fe, bulk concentrations range from 0.05 to 8.17% Mn (Site 834) and 0.19 to 11.58% Mn (Site 835), and from 4.31 to 10.51% Fe (Site 834) and 3.99 to 9.82% Fe (Site 835). Non-detrital (HCl-soluble), carbonate-free values range from 0.04 to 14.50% Mn (Site 834) and 0.013 to 13.85% Mn (Site 835), and from 0.53 to 18.63% Fe (Site 834) and 1.34 to 13.65% Fe (Site 835).

Factor analysis

At Sites 834 and 835, factor analysis of the geochemical data has resolved two factors, accounting for 80.0% (Site 834) and 89.8% (Site 835) of the data variance (Tables 1 and 2). Factor 1 shows loadings of Mn, Fe, Co, Ni, Cu,

Table 1. *Varimax rotated factor loadings for two factors obtained for the hot 50% HCl-soluble, carbonate-free phase of sediments from Site 834 (N = 219)*

Variable	Factor 1	Factor 2
Mg	0.02	0.65
Al	−0.07	0.90
Ti	0.06	0.87
V	0.68	0.07
Mn	0.92	−0.02
Fe	0.80	0.12
Co	0.69	0.01
Ni	0.85	−0.09
Cu	0.89	−0.05
Zn	0.86	0.08
Pb	0.76	0.05
P	0.88	−0.02
Variance (%)	58.4	21.6
Cumulative variance (%)	58.4	80.0

Table 2. *Varimax rotated factor loadings for two factors obtained for the hot 50% HCl-soluble, carbonate-free phase of sediments from Site 835 (N = 248)*

Variable	Factor 1	Factor 2
Mg	0.05	−0.79
Al	− 0.06	−0.86
Ti	0.08	−0.75
V	0.87	0.24
Mn	0.83	0.19
Fe	0.94	− 0.08
Co	0.80	− 0.13
Ni	0.78	0.19
Cu	0.91	0.19
Zn	0.91	0.19
Pb	0.88	0.16
P	0.84	− 0.25
Variance (%)	66.8	23.0
Cumulative variance (%)	66.8	89.8

Zn, Pb, V and P, whereas factor 2 shows loadings of Mg, Al and Ti.

As the HCl leach removes all but the resistant detrital material from the sediments, factor 2 is likely to represent a combination of several sedimentary components enriched in Mg, Al and Ti. These include partially altered volcaniclastic material, halmyrolytic products such as smectite and other minor components such as aeolian dust. Based on previous studies of Lau Basin surface sediment lithology and mineralogy (Hodkinson *et al.* 1986; Riech *et al.* 1990; Walter *et al.* 1990), it is probable that altered volcaniclastic material is quantitatively the most important sedimentary component influencing this factor and it is thus termed an 'altered volcaniclastic' factor. It is not discussed further here.

Factor 1 loadings of Mn, Fe, Co, Ni, Cu, Zn, Pb, V and P are typical of an hydrothermal oxide sediment phase, as widely reported in Lau Basin surface sediments. Factor 1 is thus taken to represent a composite 'hydrothermal oxide' sediment phase. Unlike in modern sediments from the central Lau Basin, no sulphide/ weathered sulphide factor was identified in the sediments studied here. This suggests that Sites 834 and 835 were never immediately adjacent to the source of hydrothermal discharge, where such a component is likely to be found. The hydrothermal phase at Sites 834 and 835 is considered to be a distal 'oxide plume fallout' phase.

Phase association of hydrothermal input

Geochemical partition analysis has been carried out on selected samples at both sites to assess which sediment phases individual element enrichments are associated with. Figure 3 shows patterns of partitioning for Mn and Fe at both Sites 834 and 835.

The mean bulk concentrations of both Mn and Fe show an increase in the sediments as the relative hydrothermal influence, based on factor analysis, increases. Selective leach data show that Mn occurs principally in the acid-reducible phase, where it will largely be present as amorphous oxide, and that the increase in mean bulk concentration with increasing relative hydrothermal influence occurs due to increased Mn concentrations in this phase. Little or no Mn occurs in either insoluble or HCl-soluble detrital and altered detrital phases and only relatively small amounts occur in acetic acid soluble carbonates or adsorbed phases. In the case of Fe, although a significant amount is associated with insoluble and HCl-soluble detrital and altered detrital phases, the increasing mean bulk concentrations with increasing relative hydrothermal influence occur, like Mn, due to increased concentrations in the reducible (oxide-associated) phase.

Factor scores with respect to ELSC propagation

Based on factor 1 scores (Fig. 4), variations in the hydrothermal oxide plume fallout show a similar trend at Sites 834 and 835. In general there is an increase in input uphole from the basement to a period of maximum input before a decrease towards the surface. The peak in the period of enhanced hydrothermal input occurs at around 54 to 55 mbsf* at Site 834 and 43 to 44 mbsf* at Site 835, equating to approximately 3.2 Ma at both sites. This trend is unlike the typical pattern of hydrothermal plume fallout recorded in sediments on mid-ocean ridges, which show a decrease in intensity uphole from the basement as crustal accretion continues to increase the distance between the source of hydrothermal discharge and the site of plume fallout (e.g. EPR, Lyle *et al.* 1986).

The trend observed at Sites 834 and 835 is interpreted as an increase in hydrothermal input to each site as the ELSC propagates southwards into pre-existing crust and hydrothermal discharge from it becomes more proximal to the sites. This is followed by a decrease in the hydrothermal input to both sites as crustal accretion at the newly formed spreading centre increases the distance between the hydrothermal vents associated with it and the sites. Significantly, however, the period of maximum hydrothermal input recorded at both sites is not

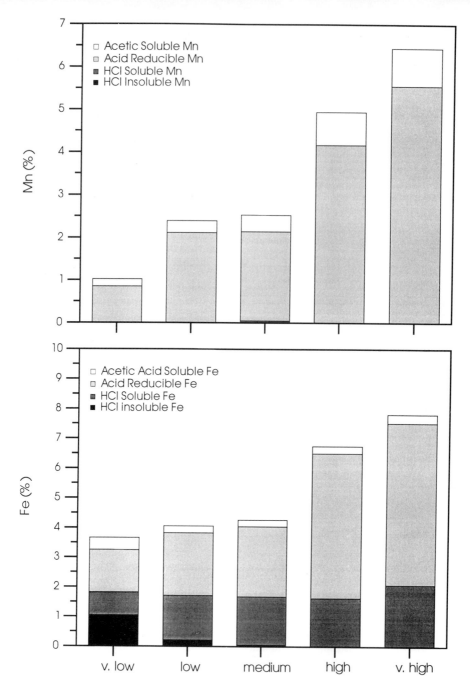

Relative Hydrothermal Influence

Fig. 3. Mean geochemical partitioning of Mn and Fe at Sites 834 and 835. Samples ($n = 85$) have been assigned a 'relative hydrothermal influence', based on the 20th, 40th, 60th and 80th percentile divisions of factor scores for the hydrothermal oxide factor. For each group, the height of the bar shows the mean bulk concentration of each element. Shading within the bar indicates the mean proportions soluble in each of the selective leaches for samples in that group. For Mn, only 'medium' hydrothermal influence samples show any HCl-soluble, non-oxidic Mn. No Mn occurs in the HCl-insoluble phase. For Fe, samples assigned to 'high' and 'very high' hydrothermal influence groups show no significant HCl-insoluble (detrital) component. Significant amounts of HCl-soluble Fe (principally altered detrital phases) occur in all hydrothermal influence groups.

Factor 1 Scores (Hydrothermal Oxide Factor)

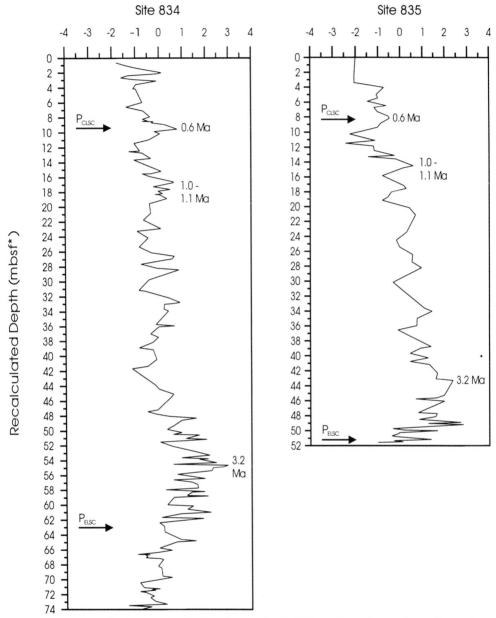

Fig. 4. Plot of factor 1 (hydrothermal oxide factor) scores for the HCl-soluble, carbonate-free sediment phase versus recalculated depth at Sites 834 and 385. P_{CLSC} = age of closest passage of southward propagating CLSC to both sites (0.5 to 0.6 Ma), P_{ELSC} = age of closest passage of southward propagating ELSC to both sites (3.6 Ma), based on Parson & Hawkins (1994) model of basin evolution. Ages of selected peaks in input are indicated (see text).

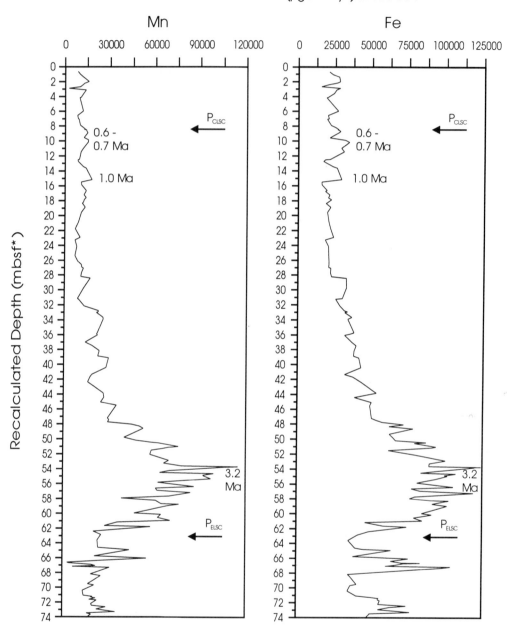

Fig. 5. Plot of non-detrital Mn and Fe accumulation rates versus recalculated depth at Site 834. P_{CLSC} = age of closest passage of southward propagating CLSC to both sites (0.5 to 0.6 Ma) and P_{ELSC} = age of closest passage of southward propagating ELSC to both sites (3.6 Ma), based on Parson & Hawkins (1994) model of basin evolution. Ages of selected peaks in input are indicated (see text).

Metal accumulation rates (μgcm⁻²ky⁻¹) at Site 835

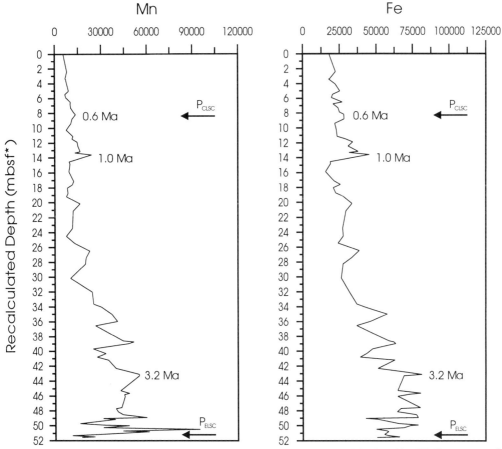

Fig. 6. Plot of non-detrital Mn and Fe accumulation rates versus recalculated depth at Site 835. P_{CLSC} = age of closest passage of southward propagating CLSC to both sites (0.5 to 0.6 Ma) and P_{ELSC} = age of closest passage of southward propagating ELSC to both sites (3.6 Ma), based on Parson & Hawkins (1994) model of basin evolution. Ages of selected peaks in input are indicated (see text).

coincident with the closest passage of the propagating ELSC past them (3.6 Ma), but instead occurs in sediments approximately 0.4 Ma younger.

Factor scores with respect to CLSC propagation

In the model of basin evolution proposed by Parson & Hawkins (1994), the southward propagation of the CLSC into crust generated by the ELSC would be expected to produce a peak in hydrothermal input followed by a subsequent decrease, analogous to that related to the ELSC.

At both Sites 834 and 835, peaks in factor scores are less clear with respect to CLSC

propagation than to ELSC propagation. The most prominent peaks occur at around 9 to 10 and 16 to 17 mbsf* at Site 834 and 8 and 14 to 15 mbsf* at Site 835, equating to approximately 0.6 Ma and 1.0 to 1.1 Ma at both sites. The ages of these peaks are difficult to reconcile with the Parson & Hawkins (1994) model of basin evolution which predicts passage of the south-wardly propagating CLSC past the latitude of Sites 834 and 835 at around 0.5 to 0.6 Ma.

Metal accumulation rates with respect to ELSC propagation

The principal feature of downhole plots of element accumulation rates for Site 834 is the peak in accumulation rates for Mn and Fe

Table 3. *Summary statistics for HCl-soluble Mn and Fe accumulation rates ($\mu g\,cm^{-2}\,ka^{-1}$) at ODP Sites 834 and 835 in the Lau Basin and comparative Mn and Fe accumulation rates from areas of known hydrothermal plume fallout*

Site/location	Mn	Fe
Site 834		
Mean	29 362	48 995
Max.	114 946	124 281
Min.	2384	15 147
3.2 Ma peak mean*	82 460	97 464
3.2 Ma peak max.*	114 946	124 281
3.2 Ma peak min.*	57 346	79 609
≤1.0 Ma mean†	11 697	24 309
≤1.0 Ma max.†	17 891	34 325
≤1.0 Ma min.†	2384	15 147
Site 835		
Mean	24 929	42 025
Max.	94 785	81 104
Min.	5803	15 445
3.2 Ma peak mean*	47 356	69 677
3.2 Ma peak max.*	54 651	81 104
3.2 Ma peak min.*	39 344	51 217
≤1.0 Ma mean†	10 987	25 691
≤1.0 Ma max.†	23 910	45 403
≤1.0 Ma min.†	5803	17 667
EPR crest[a]	5400	29 300
EPR crest[b]	5800–28 000	11 000–82 000
EPR Leg 92[c]	130–35 100	900–102 900
EPR (recent ridge proximal)[d]	36 000–86 000	120 000–240 000
Lau Basin[e]	5500	18 000
Lau Basin[f]	4700–15 000	10 500–48 500
North Fiji Basin[g]	1600–2000	
Bauer Deep[h]	4400–88 000	3900–85 000
Bauer Deep[i]	3330	9000

* Values for 3.2 Ma peak at Sites 834 (52–56 mbsf* recalculated depth) and 835 (42–46 mbsf* recalculated depth).
† Values for sediments 0–1.0 Ma at Sites 834 (0–15 mbsf* recalculated depth) and 835 (0–14 mbsf* recalculated depth).
[a]Leinen & Stakes (1979), [b]Dymond & Veeh (1975), [c]Barrett *et al.* (1987), [d]Lyle *et al.* (1986), [e]Cronan *et al.* (1984), [f]Cronan *et al.* (1986), [g]Riech (1990), [h]Boström *et al.* (1976), [i]Sayles (1975a,b).

(Fig. 5) at around 54 to 55 mbsf*, equating to an age of about 3.2 Ma. Above and below this depth interval, accumulation rates generally initially decrease rapidly (to about 42 and 67 mbsf*, respectively) followed by a more gradual decrease. Local variability does, however, occur (particularly in the case of Fe, where a pronounced peak occurs at around 67 mbsf*, equating to approximately 4.0 Ma).

Site 835 shows a similar general pattern to Site 834, with a peak in accumulation rates for Mn and Fe (Fig. 6) occurring at around 43 mbsf*, again equating to an age of 3.2 Ma. Below this both elements show a slight decrease in accumulation rates towards the base of the section, although in the case of Mn some anomalously high values occur. These are due to small fragments of Mn crust (probably not *in situ*)

included within the sediments between 50 and 51 mbsf*. The crusts comprise fragments with a purplish black metallic to submetallic lustre and massive texture with 10 Å manganite (todorokite) and δMnO_2 mineralogy. Compositionally they have high Mn (up to 35%) and low Fe and trace metal contents. Based on their morphological, mineralogical and chemical similarity with previously studied hydrothermal crusts it is concluded they are hydrothermal in origin (Rothwell *et al.* 1994a).

The pronounced decrease seen in Mn and Fe accumulation rates below the 3.2 Ma peak at Site 834 was not recognized at Site 835 due to the closer proximity of the site to the ELSC generated crust/pre-existing crust boundary (Parson & Hawkins 1994), resulting in a shorter record in the sediment section of plume fallout prior to propagation of the ELSC past the site. Above the peak in input at around 3.2 Ma, element accumulation rates decrease moderately rapidly to around 30 mbsf* (but not as rapidly as is seen at Site 834), but with significantly more local variability than at Site 834.

The trend above the peak in accumulation rates at 3.2 Ma for both sites is similar to the generally logarithmic decrease seen in DSDP Leg 92 EPR sediments away from basement, due to crustal accretion (Lyle *et al.* 1986).

In addition to trends with respect to ELSC propagation, Mn and Fe accumulation rates allow the intensity of the hydrothermal inputs to the sediment sections at Sites 834 and 835 to be quantified and compared with each other. Table 3 gives summary statistics for Sites 834 and 835 as a whole, along with the mean and maximum values for the interval of peak hydrothermal input in the cores at around 3.2 Ma (52–56 and 42–46 mbsf*, respectively), related to the passage of the ELSC propagator past the sites. For both elements, mean values for the peak in input are higher at Site 834 than at 835. In addition, mean accumulation rates are variably higher at Site 834, with Fe typically around 40% higher and Mn typically around 75% higher than at Site 835. Site 834 is presently located on pre-spreading crust approximately 90 km west of Site 835. This distance may, however, have been slightly less during the period of peak hydrothermal input described here due to continuing crustal attenuation in the western part of the Lau Basin post-ELSC initiation (Parson & Hawkins 1994). However, given the fact that Site 834 was located at a greater distance from the source of hydrothermal discharge at the ELSC than Site 835, the occurrence of significantly higher element accumulation rates at this site compared with Site 835 is the opposite of that which would be expected if simple plume dispersion away from the source of discharge had occurred.

Metal accumulation rates with respect to CLSC propagation

Factor analysis tentatively identified weak peaks in hydrothermal input at both Sites 834 and 835 that may be related to CLSC propagation. The timing of these peaks (approximately 0.6 and 1.0 to 1.1 Ma) is, however, not consistent with the proposed timing of CLSC propagation which estimates the closest passage to Sites 834 and 835 at 0.5 to 0.6 Ma. Element accumulation rate trends for Mn and Fe in the younger part of the

sediment section at Sites 834 and 835 (Figs 5 and 6) do not show as significant a peak as is seen for the ELSC propagation. At Site 834, Fe shows a slight increase in background accumulation rates above about 15 mbsf*, with peaks at depths equating to approximately 0.6 to 0.7 and 1.0 Ma, although significant variability is seen. Mn shows little significant overall increase in background accumulation, although small peaks occur at time-equivalent depths to those of Fe. At Site 835, Fe shows a more significant increase in accumulation rate at around 13 to 14 mbsf* (approximately 1.0 Ma), followed by a decrease towards the surface. Mn shows a similar but weaker trend. Smaller peaks for both Mn and Fe are also seen at approximately 0.6 Ma. The relative differences in accumulation rates between Sites 834 and 835 are reflected in mean and maximum accumulation rate values for sediments $\leqslant 1.0$ Ma (Table 3). These show significantly higher mean values at Site 835 compared with Site 834 and larger peaks (maximum values) relative to the mean values.

In terms of the trends in Mn and Fe accumulation rates with respect to the proposed model of basin evolution, it is tempting to suggest that the relatively small increase in accumulation rates seen at around 1.0 Ma at both sites may be related to CLSC propagation. The peak in Fe accumulation rate at Site 835 is particularly pronounced compared with Site 834, whereas the difference between the respective Mn peaks is less pronounced. Given the closer proximity of Site 835 to the source of hydrothermal discharge at the CLSC relative to Site 834, such a pattern is in accord with the established dispersion patterns of Fe and Mn in hydrothermal plumes, with Mn showing a greater dispersion than Fe (cf. Bignell *et al.* 1976; Cronan 1976*b*; Edmond *et al.* 1979).

Discussion

The Lau Basin has formed in response to subduction of the Pacific plate at the Tonga Trench. As summarised by Hawkins *et al.* (1994), basin evolution commenced in the late Miocene with initial crustal extension and arc rifting which developed a basin and range structure, followed by propagating seafloor spreading along the ELSC axial ridge. Subsequently, backarc spreading was, and still is being, transferred to the CLSC, in accord with the propagating rift model first described by Hey (1977) for the Cocos–Nazca spreading system. This interpretation is similar to that described for the Mariana Trough, Izu–Bonin Arc and Japan Sea (e.g. Taylor 1992; Tamaki *et al.* 1992).

The data presented in this paper support discussion of the influence of ridge propagation on the hydrothermal sedimentary record in the Lau Basin. Given the postulated similarity between crustal evolution in the Lau Basin and other Pacific backarc basins, the conclusions reached here may have some applicability to these and other backarc basins.

ELSC propagation

Metal accumulation rate trends at ODP Sites 834 and 835 agree well with the principal results of factor analysis of geochemical data on the sediments. The peak in metal accumulation rates in the sediment sections at 3.2 Ma reflects maximum hydrothermal plume fallout associated with the southward propagation of the ELSC into attenuated pre-spreading crust. This is evinced by an increase in hydrothermal input to the sediments at Sites 834 and 835 as the propagating ELSC becomes more proximal to both sites. Subsequently a decrease in input occurs after the ELSC and its associated hydrothermal system have become established and crustal accretion increases the distance between the site of hydrothermal venting and Sites 834 and 835. As in the case of the factor analysis results, timing of the peak in hydrothermal input based on metal accumulation rates (3.2 Ma) compared with the closest passage of the southwardly propagating ELSC to Sites 834 and 835 (3.6 Ma) indicates a 'time lag' of about 0.4 Ma between propagator passage and maximum hydrothermal input to the sediments. This 'time lag' is considered to result from a delay between the formation of the spreading ridge by rift propagation and either the hydrothermal system associated with it being initiated and/or becoming fully established and producing its maximum hydrothermal flux. These alternatives are not at variance with observations on the modern hydrothermal systems associated with the CLSC and ELSC, where hydrothermal fields discovered to date occur at some distance behind the propagating tips of both spreading systems (Malahoff & Falloon 1991; Fouquet et al. 1993). At the CLSC, Malahoff & Falloon (1991) found extensive sulphide deposits at a distance behind its propagating tip equating to about 0.5 to 0.6 Ma of southward propagation. In the case of the ELSC, studies of the Valu Fa Ridge (Fouquet et al. 1993) suggest that the most actively discharging vents are located on the older, more evolved parts of the ridge (the central Valu Fa Ridge segment), equating to a distance behind the propagating tip of about 0.7 Ma of southward ELSC propagation.

Although the time between ridge propagation and the initiation and/or full development of hydrothermal systems may account for the 'time lag' recorded in the sediments, it is also possible that directed plume dispersion by prevailing currents could also produce an apparent 'time lag'. A possible scenario in the case of the ELSC is that a northward flowing current would have dispersed the plume 'behind' the direction of propagation. This would result in the maximum fallout to the sediments apparently occurring after the time of ridge propagation, as recorded in the sediments here. Conversely, a southward flowing current would have dispersed the plume ahead of the propagating ridge, resulting in maximum fallout to the sediments apparently occurring prior to ELSC propagation past the sites. This was clearly not the case based on the present data.

As described, both Mn and Fe show higher mean accumulation rates at Site 834 than at Site 835 for the interval of peak hydrothermal input at 3.2 Ma associated with ELSC propagation, even though Site 834 was further from the ridge than Site 835 at the time of maximum hydrothermal input. In addition, the mean Mn accumulation rates are higher at Site 834 relative to those of Fe compared with Site 835.

Possible reasons for these differences could include: (1) a local hydrothermal source proximal to Site 834; (2) differences in oxidation conditions between the sediments at Sites 834 and 835; (3) differences in the dispersion of Mn relative to Fe from the plume source; and (4) modification of plume dispersion/ deposition patterns by currents and basin topography.

Local off-axis volcanism and associated hydrothermalism in the western part of the Lau Basin has been postulated by Cronan et al. (1986), Hodkinson & Cronan (1991) and Parson et al. (1994b). However, given the similarity in the trends of element accumulation rates at Sites 834 and 835 and their strong relationship with the proposed model of basin evolution (i.e. the peak in hydrothermal input associated with the ELSC propagation) it seems unlikely that a local off-axis hydrothermal source could account for the significantly higher accumulation rates seen at Site 834 than at Site 835. To produce such a pattern, any local hydrothermal input at Site 834 would have had to vary in discharge intensity in a similar manner to the plume fallout intensity produced by the ELSC. It is, however, possible that local hydrothermal activity may have produced some of the small-scale variation seen in downhole metal accumulation rate trends (e.g. the peaks in Mn and, in particular, Fe accumulation rates seen between 66 and 68 mbsf* at Site 834; Fig. 4).

A higher degree of oxidation at the sediment surface at Site 834 than at Site 835 could have favoured oxide precipitation at Site 834 relative to Site 835, resulting in the enhanced accumulation rates of Mn and Fe seen at Site 834. However, lithostratigraphic studies (Parson *et al.* 1992*a*) indicate similar depositional environments in terms of oxidation conditions at the two sites. Further, pore water studies at both sites (Blanc *et al.* 1994) show similar and entirely oxic interstitial water geochemistry, indicating that differential diagenetic remobilization of metal oxides, particularly Mn oxide, is unlikely to have resulted in the metal accumulation rate differences observed. It would thus seem unlikely that the higher Mn and Fe accumulation rates at Site 834 than at Site 835 are due to differences in oxidation conditions or to post-depositional diagenetic remobilization.

High Mn accumulation rates relative to Fe in the sediments could be due, in part, to the nature of the hydrothermal dispersion patterns of Mn and Fe from the hydrothermal vents. Several studies (e.g. Bignell *et al.* 1976; Cronan 1976*b*; Edmond *et al.* 1979) have shown that Mn is dispersed greater distances from the plume source than Fe as a result of its slower rate of oxidation and precipitation relative to Fe. This mechanism results in more distal deposits from the plume source being relatively enriched in Mn. Although this process could be held to account for the enrichment of Mn *relative* to Fe at Site 834 compared with Site 835, it could not account for the higher accumulation rates of both Mn *and* Fe seen at Site 834 than at Site 835.

Currents and basin topography could have had an effect on plume dispersion and fallout, and the resulting element accumulation rates in the sediments at Sites 834 and 835. Preferential plume dispersion directions due to prevailing currents have been reported by Edmond *et al.* (1982) at the EPR, with sediments to the west of the EPR receiving higher amounts of plume fallout than those to the east. In the Lau Basin, Sites 834 and 835 are both located at approximately the same latitude, to the west of the source of hydrothermal discharge. It thus seems unlikely that a prevailing current direction alone could account for the higher accumulation rates seen at Site 834 as the peaks in accumulation rates at the two sites are time-equivalent.

Crust in the western part of the Lau Basin exhibits extremely irregular topography, including discontinuous basins separated by ridges, predominantly formed before ELSC propagation (Parson *et al.* 1990; 1992*a*; Parson & Hawkins 1994). Although basin topography is likely to have changed since the period of maximum hydrothermal input at 3.2 Ma associated with ELSC propagation, this region is one in which topography is always likely to have been irregular, as evinced by sedimentation patterns at Sites 834 and 835 (Parson *et al.* 1992*a*, 1994*b*; Rothwell *et al.* 1994*a*, *b*). Such irregular topography could have influenced both hydrothermal plume dispersion and plume fallout and may in part account for the differences seen in accumulation rates between the two sites. Site 835 is located close to the eastern margin of a basin which at the present day has steeply sloping margins, whereas Site 834 is located towards the centre of a basin with relatively shallow sloping margins. Lesser amounts of plume material may have reached the sediments at Site 835 than received in the more gently sloping basin at Site 834 because of obstruction of the plume by the bathymetric high at the eastern edge of the Site 835 basin resulting in a 'shadow effect' in regard to plume dispersion and fallout.

CLSC propagation

It is evident that, by comparison with ELSC propagation into the Lau Basin, the later propagation of the CLSC appears to have produced comparatively little signature in terms of metal accumulation rate trends in the sediments. Accumulation rate values are typical of those in presently accumulating sediments in the vicinity of the CLSC (Table 3). The hydrothermal flux from sources on the CLSC thus appears to have been similar during the early stages of its evolution to that at the present day, and markedly lower than that associated with the ELSC. Surface sediment studies (Riech *et al.* 1990) support the latter conclusion, showing that the hydrothermal input to modern sediments associated with the CLSC is markedly lower than that associated with the Valu Fa Ridge (the southernmost part of the ELSC).

If the peaks in accumulation rates seen in the upper parts of the sections at Sites 834 and 835 do reflect the hydrothermal activity associated with CLSC propagation, a discrepancy exists as regards their timing relative to the model of basin evolution. Parson & Hawkins (1994) suggest that the closest passage of the southwardly propagating CLSC to Sites 834 and 835 occurred at around 0.5 to 0.6 Ma. Peaks in accumulation rates seen in Sites 834 and 835 sediments at around 1.0 Ma are about 0.4 Ma older than the period of closest passage of the CLSC to both sites. As discussed in regard to

ELSC propagation, the palaeo-current direction might contribute to the stratigraphic position of the peaks in metal accumulation rates seen in the sediments. For example, dispersion of the plume 'ahead' of the propagator by a southward flowing current would give an anomalous 'old' age for maximum hydrothermal input relative to propagator passage past the site. However, errors in the timing of the peaks in hydrothermal input, in the recalculation of non-redeposited sediment thicknesses and in the predicted age of closest passage of the CLSC propagator past the latitude of Sites 834 and 835 could all have contributed to the observed age discrepancies.

Comparison with hydrothermal sediments from other regions

Element accumulation rates provide a means of quantifying the intensity of the hydrothermal flux to sediments in different regions. Table 3 gives summary data for Sites 834 and 835, along with comparative data for other regions, principally the EPR.

For sediments $\leqslant 1.0$ Ma at Sites 834 and 835, Mn and Fe accumulation rates are similar to those of modern Lau Basin sediments associated with the CLSC at the same latitude (Table 3, Cronan et al. 1986). These sediments generally show accumulation rates within the range of most of the EPR/Bauer Deep sediments, although values reported for EPR ridge proximal sediments are significantly higher. Mn accumulation rates are also significantly higher than those in North Fiji Basin sediments.

Sites 834 and 835 sediments associated with the peak in hydrothermal input at around 3.2 Ma generally show the mean accumulation rates of Mn and Fe comparable with the highest reported values in EPR/Bauer Deep sediments. Maximum accumulation rates frequently exceed those reported from these sediments.

Compared with EPR sediments, accumulation rate data also show that Lau Basin hydrothermal plume fallout is rich in Mn relative to Fe. In particular, plume fallout associated with the ELSC exhibits a two- to four-fold enrichment of Mn relative to Fe compared with the EPR. Such findings are in accord with studies on modern Lau Basin sediments (Riech et al. 1990) which have shown that the hydrothermal oxide component, in particular that associated with the Valu Fa Ridge (ELSC), is enriched in Mn relative to Fe.

Summary and conclusions

(1) Multivariate statistical analysis (factor analysis) of data for the HCl-soluble, carbonate-free sediment phase suggests that hydrothermal input to the sediments at Sites 834 and 835 occurs primarily in an 'oxide plume fallout' phase, with no evidence of a sulphide/weathered sulphide phase. Sediment lithology, XRD and selective chemical leach data confirm this, indicating that the hydrothermal sediments at Sites 834 and 835 represent distal oxide plume fallout.

(2) Variations in factor scores for the hydrothermal oxide phase of the sediments reflect changing hydrothermal plume fallout flux to ODP Sites 834 and 835 associated with propagation of the ELSC, and possibly the CLSC, into the Lau Basin. Element accumulation rate data support this.

(3) Timing of the period of maximum hydrothermal input to the sediments at both sites suggests a 'time lag' between ELSC propagator passage and the maximum input recorded in the sediments. This may be due to a 'time lag' between ridge propagation and (a) hydrothermal systems being initiated on the newly formed ridge and/or (b) hydrothermal system becoming sufficiently established to produce maximum element flux to the sediments. However, hydrothermal plume dispersion and fallout could have been influenced by the prevailing palaeo-current directions.

(4) Higher element accumulation rates at Site 834 than at Site 835 associated with the peak in hydrothermal input from ELSC propagation could result from a combination of basin morphology affecting plume dispersion patterns and the established pattern of greater dispersion of Mn relative to Fe with increasing distance from the plume source.

(5) Propagation of the CLSC into the Lau Basin is not clearly seen in terms of factor scores or element accumulation rates in the sediments. This probably reflects a lower hydrothermal flux associated with the CLSC than with the ELSC, which is consistent with studies of modern hydrothermal sediments from them.

(6) Element accumulation rate values show that hydrothermal plume fallout associated with the CLSC is similar to modern plume fallout element accumulation rates adjacent to the CLSC and within the range typical of EPR sediments. In contrast, accumulation rates of plume fallout associated with the ELSC are significantly higher than those associated with the CLSC and are comparable with, or higher than, the highest reported values from the EPR.

(7) Plume fallout products in the Lau Basin are generally richer in Mn than in Fe relative to those associated with the EPR. This is particularly so for sediments associated with the ELSC

which show a two- to four-fold enrichment of Mn relative to Fe compared with EPR sediments.

This work was funded by the NERC, under ODP Special Topic Research Grant GST/02/442. We thank K. St Clair-Gribble for help with the chemical analysis and L. M. Parson and R. G. Rothwell (IOSDL) for their help during the research grant. Thanks are owed to the master, crew and technical staff of the *JOIDES Resolution* for their efforts during Leg 135, in which R.A.H. participated, and to the Leg 135 Scientific Party.

References

BARRETT, T. J., TAYLOR, P. N. & LUGOWSKI, J. 1987. Metalliferous sediments from DSDP Leg 92: the East Pacific Rise transect. *Geochimica et Cosmochimica Acta*, **51**, 2241–2253.

BIGNELL, R. D., CRONAN, D. S. & TOOMS, J. S. 1976. Metal dispersion in the Red Sea as an aid to marine geochemical exploration. *Transactions of the Institute of Mineralogy and Metallurgy B*, **85**, 273–278.

BLANC, G., STILLE, P. & VITALI, F. 1994. Hydrogeochemistry in the Lau Backarc Basin. *Proceedings, ODP, Scientific Results*, **135**, 677–688.

BOSTRÖM, K., JOENSUU, O., VALDES, S., CHARM, W. & GLACCUM, R. 1976. Geochemistry and origin of East Pacific Rise Sediments sampled during DSDP Leg 34. *Initial Reports of the DSDP*, **34**, 559–574.

CHESTER, R. & HUGHES, M. J. 1967. A chemical technique for the separation of ferromanganese minerals, carbonate minerals and adsorbed trace elements from pelagic sediments. *Chemical Geology*, **2**, 249–262.

CRONAN, D. S. 1976a. Basal metalliferous sediments from the eastern Pacific. *Bulletin of the Geological Society of America*, **87**, 928–934.

—— 1976b. Implications of metal dispersion from submarine hydrothermal systems for mineral exploration on mid-ocean ridges and island arcs. *Nature*, **262**, 567–569.

—— 1983. Metalliferous sediments in the CCOP/SOPAC region of the Southwest Pacific, with particular reference to geochemical exploration for the deposits. *CCOP/SOPAC Technical Bulletin*, **4**, 55pp.

——, HODKINSON, R. A., HARKNESS, D. D., MOORBY, S. A. & GLASBY, G. P. 1986. Accumulation rates of hydrothermal metalliferous sediments in the Lau Basin, S.W. Pacific. *Geo-Marine Letters*, **6**, 51–56.

——, MOORBY, S. A., GLASBY, G. P., KNEDLER, K. E., THOMPSON, J. & HODKINSON, R. A. 1984. Hydrothermal and volcaniclastic sedimentation on the Tonga-Kermadec Ridge and its adjacent marginal basins. *In*: KOKELAAR, B. P. & HOWELLS, M. F. (eds) *Marginal Basin Geology*. Geological Society, London, Special Publication, **16**, 137–149.

DYMOND, J. & VEEH, H. H. 1975. Metal accumulation values in the southeast Pacific and the origin of

metalliferous sediments. *Earth and Planetary Science Letters*, **28**, 13–22.

EDMOND, J. M., MEASURES, C. & 7 others 1979. On the formation of metal-rich deposits at ridge crests. *Earth and Planetary Science Letters*, **46**, 19–30.

——, VON DAMM, K. L., MCDUFF, R. E. & MEASURES, C. I. 1982. Chemistry of hot springs on the East Pacific Rise and their effluent dispersal. *Nature*, **297**, 187–191.

FOUQUET, Y., VON STACKELBERG, U., CHARLOU, J. L., ERZINGER, J., HERZIG, P. M., MÜHE, R. & WIEDICKE, M. 1993. Metallogenesis in back-arc environments: the Lau Basin example. *Economic Geology*, **88**, 2154–2181.

HAWKINS, J. W., PARSON, L. M. & ALLAN, J. F. 1994. Introduction to the scientific results of Leg 135: Lau Basin – Tonga Ridge drilling transect. *Proceedings, ODP, Scientific Results*, **135**, 3–5.

HEY, R. 1977. A new class of 'pseudofaults' and their bearing on plate tectonics: a propagating rift model. *Earth and Planetary Science Letters*, **37**, 321–325.

HODKINSON, R. A. & CRONAN, D. S. 1991. Geochemistry of recent hydrothermal sediments in relation to tectonic environment in the Lau Basin, southwest Pacific. *Marine Geology*, **98**, 353–366.

—— & —— 1994. Variability in the hydrothermal component of the sedimentary sequence in the Lau back-arc Basin (Sites 834–839). *Proceedings, ODP, Scientific Results*, **135**, 75–86.

——, ——, GLASBY, G. P. & MOORBY, S. A. 1986. Geochemistry of marine sediments from the Lau Basin, Havre Trough, and Tonga–Kermadec Ridge. *New Zealand Journal of Geology and Geophysics*, **29**, 335–344.

HOWARTH, R. J. 1983. *Handbook of Exploration Geochemistry (Vol. 2): Statistics and Data Analysis in Geochemical Prospecting*. Elsevier, New York.

LEINEN, M. & STAKES, D. 1979. Metal accumulation rates in the central equatorial Pacific during Cenozoic time. *Bulletin of the Geological Society of America*, **90**, 357–375.

LYLE, M. W., OWEN, R. M. & LEINEN, M. 1986. History of hydrothermal sedimentation at the EPR, 19°S. *Initial Reports of the DSDP*, **92**, 585–596.

MALAHOFF, A. & FALLOON, T. 1991. *Preliminary Report of the Akademik Mstislav Keldush/Mir Cruise 1990. Lau Basin Leg (May 7–21)*. Suva, South Pacific Applied Geoscience Commission, unpublished cruise report 137, 12pp.

PARSON, L. M. & HAWKINS, J. W. 1994. Two-stage ridge propagation and the geological history of the Lau backarc basin. *Proceedings, ODP, Scientific Results*, **135**, 819–828.

PARSON, L. M., HAWKINS, J. W., ALLAN, J. *et al.* 1992a. *Proceedings, ODP, Initial Reports*, **135**, 1230pp.

——, —— *et al.* 1994a. *Proceedings, ODP, Scientific Results*, **135**, 984pp.

——, —— & HUNTER, P. M. 1992b. Morphotectonics of the Lau Basin seafloor – implications for the opening history of backarc basins. *Proceedings, ODP, Initial Reports*, **135**, 81–82.

——, PEARCE, J. A., MURTON, B. J. & HODKINSON, R. A. 1990. Role of ridge jumps and ridge propagation in the tectonic evolution of the Lau back-arc basin, southwest Pacific. *Geology*, **18**, 470–473.

——, ROTHWELL, R. G. & MacLEOD, C. J. 1994b. Tectonics and sedimentation in the Lau Basin (Southwest Pacific). *Proceedings, ODP, Scientific Results*, **135**, 9–22.

RIECH, V. 1990. Calcareous ooze, volcanic ash, and metalliferous sediments in the Quaternary of the Lau and North Fiji Basins. *Geologische Jahrbuch*, **D92**, 109–162.

——, MARCHIG, V., SUNKEL, G. & WEISS, W. 1990. Hydrothermal and volcanic input in sediments of the Lau Back-Arc Basin, SW Pacific. *Marine Mining*, **9**, 183–203.

ROTHWELL, R. G., BEDNARZ, U. & 6 others 1994a. Sedimentation and sedimentary processes in the Lau Backarc Basin: results from ODP Leg 135. *Proceedings, ODP, Scientific Results*, **135**, 829–842.

——, WEAVER, P. P. E., HODKINSON, R. A., PRATT, C.

E., STYZEN, M. J. & HIGGS, N. C. 1994b. Clayey nannofossil ooze turbidites and hemipelagites at Sites 834 and 835 (Lau Basin, SW Pacific). *Proceedings, ODP, Scientific Results*, **135**, 101–130.

SAYLES, F. L. 1975a. Chemistry of ferromanganoan sediment of the Bauer Deep. *Bulletin of the Geological Society of America*, **86**, 1423–1431.

—— 1975b. Element accumulation values in the Bauer Deep: a correction. *Bulletin of the Geological Society of America*, **87**, 1396.

TAMAKI, K., SUYEHIRO, K., ALLAN, J., INGLE, J. C., JR. & PISCIOTTO, K. A. 1992. Tectonic synthesis and implications of Japan Sea ODP drilling. *Proceedings, ODP, Scientific Results*, **127/128**, 1333–1348.

TAYLOR, B. 1992. Rifting and the volcanic-tectonic evolution of the Izu-Bonin-Mariana Arc. *Proceedings, ODP, Scientific Results*, **126**, 627–651.

THOMPSON, M. & WALSH J. N. 1989. *The Handbook of Inductively Coupled Plasma Spectrometry*, 2nd edn. Blackie, Glasgow, 156–160.

WALTER, P., STOFFERS, P., GLASBY, G. P. & MARCHIG, V. 1990. Major and trace element geochemistry of Lau Basin sediments. *Geologische Jahrbuch, reihe D*, **92**, 163–188.

Distribution and transformation of Fe and Mn in hydrothermal plumes and sediments and the potential function of microbiocoenoses

S. M. SUDARIKOV[1], M. P. DAVYDOV[1], V. L. BAZELYAN[2]
& V. G. TARASOV[3]

[1]*Institute for Geology and Mineral Resources of the Ocean (VNIIOkeangeologia) 1, Angliysky pr., St Petersburg 190121 Russia*
[2]*Odessa University, 2 Shampansky Street, Odessa 270015, Ukraine*
[3]*Institute of Marine Biology, Far East Branch of RAN, 159 100-years road, Vladivostok 690022, Russia*

Abstract: The spatial separation of suspended Fe and Mn in plumes, recorded at Mid-Atlantic Ridge (MAR) sites, some regions on the East Pacific Rise (EPR) and in the western Pacific, is reflected in the Fe and Mn distribution in metalliferous sediments of hydrothermal origin. Investigations at 21.5–23.5°S EPR show that at a downstream distance of 10 km from the vents the total Fe and Mn contents are lower than those on the flanks. Primary suspended Mn precipitates more than 20 km from the high-temperature vents, whereas the main suspended Fe precipitates within 10 km of the active vent area. In the vicinity of the vents a zone of 'geochemical minimum' was observed. Bacterial masses were observed in plumes and sediments in hydrothermal sites of the Pacific. Microbial coenoses were studied to show the increased activity of heterotrophic and Fe–Mn bacteria at several horizons of MAR hydrothermal plumes. Near-surface plumes studied in Matupi Harbour (New Britain) were characterized by both the separation of suspended Fe and Mn horizontally and a corresponding increased presence of micro-organisms. The activity of micro-organisms accompanied differentiation of the suspended and dissolved Mn at the periphery of the plume. Evidence suggests a significant role for bacteria in removing Mn from plumes in different regions of the ocean.

Submarine hydrothermal activity is accompanied by plumes of metalliferous and non-metalliferous elements, gases and organic matter in bottom waters (Klinkhammer 1980; Baker & Massoth 1987). Only about 1% of these elements are deposited in the immediate vicinity of vents to form ore bodies. The remaining elements are distributed by bottom currents throughout an area of surrounding water and are deposited at varying distances from the vents due to differential transformation processes (Heath & Dymond 1977; Sudarikov & Ashadze 1989; Sudarikov & Cherkashev 1993).

The present work considers a potential biological explanation for the formation of Fe and Mn halos on the seafloor downstream from hydrothermal vent plumes, as reflected in the surface layer of metalliferous sediments. The work is based on the results of phase analyses of 30 metalliferous sediment samples from surface (0–2 cm) layers of cores obtained by box samplers during cruise 4 of the R/V *Geolog Fersman* at 21.5–23.5°S. Stations were worked at the flanks of the East Pacific Rise (EPR) at a distance of 10, 20 and 40 km from the rift zone axis.

Successive leaching was carried out according to the method of Chester & Hughes (1967). The method facilitates the division of the total sample content of Fe and Mn into the following components: phase 1, carbonates, ions and evaporates of mud solutions adsorbed onto the surface of particles; phase 2, amorphous and poorly crystallized oxides and hydroxides of Fe and Mn; and phase 3, crystallized ('old') oxides of Fe and the remains of the previous extracts of crystallized Mn oxides.

As a part of investigations at 21.5–23.5°S, three segments of the rift zone were examined to determine whether the hydrothermal activity becomes less intense southwards and whether the content of hydrothermal elements in metalliferous sediments is higher on the western flank than on the eastern flank because of the effects of the mostly westward currents.

The western flank of the northern segment, characterized by the most intense hydrothermal activity, revealed the widest halos of high Fe and Mn concentrations. The halos became narrower towards the central segment and fragmented on the flanks of the southern segment. However, some differences in Fe and Mn distribution were

From PARSON, L. M., WALKER, C. L. & DIXON, D. R. (eds), 1995, *Hydrothermal Vents and Processes*, Geological Society Special Publication No. 87, 249–255.

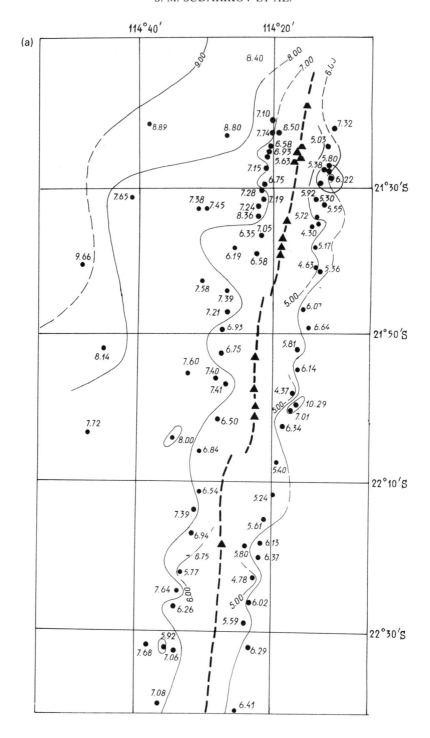

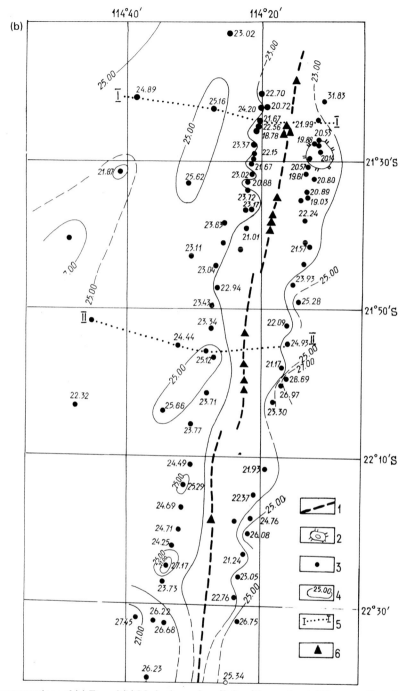

Fig. 1. Concentrations of (**a**) Fe and (**b**) Mn in the surface (0–2 cm) layer of metalliferous sediments (recalculated for non-calcareous matter). **1**, Axial part of rift zone; **2**, caldera of submarine volcano; **3**, sampling stations with values of element contents; **4**, isolines of element contents; **5**, geochemical profiles; and **6**, massive sulphide ores.

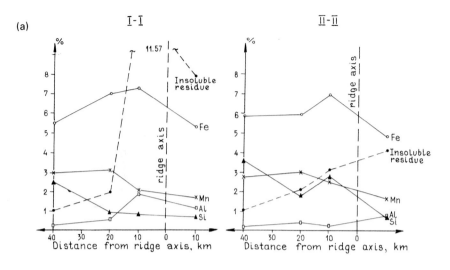

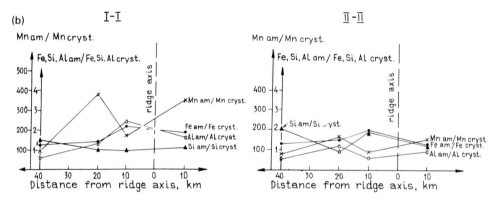

Fig. 2. (**a**) Content of amorphous and poorly crystallized forms of elements and insoluble residues and (**b**) ratio of concentrations of amorphous and poorly crystallized forms of elements and crystallized forms on Profiles I–I and II–II.

recorded at a distance of 20 km from hydro-thermally active sites in the northern segment and 10–15 km from those of the central segment (Fig. 1a). The highest concentrations of Mn were recorded 40 km from the central segment and appeared likely to occur outside the study area at the northern segment (Fig. 1b).

The distribution of suspended and dissolved Mn and Fe in plumes of high-temperature vents was also examined in a search for potential mechanisms responsible for the separate pre-cipitation of suspended forms of Fe and Mn at various distances from the rift zone axis that could be reflected in the Fe and Mn contents of the metalliferous sediment surface layer.

In the vicinity of the vents at a downstream distance of 10 km the total Fe and Mn contents were lower than those on the flanks. Such a

distribution pattern may be caused by both a dilution of volcanogenic products (as evidenced by the concentrations of insoluble residues) and by a zone of 'geochemical minimum' for Fe and Mn near high-temperature vents due to the precipitation of the primary suspended forms some distance away from the active sites (Fig. 2a).

The highest contents of secondary phases of Fe and Mn were recorded at distances of 10 and 20 km, respectively, on the western flank. The contents of these phases at a distance of 10 km were lower on the eastern flank than on the western flank. In other words, the primary suspended Mn precipitates less than 20 km from the high-temperature springs; the primary sus-pended Fe, in contrast, precipitates within the first 10 km. The zone of 'geochemical minimum'

(a)   (b)

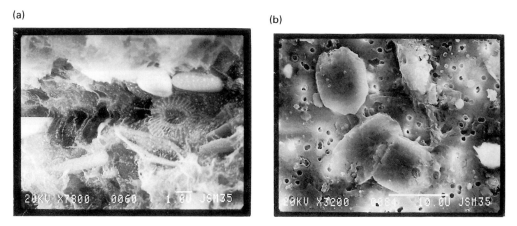

Fig. 3. Bacterial remains (including bacterial cells) in samples from EPR at 13°N. (**a**) metalliferous sediments (upper right); (**b**) suspension at ultra-fine filtres (centre). Scanning electron microscope YSM-35. Original magnification (**a**) ×7800 and (**b**) ×3200.

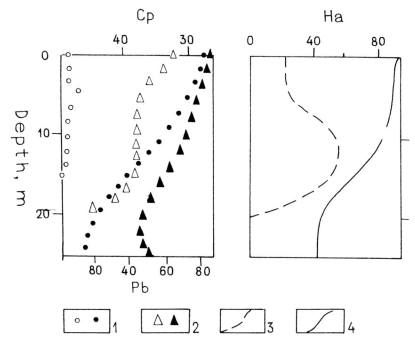

Fig. 4. Phytoplankton production (C_p) mg C m^{-3} day^{-1} (1) and total bacterial production (Pb) mg m^{-3} day^{-1} (2) near vents (open symbols) and in the centre of the Matupi Harbour (closed symbols) and relative activity of heterotrophic bacteria (Ha) near vents (3) and in the centre of the Matupi Harbour (4) in the water column.

around the high-temperature springs was also observed visually as a lighter colour of the surface layer of metalliferous sediments collected at a distance of 10 km from the axis, in contrast with the dark-coloured sediments at the periphery. Separate maxima of Fe and Mn were observed in the zone of abrupt precipitation of suspended forms of Fe and Mn where the amorphous phases dominated over crystallized phases.

Temporal variations in hydrothermal activity, the presence of hydrothermal vents on rift flanks, changes in the direction and velocity of water currents, together with a number of other

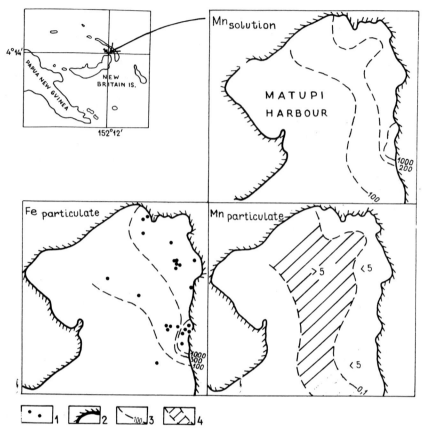

Fig. 5. Distribution of metals in the subsurface plume, Matupi Harbour. **1**, Sampling stations; **2**, coastline; **3**, isolines of metal contents (μg); and **4**, zone of bacterial activity (Fig. 4).

factors, especially early diagenesis, considerably complicate the pattern of Fe and Mn distribution by concealing and smoothing the after-effects of differentiation in the seawater which would be more prominent given a more stable system. In analysing the total concentrations of elements, we are dealing only with the averaged results of vent activity. The highest contents of Fe in metalliferous sediments on the western flank of the northern segment were recorded to the west of the highest contents of amorphous and poorly crystallized forms accumulating due to the recent activity of hydrothermal springs on the northern segment. This reflects the increasing proportion of the crystallized phase as part of the total Fe content as it moves away from the rift zone.

According to our observations the higher the temperature and flow rate of hydrothermal vents, the wider is the zone of 'geochemical minimum'. Products of low-temperature hydrothermal activity accumulate close to the vents.

The reason for the separate precipitation of Fe–Mn suspensions is of great interest (Figs 1 and 2). Some features indicate a significant role for bacteria in removing dissolved Mn from seawater. We observed bacterial masses in hydrothermal plumes and sediments in different regions of the ocean (Fig. 3). It is possible that the highest concentrations of Mn on the flanks of the EPR segment studied are due to the abrupt deposition of some of these elements due to bacterial activity (Cowen *et al.* 1986).

Microbial coenoses were studied during the cruise of R/V *Geolog Fersman* to the Snake Pit field and 15°N (Mid-Atlantic Ridge) and showed increased activity of heterotrophic and Fe–Mn bacteria at several different horizons in hydrothermal plumes on the Mid-Atlantic Ridge (Sudarikov *et al.* 1994).

Near-surface dispersion halos studied in Matupi Harbour Bay (New Britain) during the 1990 cruise of R/V *Akademik Nesmeyanov* were characterized by the spatial separation of suspended Fe and Mn (Shulkin *et al.* 1992*a*), which appeared to correlate with the activity of micro-organisms (Fig. 4). The abundance of phytoplankton and total bacterial production, as well as the activity of many microbial groups such as heterotrophic bacteria, at the dispersion horizon was maintained and even amplified at the periphery of the hydrothermal plume in the centre of Matupi Harbour Bay in parallel with the differentiation of suspended and dissolved Mn forms (Fig. 5). The spatial separation of suspended Fe and Mn in plumes was reflected in their distribution in the bottom sediments of Matupi Harbour (Shulkin *et al.* 1992*b*).

Evidence from different regions indicates that bacteria including encapsulated bacteria, may play a pivotal part in Mn concentration and suspension, whereas Fe phase transformations are much less affected by biogeochemical factors (Cowen *et al.* 1986; Tambiev *et al.* 1992; Sudarikov *et al.* 1994). A possible reason could be that about half the Fe of the hot fluids is transformed into Fe sulphides within a few seconds of release and the remaining half is rapidly oxidized to Fe oxyhydroxides which precipitate in close proximity to the vents (Rudnicki *et al.* 1992). We suggest that this differential microbial response at least partially explains the element distributions in sediments on the flanks of axial structures along oceanic rift zones.

We thank the Russian Fundamental Science Foundation for financial support.

References

BAKER, E. T. & MASSOTH, G. J. 1987. Characteristics of hydrothermal plumes from two vent fields on the Juan de Fuca Ridge, Northeast Pacific Ocean. *Earth and Planetary Science Letters*, **85**, 59–83.

CHESTER, R. & HUGHES, M. 1967. A chemical technique for separation of ferromanganese minerals and adsorbed-trace elements from pelagic sediments. *Chemical Geology*, **3**, 249–265.

COWEN, J. P., MASSOTH, G. J. & BAKER, E. T. 1986. Bacterial scavenging of Mn and Fe in a mid- to far-field hydrothermal particle plume. *Nature*, **322**, 169–171.

HEATH, G. & DYMOND, Y. 1977. Genesis and transformation of metalliferous sediments from the East Pacific Rise, Bauer Deep, and Central Basin, northwest NAZCA plate. *Geological Society of America Bulletin*, **5**, 723–733.

KLINKHAMMER, G. 1980. Observations of the distribution of manganese over the East Pacific Rise. *Chemical Geology*, **29**, 211–226.

RUDNICKI, M. D. & ELDERFIELD, H. 1992. A chemical model for the hydrothermal plume above the TAG vent field, 26°N, Mid-Atlantic Ridge. *EOS, Transactions of the American Geophysical Union*, **73**, 585.

SHULKIN, V. M., TSUKANOVA, E. V. & MAIBORODA, A. B. 1992*a*. [The influence of the modern hydrotherms on metal distributions in Matupi Bay water]. *Geohimia*, **3**, 389–399 [in Russian].

——, —— & —— 1992*b*. [The influence of the modern hydrothermal activity on metal content in bottom sediments of Matupi Bay (New Britain)]. *Litologia i Poleznie Iskopaemia*, **2**, 3–13 [in Russian].

SUDARIKOV, S. M. & ASHADZE, A. M. 1989. [Distribution of metals in hydrothermal plumes of Atlantic ocean according to ICP data]. *Doklady Akademii Nauk*, **308**, 452–456 [in Russian].

—— & CHERKASHEV, G. A. 1993. [Structure of hydrothermal plumes of the Pacific and Atlantic oceans]. *Doklady Akademii Nauk*, **330**, 757–759 [in Russian].

——, DAVYDOV, M. P., BAZELYAN, V. L. & TARASOV, V. G. 1994. [The function of microbiocoenoses in matter redistribution and transformation in hydrothermal plumes and sediments]. *In: Problems of Development of Marine Geotechnologies.* Information Science and Geoecology VNIIOkeangeologiya, St Petersburg, 131 [in Russian].

TAMBIEV, S. B., DEMINA, L. L. & BOGDANOVA, O. YU. 1992. [Biogeochemical cycles of manganese and other metals in the hydrothermal zone of Guaimas Basin (Gulf of California)]. *Geohimia*, **2**, 201–213 [in Russian].

Ecology of Mid-Atlantic Ridge hydrothermal vents

CINDY LEE VAN DOVER

Department of Marine Chemistry and Geochemistry, Woods Hole Oceanographic
Institution, Woods Hole, MA 02543, USA

[1] Present address: Institute of Marine Science, University of Alaska, Fairbanks, AK 99775,
USA

Abstract: In the past year, the number of explored deep-water vent sites on the
Mid-Atlantic Ridge has doubled. The fauna of Atlantic vents consists for the most part of a
subset of invertebrate types found elsewhere in chemosynthetic ecosystems, with taxonomic
differentiation usually at the species or genus level. Despite this similarity in taxonomic
composition, the ecology of Atlantic vents differs from the ecology of Pacific vents in ways
that highlight aspects of biogeography, trophic ecology and sensory adaptations.

The discovery of biological communities associated with active deep-water hydrothermal systems on the Mid-Atlantic Ridge lagged eight years behind the first explorations of vent fields on the Galapagos Spreading Center and East Pacific Rise. By 1985, when the high temperature TAG mound was encountered (Rona *et al.* 1986), the tubeworm, mussel and clam-dominated vents of the eastern Pacific were already textbook classics. TAG, dominated by swarms of densely packed shrimp and extensive anemone beds, seems to have ecological characteristics different from the Pacific vents. Discovery of shrimp swarms at the Snake Pit site (ODP Leg 106 Scientific Party 1986), three degrees of latitude south of TAG, reinforces the idea that the community structure of Atlantic vent systems is distinct from that of Pacific vents. Until recently, TAG and Snake Pit remained the only known vent sites in the Atlantic. However, in autumn 1992 and winter 1993, water column prospecting for active vent systems (Langmuir *et al.* 1992; Murton *et al.* 1994) led to the discovery of two new sites on the Mid-Atlantic Ridge: Lucky Strike and Broken Spur. These sites were visited by submersible in May–June 1993 and now double the number of explored deep-water hydrothermal systems on the Mid-Atlantic Ridge (Fig. 1). Discovery of a fifth deep site on the Mid-Atlantic Ridge south of the Fifteen-Twenty Fracture Zone was made by Russian scientists in February 1994 (Sudarikov and Galkin, this volume). The biological attributes of Lucky Strike and Broken Spur add interesting twists to our catalogue of vent ecologies on the Mid-Atlantic Ridge, with Broken Spur being strikingly depauperate in terms of animal biomass, and with mussels usurping what might

have been thought to be 'shrimp habitat' at Lucky Strike.

Based on observations of the fauna at TAG and Snake Pit, several invertebrate types characteristic of Pacific hydrothermal systems, including vestimentiferan tubeworms, vesicomyid clams and alvinellid polychaetes, are noticeably absent from these Atlantic vent sites (Mevel *et al.* 1989). This observation remains valid and is expanded to include the absence of these taxa at Lucky Strike (Van Dover *et al.*, submitted) and Broken Spur (Murton *et al.*, this volume). Decapod crustaceans–shrimp in the family Bresiliidae and crabs in the family Bythograeidae (Table 1) – appear to be especially (though not ubiquitously) important components of the Mid-Atlantic Ridge vent fauna, as are bathymodiolid mussels. Vestimentiferan tubeworms do occur in the Atlantic where they have been recorded from seep sites in the Gulf of Mexico (350–3266 m; Paull *et al.* 1984; MacDonald *et al.* 1990), off Guyana (500 m) and off Montevideo (300 m) as well as from a shipwreck site 30 miles west of Vigo, Spain (1160 m; Dando *et al.* 1992). Similarly, vesicomyid clams are reported from several non-hydrothermal areas, including seep sites in the Gulf of Mexico (Paull *et al.* 1984) and the Laurentian Fan (Mayer *et al.* 1988) and valves of a vesicomyid clam have been recovered from the Mid-Atlantic Ridge near the Fifteen-Twenty Transform Fault (Segonzac, pers. comm.). It seems likely that these taxa will eventually be recorded from a Mid-Atlantic Ridge hydrothermal setting as well.

Research on the biota of Mid-Atlantic Ridge vents to date directs us to three principal areas of review: site characterization and biogeography; trophic ecology; and sensory physiology. The

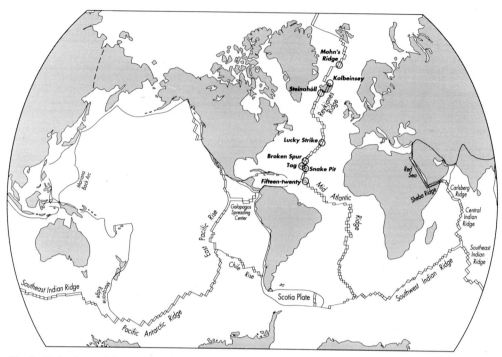

Fig. 1. Hydrothermal sites on the Mid-Atlantic Ridge.

Mid-Atlantic Ridge, with its slow spreading rate, is a natural counterpoint to the fast-spreading East Pacific Rise. With increasing exploration of Atlantic vent sites, we can begin to develop biogeographical models that help us to understand and predict patterns of diversity under hydrothermal constraints we believe are imposed by fast- or slow-spreading rates. Trophic ecology focuses on the discussion in published work of the diet of the dominant shrimp species at TAG and Snake Pit as well as the unusual endosymbionts in the mussel that occurs at Snake Pit. Finally, the motility of the vent shrimp places a premium on sensory adaptations that can detect steep chemical and physical gradients in hydrothermal environments. Although a systematic study of sensory capabilities remains to be undertaken, work on chemo- and photosensory capabilities suggests that Mid-Atlantic Ridge vent shrimp may serve as good models for this aspect of vent research.

Overview of physical settings and faunal characteristics

TAG

The TAG segment of the Mid-Atlantic Ridge is 40 km long, bounded by small offset (<10 km)

transform faults (Rona *et al.* 1976, 1986; Thompson *et al.* 1988). The hydrothermal field is complex, with low-temperature, high-temperature and relict mineralization within a 5 km × 5 km region (Fig. 2; Rona *et al.* 1993). The spreading rate on the eastern wall, which rises from 4000 m at its base to 2000 m in a series of fault steps (Temple *et al.* 1979) is $1.3 \, \text{cm a}^{-1}$ (half-rate; McGregor and Rona 1975). Active low-temperature hydrothermal activity was first located between 2400 and 3100 m on the east wall (Rona *et al.* 1984). Increased biological activity of vent-related forms (crabs, fish, ophiuroids) is noted from photo transects at two locations, the east inner wall and the east lower outer wall and occurs over 1.5 km regions (Eberhart *et al.* 1988). Bivalve-like forms, tentatively identified as vent clams in various stages of dissolution, were documented on bottom photographs in the low-temperature field (Rona *et al.* 1984), but to date no *in situ* observations by submersible confirmed this interpretation of photographic data. Biological characterization of the low-temperature field is obviously incomplete.

Relict hydrothermal zones in the TAG region located on the lower part of the east wall (3400–3500 m) between a shallower low-temperature zone (2500–3000 m) and the active

Table 1. *Dominant decapod taxa of Mid-Atlantic Ridge hydrothermal vents and related species*

Bresiliid Shrimp

Vent shrimp were originally assigned to the family Bresiliidae until Christofferson (1989) created a new family Alvinocarididae for *Rimicaris* Williams & Rona 1986 and *Alvinocaris* Williams & Chace 1982. M. de Saint Laurent (in Segonzac *et al*. 1993) lists diagnostic features of Alvinocarididae, the most readily observable being 'eyestalks reduced and attached to each other and to neighboring parts' and implicitly includes the genus *Chorocaris* Martin & Hessler 1990. These genera, although not restricted to hydrothermal systems as a group, are so far restricted to chemosynthetic habitats. *Rimicaris* is at present a monospecific genus known only from *R. exoculata* in the Atlantic. Other crustacean taxonomists have not followed Christofferson's lead, however (Chace 1992; Holthuis 1993). This review takes a conservative approach, keeping the vent taxa within the Bresiliidae, but recognizing that the taxonomy of this group is difficult.

As a first-order means of distinguishing between the three bresiliid shrimp genera which co-occur at Mid-Atlantic Ridge vents, the character of the rostrum is useful. Members of the genus *Alvinocaris* are readily identified by the well-developed, toothed rostrum that projects forward at the anterior end of the head. *Rimicaris* lacks a rostrum and has a distinctive anterior flat plate that sits where the eyes ought to be in a normal shrimp. *Chorocaris* has a stubby, smooth rostrum that barely projects forward.

Mid-Atlantic Ridge species	Authority	Known locations
Alvinocaris markensis	Williams 1988	TAG, Snake Pit, ?Broken Spur
Rimicaris exoculata	Williams & Rona 1986	TAG, Snake Pit, ?Broken Spur
?*Chorocaris* n.sp.	Segonzac *et al*. 1993	Snake Pit
Chorocaris (Rimicaris) chacei	(Williams & Rona 1986)	TAG, Snake Pit, ?Broken Spur
Chorocaris n.sp.	Martin & Christiansen in press	Lucky Strike
Other Atlantic species		
Alvinocaris muricola	Williams 1988	West Florida Escarpment
Alvinocaris stactophila	Williams 1988	Seep, N. Gulf of Mexico
Pacific species		
Alvinocaris lusca	Williams & Chace 1982	Galapagos Spreading Center, East Pacific Rise
Chorocaris vandoverae	Martin & Hessler 1990	Mariana Back Arc Basin
New genus, new species	Williams & Dobbs in press	Loihi Seamount

Bythograeid Crabs

Only a single genus of brachyuran crab, *Segonzacia* (Guinot 1989), is so far documented on the Mid-Atlantic Ridge. It is not known from other ocean basins. The character of the ocular region of bythograeid crabs seems to be one of the most useful means of distinguishing genera at present (Hessler & Martin 1989). In *Austinograea*, the most extreme reduction of eyestalks is apparent, with the eyestalks absent or perhaps reduced to an obscure plate fused to the wall of the orbital recess. In *Cyanograea*, the eyes are located in small, clearly defined pockets, while in *Bythograea* and *Segonzacia* the eyes lie in what is perhaps best described as elongate slots. *Segonzacia* is distinguished by the presence of distinctive oval patches of modified cuticle ventrolateral to the eyes. The function of these patches, if any, is unknown. They look as if they might be the cuticular expression of a sensory organ and as such, histological examination is warranted.

Mid-Atlantic Ridge species	Authority	Known locations
Segonzacia (Bythograea)	Williams 1988	Snake Pit, Broken Spur, TAG,
mesatlantica		Lucky Strike
Pacific species		
Bythograea thermydron	Williams 1980	Galapagos Spreading Center, East Pacific Rise
Bythograea microps	de St. Laurent 1984	East Pacific Rise
Bythograea intermedia	de St. Laurent 1984	East Pacific Rise
Cyanagraea praedator	de St. Laurent 1984	East Pacific Rise
Austinograea williamsi	Hessler & Martin 1989	Mariana Back Arc
Austinograea alayseae	Guinot 1989	Lau Back Arc

high-temperature mound (3625–3670 m) have been explored with the *Mir* and *Alvin* submersibles (Rona *et al*. 1993). The Mir zone is 400–600 m in diameter while the Alvin zone (formerly the North zone) forms an interrupted line of inactive chimneys and large mounds (some as large as the active TAG mound) extending for 2 km (Rona *et al*. 1993). No

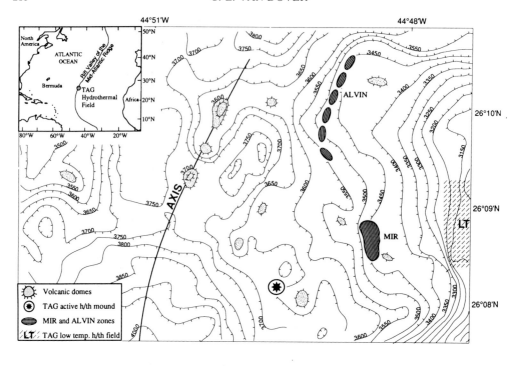

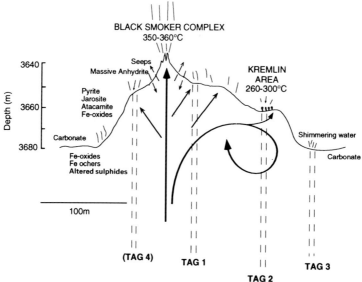

Fig. 2. Upper panel: SeaBeam bathymetric map (50 m contours) of the TAG hydrothermal field, showing the active mound and the Mir and Alvin inactive (relict) zones. Lower panel: schematic cross-section of the active TAG mound. Subsurface fluid flow illustrated is inferred from sulphide mineralogy and fluid chemistry (from Tivey in press). The relative positions of the holes proposed by the Ocean Drilling Program are shown (from Humphris 1994).

evidence of significant biological exploitation of the inactive sulphides was reported, although a curious hexagonal pattern of holes (3–7 cm diameter) tentatively identified as indicative of a living representative of the trace fossil *Paleo-dictyon nodosum* (Seilacher 1977; Rona &

Merrill 1978) is locally abundant (several specimens per square metre) on carbonate ooze surrounding the Mir relict zone (Rona *et al.* 1993).

Active high-temperature hydrothermalism has been documented at the TAG Hydrothermal Mound, located along the eastern inner wall of a U-shaped, narrow (3 km) rift valley with an 800 m wide neovolcanic zone (Rona *et al.* 1986; Thompson *et al.* 1988; Eberhart *et al.* 1988). It is located on older sedimented crust 2.4 km east of the spreading axis on the wall of the rift valley (Thompson *et al.* 1988; Rona *et al.* 1993). The TAG Mound (Fig. 2) itself is nearly circular in plan view and consists of concentric rings (Rona *et al.* 1986). The outer ring is characterized by a shallow layer of carbonate ooze (a few to tens of centimetres thick), with outcrops of basalt talus and massive sulphide blocks. This talus and ponded sediment habitat slopes up to the inner mound of constructional sulphide deposits. The junction of the outer and inner mounds marks the boundary of hydrothermal flux. At the centre of the mound, black smoke emanates from fractures in the sulphides (290–321°C; Campbell *et al.* 1988) as well as from distinct chimneys (365°C; Rona *et al.* 1993), giving the effect of a broad, impenetrable curtain of high temperature water (Fig. 3A). TAG black smoker fluids are enriched in copper, but otherwise their major element chemistry is similar to that of East Pacific Rise hydrothermal fluids (Campbell *et al.* 1988; Edmond *et al.* 1990; Edmond *et al.* this volume). The hydrothermally active inner mound wall is steep, rising 30 m above the outer mound and is composed of massive sulphide talus. The inner mound is approximately 200 m in diameter, with the gradient of hydrothermal activity increasing from near ambient at the outer margin to the maximum expression of high-temperature black smokers at the very top of the mound. A region of white smokers ($\leqslant$300°C) known as the Kremlin (Thompson *et al.* 1988; Rona *et al.* 1993) is located on the southeast portion of the inner TAG mound and emits zinc-rich, sulphide-poor fluids (Edmond *et al.* 1990).

Initial documentation of active high-temperature hydrothermalism at TAG was based on water column and imaging data (Rona *et al.* 1986). Worm tubes lying recumbent on sulphide talus and basalt at the periphery of the vent field at the juncture between the inner and outer mounds, tentatively identified as tubeworms (Rona *et al.* 1986) have since been confirmed to be an undescribed species of polychaete worm in the family Chaetopteridae (Van Dover, pers. observ.). Rona *et al.* (1986) also noted

anemones in the peripheral field and shrimp, brachyuran crabs and fish in greatest abundance in the inner mound.

Zonation in biological activity at TAG is roughly correlated with the hydrothermal gradient (Galkin & Moskalev 1990), although there is a striking paucity of biota associated with the Kremlin area (S. Humphris, pers. comm.) that is probably related to the sulphide-depleted fluid chemistry of the white smokers there. Low-temperature regions at the periphery of the mound are colonized by anemones (typically 10–20 per m^2 but reaching densities up to 100 per m^2 on the slope up to the black smoker region; Galkin & Moskalev 1990) and chaetopterid worm tubes as noted above, as well as by gastropods, a galatheid crab (*Munidopsis* sp.) and occasional eel-like (zoarcid and synaphobranchid) fish (Fig. 3A and 3B; Grassle *et al.* 1986; Galkin & Moskalev 1990). Meiofauna collected in this region include a nematode species, a cyclopoid copepod and a desmosomatid isopod (Galkin & Moskalev 1990). Photos from submersible exploration in this area show occasional small patches (1–2 m maximum dimension) of ponded sediments colonized by regularly spaced upright tubes (3–4 cm apart) and small anemones (Karson & Brown 1988; Van Dover, pers. observ.), but these sediments have not been sampled effectively for their biota. Galkin & Moskalev (1990) report collections of numerous fragments of an ampharetid polychaete species from the peripheral region, as well as four other types of undescribed organisms observed on video images. Mussels, though present at the base of sulphide mounds at Broken Spur to the north and Snake Pit to the south, are so far not documented from TAG (Galkin & Moskalev 1990), although anecdotal observations of an occasional mussel at TAG are known (e.g. J. Edmond, pers. comm.). Crustaceans (principally the bresiliid shrimp *Rimicaris exoculata*) occupy and dominate the warm waters of the black smoker field at the centre of the TAG mound (Fig. 3A; Williams & Rona 1986). Collections of these shrimp usually include the copepod, *Stygiopontius pectinatus*, which can be found in the shrimp's branchial chamber (Humes 1987; Galkin & Moskalev, 1990). *Stygiopontius pectinatus* is also known from Mariana vents, where the bresiliid shrimp, *Chorocaris vandoverae*, occurs (Hessler & Lonsdale 1991). The precise relationship between this copepod and shrimp, or the shrimp's epibiotic bacteria that cover surfaces and appendages in the branchial chamber, is unknown. Individuals of *C. chacei* co-occur with *R. exoculata* at TAG, as does a third, larger species

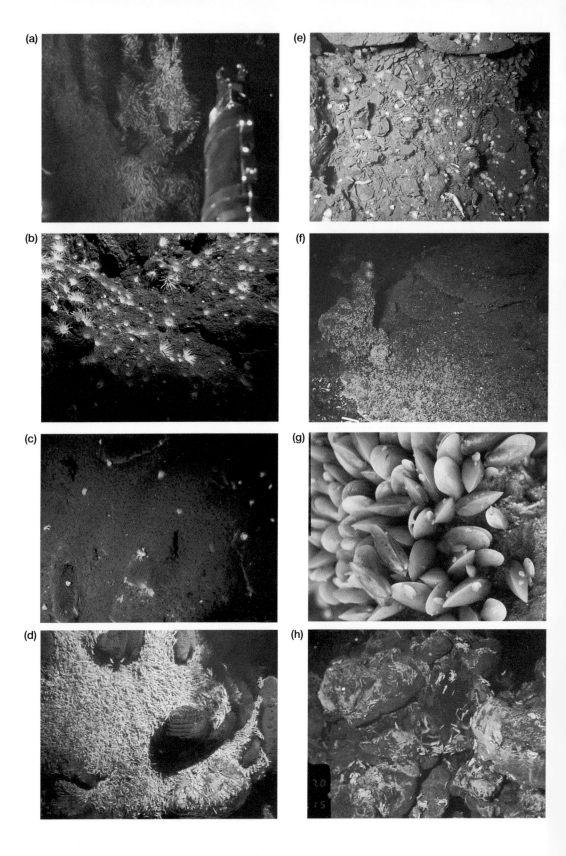

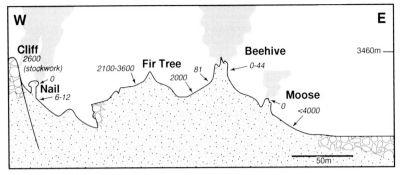

Fig. 4. East-west cross-section of the Snake Pit hydrothermal field (from Fouquet *et al*. 1993) with ages of dated sulphides (from Lalou *et al*. 1993).

observed on videos, but not collected (possibly *Alvinocaris markensis*), whereas brachyuran crabs (*Segonzacia mesatlantica*) are relatively abundant within 1–10 m of the black smoker region (Galkin & Moskalev 1990).

The Ocean Drilling Program drilled a series of holes across the active hydrothermal mound at TAG in November 1994 (Fig. 2; Humphris 1994). This provides us with a novel opportunity to study the response of motile shrimp populations to perturbations in fluid flow brought about by the drilling activity. *A priori* expectations are that the shrimp and other crustaceans will quickly redistribute themselves to occupy optimal thermal and chemical habitats, but that sessile populations of anemones and chaetopterid worms in the peripheral regions might succumb to both changes in fluid flow that leave their habitat barren of suspended food particles and to the deleterious effects of heavy burdens of drillhole debris suspended in the water column and settling on to the seafloor. Pre- and post-drilling mapping of animal distributions and fluid flow regimes on the TAG mound are planned by Japanese, UK and US biologists, together with time series observations of temperature, animal behaviour and other parameters.

Snake Pit

The TAG (26°N) and Snake Pit (23°N) ridge segments (each about 40 km long) are separated by a linear distance of about 300 km, with a 145 km offset of the Kane fracture zone. Snake Pit (also known as the MARK site – for Mid-Atlantic Ridge Kane; Fig. 1) was first discovered by seafloor photography of stained and mottled sediment (Kong *et al*. 1985) and was described in more detail as part of an effort to drill into the hydrothermal system in 1985 (ODP Leg 106 Scientific Party 1986) and a post-drilling

Fig. 3. (**A**) TAG. *Rimicaris exoculata* (bresiliid shrimp) occur in high densities only around areas within close proximity to active black smokers at the centre of the TAG mound. Photo taken by *Alvin*. (**B**) TAG. Anemones at the TAG mound dominate the diffuse flow environment. Large *Phymorhynchus* sp. gastropods and chaetopterid polychaetes co-occur. Photo taken by *Alvin*. (**C**) TAG. Peripheral microhabitats include sediment ponds colonized by small polychaetes (probably ampharetids) living in fluted tubes that are evenly spaced. Galatheid squat lobsters, a gastropod and small ?anemones are also found. Photo taken by *Alvin*. (**D**) Snake Pit. Shrimp swarms adjacent to diffuser vents at the Beehive mound. On the left side of the photograph, the swarm is dominated by the small, reddish ?*Chorocaris* sp. *Rimicaris exoculata* make up the swarm on the right. Photo by C. L. Van Dover. (**E**) Snake Pit. Mussels (*Bathymodiolus puteoserpentis*), anemones and chaetopterid polychaetes at the base of the Moose mound. Photo by *Alvin*. (**F**) Lucky Strike. Mussels (*Bathymodiolus* sp.) colonize sulphide flanges and chimneys, two urchins (*Echinus* sp.) sit on top of a sulphide spire. Photo by *Alvin*. (**G**) Lucky Strike. Mussel clump illustrating an abundance of small mussels presumed to be newly settled. Photo by C. L. Van Dover. (**H**) Broken Spur is characterized by a low biomass of invertebrates. This photograph is representative of the greatest biomass documented during the exploratory dives of June 1993 during initial dives in June 1993. The bythograeid crab, *Segonzacia mesatlantica*, and possibly two species of shrimp are visible. Photo by *Alvin* taken at the base of the Wasp site.

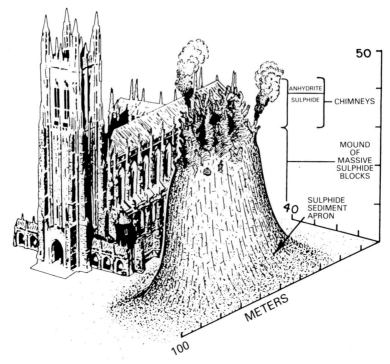

Fig. 5. Schematic illustration of a Snake Pit mound. Active chimneys are located at the top of the mound. The sides of the mounds are steep and end in an apron of hydrothermal sediment and talus at the base. The Duke University Chapel is drawn to scale. From Karson & Brown (1988).

Alvin reconnaissance in 1986 (Karson & Brown 1988). Subsequent dives by *Nautile* during the HYDROSNAKE mission in 1988 provide more details of the geological and biological setting (Mevel *et al.* 1989; Segonzac 1992; Fouquet *et al.* 1993). The known Snake Pit system (Fig. 4) consists of a series of coalescing sulphide mounds along an E–W line in the middle of the axial valley at nearly the same depths as TAG, 3480 m. There are four active mounds – Moose (L'Élan), Beehive (Les Ruches), Fir Tree (Le Sapin) and Nail (Le Clou) – as well as several inactive or low-temperature sulphide deposits (Mevel 1989; Fouquet *et al.* 1993). The high-temperature regions of each of the active mounds are separated by 50–100 m intervals, with the entire line of hydrothermal activity contained within 300 m. Relative to TAG, the Snake Pit mounds are small, 20–60 m in diameter and 20–25 m high. The active Snake Pit mounds (Fig. 5) are topped by high-temperature black smokers (325–330°C) and beehive-like diffuser vents (>70°C; Fouquet *et al.* 1993; Van Dover, pers. observ.) with gradients of temperature and hydrothermal flux going toward ambient conditions progressing down the steep

slopes and on to the sulphide talus aprons of the mounds.

The Snake Pit fauna overlaps significantly with the TAG fauna and, like TAG, the Snake Pit Mounds are dominated by densely packed (up to 2500 per m²) shrimp in the warm waters adjacent to black smokers and beehive diffusers (Fig. 3D; ODP Leg 106 Shipboard Scientific Party 1986; Williams 1987; Mevel *et al.* 1989; Segonzac 1992). Although the shrimp are typically described as 'swarms' or 'swarming', Segonzac *et al.* (1993) emphasize that we use these terms in this context to mean high densities of active individuals with well-defined boundaries, without meaning to infer social behavior on the swarming activity. Segonzac *et al.* (1993) and dives to Snake Pit in 1993 documented clusters of a small, orange ?*Chorocaris* species within or adjacent to the main swarm of *R. exoculata*. The orange colouration is derived from an oily, lipid-rich tissue of the hepatopancreas immediately beneath the carapace in the thoracic region of the body and appears to be characteristic of juveniles of both *R. exoculata* and this ?*Chorocaris* species (Casanova *et al.* 1993). The shape of the photoreceptors (oval in

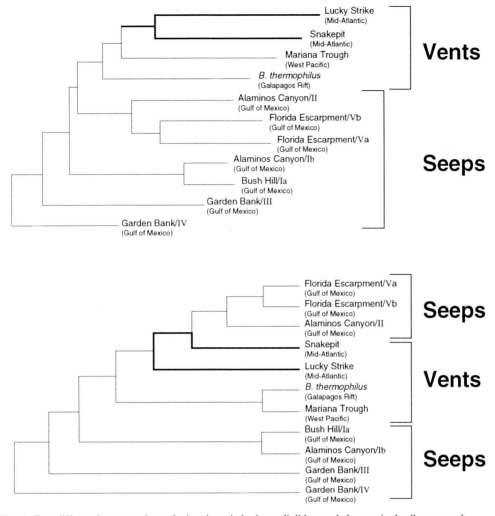

Fig. 6. Two different interpretations of relatedness in bathymodiolid mussels from a single allozyme and morphological data set. In both cases, the Mid-Atlantic Ridge mussel species from Lucky Strike and Snake Pit are more closely allied to each other than to mussels from vent sites in the Pacific. From Craddock *et al*. (in press).

the small *?Chorocaris* sp. and angel wing-shaped in *R. exoculata*) readily allows the observer to distinguish juveniles of these two genera (Van Dover pers. observ.) in *in situ* images or fresh and frozen material. Although the proportion of small *Chorocaris* sp. to *R. exoculata* in any given habitat has not been determined quantitatively, the visual impression is of dominance by *R. exoculata* (at least 70%). Preliminary identification of these *?Chorocaris* sp. individuals suggests that they may be a new species, closely related to *C. chacei* (Segonzac *et al*. 1993). *Chorocaris chacei* and *A. markensis* are found beneath but close

to the main body of the shrimp swarms, but seem to live a more solitary life in cooler waters (Segonzac *et al*. 1993).

Snake Pit supports a cryptic population of a new species of mussel, *Bathymodiolus puteoserpentis* (Fig. 3E; von Cosel *et al*. 1994; Craddock *et al*. in press), at Moose and Nail (Segonzac 1992). The mussels are not readily observed with remote imaging devices (ODP 106 Leg 106 Scientific Party 1986), but were discovered during the first exploration dive in 1985 (Karson & Brown 1988). The mussels dwell on the sulphide structure, principally beneath the region dominated by shrimp, often tucked in

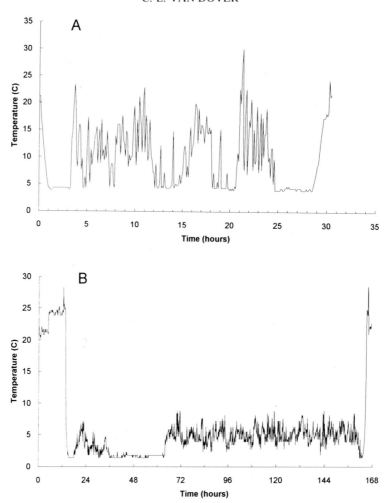

Fig. 7. Time series fluctuations in temperature at mussel-covered clumps. Both temperature records illustrate the characteristic 'flickering' of temperature in environments colonized by invertebrates. From Fornari *et al.* (1994). (**A**) Lucky Strike mussels. The probe was positioned on a mussel-covered flange at *c.* 3 h and recovered at 25 h. (**B**) Snake Pit mussels. The start of the trace is the on-deck temperature. Records at the clump begin at about 18 hours. At about 36 h, the probe was dislodged from the clump; at *c.* 65 h the probe was repositioned. The probe was recovered at *c.* 162 h.

crevices or beneath outcrops of sulphides. Allozyme analysis distinguishes the Snake Pit and Lucky Strike mussels as different species, with some uncertainty in their degree of related-ness to Gulf of Mexico seep mussel species and Pacific vent species (Fig. 6; Craddock *et al.* in press). According to von Cosel *et al.* (1994), *B. puteoserpentis* has much closer morphological affinities to the western Pacific *Bathymodiolus* species than it does to *B. thermophilus* from the East Pacific Rise and Galapagos Spreading Center. Temperature data measured at a single point at a Snake Pit mussel clump (Fig. 7) during

a six-day deployment illustrates the low-temperature (~5°C) environment occupied by these mussels (Fornari *et al.* 1994).

Segonzac (1992) provides the most complete description of the Snake Pit faunal assemblage (Fig. 8), which includes the four species of shrimp mentioned above (*R. exoculata, C. chacei, ?C.* sp. and *A. markensis*). Other decapod crustaceans present include the brachy-uran crab, *S. mesatlantica*, and the galatheid crab, *Munidopsis crassa*. The copepod (*S. pectinatus*) that lives in the branchial chamber of *R. exoculata* is present, as well as several other

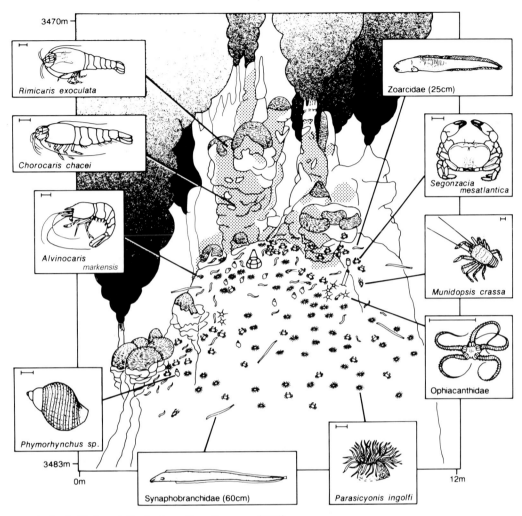

3470m —

Rimicaris exoculata

Chorocaris chacei

Alvinocaris markensis

Phymorhynchus sp.

3483m —

0m

Synaphobranchidae (60cm)

Parasicyonis ingolfi

Zoarcidae (25cm)

Segonzacia mesatlantica

Munidopsis crassa

Ophiacanthidae

12m

Fig. 8. Snake Pit faunal distributions. From Segonzac (1992).

undescribed representatives of this copepod genus. The polychaete, *Branchipolynoe seepensis*, a polynoid polychaete described as a commensal with a different mussel species from the Florida Escarpment (Gulf of Mexico), is found in the mantle cavity of *B. puteoserpentis*. Karson & Brown (1988) illustrate fragments of a maldanid polychaete collected from Snake Pit. The predatory *Phymorhynchus* sp., a large turrid gastropod, occurs among chaetopterid tubes on the lower slopes of Snake Pit mounds and is attracted to baited fish traps (Van Dover, pers. observ.). Less conspicuous invertebrate species occur, including a variety of nematodes, polychaetes and gastropods. A limpet, *Pseudorimula midatlantica*, has as its closest known

relative *P. marianae* from the Mariana Back Arc vents (McLean 1992). Tyler *et al.* (in press) describe a new genus of ophiuroid that is now known from Snake Pit (cf. Segonzac 1992) and Broken Spur. This new ophiuroid type has unusual oral and dental features that suggest this it is adapted for feeding on filamentous bacteria. Two eel-like fish – one a zoarcid, the other a synaphobranchid (Segonzac 1992) – give Snake Pit its name (ODP Leg 106 Scientific Party 1986). These fish are observed in cooler waters on the sides of the sulphide mounds, together with occasional individuals of *Nematonurus armatus* (Macrouridae; Segonzac 1992). Sudarikov & Galkin (this volume) used photographic surveys to provide additional information about the Snake Pit and nearby benthic faunas. These workers note the predominance of suspension-feeding invertebrate types in the nearby area, including hexactinellid sponges, a variety of coelenterates (hydroids, crinoids, gorgonians) and serpulid polychaetes. Along transects running 500 m from active vent sites, Sudarikov & Galkin (this volume) found maxima in non-vent faunal types at about 120–180 m away from the vents. Bacterial mats are also reported in this region and are indicative of local diffuse hydrothermal venting.

Lucky Strike

In 1992, a dredge pulled across what became known as the Lucky Strike Seamount at 37°18′N on the Mid-Atlantic Ridge (Fig. 1) brought up fragments of fresh sulphide material colonized by invertebrates – mussels with commensal polychaetes, shrimp, and limpets (Langmuir et al. 1992, 1993). A subsequent Alvin dive series in 1993 located active hydrothermal sites in a depression formed between three cones that make up the summit of the seamount (Langmuir et al. 1993). Seven more or less discrete active sites occur over an area that extends 700 m in length at depths of 1630–1710 m (Fig. 9). Venting styles ranged from low-temperature (a few to tens of degrees above ambient) diffuse flows from cracks in a silicified conglomerate of volcanic and hydrothermal debris and from 'leaky' surfaces of sulphide chimneys and a flange, to pooled high-temperature (200°C) water beneath a sulphide flange, to turbulent jets of hot water (292–333°C) from black smoker chimneys (Humphris et al. 1993). Fluid compositions generally fall within the ranges established for other hydrothermal systems (Colodner et al. 1993). The high-temperature fluids are enriched in CH_4 and H_2 relative to TAG fluids and have low sulphide concentrations (<3.3 mmol/l). Attributes of the fluid chemistry suggest to Colodner et al. (1993) that Lucky Strike has recently been reactivated. Other evidence cited to support the idea of hydrothermal episodicity at Lucky Strike is the situation of fresh sulphides on top of older, more weathered and silicified debris (Humphris et al. 1993).

Although most of the faunal components of the Lucky Strike vent communities are familiar taxa from other hydrothermal systems (bathymodiolid mussels, bresiliid shrimp, bythograeid crabs), the visual impression of the vent community is strikingly different from the other known hydrothermal sites on the Mid-Atlantic Ridge or elsewhere (Van Dover et al. 1993,

submitted). Mussels predominate (Fig. 3F), occupying much of the available sulphide surface on active chimneys and flanges as well as crowding into cracks in the sulphide conglomerate that paves the seafloor in the nearby vicinity. On what appear to be the older, more stabilized sulphide mounds, mussels are absent only in small areas where the temperature is too hot or at the boundary between leaky sulphides and non-hydrothermally active substrates. Allozyme analysis of this mussel indicates it is a species distinct from other mussels analysed, including mussels from Snake Pit, the Gulf of Mexico and Pacific vent sites (Fig. 6; Craddock et al. in press). Its taxonomic relationship with the Broken Spur mussel is unknown at present. Mussel shell hash typically marks the periphery of the vent sites. Overcrowding of mussels and predation by fish may dislodge mussel clumps from suitable habitats to form the shell hash; collapse or other types of mass wasting of mussel-bearing sulphides may also contribute to the accumulations of broken shells. The temperature of a single point immediately above a region of mussels on the top of a sulphide flange was monitored for 20 hours, providing a synoptic view of the thermal environment of this species (Fig. 7; Fornari et al. 1994). As elsewhere (e.g. Johnson et al. 1988), the mussels live in a zone of 'flickering' water, the mixing zone between cold oxygenated seawater and warmer hydrothermal fluids. Although the temperature data from mussel habitats at Snake Pit and Lucky Strike suggest that the Lucky Strike mussels occupy warmer water (up to 30°C), spatial gradients in temperature, both horizontally and vertically, are steep and need to be taken into consideration when comparing single-point measurements between populations. A high percentage (>95%) of the mussels dissected shipboard contained a commensal species of the polynoid polychaete B. seepensis (Van Dover et al. unpublished). The minimum mussel size invaded by the polychaete is unknown at present.

Mussel size-frequency data (Fig. 10) illustrate multiple size-class distributions. The smallest mussels (less than 5 mm in length; Fig. 3G) suggest a recent recruitment event based on size and rapid growth rates of mussels determined at other hydrothermal systems (Rhoads et al. 1981). The possibility that one or more of these size classes represents interspecific competition and/or differential growth rates cannot be excluded at present; this may be resolved when the site is reoccupied by a submersible programme. If the mussel size classes do reflect recruitment events rather than stunted growth at some sites, some unrecognized factor must

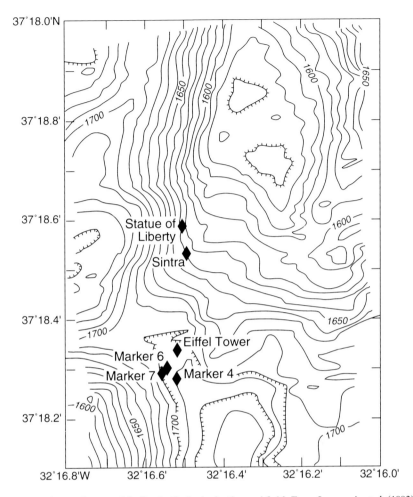

Fig. 9. SeaBeam bathymetric map of the Lucky Strike hydrothermal field. From Langmuir *et al.* (1993).

control reproductive synchrony and/or the physical concentration of larvae to account for the 'waves' of recruitment. One possibility is that the reactivation of Lucky Strike discussed above has taken place very recently, with the most abundant smaller size classes (with fresh, unstained periostracums) arriving as early colonists of new sulphides, while the older mussels (manganese-stained) represent a residual population that had been eking out a living on suspended particles and very low fluxes of reduced compounds.

The bresiliid shrimp at Lucky Strike is a previously unknown species in the genus *Chorocaris* (Martin & Christiansen in press). It is a small species (carapace length of gravid female 6.4 mm) relative to shrimp species found at TAG, Snake Pit and Broken Spur. The shrimp occur in localized regions of warm water where fresh sulfides are exposed and are only patchily

abundant (20–30 per m^2), nowhere reaching densities observed for *R. exoculata* at TAG or Snake Pit. There was a relatively large abundance of gravid females within the small sample collected.

Sea urchins in the genus *Echinus* (Sibuet, pers. comm.) represent the most unusual component of the Lucky Strike faunal assemblage (Langmuir *et al.* 1993). The urchins, with test diameters on the order of 7 cm, occur at the edges of the mussel beds and on sulphides at the edge between active and inactive regions of the sites (Fig. 3F). Densities were on the order of ten or fewer urchins per site. The urchins were not observed outside areas of active venting, even though transits in non-vent terrain comprised much of the bottom time on the first two dives at Lucky Strike. Urchins have not previously been recorded from any deep-water hydrothermal

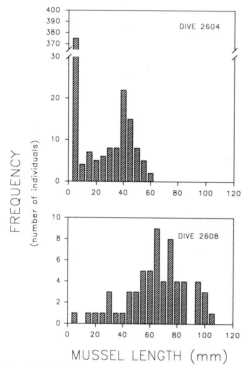

Fig. 10. Mussel size frequency histograms from collections during *Alvin* dives 2604 and 2608, illustrating three abundant size classes: <5 mm, *c.* 40–45 mm, *c.* 60–80 mm. From Van Dover *et al.* unpublished data.

site. Their diets are unknown but, as they were most abundant at a site where small mussels were very abundant, it seems plausible that they are foraging on the smallest mussel size classes.

Other taxa collected from Lucky Strike include a bythograeid crab, ampharetid and free-living polynoid polychaetes, limpets and other gastropods, hydroids, nemerteans, tanaids, isopods, ostracods, amphipods and barnacles (Van Dover *et al.* unpublished).

Langmuir *et al.* (1992), using water column techniques, surveyed the Mid-Atlantic Ridge rift valley between 34° and 41°N, an area which includes Lucky Strike, and found evidence for hydrothermal activity in seven of 17 segments visited. Strong plume signals were detected along the Lucky Strike segment at 37°N in accordance with the discovery of hydrothermal activity there, and Wilson *et al.* (1993) predict from plume data that there are additional hydrothermal sites northwest of those visited by *Alvin* in 1993. In a comparison of plume signals at Lucky Strike, TAG, Snake Pit and Broken Spur, lower particle densities as measured by

transmissometry and nephelometry at Lucky Strike are consistent with lower end-member concentrations and lower temperatures of venting fluids observed at Lucky Strike (Chin *et al.* 1993). Two strong manganese signals, comparable with TAG–Snake Pit–Broken Spur signals, were observed at the southern end of the AMAR segment and the northern end of the adjacent South AMAR segment (Chin *et al.* 1993). As exploration of these areas progresses, it should be possible to obtain a regional measure of the pattern of distribution of vent faunas with respect to depth and other variables.

Broken Spur

Broken Spur (Fig. 1) was located in 1993 using a combination of remote geophysical and geochemical sensors towed along a first-order mid-ocean ridge segment between 27° and 30°N on the Mid-Atlantic Ridge (Murton *et al.* 1994). Plume signals (manganese anomalies, turbidity, temperature) were at a maximum over an area near 29°N, which became the target area for more intense survey work to provide a best location for the hydrothermal field. Active, waning and inactive hydrothermal mounds (Fig. 11) were subsequently located at 3090 m during the first of two *Alvin* dives there (Alvin 2624 and 2625; Murton & Van Dover 1993). Mound topography is similar in scale and morphology to that of Snake Pit, with individual sulphide structures reaching heights from 4 to 36 m and with diffuser-type vents (beehives) as well as chimneys with turbulent jets of hot water.

One mound described as fundamentally 'inactive', but probably with some very low level of hydrothermal flux, was colonized by a few live mussels and many dead valves. The impression was of fugitive mussels able to survive redirection of a plumbing system long after other invertebrates died or moved off to more active area.

The active sites – The Spire, Saracen's Head and Wasp's Nest – each topped by black smoker chimneys and beehive, diffuser-type structures, were colonized by at least three species of shrimp, a bythograeid crab and a new species of ophiuroid (Murton *et al.* this volume). The most abundant shrimp appears to be closely related or identical to *R. exoculata*, both taxonomically and in its habitat, but it did not occur in high densities. Mussels grow in a small cluster at the base of The Spire; the bases of the other sites are less well explored. The taxonomic relationships of Broken Spur fauna to faunas of other Mid-Atlantic Ridge vent sites are under investigation.

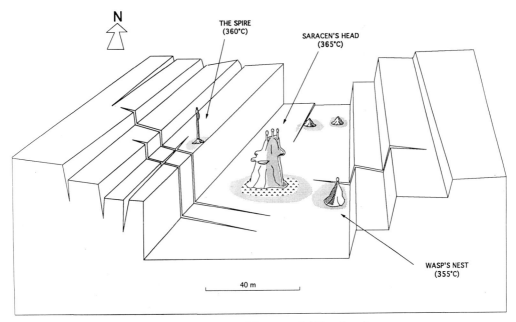

Fig. 11. Interpretation of Broken Spur setting. From Murton *et al.* (this volume).

As at the base of TAG and Snake Pit mounds, Broken Spur supports a peripheral fauna that includes zoarcid fish, chaetopterid(?) worm tubes and anemones as well as shrimp and ophiuroids that appear to occupy both the base of mounds as well as warm water habitats adjacent to the black smokers. Sediment ponds at the base of Saracen's Head contain evenly spaced fluted tubes occupied by an ampharetid polychaete as well as other infaunal species. A similar-looking sediment community has been photographed at TAG (e.g. Fig. 3C).

The most conspicuous feature of Broken Spur is the lowest biomass at any of the sites encountered (Fig. 3H). Large sulphide deposits indicate that Broken Spur is not a new hydrothermal system (Murton *et al.* this volume), so the paucity of invertebrate biomass cannot be accounted for in this manner. Vigorous black smoker activity and the presence of low-temperature microhabitats that seem suitable for invertebrate colonization suggest that the system is not in a dying stage. Water chemistry indicates that there is nothing particularly unusual about the sulphide concentration or major element chemistry of this site to preclude the accumulation of a large biomass (Murton *et al.* this volume). A testable hypothesis is that Broken Spur has recently been reactivated and that invertebrates are only just arriving to recolonize the system.

Surviving mussels suggest that deactivation was of sufficient quality and duration for motile organisms to emigrate or die off, but not so long nor so complete as to wipe out the mussel populations. If Broken Spur has been reactivated, then subsequent visits to Broken Spur and remapping of animal communities on sulphide edifices should detect an increase in biomass as reproduction and immigration proceed.

As at Lucky Strike, plume mapping indicates that there is another site of active venting within the Broken Spur region (Elderfield *et al.* 1993). The plume associated with the known Broken Spur vents is relatively weak; a plume more intense by an order of magnitude (comparable with that associated with the TAG mound) lies to the south of the explored vents. Of three possible plume anomalies detected during continuous coverage of the ridge axis between 27° and 29°N, only the Broken Spur plumes were thought to necessarily signify high-temperature hydrothermal activity (Murton *et al.* 1994).

14°45'N vent site

In February 1994, Russian scientists aboard the R/V *Professor Logatchev* located a new hydrothermal field (2900–3000 m) on the Mid-Atlantic Ridge south of the Fifteen-Twenty Fracture Zone (Batuyev *et al.* 1994; Sudarikov & Galkin, this volume). The geological setting resembles Snake Pit, but evidence to date suggests that hydrothermal activity is lower and that the sulphide mounds are older at the new site. Sudarikov & Galkin (this volume) report the presence of bacterial mats, bathymodiolid mussels and large gastropods.

Shallow water sites

The Mid-Atlantic Ridge shoals north and south of Iceland (Jan Mayen and Reykjanes Ridges, respectively) as well as in the vicinity of the Azores. A detailed water column survey designed to locate a series of hydrothermal systems along the depth gradient of the Reykjanes Ridge (between 57°45′N and 63°09′N) failed to locate more than a single site at Steinahóll (250–350 m) along 750 km of ridge crest (German *et al.* 1993). The Steinahóll site (Fig. 1) had been known previously, originally by fishermen as a productive fishing hole and subsequently by scientific investigators. Descriptions of the associated fauna are not published, but bubble-rich plumes have been imaged at Steinahóll with a 38 kHz echosounder (Ólafsson *et al.* 1989; Walker 1992; German *et al.* 1993). High dissolved gas concentrations (up to 18 nmol/l CH_4 and 30 nmol/l H_2) are reported from these plumes (German *et al.* 1993).

Fricke *et al.* (1989) explored shallow hydrothermal vents (100–106 m) off Kolbeinsey Island (Ólafsson *et al.* 1989) on the Jan Mayen Ridge north of Iceland (Fig. 1), where hydrothermal fluids are emitted from fissures, small chimneys (30 cm high) and large 'crater-like dips' in the seafloor. Boiling water, identified by the generation of bubbles, was observed at the craters, implying a temperature of 180–182°C. At the periphery of the craters, where an increased abundance of organisms occurred, the water temperature was between 5 and 20°C (ambient 2.6°C). Extensive mats of bacteria were found at the two active sites explored. One type of hyperthermophilic methanogenic bacterium isolated from high temperature regions (59–180°C) of these vents (106 m) belongs to a new species that also occurs at 2000 m depth in hydrothermal systems of Guaymas Basin, Gulf of California (Kurr *et al.* 1991). Invertebrate biomass is enriched at the Kolbeinsey sites, but species diversity is lower than in the surrounding ambient seafloor environment. The fauna is not vent-specific and contains none of the taxa normally associated with deep-water hydrothermal systems. Instead, a solitary sponge, *Scypha (Sycon) quadrangulatum*, dominates one site where another sponge, *Tethya*

aurantium, also occurs. A large solitary hydroid polyp, *Corymorpha groenlandica*, dominates the second site. The sponges and hydroid are suspension feeders and are known from non-vent regions of the North Atlantic and Arctic Sea. No evidence for bacterial endosymbionts in any of the species was identified. Neither the known shallow-water Atlantic vents nor the shallow vents of the Kraternaya Caldera (shallow subtidal to 20 m depth; Kurile Islands, NW Pacific; Tarasov et al. 1986) support invertebrate taxa considered to be specific to deep-water hydrothermal vent or other sulphitic environments. The Kraternaya vents (maximum temperatures 85–87°C) support bacterial mats and a large faunal biomass that includes solitary corals (*Cerianthus* sp.) reaching densities of 200 individuals per m^2, bivalve molluscs (*Mya priapus*, *Macoma* sp. and *Hiatella arctica*) and polychaetes as well as populations of holothurians. Unlike the Kolbeinsey fauna, the Kraternaya fauna is reported to be distinct from that of the immediate surrounding, non-vent seafloor environment, although the species are known from other locales. Shallow warm water vents in subtidal areas of White's Point, near Los Angeles, California (Kleinschmidt and Tschauder 1985) are colonized by bacteria and are not characterized by an enriched biomass, although gastropods do graze on the bacterial material (Stein 1984). In contrast, at shallow (80 and 100 m) fumarole sites in Kagoshima Bay (Japan), Hashimoto et al. (1993) report dense thickets of an undescribed species of *Lamellibrachia* in the Vestimentifera in warm water effluents (28–30°C; ambient 16°C). As at Steinahóll, the fumarole sites were known to fishermen who referred to them as Tagiri, which means bubbling or boiling, in reference to gas bubbles that rise to the surface above the vents. Before the 1993 discovery of tubeworms at the Tagiri fumaroles, two earlier surveys located only areas of bacterial mats at 200 m depth (Naganuma et al. 1991). This suggests that vent taxa at shallow vents can have patchy distributions and that the absence of vent taxa in one area does not preclude their presence in another.

The apparent restriction of vent taxa to deeper waters of the Mid-Atlantic Ridge and the occurrence of non-vent species at vents in shallow waters raises the issue of where the restriction begins and what causes the separation. Van Dover et al. (1993) propose that depth (pressure) may play a part in controlling the disparate fauna of Lucky Strike (1600 m) compared with deeper sites on the Mid-Atlantic Ridge (>3000 m), although the affinities of

Broken Spur fauna (3300 m) to the Lucky Strike fauna may be greater than initially supposed (E. Southward, pers. comm.). Within the deep sea in general, some species have very narrow depth distributions while others are broadly distributed across wide bathymetric ranges (Carney *et al.* 1983). We expect to find narrow depth tolerances in some vent species, but compelling evidence for this remains to be gathered. A related hypothesis is that larval stages of vent species will be more tolerant of broad ranges of environmental parameters (e.g. pressure and temperature) than adult stages. Discovery and exploration of vent sites at intermediate to shallow depths in the vicinity of the Azores and the Lucky Strike hydrothermal area may help to resolve these issues.

Other hydrothermal indications on the Atlantic Ridge System

The Mohn's Ridge. A 1987 ARGO Survey (Schwab *et al.* 1992) provided sidescan sonar, video and still images of the flank and crest of the topographic high on a segment of the Mohn's Ridge (Fig. 1) in the Norwegian–Greenland Sea (71–72°N, 0–4°W). Although no active venting was observed, hydrothermal activity associated with relatively fresh sheet and pillow basalts (low sediment cover) in the vicinity of 71°57′N, 0°48.5′W (2600–2660 m) was indicated on the basis of a density plume (observed with a SeaMarc 4.5 kHz sub-bottom profiler), a limited area of hydrothermal sediments, minor temperature anomalies and unusual abundances of sponges. In addition, a series of features interpreted to be sediment elutriation structures (blowouts) were found in nearby areas of nearly 100% sediment cover.

Serpentinization and methane. Hydrothermal activity occurs in association with serpentinized ultramafic (mantle) outcrops that occur at ridge–transform intersections (e.g. Rona *et al.* 1992; Bougault *et al.* 1993) and elsewhere on slow-spreading ridge systems. This kind of hydrothermal system generates a high methane signal in the overlying water column, but low manganese and suspended particulate concentrations relative to black smoker hydrothermal systems (Rona *et al.* 1987; Charlou *et al.* 1991; Bougault *et al.* 1993). Elevated δ^3He values of water combined with high methane (low manganese) signals overlying an ultramafic outcrop near the Fifteen-Twenty Fracture Zone on the Mid-Atlantic Ridge (Fig. 1) indicate an ongoing and well-developed hydrothermal discharge that is thermally buoyant (Rona *et al.* 1987). Methane anomalies associated with ultramafic outcrops can be generated by the chemical process of serpentinization, a process which is exothermic (Charlou *et al.* 1991) and does not require an external heat source to maintain circulation of low-temperature fluids (Bougault *et al.* 1993). Mantle degassing may also contribute to the methane anomalies. This kind of chronic hydrothermalism is reported to be detectable everywhere along the Mid-Atlantic Ridge from 12 to 26°N (Fig. 12), but is maximal (3.6 nmol/kg) near fracture zones (Kane and Fifteen-Twenty) and is correlated with ultramafic rocks and domed regions of fault intersection corners (Rona *et al.* 1992; Bougault *et al.* 1993). Submersible exploration of ultramafic outcrops at 15°N detected very diffuse hydrothermal activity, barely recognizable, but the large overlying methane plume in the regions suggests that there must be a source of highly focused, intense flow (Charlou *et al.* 1992). Eberhart *et al.* (1988) describe a plateau of the inner east wall of the rift valley at 14°54′N covered by a 'knurly and frothy deposit' along a 400 m extent. This deposit is interpreted to be hydrothermal, reminiscent of TAG deposits.

The availability of methane along long stretches of the ridge crest raises the question of whether organisms are able to take advantage of this reduced carbon source. Sulphide is the reduced compound typically associated with chemosynthetic production at vents, but mussels from Mid-Atlantic Ridge vents do contain methylotrophic endosymbiotic bacteria that can use methane as both an energy and carbon source (discussed below). To date, the only biological material collected close to areas of serpentinized ultramafic outcrops (15°07.07′N, 44°50.64′W; 3424 m) are fragments of two manganese-coated shells belonging to the family Vesicomyidae (Segonzac, pers. comm.).

Controls on the nature of Mid-Atlantic Ridge vent communities

Although we do not know all the details of microhabitat variability at Mid-Atlantic Ridge versus East Pacific Rise hydrothermal systems, the gestalt sense from having worked both systems is that habitat availability in terms of substrate types, fluid flow and fluid composition is remarkably similar in the two systems, yet the characters of the vent communities and of species that exploit similar microhabitats in the two systems are distinct. It is this admittedly qualitative observation that obliges us to

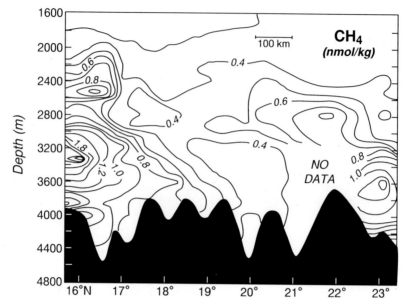

Fig. 12. Methane plumes between 1600 m and the seafloor along the Mid-Atlantic Ridge from 15°36′N to 23°N. Ridge axis bathymetry is expanded and schematic. Buoyant plumes are evident near the Fifteen-Twenty and Kane (23°N) Transform Faults. From Charlou & Donval (1993).

consider alternative explanations for differences between eastern Pacific and Mid-Atlantic Ridge vent faunas. In the following discussion, a descriptive model is used to consider how temporal and spatial aspects of fast- and slow-spreading centres may influence the community structure and biogeography of vent communities along ridge axes. The model does not seek to explain why a particular species should or should not occur on a fast- or slow-spreading ridge axis; instead, the model considers beta diversity – regional patterns of diversity – in the two types of systems.

Community structure on slow versus fast spreading centres. Controls influencing the nature of Atlantic vent communities are not yet known, but reports of their physical and temporal settings encourage speculation. Spatial and temporal contexts of hydrothermal systems on the Mid-Atlantic Ridge are influenced by the slow rate of crustal accretion here relative to Pacific spreading centres (c.25 mm a⁻¹ versus 180 mm a⁻¹ full spreading rates). Throughout its existence, the Mid-Atlantic Ridge has always been slow-spreading (Fowler 1990). Unlike fast-spreading ridge systems of the East Pacific Rise, where only active or very recent hydrothermal fields occur in the axial valley, slow-spreading ridges can co-host fossil and active fields. This is expressed as accumulations of

massive sulphide blocks and talus which survive as relicts beneath and beside recent and hydrothermally active sulphide edifices. Lalou *et al.* (1990, 1993) present radiochronological data of relict sulfides (Fig. 13) that suggest the TAG field is at least 125 000 years old, with the present episode of venting beginning about 50 years ago and with intermittent pulses of activity every 4000–6000 years over the past 20 000 years. Snake Pit similarly has seen episodic activity, though the initial onset of hydrothermalism is more recent, the oldest sulphides collected dating to ages of 4000 years (Fig. 4); the present episode of venting began about 80 years ago (Lalou *et al.* 1993). Thus on slow-spreading ridges, major fracture systems are believed to focus hydrothermal discharges that undergo low frequency, long wavelength cyclic activity, with new sulphide deposits precipitated on top of older ones. In contrast, on the East Pacific Rise each pulse of activity is separated in time and space, resulting in discrete, smaller deposits (Lalou *et al.* 1993). Murton *et al.* (1994) suggest a maximum average spatial frequency of occurrence of high-temperature hydrothermal activity on a stretch of the Mid-Atlantic Ridge to be on the order of one site every 360 km; German *et al.* (1993) cite a higher frequency of occurrence, on the order of one site every 100–150 km based on surveys conducted between 11° and 40°N. In contrast, on the East Pacific Rise, the spatial

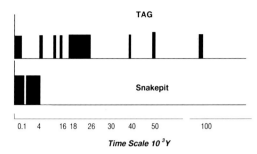

Fig. 13. Episodicity of high-temperature hydrothermal activity at TAG and Snake Pit from 0 to about 130 000 years. From Lalou *et al.* (1993).

frequency can be as high as one active high temperature site for every 5 km of ridge axis (Haymon *et al.* 1991). Hydrothermal activity can be severely affected by volcanic and tectonic events, with a range of consequences, including capping and/or diversion of active vents by overrunning lava flows, creation of new hydrothermal systems or refreshment of old fields. The frequency of eruptive events is greater on fast-spreading centres, but the potential for basalts to be stressed by cooling and extensional processes may be greater at slow-spreading centres (Fornari & Embley in press).

Differences in timing and spacing of hydrothermal systems seem likely to affect the biological and evolutionary attributes of vent systems. Structural details of ridge axes (e.g. relief of the axial valley, Fig. 14) will influence current regimes and fluid flow along ridge crests and thereby also have an influence on the ecology and perhaps evolution of vent communities. On the Mid-Atlantic Ridge, rift valleys are wide (30–45 km) and deep (1–2 km), deeper than the height of buoyant plumes generated by high-temperature vents (Rudnicki & Elderfield 1993). This containment of plumes may have a significant effect on the dispersal of larvae, specifically in increasing the retention of larvae within the rift valley. Containment may be abetted by complex topography within the rift valley created by volcanic ridges (Smith & Cann 1993), including cross-axis sills that may effectively isolate hydrothermal systems in basins (B. Murton, pers. comm.). This potential for pooling and restricted horizontal exchange of a hydrothermally influenced water column found in certain parts of the Mid-Atlantic Ridge is considerably reduced in the low-relief axial environment of the East Pacific Rise where lateral and along-axis flow of water is relatively unrestricted by seafloor morphology.

The actual distribution of vents in space and time on the global ridge system is already proving to be complex, but for the sake of having an *a priori* argument, two theoretical end-members may be considered: type I, vent distribution is high frequency, short wavelength in time and space; and type II, vent distribution is low frequency, long wavelength in time and space.

We are arguably justified at present in considering the fast-spreading East Pacific Rise hydrothermal systems to be our best empirical model for end-member type I, whereas the slow-spreading Mid-Atlantic Ridge systems are our empirical model for the type II end-member. Although it is artificial to consider these models as operating exclusive of any other controls on community structure, these models serve to emphasize some of the factors that might influence patterns of diversity that are observed in hydrothermal systems. The discussion below focuses on diversity measured simply as the number of species in entire collections from a site. Community structure is used here as a qualitative measure of the types of species and their relative dominance within a community.

High frequency, short wavelength cycles of activity in type I systems are short relative to the life span of the species colonizing those systems. Community development is sporadically interrupted, having the effect of influencing the species diversity of a site in a manner analogous to the intermediate disturbance hypothesis in shallow-water communities: along a gradient of increasing disturbance (in this case the gradient is defined by decreasing cycle durations, increasing cycle frequency), species richness oscillates, first increasing, then decreasing as the strength (duration or frequency) of the disturbances becomes too severe to accommodate most species (Sousa 1979). Changes in community structure in type I systems are rapid (e.g. Fustec *et al.* 1987; Hessler *et al.* 1988; Lutz & Fornari 1992; Jollivet 1993) and can occur within the life span of dominant species. Differences in community structure can be found over extremely short spatial scales (tens of metres; Hessler & Smithey 1983), although long extents of ridge axis share the same pool of invertebrate species (Van Dover & Hessler 1990). In the type I case, the community structure of a given site reflects the particular stage of the cycle of that site (Tunnicliffe 1991). Stochastic events related to colonization will also be important in determining the community structure given the complex mosaic in time and space of colonization opportunities.

In type II systems, cycles of activity are long relative to the life span of individuals colonizing

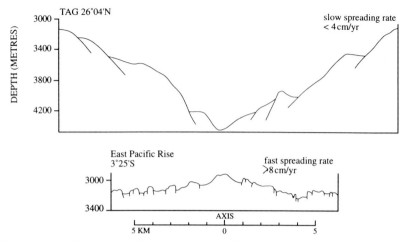

Fig. 14. Comparison of ridge crest relief between slow-spreading (top) and fast-spreading (bottom) ridge systems. From Macdonald (1982).

these systems, with vent systems lasting decades. Community development is not as frequently interrupted by volcanic and tectonic activity and there is the possibility of 'climax' communities, at least in the broad sense, where dominant species in a system become established and remain dominant for longer than the life span of the species, and community structure does not change rapidly over time (cf. shrimp-dominated systems at TAG which remained essentially unchanged in terms of species composition and dominance from 1985 to 1993; Van Dover, pers. observ.). The smaller number of sites, the geographical and geomorphological isolation and longer duration of cycles in type II systems have the effect of minimizing local geographic differences in community structure in this kind of system.

In the simplistic model of a type I ridge system, vents are close enough in space and time relative to a species' or individual's dispersal capabilities and life history attributes, the geographical range of a species becomes virtually unlimited until a major biogeographical barrier is encountered (France et al. 1992) and, as a consequence, populations are well mixed (Black et al. in press; Vrijenhoek et al. in press). What is important here is that, within the scale of life-history attributes such as generation time and dispersability of vent species, distance and time per se are typically not biogeographical barriers in type I vent systems. Colonization opportunities are usually available within the dispersal shadow of a species and, although a species' demographics may wax and wane along

any particular region of the ridge axis over short time frames (years) due to the vagaries of biological processes and the dynamic response of biota to cycles of hydrothermal activity, the species itself is unlikely to undergo local extinction unless it is adapted to a specific microhabitat or resource requirement that becomes lost from the system. In this model, species will tend to accumulate to form one or at most a few biogeographical centres of higher diversity, with decreasing numbers of species as one approaches biogeographical barriers at either end of the linear system of vents. Competitive exclusion in the type I system will be a local phenomenon in both space and time, in the sense that no one species will be able to exclude another species on a regional scale because of the stochastic nature of the timing and spacing of new vent systems. Instead, the dominant character of a particular community at a given phase of a cycle will be controlled by lottery effects related to larval availability and successful colonization. Thus competitive exclusion will not be as powerful an evolutionary forcing function as it will be in a type II ridge system. In the type I end-member, the evolutionary premium on long-distance dispersal abilities is reduced. Where vents are relatively close and short-lived, it is harder for the shut-down of one system to cause the extinction of a species because that species has had a high likelihood of dispersing to an adjacent 'stepping stone' where it finds temporary refuge and from which it can make another step.

Under type II end-member conditions, the distance between suitable habitats in space and time is large relative to the dispersal capabilities and life-history attributes of species or individuals, making distance and time significant biogeographical barriers (cf. arguments for genetic isolation of seep environments; Craddock *et al.* in press). Under this regime, geographical ranges of species are in general limited and their populations are not well mixed. Colonization opportunities are relatively rare in space and time and species' demographics, although subject to control by cycles of hydrothermalism, are overall more stable than those of species of type I ridge systems because of the longevity of the hydrothermal activity at a given vent field. The fauna of the type II ridge system would be characterized by lower regional diversity and a multiplicity of biogeographical centres relative to the faunal communities of type I ridge systems. Competitive exclusion is more likely to be an important evolutionary factor in the type II system, given the possibility that successful exclusion from a local area could result in regional extinction, and there is an evolutionary premium for having good long distance dispersal and colonizing abilities. Successful species in type II systems might be those that hedge their dispersal bets by remaining capable of dispersal throughout their life cycle (cf. the dominance of motile crustaceans at TAG, Snake Pit, Broken Spur).

A rigorous theoretical analysis of the consequences of these two kinds of end-member vent distributions in space and time is called for to substantiate, revise and expand the descriptive model for comparative community structure presented here. Empirical observations of biogeographical parameters at other fast- and slow-spreading systems (e.g. the slow-spreading Southwest Indian Ridge and the adjacent fast spreading Southeast Indian Ridge) will provide additional field tests of this model.

Palaeotectonic controls on Mid-Atlantic Ridge biogeography. The palaeotectonic history of the Atlantic basins (Fig. 15) and its possible role as a first-order control of Atlantic vent biogeography was considered at the DOVE Workshop (Cann *et al.* 1994). That the Mid-Atlantic Ridge fauna seems to be a subset of the Pacific fauna and not vice versa is at least consistent with the fact that the Atlantic basin is much younger than the Pacific. The Central Atlantic ($c. 0°$ to 30°N) formed first during the middle Jurassic ($c. 175–180$ Ma), with likely communication both to the Pacific (probably via transform faults) between North and South America and to the now vanished Tethys Sea between

Eurasia and Africa (Fowler 1990). Whether these corridors were barriers or conduits for the dispersal of vent faunas is not known. A second Atlantic mid-ocean ridge system began to form as Africa and South America began to move apart during the early Cretaceous ($120–135$ Ma); this system was continuous with a spreading centre that connected with the proto-Indian Ocean. The central and south Atlantic spreading systems joined in the late Cretaceous ($c. 95–100$ Ma). Eurasia and North America did not begin to move apart to form the North Atlantic (north of the Azores) until the Late Cretaceous (90 Ma). Thus the Atlantic vent faunas could have been seeded from the Pacific as early as 175 Ma via long conduits to the Central Atlantic from east or west. The putative southern Atlantic vent fauna might have been seeded from the proto-Indian Ocean as early as $c. 135$ Ma, with exchange between central and southern Atlantic ridge systems possible at $c. 100$ Ma. The North Atlantic was colonized at the earliest 90 Ma from the Central Atlantic. Palaeotectonic influences on Atlantic biogeography may only become apparent as putative vent faunas of South and North Atlantic ridge systems are compared with the known faunas of the Central Atlantic.

The Arctic Ocean opened about 65–70 Ma, with Iceland forming a subaerial break in the mid-ocean ridge system since $c. 55$ Ma. Isolation of the ridge system north of Iceland is abetted by the shallow sills of the Greenland–Iceland and Iceland–Faeroes Rises. The Scotia Ridge in the southern Atlantic is young (5 Ma) and is at present geographically isolated from the Mid-Atlantic Ridge. Characterization of putative Arctic and Scotia vent faunas in the context of their palaeotectonic and geographical situations are certain to provide critical insight into the biogeography of Atlantic and global vent biogeography.

The present day pathway for the most direct stepwise dispersal of vent fauna to and from the Mid-Atlantic Ridge is via the ultra-slow spreading (<2 cm per year full rate) Southwest Indian Ridge. Martin & Hessler (1990) note faunal affinities between Mariana Back Arc hydrothermal communities in the Pacific and Mid-Atlantic Ridge faunas (e.g. the copepod *S. pectinatus*, which is known from TAG and Snake Pit as well as from Mariana vents; the shrimp genus *Chorocaris* and limpet genus *Pseudorimula*, which are so far represented only at western Pacific and Mid-Atlantic Ridge vents), but there is no way of knowing the direction or timing of the faunal communication at present.

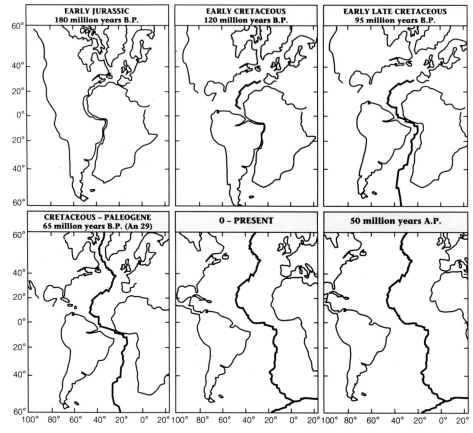

Fig. 15. Formation of the Mid-Atlantic Ridge. Note the formation of the Central Atlantic first, with possible connections to the Pacific to the east and west, followed by spreading in the southern Atlantic and then the northern Atlantic. From Emery and Uchupi (1984).

Transform faults as barriers to dispersal. The role of deep, lateral offsets of hydrothermally active ridge segments as barriers to the dispersal of vent invertebrates has been postulated (e.g. Van Dover 1990; Craddock *et al.* in press), but the significance of transform faults may depend on the dispersal strategies of individual species. France *et al.* (1992) found support for the Hess Deep as a topographic barrier to the dispersal of an amphipod species that broods its young, but populations of vent taxa with dispersive larval stages (mussels, clams, tubeworms) appear to mix easily across the same feature (Black *et al.* 1994; Craddock *et al.* in press; Vrijenhoek *et al.* 1994). There are several major transform faults in the Atlantic (Fig. 16) that might serve as barriers to dispersal (see Table 2). The Kane Transform Fault, with an offset of 145 km, separates TAG from Snake Pit but does not appear to represent a major barrier for the dispersal of the dominant shrimp species at these

sites. The bathymodiolid mussel species that occurs at Snake Pit is so far absent or extremely rare at TAG; if the Broken Spur bathymodiolid proves to be a new species, then the Kane might be implicated as a possible filter to dispersal for the mussel. The relationship between TAG and Snake Pit shrimp species needs to be examined at the population level to determine how much genetic mixing there is across the Kane. The Oceanographer Transform Fault also represents a moderate offset (120 km) of the ridge axis and separates the Lucky Strike hydrothermal field from the other deep-water vents to the south. Careful assessment of species identities and genetic affinities between Broken Spur and Lucky Strike is critical if we are to evaluate the potential role of the Oceanographer as a barrier to dispersal. Far greater offsets are found at the Charlie Gibbs and Vema Transforms. The Romanche Transform Fault appears as a major ridge discontinuity near the equator. As

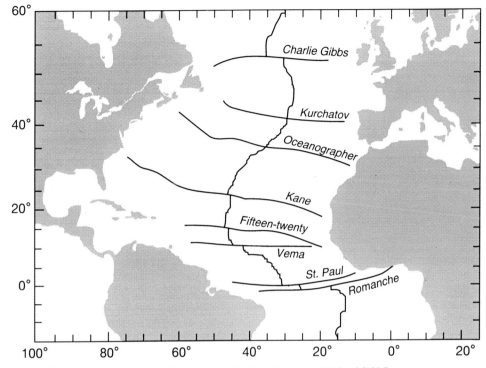

Fig. 16. Major transform faults in the Central Atlantic. From Emery and Uchupi (1984).

Table 2. *Transform faults in the Atlantic (from Fowler, 1990)*

Name	Location	Offset (km)	Age (ma)	Relief (km)
Charlie Gibbs	52°N	260	20	2.5
Kurchatov	40°N	22	2	1.5
FAMOUS A,B	37°N	20	2	1–1.5
Oceanographer	35°N	120	13	4
Kane	24°N	145	10	3
Vema	11°N	315	20	3
Romanche	0°	935	60	4

the hydrothermal systems of the Atlantic are located and characterized, faunal discontinuities associated with these offsets should be catalogued. The major transform faults in the Atlantic in general have greater offsets, are older, and are deeper than the major transforms of the East Pacific Rise (Fowler 1990) and may therefore have served as stronger barriers or filters to dispersal. The biogeographical expression of this may be that more discrete biogeographical provinces – bounded by major transform faults – occur on the Mid-Atlantic Ridge than will be found on the East Pacific Rise. Less testable is the hypothesis that the nature of transform discontinuities in the Atlantic also selects for species that are dispersive both as larvae and as adults. East Pacific Rise transform faults may contain small spreading axes within them, which, if hydrothermally active, may facilitate the stepwise dispersal of vent taxa. Although these spreading relays are thought to be absent in Mid-Atlantic Ridge transform faults, ridge-transform intersections in the Atlantic do have associated serpentinization and consequent methane production (discussed earlier), which might assist the stepwise dispersal of methanotrophic taxa.

Trophic studies

Primary production

One of the most striking ecological differences between the TAG and Snake Pit vent communities in the Atlantic and tubeworm or bivalve-dominated vents in the Pacific is that the bulk of the chemoautotrophic primary production at Pacific vent sites arguably is undertaken by sulphide-oxidizing endosymbionts within invertebrate host tissues (Wirsen *et al.* 1993), whereas at TAG and Snake Pit, the dominant invertebrate biomass appears to be maintained by some combination of primary production associated with free-living (suspended or growing on abiotic surfaces) and epibiotic micro-organisms. This is a misleading dichotomy, however, as certain microhabitats at Pacific vents, such as alvinellid colonies on chimneys may be close trophic analogues to shrimp-dominated TAG and Snake Pit environments (Segonzac *et al.* 1993; Wirsen *et al.* 1993). Nevertheless, the apparent lack of endosymbionts in bresiliid shrimp at TAG and Snake Pit, their abundance and their high level of activity all focus attention on primary production of non-endosymbiotic bacteria at these sites.

Wirsen *et al.* (1993) report high *in situ* CO_2 fixation rates for suspensions of micro-organisms collected in samples of warm water from TAG, comparable with or greater than rates measured by Tuttle *et al.* (1983) and Wirsen *et al.* (1986) for Galapagos and East Pacific Rise vents and Tunnicliffe *et al.* (1985) for Juan de Fuca vents. The organic carbon contents of surface and subsurface sulphide deposits from TAG were in some samples comparable with values measured for coastal sediments (as much as 2.75%) and large resident populations of microbial cells were observed on the surfaces of sulphide deposits using scanning electron microscopy (Wirsen *et al.* 1993). These observations, combined with high levels of chemosynthetic activity indicated by CO_2-fixation experiments on scrapings of sulphide material from areas of active shrimp swarms, indicate that high primary productivity associated with sulphide surfaces might well serve as a renewable resource for grazing organisms (Van Dover *et al.* 1988*a*; Wirsen *et al.* 1993). No bacteria were observed on sulphide samples from a vent at Snake Pit where shrimp were absent (Segonzac *et al.* 1993). Wirsen *et al.* (1993) also demonstrate the potential for microbial primary production through lithoautotrophic oxidation of pyrite minerals at TAG, in addition to oxidation of dissolved sulphide.

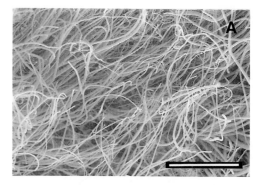

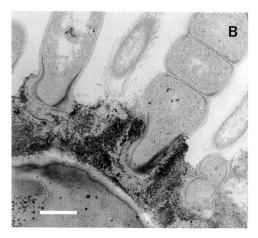

Fig. 17. (A) Filamentous bacteria comprising 'pads' on undersurface of cephalothorax of *Rimicaris exoculata*. Photo by J. Kluft. Scale bar = 10 μm. (B) Transmission electron micrograph of polysaccharide holdfast of bacteria on a section of shrimp. (*R. exoculata*) maxilla; scale bar =0.5 μm. From Wirsen *et al.* (1993).

The chemosynthetic primary production potential of epibiotic bacteria on shrimp is also significant. Both filamentous and rod/coccoid forms have been illustrated using light and scanning electron microscopy (Van Dover *et al.* 1988*b*; Segonzac 1992; Gebruk 1993; Wirsen *et al.* 1993; Segonzac *et al.* 1993). These epibiotic bacteria cover the legs, gills, mouthparts, carapace and antennae of all the species of bresiliid shrimp collected at Mid-Atlantic Ridge vents (Wirsen *et al.* 1993; Van Dover, pers. observ.), but they are most conspicuous on *R. exoculata* and *C. chacei* (Segonzac *et al.* 1993; Casanova *et al.* 1993). In fresh material, bacteria are especially noticeable as cream-coloured pads on the under surface of the cephalothorax; under scanning microscopy this pad proves to be a

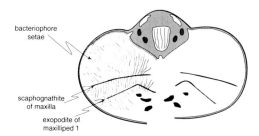

Fig. 18. Schematic cross-section of the shrimp *R. exoculata* illustrating the inflated branchial chambers formed by the carapace that folds around the body (shaded area) to almost meet at the ventral mid-line of the shrimp. Bacteria attach to the bacteriophore setae of the mouthparts (especially the maxillary scaphognathite and the exopodite of maxilliped I). Dense carpets of bacteria also formed in the regions denoted as 'fringed'. From Segonzac *et al.* (1993).

dense lawn of filamentous bacteria (Figs 17 and 18) within an understory of rods and coccoid forms. CO_2-fixation rates of shrimp epibionts (0.5–0.7 nmol CO_2 fixed/mg protein/min; Wirsen *et al.* 1993) are comparable with estimated CO_2 fixation rates of *Beggiatoa* sp. from Guaymas Basin (Nelson *et al.* 1989) and endosymbionts of *Riftia pachyptila* (*c.*45.8 nmol/mg protein/min; Childress *et al.* 1991). It seems likely that high growth rates of bacteria in the branchial chamber of the shrimp are facilitated by the habit of occupying areas of continuous mixing of sulphide-enriched vent water and oxygenated seawater (Gebruk *et al.* 1993; Segonzac *et al.* 1993; Wirsen *et al.* 1993).

Shrimp

The striking 'swarming' behaviour and high densities of shrimp at TAG and Snake Pit raise questions about how these populations are sustained by chemosynthetic production. Van Dover *et al.* (1988*b*) found no evidence for endosymbiosis, although they cautioned that the work is not conclusive in this regard, a caution that is still valid. These workers found strong evidence for the ingestion of large volumes of sulphide minerals. Foreguts of all individuals examined were filled with sulphide minerals precipitated at high temperatures. Scoop-shaped chelipeds and file-like dactyls on the pereiopods could facilitate ingestion of sulphide deposits from the walls of chimneys and associated microbes. Bulk stable isotope values of a limited number of specimens ($\delta^{13}C = -11.8‰$, $n = 3$) indicate, not surprisingly, that the shrimp are dependent on a non-photosynthetic source

of organic carbon (Van Dover *et al.* 1988*b*). Unlike species known to rely to a large extent on endosymbiotic bacteria, where $\delta^{15}N$ values are typically depleted in ^{15}N (<4‰), the nitrogen stable isotope composition of shrimp tissues was +7.6‰. Van Dover *et al.* (1988*b*) also note the unusually enlarged gill chamber of *R. exoculata* and expansion of the exopods of the first maxillipeds and second maxillae, as did Williams and Rona (1986) in their original description. Dense mats of DAPI-staining, filamentous bacteria occur as epibionts on these mouthparts as well as on other regions of the shrimp's carapace. Although ingestion of epibionts was deemed likely, Van Dover *et al.* (1988*b*) suggested that the major source of diet items for the shrimp is likely to be micro-organisms growing on surfaces of chimney sulphides. Patches of bacteria are observed on other crustaceans as well, such as bythograeid crabs collected from East Pacific Rise vents where the patches are conspicuous, attached to setae on various regions of the chelipeds, walking legs and carapace (Van Dover, pers. observ.). Attachment of epibionts to shrimp epicuticle is by a polysaccharide holdfast (Fig. 17; Wirsen *et al.* 1993). Comparative microbial genetic studies are needed to determine if the epibionts of *R. exoculata* and other bresiliid shrimp are host-specific or whether they occur elsewhere in the vent environment. Are these epibionts best viewed as a type of generalist fouling organism, or is the relationship between the shrimp and its epibionts an exclusive and co-evolving one? The distinction between fouling and a mutualistic symbiosis is important ecologically as in one case the host might have to expend energy to remove fouling material that interferes with, for example, gill activity, whereas in the mutualistic symbiosis the host and the symbiont benefit from attributes of each other.

Gal'chenko *et al.* (1988, 1989) also examined the question of nutrition in *R. exoculata*. Working with fresh animals, they looked for CO_2-uptake by various shrimp tissues and found it in a homogenate of gill, hepatopancreas, abdominal muscle and intestine. Surprisingly, given the high chemoautotrophic activity in filamentous bacteria associated with shrimp mouthparts (Wirsen *et al.* 1993), Gal'chenko report an order of magnitude less CO_2-uptake in an homogenate of mouthparts (second maxillae and first maxillipeds). Because of the dense populations of epibiotic bacteria associated with the various tissues chosen for analysis, it is difficult to agree with these workers' interpretation of these data as evidence for endosymbiotic autotrophic activity in the gills.

Table 3. *Stable isotope compositions ‰ reported for bresiliid shrimp. MAR = Mid Atlantic Ridge*

Species/location	Tissue	$\delta^{13}C$	$\delta^{15}N$	$\delta^{25}S$	Reference
Rimicaris exoculata TAG, Snake Pit (MAR)	Abdominal muscle	-11.8 ± 0.2 (3)	$+7.6 \pm 0.2$ (3)	$+9.7$ (5 pooled)	Van Dover et al. 1988; Van Dover, unpublished data
	Entire	-11.5	—	—	Gal'chenko et al. 1989
	Entire	-8.5 to -14.0	—	—	Pimenov et al. 1992
	Entire	-10.5 to -12.1	—	—	Gebruk et al. 1993
	Abdominal muscle and mouthparts	-15.1	—	—	Pimenov et al. 1992; Gebruk et al. 1993
	Mouthparts	-16	—	—	Gebruk et al. 1993
	Abdominal muscle and gills	-13.1	—	—	Gebruk et al. 1993
Chorocaris chacei TAG (MAR)	Entire	-11.6 to -12.5	—	—	Gebruk et al. 1993
Chorocaris sp. 1 Snake Pit (MAR)	Abdominal muscle	-15.8 ± 0.6 (10)	4.9 ± 0.2 (10)	—	Van Dover, unpublished data
Chorocaris sp. 2 Lucky Strike (MAR)	Abdominal muscle	$-17.9, -16.2$	$+1.4, +3.0$	—	Van Dover, unpublished data
Chorocaris vandoverae Mariana Back Arc (W. Pacific)	Abdominal muscle	$-16.7, -16.4$	$+8.9, +8.6$	—	Van Dover & Fry 1989
Alvinocaris lusca Galapagos Spreading Center, (E. Pacific)	Abdominal muscle	-21.7 (11) (bimodal: 3 ind.: -27.8 to -29.3 7 ind.: -16.2 to -22.6)	$+5.9$ (7) (range: $+3.0$ to $+8.3$)	—	Fisher et al. 1994

Gal'chenko *et al.* also found the shrimp tissues to have a carbon isotope ratio of −11.5‰.

Comparison of carbon isotope ratios of bresiliid shrimp (*R. exoculata, C. chacei, Chorocaris* spp.) from Mid-Atlantic Ridge vents (Table 3) with bresiliid shrimp (*Alvinocaris lusca*) from Rose Garden on the Galapagos Spreading Center show that the Mid-Atlantic Ridge shrimp have a unimodal range in carbon isotope values (*c.* −8 to −18‰) whereas the Rose Garden shrimp $\delta^{13}C$ values are bimodal (*c.* −16 to −22 and *c.* −28 to −30‰). This is consistent with the idea that the Mid-Atlantic Ridge shrimp consume food resources that are relatively homogeneous in carbon isotope composition compared with the Rose Garden shrimp. Nitrogen isotope (Table 3) compositions of Mid-Atlantic Ridge vent shrimp are consistent with the role of the shrimp as primary consumers whereas the sulphur isotope composition of *R. exoculata* (Table 3) indicates that a major portion of the shrimp's organic sulphur is ultimately derived from hydrothermal sources rather than seawater sulphate (Van Dover *et al.* 1988*b*).

Feeding in *R. exoculata* was revisited by Pimenov *et al.* (1992) and Gebruk *et al.* (1993), who repeated all of the above observations and conclude, with little new evidence, that the shrimp feed primarily on their epibionts. Segonzac *et al.* (1993) undertook a comparative study of shrimp behaviour and feeding biology based on material collected from Snake Pit during the HYDROSNAKE mission. They document differences in the morphology of feeding appendages and digestive systems of *A. markensis, R. exoculata* and *C. chacei* and discuss how these morphologies relate to the feeding biology of each species. *Rimicaris exoculata* and *A. markensis* appear to be two extremes of morphological differentiation. *Alvinocaris markensis*, which lives in the periphery of the high-temperature vents, has normal caridean mouthparts and is attracted to baited traps, where it feeds readily on the bait, and has the most capacious stomach of the three shrimp species (Fig. 19). *Rimicaris exoculata* has several morphological peculiarities that may relate to its trophic ecology, including 'bacteriophore' setae covering the enlarged exopodites of the maxillae and first maxillipeds and the inflated branchial chamber as has been noted elsewhere (e.g. Van Dover *et al.* 1988*b*; Figs 17 and 18). *Rimicaris exoculata* has a small stomach (Fig. 19) relative to *A. markensis* which, to Segonzac *et al.* (1993), is reminiscent of the reduction of digestive systems in symbiont-bearing invertebrates; these workers do not address whether differences in

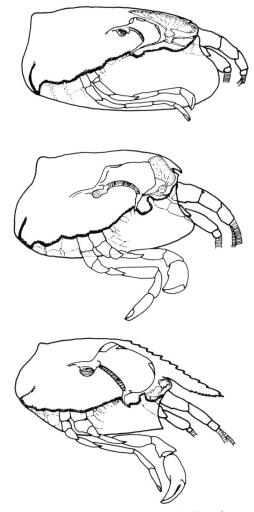

Fig. 19. Comparison of shrimp stomach size and location: *Rimicaris exoculata* (top), *Chorocaris chacei* (middle), *Alvinocaris markensis* (bottom). Redrawn from Segonzac *et al.* (1993).

stomach size could be accounted for by differences in rates or style of digestive processes depending on the type of presumptive prey (i.e. animal for *A. markensis* and bacterial for *R. exoculata*). The chelipeds of *R. exoculata* are small and carried within the enclosure created by the unusual nature of the carapace, which wraps from either side around the body of the shrimp, almost meeting ventrally (Williams and Rona 1986). Accordingly, Segonzac *et al.* (1993) argue that these chelipeds cannot be used effectively to scrape or scoop sulphides, as suggested by Van Dover *et al.* (1988*b*), but they might be very effective at cleaning epibiotic bacteria from the mouthparts and inner walls of the carapace. Despite its inadequate chelipeds, Segonzac *et al.* (1993) suggest *R. exoculata* may have access to chimney sulphides by scraping and loosening the particles with their walking legs (as described in Van Dover *et al.* 1988*b*) and subsequently ingesting the particles. In concurrence with Gebruk *et al.* (1992), Segonzac *et al.* (1993) hypothesize that the behaviour and morphology of *R. exoculata* are co-evolved as part of a symbiotic relationship with the epibiotic bacteria from which they are presumed to derive most of their nutrition.

Thus a strong circumstantial case for the role of epibiotic organisms in the nutrition of the shrimp has been developed (Van Dover *et al.* 1988*b*; Gebruk *et al.* 1992; Segonzac *et al.* 1993) but the relative importance of this role has not been determined. An alternative hypothesis that cannot be rejected yet is that behavioural and morphological characteristics of *R. exoculata* have been driven primarily by its invasion of a high-temperature, low-oxygen zone (>30°C) to graze on microbial primary production associated with chimney sulphides. The inflated branchial chamber formed by the carapace which wraps around the body of the shrimp may have evolved primarily to maximize oxygen exchange over the gills, with fouling and subsequent ingestion of bacteria a secondary consequence. Epibiotic bacteria foul many parts of each of the shrimp species, not just the mouthparts of *R. exoculata* (Segonzac 1993; Van Dover, pers. observ.); they are also found on other invertebrates at vents. Even so, the case that there is a special relationship between the shrimp and its epibiotic bacteria is compelling, reminiscent of similar epibiont–invertebrate associations of nematodes described by Polz *et al.* (1992). Resolution of the relative importance of epibiotic versus free-living micro-organisms in the diet of the shrimp requires new approaches such as molecular biomarker techniques. Lipid and genetic biomarker approaches are currently underway (D. Hedrick, C. Cavanaugh & M. Polz, pers. comm.). Because certain bacterial lipids can become incorporated in animal tissue with little change, lipid biomarkers hold the most promise as a tool to measure the relative importance of epibionts versus bacteria growing on sulphides in the diet of the shrimp. Rieley *et al.* (this volume) show that an invertebrate that derives its nutrition from endosymbionts has a relatively homogeneous lipid composition, whereas an invertebrate that grazes on free-living bacteria in the same locale has a heterogeneous lipid composition, reflecting the heterogeneity of functional bacterial groups in the environment. As Rieley *et al.* point out, this simple level of analysis may prove to be useful in determining the relative importance of epibionts in the diet of *R. exoculata* in as much as there appears to be little morphologic or genetic variation in the epibiotic population (Segonzac *et al.* 1993; Polz, pers. comm.). Thus if *R. exoculata* lipids are homogeneous, this would provide strong evidence that their diet is derived principally from a single source.

Mussels

Bathymodiolid mussels from the Galapagos Spreading Center and the East Pacific Rise were among the first invertebrate–bacterial symbioses to be studied and, in Pacific hydrothermal settings, endosymbiotic bacteria associated with these bivalves are sulphide-oxidizing chemoautotrophs that obtain their carbon from CO_2 (Felbeck *et al.* 1981; Cavanaugh 1983). However, in the Atlantic, hydrothermal bathymodiolid mussels from Snake Pit have two types of endosymbionts, one of which is implicated in the activity of methanol dehydrogenase (MeDH), an enzyme diagnostic of methylotrophy (Cavanaugh *et al.* 1992) and assimilation of carbon from CH_4. Methanotrophy and the co-occurrence of two endosymbiont morphs is also known in mussels from brine and seep habitats in the Atlantic (Gulf of Mexico) and elsewhere (Childress *et al.* 1986; Cavanaugh *et al.* 1987; Fisher *et al.* 1993).

In studies of mussels collected from Snake Pit, Cavanaugh *et al.* (1992) examined the morphology of the endosymbionts using transmission electron microscopy (TEM) and conducted enzyme assays for ribulose-1,5-bisphosphate carboxylase/oxygenase (RuBisCO), an enzyme diagnostic of Calvin–Benson CO_2 fixation (autotrophy) and MeDH, diagnostic for CH_4 assimilation (methylotrophy). Their TEM data from gill tissues of the mussel reveals the presence of two kinds of endosymbiont ultrastructure within gill epithelial cells (Fig. 20): (1) a large (1.2 μm diameter) oval symbiont with stacked intracytoplasmic membranes and (2) a smaller, rod- or coccoid-shaped symbiont (0.4 μm diameter) lacking internal membranes. Stacked internal membranes such as those found in the large cells are not known in sulphide-oxidizing bacteria, but are known from other metabolic types, including methanotrophs (Lidstrom 1992). Methanol dehydrogenase activity of the mussel gill tissue supports the idea that the large symbiont cells are methylotrophs; no RuBisCO activity was observed in the mussel gill tissue preparations (Cavanaugh *et al.* 1992).

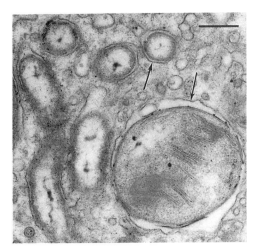

Fig. 20. Transmission electron micrograph of the Snake Pit mussel gill tissue. Arrows point to two different bacterial cell types. The large symbiont has stacked intracytoplasmic membranes characteristic of methylotrophic bacteria. Scale bar = 0.5 μm. From Cavanaugh *et al.* (1992).

Stable isotope compositions of the Snake Pit mussels were also reported by Cavanaugh *et al.* (1992), but provide few clues about the symbiont inorganic carbon or nitrogen sources. Mussel $\delta^{13}C$ values of -32.7 to $-35.6‰$ (Table 4) fall within the range of values observed in Pacific vent bivalve–sulphide-oxidizing bacteria symbioses (Fisher 1990; Conway *et al.* 1994). Carbon derived from biogenic or thermogenic sources would be expected to give the mussel tissue a very ^{13}C-depleted (more negative) value, as observed, for example, in methanotrophic mussel endosymbioses in the Gulf of Mexico [$\delta^{13}C = -45.7$, $-51.8‰$ (mantle tissue); Fisher *et al.* 1993]. Snake Pit mussel $\delta^{13}C$ values do not preclude the use of methane in vent fluids, however. The isotope composition of Snake Pit methane has not been reported, but at Pacific vents, methane has a $\delta^{13}C$ composition of -15 to $-17‰$ (Welhan and Craig 1983). Snake Pit vent fluids are not especially enriched in methane relative to vent fluids of Pacific hydrothermal systems, with concentrations in the range of $50–100$ μmol/l (Lilley *t al.* 1983; Welhan & Craig 1983; Jean Baptiste *et al.* 1991). The carbon isotope composition of the Lucky Strike mussel ($-24.1‰$) falls outside the range measured for any other vent bivalve that hosts endosymbionts (Van Dover *et al.* unpublished); characterization of the host-symbiont association is in progress (Fiala-Medioni, pers. comm.). Nitrogen isotope compositions of Snake Pit and

Lucky Strike mussel tissues are similar (-4.2 to $-10.5‰$; Table 4) and are within the range reported for symbiont-containing bivalve tissues and are depleted in ^{15}N relative to non-vent invertebrates ($>9.8‰$). Sulphur isotope composition of the Lucky Strike mussel is low ($+3.1$ to $+4‰$), consistent with a hydrothermal source for much of its organic sulphur via sulphide oxidation (Van Dover *et al.* unpublished).

Trophic ecology

Complete subordination of endosymbiont–host relationships to putative grazing taxa at some Mid-Atlantic Ridge vents emphasizes the potential for divergent evolution of hydrothermal vent trophic ecologies. Vent food webs can operate independently of endosymbioses, as has been suggested by Van Dover & Fry (1994) for situations where endosymbiotic primary production is conspicuous in terms of biomass, but relatively unimportant as a food resource for non-host grazers and consumers within the same microhabitat. Is there a difference in food web efficiency (in terms of biological oxidation of reduced compounds emitted from vents) in the presence or absence of endosymbioses? The standing crop of shrimp at one of the Snake Pit mounds is estimated to be of the order of 1.2 kg/m^2 (Segonzac 1992). This is an order of magnitude lower than has been reported for invertebrate–endosymbiont-dominated systems of the eastern Pacific where, for example, one clump of mussels (*B. thermophilus*) can have a biomass of 10.1 kg/m^2 (Hessler & Smithey 1983), or where a biomass of $10–15$ kg/m^2 in one clump of tubeworms (*R. pachyptila*) has been reported (Somero *et al.* 1983). Alvinellid polychaetes, thought to have a close trophic relationship with epibiotic bacteria (Desbruyéres *et al.* 1983) in much the same way as postulated for the shrimp, have a relatively low biomass of 0.55 kg/m^2. Standing crops of invertebrates are a poor measure of primary production, however, especially given the apparent difference in energetic demands between sessile and motile species. It remains to be determined whether primary productivity in an endosymbiosis-dominated community is more efficient at using reduced compounds in hydrothermal fluids than communities where endosymbiont populations are unimportant.

In the apparently simple food chains of shrimp-dominated Atlantic vents, do trophic interactions play any part in determining the distributions and abundances of organisms? Physical gradients of temperature and chemistry seem to impose a corresponding zonation in the

Table 4. *Stable isotope compositions (‰) reported for bathymodiolid mussels. MAR = Mid-Atlantic Ridge*

Species/location	Tissue	δ^{13}C	δ^{15}N	δ^{25}S	Reference
Bathymodiolus puteoserpentis Snake Pit (MAR)	Gill	−34.5 ± 1.5 (3)	+7.6 ± 3.0	–	Cavanaugh *et al.* 1992
	Foot/ mantle	−33.7 ± 1.1 (3)	−8.3 ± 3.2	–	Cavanaugh *et al.* 1992
Bathymodiolus thermophilus Galapagos Spreading Center (E. Pacific)	Foot/ mantle/gill	−30.5 to −37.1	−8.7 to +6.8	–	Rau & Hedges 1979 Fisher *et al.* 1987, 1988
Mytilid sp. 3 Mariana Back Arc (W. Pacific)	Foot	−32.8	−0.5	–	Van Dover & Fry 1989

biota, with mussels and anemones and chaetopterid polychaetes among the dominants in cooler waters at the base of sulphide mounds, and with shrimp abundant in warmer waters. However, the lessons of intertidal ecology, where the upper limits of some species and lower limits of others may be determined by trophic interactions (e.g. Connell 1961; Dayton 1971; Paine 1974), should not be forgotten. It seems plausible, for example, that *R. exoculata* excludes other invertebrates from its 'feeding turf', with the lower limits of the shrimp swarms defining the upper limits of successful colonization by sessile invertebrates. This kind of hypothesis is amenable to experimentation, but remains to be tested.

Sensory adaptations

Vision

Dominance of vent communities by highly active, swimming shrimp at TAG and Snake Pit focuses attention on sensory adaptations and ways in which these organisms perceive their environment. Van Dover *et al.* (1989) demonstrated that the 'eyeless' shrimp, *R. exoculata* has a novel photoreceptor presumably derived from the normal shrimp eye. Although *R. exoculata* lacks an externally differentiated eye, a pair of anteriorly fused dorsal organs located within the cephalothroax and immediately beneath the carapace represent highly derived 'eyes' (Fig. 21). There are no image-forming optics associated with these eyes, but they do have a cellular organization reminiscent of ommatidia, they are connected to the brain of the shrimp by a pair of large nerve tracts and they contain a visual pigment with the absorption characteristics of rhodopsin, with an absorption maximum at about 500 nm (Fig. 21). O'Neill *et al.* (in press), examine the histological structure of the *R. exoculata* eye and found that it is composed of between 5000 and 10 000

quasi-ommatidia, each containing five photoreceptor cells. Instead of a reflective tapetal layer (Van Dover *et al.* 1989), O'Neill *et al.* found that the ommatidia are embedded in a white, honeycomb-like matrix that acts as a diffuser, scattering incident light to maximize photon absorption by the retina. It is this matrix that gives the eyes their apparent reflective properties under artificial lighting conditions. The photoreceptor cells of *R. exoculata* are unusual in that they consist of an hypertrophied region containing an exceptionally high volume of rhabdomeric membrane (which contains the rhodopsin) and an extremely attenuated arhabdomeral region that appears to lack the cellular apparatus for cyclic membrane turnover (Van Dover *et al.* 1989; O'Neill *et al.* in press). Exposure to operating lights of submersibles is thought to cause the severe light damage observed in the microvillar array of the rhabdom, which is found even in animals preserved on the seafloor immediately after collection.

Van Dover *et al.* (1989) interpret the organs as being adapted for detection of low-level illumination associated with the environment around hydrothermal vents and, although they do not eliminate the possibility that the eye is used for detecting bioluminescence, they suggest that the extreme modification of the eye argues for a strong selective pressure not encountered by non-vent deep-sea shrimp. One possibility which remains to be tested is that the shrimp use their visual organs to detect black-body radiation from high-temperature (350°C) vents around which they cluster. Pelli and Chamberlain (1989) take a theoretical approach to address this question, using attributes of the absorption spectrum of the *R. exoculata* rhodopsin and the spectral properties of black-body radiation and attenuation in seawater. They conclude that it is theoretically possible that the shrimp could detect black-body radiation but, contrary to scattered reports in the popular literature, this is unproved. Segonzac *et al.*

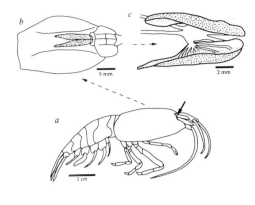

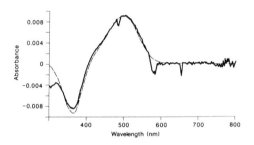

Fig. 21. (a) *Rimicaris exoculata*: arrow points to position where eyes are located in typical shrimp. **(b)** Position of the modified eye beneath the transparent carapace of the cephalothroax. **(c)** Dissected lobes of the eye show attachment to supraesophageal ganglion; lobes are fused anteriorly (not shown). The graph illustrates the absorption difference spectrum of the shrimp rhodopsin (solid line) compared with that of frog rhodopsin (broken line). Note the absorption maximum for the shrimp rhodopsin at about 500 nm. From Van Dover *et al.* (1989).

(1993) do not agree that the eye of *R. exoculata* plays a sensory role, based on their observations of the shrimp behaviour under submersible operating conditions. Attempts to examine the functional physiology of the photoreceptor on live material have so far proved futile, presumably due to irreversible bleaching of the photoreceptive pigment by the submersible lights during collection (S. Chamberlain, pers. comm.).

Ambient light fields associated with high-temperature black smoker chimneys have been demonstrated (Van Dover *et al.* 1988*a*; Smith & Delaney 1989) on the Juan de Fuca Ridge, but the spectral quality of this observed light is only coarsely resolved and predominantly of longer wavelengths perceived by the charge coupled device detector (i.e. >750 and <1050 nm).

Chave *et al.* (1993) report intriguing evidence for bursts of visible and near-infrared wavelengths (≤800 nm) of light during a single set of observations using a photodiode array at the Snake Pit site on the Mid-Atlantic Ridge. The spectral quality and intensity of ambient light at vents, its variability, its causal mechanisms and its role in photobiochemical reactions are all issues that remain unresolved and have been identified as priorities for research (LITE Workshop Participants 1993; Van Dover *et al.* 1994).

Rimicaris exoculata is not the only bresiliid with an unusual eye structure, though it represents the most extreme modification recognized. Segonzac (1993) reviewed the eye structure readily observed in live and frozen specimens of Snake Pit shrimp genera and noted that in *A. markensis*, 'the eyes are reduced, located entirely in the eyestalks which are relatively well-delimited and possess a cornea', in *C. chacei* the 'eyes are modified to form a bilobate ocular organ extending dorsally over the eyestalk region and back to the anterior region of the stomach', and in *R. exoculata* 'the two lobes of the ocular organ are coalesced at the level of the ocular plate and form a long V extending up to one-third the cepahlothroax length'. On preservation in formalin or alcohol, it is extremely difficult to detect regions of the eyes of bresiliid shrimp that lie beneath the carapace, but in fresh or frozen material or on videotapes and photographs of *in situ* animals, the outline of the dorsal surface of the eyes and the degree of their 'invasion' of the cephalothorax is readily apparent (Van Dover, pers. obs.). Martin & Hessler (1990) consider the eyestalk of *R. exoculata* to be greatly reduced and modified to form a flattened transverse plate in the orbital area of the shrimp, a step beyond the transverse, non-functional (in a mechanical sense) fused eyestalk of *C. vandoverae*. Chamberlain and co-workers are undertaking a comparative atlas of eye morphology/histology within the Bresiliidae.

Chemoreception

Renninger *et al.* (1994) examined the ability of *R. exoculata* and an unidentified species of *C.* from Snake Pit to detect sulphide and other odorants. Using neurophysiological preparations of excised antennae from live shrimp, these workers found a pronounced response to sulphide, with a dose–response curve indicating a 1% response at sulphide concentrations as low as 3 μmol/lM. This 1% response level is interpreted as physiologically significant and the effective concentration is well within the range of environmentally realistic sulphide concentrations (black smoker fluids from Snake Pit have sulphide concentrations of 18–20 mmol/l). Because sulphide is rapidly oxidized, it is believed to be involved only in near-field detection (tens of centimetres) of sulphide gradients. Other compounds such as methane which can be detected in plumes well away from the source might better serve as long-distance tracers of plumes for shrimp and need to be tested for their physiological activity.

Sensory adaptations and dispersal

Dispersal *per se* is not so often the problematic part of the life cycle of a vent organism – being carried away from a vent site is probably easy. Of paramount importance to dispersive life-history stages are the hydrodynamic conditions that allow propagules to become entrained and retained in hydrothermal systems and the physiological responses and consequent behaviours that allow propagules to recognize and respond to potential hydrothermal habitats. The motility of Mid-Atlantic Ridge vent shrimp and their amenability to physiological investigations suggest that they make a good model in which to address these kinds of questions although, ultimately, models based on more passive propagules (larvae rather adults) will need to be developed.

Conclusions

The ecology of Mid-Atlantic Ridge hydrothermal systems juxtaposed with the ecology of East Pacific Rise hydrothermal systems suggests that the two systems operate under very different biogeographical constraints. Whether this will ultimately prove to be a primary consequence of the spacing and timing of hydrothermal activity remains to be determined; at the moment, this hypothesis seems plausible. Comparative theoretical models that incorporate hydrodynamic, geographical and temporal parameters characteristic of fast- and slow-spreading end-members need to be developed to provide a context in which to interpret empirical observations, and vice versa.

Shrimp-dominated Mid-Atlantic Ridge vent communities seem likely to have trophic controls of invertebrate distributions superimposed on zonations determined by physiological tolerances to hydrothermal conditions. This question needs to be addressed in a manner analogous to that used in studies of shallow-water intertidal hard substrate communities. Details of the

trophic ecology and community energetics of Mid-Atlantic Ridge vents, including the role of methane in the nutrition of the mussels at Snake Pit and Lucky Strike and predominant modes of primary production and consumption suggest fundamental differences in food web structure between Mid-Atlantic Ridge and East Pacific Rise hydrothermal systems.

The motility of dominant species in the hydrothermal regime of steep gradients in temperature and chemistry places a premium on sensory adaptations. Shrimp-dominated communities of the Mid-Atlantic Ridge bring our attention to these adaptations more readily than the sessile invertebrate-dominated communities of the East Pacific Rise. Shrimp can serve as models of sensory responses and consequent behaviours that allow dispersive propagules to recognize and respond to hydrothermal habitats.

The author gratefully acknowledges reviews of the manuscript by Fred Grassle and Robert Hessler. Various scientists provided unpublished materials or other assistance, including M. Segonzac, A. Gebruk, C. Wirsen, C. Walker, R. Gustafson, B. Metivier, C. Cavanaugh, Wirsen, J. Karson, P. Rona, S. Humphris, D. Desbruyeres, J. Cann, B. Murton, M. Tivey, D. Fornari, C. Langmuir and A. Fiala-Medioni. Special thanks to C. Walker, D. Dixon and L. Parson for their invitation to share this work with the BRIDGE scientific community at the Hydrothermal Vents and Processes Meeting and for their editorial assistance. J. Edmond provided critical guidance when needed most. This work was supported by a grant from the Biological Oceanography Program of the US National Science Foundation.

References

BATUYEV, B. N., KROTOV, A. G., MARKOV, V. F., CHERKASHEV, G. A., KRASNOV, S. G. & LISITSYN, Y. D. 1994. Massive sulfide deposits discovered at 14°45′N, Mid-Atlantic Ridge. *BRIDGE Newsletter*, **6**, 6–10.

BLACK, M. B., LUTZ, R. A. & VRIJENHOEK, R. C. 1994. Gene flow among vestimentiferan tube worm (*Riftia pachyptila*) populations from hydrothermal vents of the eastern Pacific. *Marine Biology*, **120**, 33–39.

BOUGAULT, H., CHARLOU, J.-L. & 8 others 1993. Fast and slow spreading ridges: structure and hydrothermal activity, ultramafic topographic highs, and CH4 output. *Journal of Geophysical Research*, **98** (B6), 9643–9651.

CAMPBELL, A. C., PALMER, M. R. & 9 others 1988. Chemistry of hot springs on the Mid-Atlantic Ridge. *Nature*, **335**, 514–519.

CANN, J. R., VAN DOVER, C. L., WALKER, C. L., DANDO, P. & MURTON, B. J. 1994. *Diversity of Vent Ecosystems (DOVE)*. BRIDGE Workshop Report No. 4.

CARNEY, R. S., HAEDRICH, R. L. & ROWE, G. T. 1983. Zonation of fauna in the deep sea. *In*: ROWE, G. T. (ed.) *The Sea*, Vol. 8, Wiley-Interscience, New York, 371–398.

CASANOVA, B., BRUNET, M. & SEGONZAC, M. 1993. L'impact d'une épibiose bactérienne sur la morphologie fonctionelle de crevettes associées à l'hydrothermalisme médio-atlantique. *Cahiers de Biologie Marine*, **34**, 573–588.

CAVANAUGH, C. M. 1983. Symbiotic chemoautotrophic bacteria in marine invertebrates from sulphide-rich habitats. *Nature*, **302**, 58–61.

——, LEVERING, P. R., MAKI, J. S., MITCHELL, R. & LIDSTROM, M. 1987. Symbiosis of methylotrophic bacteria and deep-sea mussels. *Nature*, **325**, 346–348.

——, WIRSEN, C. O. & JANNASCH, H. W. 1992. Evidence for methylotrophic symbionts in a hydrothermal vent mussel (Bivalvia: Mytilidae) from the Mid-Atlantic Ridge. *Applied and Environmental Microbiology*, **58**, 3799–3803.

CHACE, F. A. 1992. On the classification of the Caridea (Decapoda). *Crustaceana*, **63**, 70–80.

CHARLOU, J.-L. & DONVAL, J.-P. 1993. Hydrothermal methane venting between 12° and 26°N along the Mid-Atlantic Ridge. *Journal of Geophysical Research*, **98** (B6), 9625–9642.

——, BOUGAULT, H., APPRIOU, P., NELSON, T. & RONA, P. A. 1991. Different TDM/CH4 hydrothermal plume signatures: TAG site at 26°N and serpentinized ultramafic diapir at 15°05'N on the Mid-Atlantic Ridge. *Geochimica et Cosmochimica Acta*, **55**, 3209–3222.

——, & FARANAUT 15°N Scientific Party. 1992. Immense CH4 plumes in seawater associated to ultramafic outcrops at 15°N on the MId-Atlantic Ridge. *Eos, Transactions of the American Geophysical Union*, **73**, 585.

CHAVE, A. D., VAN DOVER, C. L., TYSON, J. A., BAILEY, J. W., PETIT, R. A., FILLOUX, J. H. & MOELLER, H. H. 1993. Ambient light measurements from the Snake Pit area. *Eos, Transactions of the American Geophysical Union*, **74**, 100.

CHILDRESS, J. J., FISHER, C. R., BROOKS, J. M., KENNICUTT, M. C. I, BIDIGARE, R. & ANDERSON, A. E. 1986. A methanotrophic marine molluscan (Bivalvia, Mytilidae) symbiosis: mussels fueled by gas. *Science*, **233**, 1306–1308.

——, —— FAVUZZI, J. A., SANDERS, N. K. & ALAYSE, A. M. 1991. Sulfide-driven autotrophic balance in the bacterial-symbiont-containing hydrothermal vent tubeworm, *Riftia pachyptila*, Jones. *Biological Bulletin*, **180**, 135–153.

CHIN, C. S., KLINKHAMMER, G. P. & WILSON, C. 1993. Comparison of hydrothermal plume signals, 23°N to 40°N on the Mid-Atlantic Ridge. *Eos, Transactions of the American Geophysical Union*, **74**, 360.

CHRISTOFFERSON, M. L. 1989. Phylogenetic relationships among Oplophoridae, Atyidae, Pasiphaeidae, Alvinocarididae Fam. N., Bresiliidae, Psalidopodidae and Disciadidae (Crustacea Caridea Atyoidea). *Boletim de Zoologia, Universidade de Sao Paulo*, **10**, 273–281.

COLODNER, D., LIN, J. & 7 others 1993. Chemistry of Lucky Strike hydrothermal fluids: initial results. *Eos, Transactions of the American Geophysical Union*, **74**, 99.

CONNELL, J. H. 1961. Effects of competition, predation by *Thais lapillus* and other factors on natural populations of the barnacle *Chthalamus stellatus*. *Ecology*, **42**, 710–723.

CONWAY, N., KENNICUTT, M. C. & Van Dover, C. L. 1994. Stable isotopes in the study of marine chemosynthesis-based ecosystems. *In*: LAJTHA, K. & MICHENER, R. (eds) *Stable Isotopes in Ecology*. Blackwell Scientific, New York, 158–186.

CRADDOCK, C., HOEH, W. R., GUSTAFSON, R. G., LUTZ, R. A., HASHIMOTO, J. & VRIJENHOEK, R. C. Evolutionary relationships among deep-sea mytilids (Bivalvia: Mytilidae) from hydrothermal vents and cold-water methane/sulfide seeps. *Marine Biology*, in press.

DANDO, P. R., SOUTHWARD, A. J., SOUTHWARD, E. C., DIXON, D. R., CRAWFORD, A. & CRAWFORD, M. 1992. Shipwrecked tube worms. *Nature*, **356**, 667.

DAYTON, P. K. 1971. Competition, disturbance and community organization: the provision and subsequent utilization of space in a rocky intertidal community. *Ecological Monographs*, **41**, 351–389.

DE SAINT LAURENT, M. 1984. Crustacés décapodes d'un site hydrothermal actif de la dorsale du Pacifique oriental (13°N), en provenance de la campagne française *Biocyatherm*. *Comptes Rendues de l'Académie des Sciences, Paris, Series III*, **299**, 355–360.

DESBRUYÉRES, D., GAILL, F., LAUBIER, L., PRIEUR, D. & RAU, G. 1983. Unusual nutrition of the 'Pompeii' worm *Alvinella pompejana* (polychaetous annelid) from a hydrothermal vent environment: SEM, TEM, ^{13}C and ^{15}N evidence. *Marine Biology*, **75**, 201–205.

EBERHART, G.L., RONA, P.A. & HONNOREZ, J. 1988. Geologic controls of hydrothermal activity in the Mid-Atlantic Ridge rift valley: tectonics and volcanics. *Marine Geophysical Researches*, **10**, 233–259.

EDMOND, J. M., CAMPBELL, A. C., PALMER, M. R. & GERMAN, C. R. 1990. Geochemistry of hydrothermal fluids from the Mid-Atlantic Ridge: TAG and MARK. *Eos, Transactions of the American Geophysical Union*, **71**, 1650–1651.

——, —— & 7 others. Time series studies of vent fluids from the TAG and MARK sites (1986, 1990) Mid-Atlantic Ridge: a new solution on chemistry model and a mechanism for Cu/Zn zonation in massive sulphide orebodies, this volume.

ELDERFIELD, H., GERMAN, C. & 11 others 1993. Preliminary geochemical results from the Broken Spur Hydrothermal Field, 29°N, Mid Atlantic Ridge. *Eos, Transactions of the American Geophysical Union*, **74**, 99.

EMERY, K.O. & UCHUPI, E. 1984. *The Geology of the Atlantic Ocean*. Springer-Verlag, New York, 1050 pp.

FELBECK, H., SOMERO, G. N. & CHILDRESS, J. J. 1981. Calvin–Benson cycle and sulphide oxidation enzymes in animals from sulphide rich habitats. *Nature*, **293**, 291–293.

FISHER, C. R. 1990. Chemoautotrophic and methanotrophic symbioses in marine invertebrates. *Reviews in Aquatic Sciences*, **2**, 399–436.

——, BROOKS, J. M., VODENICHAR, J. S., ZANDE, J. M., CHILDRESS, J. J. & BURKE, R. A. 1993. The co-occurrence of methanotrophic and chemoautotrophic sulfur-oxidizing bacterial symbionts in a deep-sea mussel. *P.S.Z.N.I. Marine Ecology*, **14**, 277–289.

——, CHILDRESS, J. J. & 12 others 1988. Microhabitat variation in the hydrothermal vent mussel *Bathymodiolus thermophilus*, at Rose Garden on the Galapagos Rift. *Deep-Sea Research*, **35**, 1769–1791.

——, ——, MACKO, S. A. & BROOKS, J. M. 1994. Nutritional interactions in Galapagos Rift hydrothermal vent communities: inferences from stable carbon and nitrogen isotope analyses. *Marine Ecology Progress Series*, **103**, 45–55.

——, J.J., Oremland, R.S. & Bidigare, R.R. 1987. The importance of methane and thiosulfate in the metabolism of the symbionts of two deep-sea mussels. *Marine Biology*, **96**, 59–71.

FORNARI, D. & EMBLEY, R. W. Tectonic and volcanic controls on hydrothermal processes at the Mid Ocean Ridge: an overview based on near-bottom submersible studies. *In*: HUMPHRIS, S. *ET AL.* (eds) *Physical, chemical, biological and geological interactions within seafloor hydrothermal systems.* AGU Monograph, in press.

——, VAN DOVER, C. L., SHANK, T., LUTZ, R. & OLSSON, M. 1994. A versatile, low-cost temperature-sensing device for time-series measurements at deep-sea hydrothermal vents. *BRIDGE Newsletter*, **6**, 40–47.

FOUQUET, Y., WAFIK, A., CAMBON, P., MEVEL, C., MEYER, G. & GENTE, P. 1993. Tectonic setting and mineralogical and geochemical zonation in the Snake Pit sulfide deposit (Mid-Atlantic Ridge at 23°N). *Economic Geology*, **88**, 2018–2036.

FOWLER, C. M. R. 1990. *The Solid Earth: An Introduction to Global Geophysics*. Cambridge University Press, Cambridge, 472 pp.

FRANCE, S. C., HESSLER, R. R. & VRIJENHOEK, R. C. 1992. Genetic differentiation between spatially-disjunct populations of the deep-sea hydrothermal vent-endemic amphipod *Ventiella sulfuris*. *Marine Biology*, **114**, 551–556.

FRICKE, H., GIERE, O., STETTER, K, ALFREDSSON, G. A., KRISTJANSSON, J. K., STOFFERS, P. & SVAVARSSON, J. 1989. Hydrothermal vent communities at the shallow subpolar Mid-Atlantic Ridge. *Marine Biology*, **102**, 425–429.

FUSTEC, A., DESBRUYÈRES, D. & JUNIPER, S. K. 1987. Deep-sea hydrothermal vent communities at 13°N on the East Pacific Rise: microdistribution and temporal variations. *Biological Oceanography*, **4**, 121–164.

GAL'CHENKO, V. F., GALKIN, S. V., LEIN, A. Y., MOSKALEV, L. I. & IVANOV, M. V. 1988. Role of bacterial symbionts in nutrition of invertebrates from areas of active underwater hydrothermal systems. *Oceanology*, **28**, 786–794.

——, PIMENOV, N. V., LEIN, A. Y., GALKIN, S. V., MOSKALEV, L. I. & IVANOV, M. V. 1989. Autotrophic CO_2 assimilation in tissues of the shrimp *Rimicaris exoculata* from a hydrothermal region on the Mid-Atlantic Ridge. *Doklady Akademia Nauk*, **308**, 1478–1481.

GALKIN, S. V. & MOSKALEV, L. I. 1990. Hydrothermal fauna of the Mid-Atlantic Ridge. *Okeanologia*, **30**, 842–847.

GEBRUK, A. V., PIMENOV, N. V. & Savvichev, A. S. 1993. Feeding specialization of bresiliid shrimps in the TAG site hydrothermal community. *Marine Ecology Progress Series*, **98**, 247–253.

GERMAN, C., BRIEM, J. & 11 others 1993. Hydrothermal activity on the Reykjanes Ridge: the Steinahóll vent field at 63°06′N. *Eos, Transactions of the American Geophysical Union*, **74**, 360.

GRASSLE, J. F., HUMPHRIS, S. E., RONA, P. A., THOMPSON, G. & VAN DOVER, C. L. 1986. Animals at Mid-Atlantic Ridge vents. *Eos, Transactions of the American Geophysical Union*, **67**, 1022.

GUINOT, D. 1988. Les crabes des sources hydrothermales de la dorsale du Pacifique oriental (campagne *Biocyarise*, 1984). *Oceanologica Acta (special issue)*, **8**, 109–118.

—— 1989. Description de *Segonzacia* gen. nov. et remarques sur *Segonzacia mesatlantica* (Williams): campagne HYDROSNAKE 1988 sur la dorsale médio-Atlantique (Crustacea Decapods Brachyura). *Bulletin du Muséum National d'Histoire Naturelle Paris, Zoologie, Section A*, **11**, 203–231.

HASHIMOTO, J., MIURA, T., FUJIKURA, K. & OSSAKA, J. 1993. Discovery of vestimentiferan tube-worms in the euphotic zone. *Zoological Science*, **10**, 1063–1067.

HAYMON, R. M., FORNARI, D. J., EDWARDS, M. H., CARBOTTE, S., WRIGHT, D. & MACDONALD, K. C. 1991. Hydrothermal vent distribution along the East Pacific Rise crest (9°09′–54′N) and its relationship to magmatic and tectonic processes on fast-spreading mid-ocean ridges. *Earth and Planetary Science Letters*, **104**, 513–534.

HESSLER, R. R. & LONSDALE, P. 1991. Biogeography of Mariana Trough hydrothermal vent communities. *Deep-Sea Research*, **38A**, 185–199.

—— & MARTIN, J. W. 1989. *Austinograea williamsi*, new genus, new species, a hydrothermal vent crab (Decapoda: Bythograeidae) from the Mariana back-arc basin, Western Pacific. *Journal of Crustacean Biology*, **9**, 645–651.

—— & SMITHEY, W. M. 1983. The distribution and community structure of megafauna at the Galapagos Rift hydrothermal vents. *In*: RONA, P. A. *Hydrothermal Processes at Seafloor Spreading Centers*. Plenum Press, New York, 735–770.

——, ——, BOUDRIAS, M. A., KELLER, C. H., LUTZ, R. A. & CHILDRESS, J. J. 1988. Temporal change at the Rose Garden hydrothermal vent. *Deep-Sea Research*, **35A**, 1681–1709.

HOLTHUIS, L. B. 1993. *The recent genera of the Caridean and Stenopodidean shrimps (Crustacea, Decapoda): with an appendix on the order Amphionidacea*. National Natuurhistorisch Museum, Leiden, Netherlands.

HUMES, A. G. 1987. Copepoda from deep-sea hydrothermal vents. *Bulletin of Marine Science*, **41**, 645–788.

HUMPHRIS, S. E. 1994. Drilling will TAG active hydrothermal system on the MAR. *JOI/USSAC Newsletter*, **7**, 1–8.

——, FOUQUET, Y. & THE LUCKY STRIKE TEAM. 1993. Comparison of hydrothermal deposits at the Lucky Strike Vent Field with other Mid-Ocean Ridge vent sites. *Eos, Transactions of the American Geophysical Union*, **74**, 100.

JEAN-BAPTISTE, P., CHARLOU, J.-L., STIEVENARD, M., DONVAL, J. P., BOUGAULT, H. & MEVEL, C. 1991. Helium and methane measurements in hydrothermal fluids from the Mid-Atlantic Ridge: the Snake-Pit site at 23°N. *Earth and Planetary Science Letters*, **106**, 17–28.

JOHNSON, K. S., CHILDRESS, J. J., & BEEHLER, C. L. 1988. Short term temperature variability in the Rose Garden hydrothermal vent field. *Deep-Sea Research*, **35A**, 1711–1722.

JOLLIVET, D. 1993. *Distribution et evolution de la faune associee aux sources hydrothermales profondes a 13N sur la dorsale du Pacifique Oreintal: Le cas particulier des polychaetes Alvinellidae*. These de Doctorat de l'Universite de Bretagne Occidentale.

KARSON, J. A. & BROWN, J. R. 1988. Geologic setting of the Snake Pit hydrothermal site: an active vent field on the Mid-Atlantic Ridge. *Marine Geophysical Researches*, **10**, 91–107.

KLEINSCHMIDT, M. & TSCHAUDER, R. 1985. Shallow-water hydrothermal vent systems off the Palos Verdes Peninsula, Los Angeles County, California. *Bulletin of the Biological Society of Washington*, **6**, 485–488.

KONG, L., RYAN, W. B. F., MAYER, L., DETRICK, R. S., FOX, P. J. & MANCHESTER, K. 1985. Bare-rock drill sites, ODP Legs 106 and 109: evidence for hydrothermal activity at 23°N in the Mid-Atlantic Ridge. *Eos, Transactions of the American Geophysical Union*, **66**, 1106.

KURR, M., HUBER, R. & 6 others 1991. *Methanopyrus kandleri*, gen. and sp. nov. represents a novel group of hyperthermophilic methanogens, growing at 110°C. *Archives of Microbiology*, **156**, 239–247.

LALOU, C., REYSS, J.-L., BRICHET, E., ARNOLD, M., THOMPSON, G., FOUQUET, Y. & RONA, P. A. 1993. New age dating for Mid-Atlantic Ridge hydrothermal sites: TAG and Snake Pit chronology revisited. *Journal of Geophysical Research*, **98**(B3), 9705–9713.

——, THOMPSON, G., ARNOLD, M., BRICHET, E., DRUFFEL, E. & RONA, P. A. 1990. Geochronology of TAG and Snakepit hydrothermal fields, Mid-Atlantic Ridge: witness to a long and complex hydrothermal history. *Earth and Planetary Science Letters*, **97**, 113–128.

LANGMUIR, C. H. & the FAZAR CRUISE PARTICIPANTS. 1992. *Evaluation of the Relationships among Segmentation, Hydrothermal Activity and Petrologic Diversity on the Mid-Atlantic Ridge*. Lamont Doherty Earth Observatory of Columbia University, Technical Report No. LDEO–92–3, Palisades.

——, FORNARI, D. & 17 others 1993. Geological setting and characteristics of the Lucky Strike Vent Field at 37°17'N on the Mid-Atlantic Ridge. *Eos, Transactions of the American Geophysical Union*, **74**, 99.

LIDSTROM, M. E. 1992. The aerobic methylotrophic bacteria. *In*: BALOWS, A., TRÜPER, A. G., DWORKIN, M., HARDER, W. & SCHLEIFER, K. H. (eds) *The Prokaryotes*, 2nd edn. Springer-Verlag, Berlin, 431–445.

LILLEY, M. D., BAROSS, J. A. & GORDON, L. I. 1983. Reduced gases and bacteria in hydrothermal fluids: the Galapagos Spreading Center and 21°N East Pacific Rise. *In*: RONA, P., BÖSTROM, K., LAUBIER, L. & SMITH, K. (eds) *Hydrothermal processes at seafloor spreading centers*. Plenum Press, New York.

LITE WORKSHOP PARTICIPANTS 1993. *Light in Thermal Environments*. Workshop Report. RIDGE Technical Report. RIDGE Office, Woods Hole Oceanographic Institution, Woods Hole.

LUTZ, R. A. & FORNARI, D. J. 1992. A system for monitoring biologic and geologic changes on a fast-spreading mid-ocean ridge crest. *EOS, Transactions of the American Geophysical Union*, **73**, 278.

MACDONALD, K. C. 1982. Mid-ocean ridges: fine scale tectonic, volcanic and hydrothermal processes within the plate boundary zone. *Annual Reviews of Earth and Planetary Sciences*, **10**, 155–190.

MACDONALD, I. R., GUINASSO, N. L., REILLY, J. F., BROOKS, J. M., CALLENDER, W. R. & GABRIELLE, S. G. 1990. Hydrocarbon seep communities VI: species composition and habitat characteristics. *Geo-Marine Letters*, **10**, 244–252.

MARTIN, J. W. & CHRISTIANSEN, J. C. A new species of the shrimp genus *Chorocaris* Martin and Hessler, 1990, from hydrothermal vents along the Mid-Atlantic Ridge. *Proceedings of the Biological Society of Washington*, in press.

—— & HESSLER, R. R. 1990. *Chorocaris vandoverae*, a genus and species of hydrothermal vent shrimp (Crustacea, Decapoda, Bresiliidae). *Contributions in Science, Los Angeles*, **417**, 1–11.

MAYER, L. A., SHOR, A. N., CLARKE, J. H. & PIPER, D. J. W. 1988. Dense biological communities at 3850 m on the Laurentian Fan and their relationship to the deposits of the 1929 Grand Banks earthquake. *Deep-Sea Research*, **35**, 1235–1246.

MCGREGOR, B. A. & RONA, P. A. 1975. Crest of the Mid-Atlantic Ridge at 26°N. *Journal of Geophysical Research*, **80**, 3307–3314.

MCLEAN, J. H. 1992. A new species of *Pseudorimula* (Fissurellacea: Clyposectidae) from hydrothermal vents of the Mid-Atlantic Ridge. *The Nautilus*, **106**, 115–118.

MEVEL, C., AUZENDE, J. & 9 others 1989. La ride du Snake Pit (dorsale medio-Atlantique, 23°22′N): résultats préliminaires de la campagne HYDRO-SNAKE. *Comptes Rendus Hebdomadaires des Séances de l'Académie des Sciences, Paris*, **308**(II), 545–552.

MURTON, B. J. & VAN DOVER, C. L. 1993. ALVIN dives on the Broken Spur Hydrothermal Vent Field at 29°10′N on the Mid-Atlantic Ridge. *BRIDGE Newsletter*, **5**, 11–14.

——, KLINKHAMMER, G. & 11 others 1994. Direct evidence for the distribution and occurrence of hydrothermal activity between 27–30°N on the Mid-Atlantic Ridge. *Earth and Planetary Science Letters*, **125**, 119–128.

——, —— & 8 others 1993. Direct measurements of the distribution and occurrence of hydrothermal activity between 27°N and 37°N on the Mid-Atlantic Ridge. *Eos, Transactions of the American Geophysical Union*, **74**, 99.

——, VAN DOVER, C. L. & SOUTHWARD, E. Geological setting and ecology of the Broken Spur hydrothermal vent field at 29°10′N on the Mid-Atlantic Ridge, this volume.

NAGANUMA, T., DEEP SEA RESEARCH GROUP & DEEP STAR GROUP 1991. Collection of chemosynthetic sulfur bacteria from a hydrothermal vent and submarine volcanic vents. *Proceedings of the JAMSTEC Symposium on Deep-Sea Research*, **7**, 201–207.

NELSON, D. C., WIRSEN, C. O. & JANNASCH, W. W. 1989. Characterization of large autotrophic *Beggiatoa* abundant at hydrothermal vents of the Guaymas Basin. *Applied Environmental Microbiology*, **55**, 2909–2917.

ODP LEG 106 SCIENTIFIC PARTY 1986. Drilling the Snake Pit hydrothermal sulfide deposit on the Mid-Atlantic Ridge, lat 23°22′N. *Geology*, **14**, 1004–1007.

ÓLAFSSON, J., HONJO, S., THORS, K., STEFÁNSSON, U., JONES, R. R. & BALLARD, R. D. 1989. Initial observations, bathymetry and photography of a geothermal site on the Kolbeinsey Ridge. In: AYALA-CASTAÑAREA, A. WOOSTER, W. & YÁÑEZ-ARANCIBIA, A. (eds) *Oceanography 1988*. UNAM Press, Mexico, 208 pp.

O'NEILL, P. J., JINKS, R. N., HERZOG, E. R. &. CHAMBERLAIN, S. C. Is photoreceptive membrane shedding ubiquitous? Negative evidence from an ultra-high sensitivity photoreceptor. *Investigative Ophthamology and Visual Science*, in press.

PAINE, R. T. 1974. Intertidal community structure: experimental studies on the relationship between a dominant competitor and its principal predator. *Oecologia*, **15**, 93–120.

PAULL, C. K., HECKER, B. & 8 others 1984. Biological communities at the Florida Escarpment resemble hydrothermal vent taxa. *Science*, **226**, 637–654.

PELLI, D. G. & CHAMBERLAIN, S. C. 1989. The visibility of 350°C black-body radiation by the shrimp *Rimicaris exoculata* and man. *Nature*, **337**, 460.

PIMENOV, N. V., SAVVICHEV, A. S., GEBRUK, A. V., MOSKALEV, L. I., LEIN, A. J. & IVANOV, M. V.

1992. Trophic specialisation of shrimp Bresiliidae from hydrothermal regions of TAG. *Doklady Akademy Nauk*, **323**, 567–571 [in Russian].

POLZ, M. F., FELBECK, H., NOVAK, R., NEBELSICK, M. & OTT, J. A. 1992. Chemoautotrophic, sulfur-oxidizing bacteria on marine nematodes: morphological and biochemical characterization. *Microbial Ecology* **24**, 313–329.

RAU, G. H. & HEDGES, J. I. 1979. Carbon-13 depletion in a hydrothermal vent mussel: suggestion of a chemosynthetic food source. *Science*, **203**, 648–649.

RENNINGER, G. H., GLEESON, R. A., KASS, L. & VAN DOVER, C. L. 1994. Sulfide responses in antennal axons of shrimp from mid-Atlantic ridge hydrothermal vent sites. *Society of Neurosciences Abstracts*, **20** (Part 1), 774.

RHOADS, D. C., LUTZ, R. A., REVELAS, E. C. & CERRATO, R. M. 1981. Growth rates of bivalves at deep-sea hydrothermal vents along the Galapagos Rift. *Science*, **214**, 911–913.

RIELEY, G., VAN DOVER, C. L., HEDRICK, D. B., WHITE, D. C., & EGLINTON, G., Lipid characteristics of hydrothermal vent organisms from 9°N, East Pacific Rise, this volume.

RONA, P. A. & MERRILL, G. F. 1978. A benthic invertebrate from the Mid-Atlantic Ridge. *Bulletin of Marine Science*, **28**, 371–375.

——, BOGDANOV, Y. A., GURVICH, E. G., RIMSKI-KORSAKOV, N. A., SAGALEVITCH, A. M., HANNINGTON, M. D. & THOMPSON, G. 1993. Relict hydrothermal zones in the TAG Hydrothermal Field, Mid-Atlantic Ridge 26°N, 45°W. *Journal of Geophysical Research*, **98**(B6), 9715–9730.

——, BOUGAULT, H. & 7 others 1992. Hydrothermal circulation, serpentinization, and degassing at a rift valley-fracture zone intersection: Mid-Atlantic Ridge near 15°N, 45°W. *Geology*, **20**, 783–786.

——, HARBISON, R. N., BASSINGER, B. G., SCOTT, R. B. & NALWALK, A. J. 1976. Tectonic fabric and hydrothermal activity of Mid-Atlantic Ridge crest (latitude 26°N). *Geological Society of America Bulletin*, **87**, 661–674.

——, KLINKHAMMER, G., NELSEN, T. A., TREFRY, J. H. & ELDERFIELD, H. 1986. Black smokers, massive sulphides and vent biota at the Mid-Atlantic Ridge. *Nature*, **321**, 33–37.

——, THOMPSON, G. & 7 others 1984. Hydrothermal activity at the Trans-Atlantic Geotraverse hydrothermal field, Mid-Atlantic Ridge Crest at 26°N. 1984. *Journal of Geophysical Research*, **89**(B13), 11 365–11 377.

——, WIDENFALK, L. & BOSTRÖM, K. 1987. Serpentinized ultramafics and hydrothermal activity at the Mid-Atlantic Ridge Crest near 15°N. *Journal of Geophysical Research*, **92** (B2), 1417–1427.

RUDNICKI, M. D. & ELDERFIELD, H. 1993. A chemical model of the buoyant and neutrally buoyant plume above the TAG vent field, 26 degrees N, Mid-Atlantic Ridge. *Geochimica et Cosmochimica Acta*, **57**, 2939–2957.

SCHWAB, W. C., UCHUPI, E., HOLCOMB, R. T. & DANFORTH, W. W. 1992. Geologic interpretation of SeaMARC 1B and ARGO imagery in the median valley of the Mohn's Ridge, Norwegian-Greenland Sea. *Miscellaneous Field Studies, Map MF–2197*. US Department of the Interior, US Geological Survey.

SEGONZAC, M. 1992. The hydrothermal vent communities of Snake Pit area (Mid Atlantic Ridge; 23°N, 3480 m): megafaunal composition and microdistribution. *Comptes Rendus Hebdomadaires des Séances de l'Académie des Sciences, Paris*, **314**, 593–600.

——, DE SAINT LAURENT, M. & CASANOVA, B. 1993. L'énigme du comportement trophique des crevettes Alvinocarididae des sites hydrothermaux de la dorsale médio-atlantique. *Cahiers de Biologie Marine*, **34**, 535–571.

SEILACHER, A. 1977. Pattern analysis of *Paleodictyon* and related trace fossils. *Geological Journal Special Issue*, **9**, 289–334.

SMITH, D. K. & CANN, J. R. 1993. Building the crust at the Mid-Atlantic Ridge. *Nature*, **365**, 707–715.

SMITH, M. O. & DELANEY, J. R. 1989. Variability of emitted radiation from two hydrothermal vents. *Eos, Transactions of the American Geophysical Union*, **70**, 1161.

SOMERO, G. N., SIEBENALLER, J. F. & HOCHACHKA, P. W. 1983. Biochemical and physiological adaptations of deep-sea animals. *In*: ROWE, G. T. (ed.) *Deep-Sea Biology, The Sea*, Vol. 8. Wiley, New York, 261–330.

SOUSA, W. P. 1979. Disturbance in marine intertidal boulder fields: the non-equilibrium maintenance of species diversity. *Ecology*, **60**, 1225–1239.

STEIN, J. L. 1984. Subtidal gastropods consume sulphur-oxidizing bacteria: Evidence from coastal hydrothermal vents. *Science*, **223**, 696–698.

SUDARIKOV, S. M. & GALKIN, S. V. Geochemistry of the Snake Pit vent field and its implications for vent and non-vent fauna, this volume.

TARASOV, V. G., PROPP, M. V., PROPP, L. N., BLINOV, S. V. & KAMENEV, G. 1986. Shallow hydrothermal vents of Kraternaya Caldera and the ecosystem of Kraternaya Caldera, Kurile Islands. *Biologiya Morya*, **2**, 72–74. [in Russian, with English abstract].

TEMPLE, D. G., SCOTT, R. B. & RONA, P. A. 1979. Geology of a submarine hydrothermal field, Mid-Atlantic Ridge 26°N latitude. *Journal of Geophysical Research*, **84**, 7453–7466.

THOMPSON, G., HUMPHRIS, S. E., SCHROEDER, B., SULANOWSKA, M. & RONA, P. A. 1988. Hydrothermal mineralization on the Mid-Atlantic Ridge. *Canadian Mineralogist*, **26**, 697–711.

TIVEY, M. K. The influence of hydrothermal fluid composition and advection rates on vent deposit mineralogy and texture: insights from modelling and non-reactive transport. *Geochimica et Cosmochimica Acta*, in press.

TUNNICLIFFE, V. 1991. The biology of hydrothermal vents: ecology and evolution. *Oceanography and Marine Biology Annual Reviews*, **29**, 319–407.

——, JUNIPER, S. K. & DEBURGH, M. E. 1985. The hydrothermal vent community on Axial Seam-

ount, Juan de Fuca Ridge. *Bulletin of the Biological Society of Washington*, **6**, 453–464.

TUTTLE, J. H., WIRSEN, C. O. & JANNASCH, H. W. 1983. Microbial activities in emitted hydrothermal waters of the Galapagos Rift vents. *Marine Biology*, **73**, 293–299.

TYLER, P. A., PATERSON, G. J. L., SIBUET, M., GUILLE, A., MURTON, B. & SEGONZAC, M. A new genus of ophiuroid (Echinodermata: Ophiuroidea) from hydrothermal mounds along the Mid-Atlantic Ridge, *Journal of Marine Biological Association*, in press.

VAN DOVER, C. L. 1990. Biogeography of hydrothermal vent communities along seafloor spreading centers. *Trends in Ecology and Evolution*, **5**, 242–246.

—— & FRY, B. 1989. Stable isotopic compositions of hydrothermal vent organisms. *Marine Biology*, **102**, 257–263.

—— & —— 1994. Microorganisms as food resources at deep-sea hydrothermal vents. *Limnology and Oceanography*, **39**, 51–57.

—— & HESSLER, R. R. 1990. Spatial variation in faunal composition of hydrothermal vent communities on the East Pacific Rise and Galapagos Spreading Center. *In*: McMURRAY, G. R. (ed.) *Gorda Ridge: A seafloor spreading center in the United States Exclusive Economic Zone*. Springer Verlag, New York, 253–264.

——, CANN, J. R. & 8 others 1994. Light at deep sea hydrothermal vents. *Eos, Transactions of the American Geophysical Union*, **75**, 44–45.

——, DELANEY, J. R., SMITH, M. & CANN, J. R. 1988a. Light emission at deep-sea hydrothermal vents. *Eos, Transactions of the American Geophysical Union*, **69**, 1498.

——, DESBRUYÉRES, D., SALDANHA, L., FIALA-MÉDIONI, A., LANGMUIR, C. & THE LUCKY STRIKE TEAM. 1993. A new faunal province at the Lucky Strike Hydrothermal Field. *Eos, Transactions of the American Geophysical Union*, **74**, 100.

——, FRY, B., GRASSLE, J. F., HUMPHRIS, S. & RONA, P. A. 1988b. Feeding biology of the shrimp *Rimicaris exoculata* at hydrothermal vents on the Mid-Atlantic Ridge. *Marine Biology*, **98**, 209–216.

——, SZUTS, E. Z., CHAMBERLAIN, S. C. & CANN, J. R. 1989. A novel eye in 'eyeless' shrimp from hydrothermal vents of the Mid-Atlantic Ridge. *Nature*, **337**, 458–460.

VON COSEL, R., MÉTIVIER, B. & HASHIMOTO, J. 1994. Three new species of *Bathymodiolus* (Bivalvia: Mytilidae) from hydrothermal vents in the Lau Basin and the North Fiji Basin, Western Pacific, and the Snake Pit area, Mid-Atlantic Ridge. *Veliger*, **37**, 374–392.

VRIJENHOEK, R. C., SCHUTZ, S. J., GUSTAFSON, R. G. & LUTZ, R. A. 1994. Cryptic species of deep-sea clams (Mollusca: Bivalvia: Vesicomyidae) from hydrothermal vent and cold-water seep environments. *Deep-Sea Research*, **41**, 1171–1189.

WALKER, C. L. 1992. *The Volcanic History and Geochemical Evolution of the Hveragerdi Region, S.W. Iceland*. PhD Thesis, University of Durham, 365 pp.

WELHAN, J. A. & CRAIG, H. 1983. Methane, hydrogen and helium in hydrothermal fluids at 21°N on the East Pacific Rise. *In*: RONA, P., BÖSTROM, K., LAUBIER, L. & SMITH, K. (ed.) *Hydrothermal Processes at Seafloor Spreading Centers*. Plenum Press, New York, 391–409.

WILLIAMS, A. B. 1987. More records for shrimps of the genus *Rimicaris* (Decapoda: Caridea: Bresiliidae). *Journal of Crustacean Biology*, **7**, 105.

—— 1980. A new crab family from the vicinity of submarine thermal vents on the Galapagos Rift (Crustacea: Decapoda: Brachyura). *Proceedings of the Biological Society of Washington*, **93**, 443–472.

—— 1988. New marine decapod crustaceans from waters influenced by hydrothermal discharge, brine and hydrocarbon seepage. *Fisheries Bulletin*, **86**, 263–287.

—— & CHACE, F. A. 1982. A new caridean shrimp of the family Bresiliidae from thermal vents of the Galapagos Rift. *Journal of Crustacean Biology*, **2**, 136–147.

—— & DOBBS, F. C. A new genus and species of caridean shrimp (Bresiliidae) from hydrothermal vents on Loihi Seamount, Hawaii. *Proceedings of the Biological Society of Washington*, in press.

—— & RONA, P. A. 1986. Two new caridean shrimps (Bresiliidae) from a hydrothermal field on the Mid-Atlantic Ridge. *Journal of Crustacean Biology*, **2**, 136–147.

WILSON, C., KLINKHAMMER, G., SPEER, K. & CHARLOU, J.-L. 1993. Hydrothermal venting at the Lucky Strike segment (37 degrees 17′N) of the Mid-Atlantic Ridge. *Eos, Transactions of the American Geophysical Union*, **74**, 360.

WIRSEN, C. O., JANNASCH, H. W. & MOLYNEAUX, S. J. 1993. Chemosynthetic microbial activity at Mid-Atlantic Ridge hydrothermal vent sites. *Journal of Geophysical Research*, **98**(B6), 9693–9703.

——, Tuttle, J. H. & JANNASCH, H. W. 1986. Activities of sulfur-oxidizing bacteria at the 21°N East Pacific Rise vent site. *Marine Biology*, **92**, 449–456.

Composition and morphogenesis of the tubes of vestimentiferan worms

BRUCE SHILLITO[1], JEAN-PIERRE LECHAIRE[1], GÉRARD GOFFINET[2] & FRANCOISE GAILL[1]

[1]UPR CNRS 4601, URM IFREMER N° 7, Laboratoire de Biologie Marine, UPMC, 7 Quai St Bernard, Bat A, Paris, France

[2]Institut de Zoologie, Université de Liège, 22 Quai Van Beneden, B-4020 Liège, Belgium

Abstract: The structure and chitin–protein content of the tubes of different vestimentiferan species are compared. *Riftia pachyptila* and *Tevnia jerichonana* were collected at a hydrothermal vent site at 13°N on the East Pacific Rise; *Ridgeia piscesae* was from a hydrothermal vent site on the Juan de Fuca Ridge and *Lamellibrachia* sp. came from a cold seep on the Louisiana continental slope in the Gulf of Mexico. The tube ultrastructure shows chitin microfibrils organized in parallel bundles with various orientations. The chitin content varies considerably between species and is unrelated to seep or vent origin, whereas the protein content is much higher in the cold seep species than the hydrothermal vent species. All the species have specialized chitin secreting glands, as described for *R. pachyptila*. Based on this investigation, a tentative model of tube morphogenesis under cell control is proposed.

Vestimentiferans are large tubeworms characteristic of hydrothermal vent and cold seep animal communities (Gardiner and Jones 1993). The tubes protect the animals from the aggressive vent environment and from predators (Gaill & Hunt 1986; Tunnicliffe 1991; Childress & Fisher 1992). It has been proposed that the possession of a protective layer is a prerequisite for vent colonization (Tunnicliffe 1992). The tubes of vestimentiferans also act as external skeletal structures to maintain the worm's shape and the position of the plume relative to the hydrothermal vent fluid (Gaill 1993).

The tube composition has already been determined in two vent species from the 13°N site *Riftia pachyptila* (Gaill & Hunt 1986) *Tevnia jerichonana* (Gaill & Hunt 1991; Gaill *et al.* 1992*a*). In both cases the tube is made of a chitin–protein complex which differs from the chitinous structures of other animals. The high organic content (Gaill & Hunt 1986; Gaill *et al.* 1992*b*) indicates that an important percentage of the worm's metabolism is devoted to tube construction. Estimations of the annual chitin production of *Riftia* and *Tevnia* species have shown that production is about 1000 times higher at the 13°N site than in any other marine ecosystem (Gaill *et al.* 1992*b*). Furthermore, these worms exhibit a unique chitin secretion system (Gaill *et al.* 1992*c*) and data indicate that the chitin production efficiency originates in the degree of differentiation which occurs from the cellular to the anatomical level of the chitin production system (Shillito *et al.* 1993).

The length of the vestimentiferan tube may be up to 2 m in both the vent (Jones 1985; Tunnicliffe 1990; Gaill 1993) and cold seep areas (MacDonald *et al.* 1989). However, these similar lengths might result from different tube growth rates. Observations indicate that tube growth is fast in vent environments (Tunnicliffe *et al.* 1990; Tunnicliffe 1991) and very slow in cold seep habitats (Fisher pers. comm.). Growth of more than 1.2 m in a period of ten years or less was reported in the particular case of tubeworms discovered in a shipwreck (Dando *et al.* 1992).

To look at the specificity of the tube structure and composition related to environmental conditions, we analysed the tubes from two vestimentiferan species living in vent and non-vent areas. One is from the Juan de Fuca Ridge, a vent environment (Tunnicliffe *et al.* 1985) and the second is from the cold seep habitat of the Gulf of Mexico (MacDonald *et al.* 1989, 1990; Fisher 1990). In addition, for *Riftia pachyptila* we discuss tube morphogenesis on the basis of observations of the chitin-secreting system. Starting from the cellular process of microfibril manufacture we propose a model for microfibril packing which allows the transit of tube material before its formation outside the organism.

From PARSON, L. M., WALKER, C. L. & DIXON, D. R. (eds), 1995, *Hydrothermal Vents and Processes*, Geological Society Special Publication No. 87, 295–302.

Materials and methods

Biological material

Riftia pachytila and *Tevnia jerichonana* were collected during the Hero cruises at the 13°N site of the East Pacific Rise (2600 m depth) with the submersibles *Nautile* (Hero I 1991) and *Alvin* (Hero II 1992). *Ridgeia piscesae* samples were collected at 1500 m depth at the Juan de Fuca Ridge with the submersible *Pisces IV* in 1988 and provided by V. Tunnicliffe. *Lamellibrachia* sp. from the Louisiana slope of the Gulf of Mexico (700 m depth) were collected by the submersible *Johnson Sea-Link* during the JSL cruise logs (1992).

Microscopy

Tube fragments and animal tissues were fixed shortly after retrieval from the submersible containers with a 2.5% or 3% glutaraldehyde – sodium cacodylate buffered solution and later post-fixed in osmium tetroxide. For the light micrographs, the chitin-secreting glands were not post-fixed and were observed with a Nikon Optiphot. For transmission electron microscopy, samples were finally dehydrated in ethanol and propylene oxide solutions and embedded in Araldite resin. Thin sections were obtained from a Reichert-Jung ultramicrotome and were laid on 200 mesh copper grids. They were stained with uranyl acetate and lead citrate. Observations were carried out on a Philips 201 TEM or a Zeiss EM 910, both operating at 80 kV.

Chitin–protein content determination

Pieces of air-dried tube were carefully washed three times in distilled water, then dried *in vacuo* over NaOH and weighed. After treatment for four hours at 20°C with 0.5 M HCl solution and washing in distilled water, they were dried and subsequently subjected to two successive three-hour treatments with 0.5 M NaOH at 100°C. Proteins were assayed in NaOH extracts by the Lowry method (Lowry *et al.* 1951). Chitin was enzymatically assayed in residues after HCl and NaOH extractions according to the method of Jeuniaux (Jeuniaux 1963).

Results and discussion

Tube composition

The chitin–protein content was determined for different places along the tube length for at least three tubes of each species. For each individual,

Table 1. *Mean values of chitin and protein contents as percentage of dry weight of vestimentiferan tubes from hydrothermal vents and cold seeps*

Species	Chitin	Protein
*Riftia pachyptila**	24	37
*Tevnia jerichonana**	45	36
Ridgeia piscesae†	32	45
Lamellibrachia sp‡	28	75

* 13°N East Pacific Rise (hydrothermal vent).
† Juan de Fuca (hydrothermal vent).
‡ Gulf of Mexico (cold seep).

the deviation from values in Table 1 did not exceed 10%.

It is known that chitin is always associated with proteins, thus forming a chitin–protein complex (Neville 1975, 1993). The protein content shown in Table 1 corresponds to the proteins which are more specifically involved in this complex. The chitin–protein complex makes up at least 60% of the dry weight of the *R. pachyptila* tube, about 80% for the two vent species *R. piscesae* and *T. jerichonana* tubes and the whole of the tube for the cold seep species. The chitin–protein content did not show significant variations along the tube length. In the tube of *Lamellibrachia* sp., the whole protein content seems to be associated with the chitin microfibrils. This could possibly be related to its wood-like consistency, which is unlike the tubes of the vent species. The tube here is a very resistant structure which may be related to the longer lifetime, and also the slower tube growth rate than in the vent species (Fisher pers. comm.).

Table 1 also shows that the lowest chitin content is found in *R. pachyptila* and the highest in *T. jerichonana*. *Lamellibrachia* sp. and *R. piscesae* have a similar chitin content, which is close to that of pogonophoran species (Foucart *et al.* 1965). The vent species from the 13°N site exhibit the lowest protein content and the highest content is observed in the tube of cold seep *Lamellibrachia* sp. These different tubes are non-calcified and contain a low inorganic content. In the case of *R. pachyptila*, the inorganic content is low (3% ash) and the neutral hexose content of the tube material is about 3.5% (Gaill & Hunt 1986). The unidentified remainder most probably consists of sugar, glycoproteins and additional proteins, such as HCl-soluble proteins, which are not linked to the chitin system.

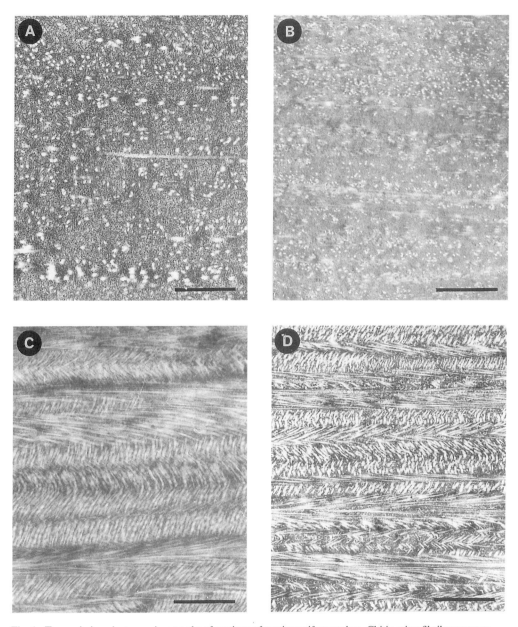

Fig. 1. Transmission electron micrographs of sections of vestimentiferan tubes. Chitin microfibrils appear as clear spots when cut transversally and as clear strips when cut obliquely or longitudinally. (**A**) *Riftia pachyptila* transverse section, parallel to the tube axis. (**B**) *Lamellibrachia* sp. transverse section, parallel to the tube axis. (**C**) *Tevnia jerichonana*, oblique section. (**D**) *Ridgeia piscesae*, oblique section. All scale bars 1 mm.

Tube ultrastructure

The tubes are always colonized at their outer surface by filamentous bacteria (not shown). The inner tube surface is always free of micro-organisms. The tube wall is made of a fibrillar system embedded in an amorphous matrix (Fig. 1). These fibrils, called microfibrils, appear as electron-lucent structures. Microfibrils are elliptical-shaped in transverse section. Their diameter is about 50 nm, with a length of several micrometres. Complementary studies have shown that each microfibril is a chitin crystallite (Gaill *et al.* 1992c). Proteins are located in the electron-dense matrix surrounding the crystallites.

In the different tube species, chitin chains assemble in microfibrils of similar diameter which adopt similar plywood configurations. From Fig. 1 it can be seen that the microfibrils are packed in successive parallel layers. Their orientation is parallel in one layer, but varies from one layer to the next, without any significant periodicity (Gaill *et al.* 1992a). The supramolecular organization of this chitin differs from that in crustacea, for example, where most of the chitin microfibrils are organized in a regular twisted plywood system, with long-range distance periodicity (Neville 1965; Bouligand 1971).

Tube morphogenesis

Data have shown that vestimentiferans have a unique system of chitin secretion. Such a system was first described in *T. jerichonana* (Gaill *et al.* 1992c) and more recently in *R. pachyptila* (Shillito *et al.* 1993). Dissections of *Riftia* and *Tevnia* samples reveal the existence of chitin-secreting glands scattered over the body wall surface, apart from the plume. Similar glands are also observed in *R. piscesae* and *Lamellibrachia* sp. (not shown). The exocrine glands of *R. pachyptila* are the largest. This fact may be related to the large size and high growth rate of the animal (Fisher 1990; Childress & Fisher 1992). In relation to the tube structure, it seems that each parallel bundle of microfibrils is produced by one gland. In all the species, except *R. pachyptila*, the animal inhabits the whole length of the tube, its hind end being located at the base of the tube (Gardiner & Jones 1993). By moving its body inside the tube, *R. pachyptila* can produce new material at both top and base ends. The magnitude of the structures which are involved in the tube production of *R. pachyptila* allows a more precise analysis of the steps of tube morphogenesis.

Microfibril synthesis. In *Riftia pachyptila*, the lumen of the glands is surrounded by subluminal cavities in the gland wall, composed of epithelial cells. The presence of chitin microfibrils was demonstrated in the lumen and sublumina (Gaill *et al.* 1992c). Each sublumen seems to correspond to a single cell. At the cell–sublumen interface, cup-shaped microvilli, 350 nm wide for about 800 nm long, occur in densities of about 4 mm^2. They are covered by a dense cell coat which in some cases appears to be filamentous (Fig. 2C). In some of the observed sublumina, the microvilli harbour a microfibril in their central cavity (Fig. 2D). Our previous findings support the hypothesis that these cups are the site of an end-synthesis process for chitin microfibrils (Shillito *et al.* 1993). Similar glandular structures, comprising subluminal cavities and associated cup-shaped microvilli, were also reported for perviate pogonophora (Southward 1984) and it was suggested that these microvilli secreted chitin microfibrils (Southward 1993).

Origin of ultrastructural organization: comparison with other systems. The origin of the other tube components remains unknown. Our observations show that the tube is composed of parallel bundles of microfibrils embedded in a dense matrix, which most probably acts as a cement by maintaining the microfibrils together. The similarity of supramolecular order in the tube and in the glands indicates that the final locally unidirectional structure of the tube is pre-determined in the glands themselves (Fig. 2B and 2C). Moreover, the nascent microfibril tips anchored in the cup structures of a particular area also adopt a common orientation, suggesting that microfibril parallelism is under cell control (Fig. 2D). An electron-dense filamentous network runs in between the microfibrils and seems to be connected to them. This network is partly composed of proteins, as indicated by the intense staining of thin sections. Additional proteoglycans or glycosaminoaglycans may participate to this network. Figure 2A shows that the chitinous product is a single coherent ribbon.

Fig. 2. *Riftia pachyptila*'s chitin secreting system. (**A**) Light micrograph of a chitin-secreting gland. During its transit through the gland neck (N), the chitinous product (P) appears as a rigid, translucid ribbon. (**B**) Electron micrograph of chitin microfibril transverse sections (clear spots) in the gland lumen. Electron-dense filaments run through the lumen and seem to interconnect microfibril sections (arrows). (**C**) Electron micrograph of the cell/sublumen interface. The cup-shaped microvilli (C) are covered by an electron-dense filamentous layer (L). Further away from the cell, the sublumen (SL) displays a stripy aspect due to longitudinal sections of microfibrils. (**D**) Electron micrograph of the cell/sublumen interface. Four sections of cup-shaped microvilli (C) harbour chitin microfibrils (MF). The four microfibril sectons have the same transversal orientation. Scale bars: (**A**) 200 mm; (**B**) 200 nm; (**C**) 400 nm; and (**D**) 300 nm.

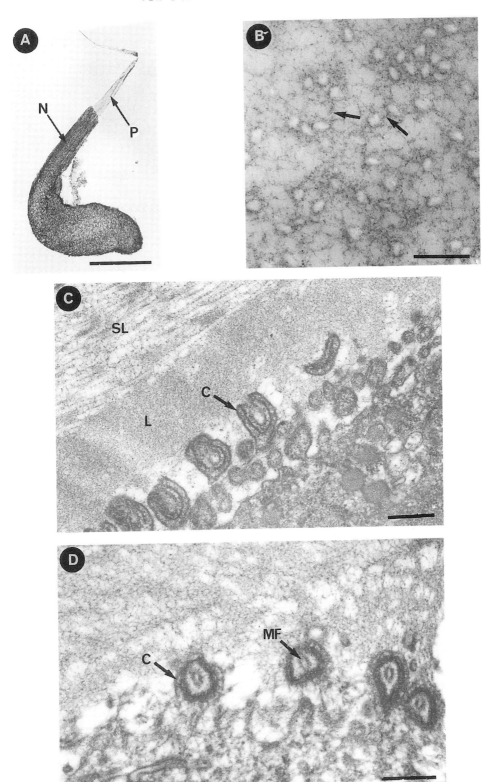

Such a system seems to differ from what is believed to occur in the chitinous arthropod cuticle. Firstly, the organization of chitin microfibrils is a twisted plywood organization, with long-range periodicity (Neville 1965; Bouligand 1971), as opposed to local parallelism in vestimentiferan tubes. Secondly, this organization is thought to result from a self-assembly process, similar to that observed in liquid crystalline phases (Neville 1975, 1993). Indeed, the aptitude of chitin and cellulose microfibrils to self-assemble in ordered phases has been demonstrated *in vitro* (Marchessault *et al.* 1959; Revol and Marchessault 1993).

Our observations of unidirectional order in the early stages of microfibril synthesis (Fig. 2D) recalls that of the secondary cell wall in green algae, in which the locally parallel arrangement of cellulose microfibrils is dependent on the organization of the membranous synthesizing complexes (Giddings & Staehelin 1988). These terminal complexes are thought to be linked to the underlying cytoskeleton, which therefore acts as a template for cellulose microfibrils in the morphogenesis of the cell wall. Still, when considering spatial constraints, the task of generating microfibril parallelism seems to be different in the previous case of algae cells. Indeed, to obtain the locally unidirectional ultrastructure of the secondary cell wall in algae, the cell produces microfibrils parallel to the plasma membrane and the pre-existing primary cell wall. Control of microfibril parallelism occurs within the two dimensions of the cell surface, whereas *R. pachyptila*'s secreting cell would have to control the parallelism of nascent microfibrils that are directed towards a large three-dimensional space, the sublumen.

Model for microfibril orientation

One simple possibility to explain the parallelism of microfibrils could be that the filamentous cross-linking between chitin microfibrils is achieved as soon as these microfibrils are synthesized. Indeed, Fig. 3 suggests how nascent microfibrils originating from neighbouring cup-structures can gradually converge towards parallelism, provided that they are linked mechanically and that the cup structures have a flexible base. When a chitin microfibril is synthesized, it seems that it has to go through the dense layer that covers the area where cup structures occur (Fig. 2C). This layer may contain densely packed polymeric proteins with an affinity for chitin. By going through this layer, at a speed corresponding to that of the end-synthesis process, all the nascent microfibrils would

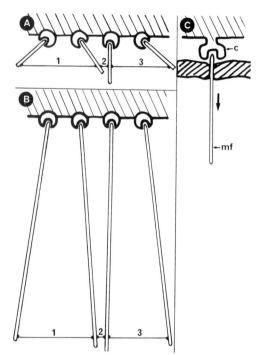

Fig. 3. (A) Schematic drawing representing four cup structures at the surface of a secreting cell (light-hatched domain), each synthesizing a microfibril in a random direction. Segments 1, 2 and 3 show material linkages established between the specific anchorage points (black dots) of adjacent microfibrils, at a given time, shortly after the beginning of synthesis. **(B)** As the end-synthesis process continues, nascent microfibrils elongate. If segments 1, 2 and 3 bear enough mechanical resistance, i.e. do not stretch, then the microfibril orientations, will gradually converge towards parallelism, provided that the cup structures can rotate freely. **(C)** Schematic representation of a cup structure (c) synthesizing a microfibril (mf). During the process the microfibril passes through a dense filamentous layer (bold-hatched domain).

simultaneously connect to these proteins and carry them away in the form of a filamentous network.

In this model, the cells can generate microfibril parallelism simply by synthesizing in the same area (the apical region where the cups occur) both these putative chitin-binding polymeric proteins and the chitin microfibrils. Thus, synthesis and supramolecular organization could be interrelated.

Conclusions

In conclusion, the structure of the tube is broadly similar in vestimentiferans from different environments. The vestimentiferan tube is always made of large chitin microfibrils embedded in a proteinaceous matrix. Variations are observed in the chitin–protein complex which constitutes the tube. This complex represents the whole tube material of the cold seep *Lamellibrachia* sp., whereas additional components are present in the tube material of the vent species. The chitin content is not closely controlled by the environment as the lowest and the highest values are observed in the vent species from the 13°N site. The percentage of proteins which are associated with chitin is higher in cold seep areas.

All the species seem to have the same system of tube secretion. This system is made of chitin-secreting glands which are similar in the different species, but most details have been observed in *R. pachyptila*. Analysis of the *R. pachyptila* system indicates that the locally parallel microfibril organization is cell-controlled and occurs in the early stages of microfibril synthesis. This, together with the observation of filamentous cross-links between microfibrils, leads to a model in which polymeric proteinaceous molecules, capable of binding to chitin, are associated to the nascent microfibrils in the proximity of the plasma membrane. With respect to the morphogenesis of chitinous extracellular matrix in general, such a model would distinguish vestimentiferans, and perhaps pogonophorans, from arthropods, for which available models suggest self-assembly processes (Neville 1993).

We thank V. Tunnicliffe for providing the vent material from Juan de Fuca and C. R. Fisher and I. R. MacDonald for the cold seep material from the Gulf of Mexico. We also thank J. J. Childress, D. Desbruyères and H. Felbeck (Hero cruises) for the material collected from 13°N EPR. Finally, we thank the C.I.M.E. Jussieu, where microscopy work was carried out, and D. Touret, F. Devienne and V. Hosansky for photographic assistance.

References

BOULIGAND, Y. 1971. Les orientations fibrillaires dans le squelette des Arthropodes. *Journal de Microscopie (Paris)*, **11**, 441–472.

CHILDRESS, J. J. & FISHER, C. R. 1992. The biology of hydrothermal vent animals: physiology, biochemistry and autotrophic symbioses. *Marine Biology and Oceanography: An Annual Review*, **30**, 337–441.

DANDO, P. R., SOUTHWARD, A. J., SOUTHWARD, E. C., DIXON, D. R., CRAWFORD, A. & CRAWFORD, M. 1992. Shipwrecked tube worms. *Nature*, **356**, 667.

FISHER, C. R. 1990. Chemoautotrophic and methanotrophic symbioses in marine invertebrates. *Review in Aquatic Sciences*, **2**, 399–436.

FOUCART, M. F., BRICTEAUX-GREGOIRE, S. & JEUNIAUX, C. 1965. Composition chimique du tube d'un pogonophore (*Siboglinum sp*) et des formations squelettiques de deux pterobranches. *Sarsia*, **20**, 35.

GAILL, F. 1993. Aspects of life development at deep sea hydrothermal vents. *The FASEB Journal*, **7**, 558–565.

—— & HUNT, S. 1986. Tubes of deep-sea hydrothermal vent worms *Riftia pachyptila* (Vestimentifera) and *Alvinella pompejana* (Annelida). *Marine Ecology Progress Series*, **3**, 267–274.

—— & —— 1991. The biology of Annelid worms from high temperature hydrothermal vent regions. *Reviews in Aquatic Sciences*, **4**, 107–137.

——, PERSSON, J., SUGIYAMA, J., VUONG, R. & CHANZY, H. 1992a. The chitin system in the tubes of deep-sea hydrothermal vent worms. *Journal of Structural Biology*, **109**, 116–128.

——, SHILLITO, B., LECHAIRE, J. P., CHANZY, H. & GOFFINET, G. 1992c. The chitin secreting system from deep-sea hydrothermal vent worms. *Biology of the Cell*, **76**, 201–204.

——, VOSS-FOUCART, M. F., GERDAY, C., COMPERE, P. & GOFFINET, G. 1992b. Chitin and protein content in the tubes of vestimentiferans from hydrothermal vents. *In*: BRINE, C. J., STANDFORD, P. A. & ZIZAKIS, J. P. (eds) *Advances in Chitin and Chitosan*. Elsevier Applied Science, Barking, 232–236.

GARDINER, S. L. & JONES, M. L. 1993. Vestimentifera. *In*: HARRISON, F. W. & RICE, M. E. (eds) *Microscopic Anatomy of Invertebrates*. Vol. 12. Wiley-Liss, New York, 371–460.

GIDDINGS, T. H. & STAEHELIN, L. A. 1988. Spatial relationship between microtubules and plasmamembrane rosettes during the deposition of primary wall microfibrils in *Closterium sp*. *Planta (Berlin)*, **173**, 22–30.

JEUNIAUX, CH. 1963. *Chitine et chitinolyse*. Masson, Paris, 181 pp.

JONES, M. L. 1985. On the Vestimentifera, new phylum: six new species, and other taxa, from hydrothermal vents and elsewhere. *In*: JONES, M. L. (ed.) *The Hydrothermal Vents of the East Pacific: an Overview. Bulletin of the Biological Society of Washington*, **6**, 117–158.

LOWRY, O. H., ROSEBROUGH, N. J., FARR, A. L. & RANDALL, R. J. 1951. Protein measurement with the phenol reagent. *Journal of Biological Chemistry*, **193**, 265–275.

MACDONALD, I. R., BOLAND, G. S., BAKER, J. S., BROOKS, J. M., KENNICUT II, M. C. & BIDIGARE, R. R. 1989. Gulf of Mexico hydrocarbon seep communities II. Spatial distribution of seep organisms and hydrocarbon at Bush Hill. *Marine Biology*, **101**, 235–247.

——, GUINASSO, N. L., REILLY, J. F., BROOKS, J. M., CALLENDER, W. R. & GABRIELLE, S. G. 1990. Gulf of Mexico hydrocarbon-seep communities VI. Patterns in community structure and habitat. *Geo-Marine Letters*, **10**, 244–252.

MARCHESSAULT, R. H., MOREHEAD, F. F. & WALTER,

N. M. 1959. Liquid crystal systems from fibrillar polysaccharides. *Nature*, **184**, 632–633.

NEVILLE, A. C. 1975. *Biology of the Arthropod Cuticle*. Springer-Verlag, Berlin, Heidelberg, New York, 448 pp.

—— 1993. *Biology of Fibrous Composites. Development Beyond the Cell Membrane*. Cambridge University Press, Cambridge, 214 pp.

REVOL, J. F. & MARCHESSAULT, R. H. 1993. *In vitro* chiral nematic ordering of chitin crystallites. *International Journal of Biological Macromolecules*, **15**, 329–335.

SHILLITO, B., LECHAIRE, J. P. & GAILL, F. 1993. Microvilli-like structure secreting chitin crystallites. *Journal of Structural Biology*, **111**, 59–67.

SOUTHWARD, E. C. 1984. Pogonothora. *In*: BEREITER-HAHN, J., MATOLTSY, A. G. & RICHARDS, K. S. (eds) *Biology of the Integument: Invertebrates*, Vol. I. Springer-Verlag, Berlin, Heidelberg, New York, Toronto, 376–388.

—— 1993. Pogonophora. *In*: HARRISON, F. W. & RICE, M. E. (eds) *Microscopic Anatomy of Invertebrates*, Vol. 12. Wiley-Liss, New York, 327–369.

TUNNICLIFFE, V. 1991. The biology of hydrothermal vents: ecology and evolution. *Marine Biology and Oceanography: An Annual Review*, **29**, 319–407.

—— 1992. The nature and origin of the modern hydrothermal vent fauna. *Palaios*, **7**, 338–350.

——, GARRETT, J. F. & JOHNSON, H. P. 1990. Physical and biological factors affecting the behaviour and mortality of hydrothermal vent tubeworms (vestimentiferans). *Deep-Sea Research*, **37**, 103–125.

——, JUNIPER, S. K. & DE BURGH, M. E. 1985. The hydrothermal vent community on Axial Seamount, Juan de Fuca Ridge. *In*: JONES, M. (ed.) *The hydrothermal vents of the eastern Pacific Ocean: an Overview. Bulletin of the Biological Society of Washington*, **6**, 453–464.

Preliminary observations on biological communities at shallow hydrothermal vents in the Aegean Sea

P. R. DANDO[1,4], J. A. HUGHES[2] & F. THIERMANN[3]

[1]*Marine Biological Association of the United Kingdom, Citadel Hill, Plymouth PL1 2PB, UK*

[2]*Institute of Marine Biology of Crete, PO Box 2214, 71003 Iraklio, Crete, Greece*

[3]*Zoological institute and Zoological Museum, University of Hamburg, Martin-Luther-King Platz 3, D–20146 Hamburg, Germany*

[4]*Present address: School of Ocean Sciences, University of Wales, Bangor, Menai Bridge, Gwynedd LL59 5EY, UK*

Abstract: Hydrothermal vents, at depths varying from the littoral zone to water depths of 115 m, have been explored around the islands of Milos and Santorini in the Hellenic Volcanic Arc. The biota were surveyed by scuba divers and with a remote operated vehicle as well as by soft-bottom sampling with grabs and corers. No species specific to hydrothermal areas were found at any of the sites investigated. Animals in the immediate vicinity of the vents belonged to opportunistic species, such as the polychaetes *Capitella capitata*, *Microspio* sp. and *Spio decoratus*. The nassariid gastropod *Cyclope neritea* was the dominant macrofaunal species found in the bacterial mat areas at Milos which overlay hot brine seeps. At all rocky hydrothermal sites deeper than 35 m water depth the echiuran *Boniellia* cf. *viridis* was observed. Around the periphery of the seeps the meiofaunal community was dominated by the nematode *Onchlaimus camplyoceroides*. Few nematodes were found in the centre of the hydrothermal brine areas. At Santorini the bacteria at the venting sites were dominated by iron bacteria, whereas at Milos large globular sulphur bacteria, *Achromatium volutans*, covered the hydrothermal brine seeps. Unlike deep-sea vents, little of the biomass production was due to symbiotic associations between animals and chemoautolithotrophic bacteria, although a few stilbonematine nematodes were found at the periphery of the brine seeps at Milos. Macrofaunal biomass reached a maximum of $80 \, \mathrm{g \, m^{-2}}$ around the vents, compared with $\geqslant c \, 500 \, \mathrm{g \, m^{-2}}$ at hydrothermal vents elsewhere. Much of the bacterial biomass appeared to be exported from the sites.

Shallow water hydrothermal vents have been described from White Point, California (Stein 1984; Jacq *et al.* 1987; Trager & DeNiro 1990), Baja California (Vidal *et al.* 1978), New Britain (Ferguson & Lambert 1972; Sorokin 1991), the Bay of Plenty, New Zealand (Sarano *et al.* 1989; Sorokin 1991; Kamenev *et al.* 1993), the Kurile Islands (Tarasov *et al.* 1990), southern Japan (Naganuma 1991; Hashimoto *et al.* 1993) and off Iceland at Kolbeinsey, to the north (Fricke *et al.* 1989), and to the south at Steinahóll, south of the Reykjanes Peninsular (Ólafsson *et al.* 1991). One of the physicochemical differences between shallow water hydrothermal venting and that in the deep sea is the frequent presence of streams of gas bubbles issuing from the shallow vents. This has been described by Tarasov et al. (1990) as gasohydrothermal activity.

In the Aegean Sea venting of hot water has been described in the Caldera of Santorini (Holm 1987; Varnavas *et al.* 1990) and off the coast of Kos and Yali (Varnavas & Cronan 1991). Venting of gas bubbles, containing mainly carbon dioxide, has been described from the submerged caldera at Santorini (Holm 1987) and from several areas around Milos (Dando *et al.* in press). Gas bubbles have been noted escaping from the seafloor in the hydrothermal area of Volcania on Kos (Bardintzeff *et al.* 1989). Hydrothermal waters on the Greek mainland in the coastal zone, as well as on the islands, are characterized by high NaCl contents with the Na to Cl ratio being identical or close to that in seawater (Dominco & Papastamatoki 1975; Bardintzeff *et al.* 1989; Minissale *et al.* 1989). This indicates that seawater has entered the reservoirs and suggests that submarine hydrothermal venting is likely to be found in all the coastal geothermal areas identified.

Studies on hydrothermal activity in the Aegean Sea have largely been confined to metal fluxes and sedimentary deposits from submarine

From PARSON, L. M., WALKER, C. L. & DIXON, D. R. (eds), 1995, *Hydrothermal Vents and Processes*, Geological Society Special Publication No. 87, 303–317.

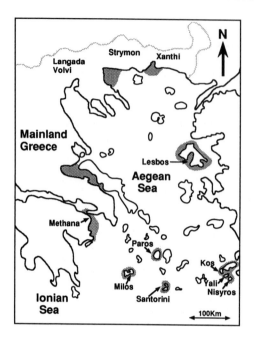

Fig. 1. Areas of the Aegean believed to have submarine hydrothermal vents.

activity (Varnavas 1989; Varnavas *et al.* 1990; Varnavas & Cronan 1991). The present paper presents a preliminary study on the biological communities at hydrothermal sites in the Aegean Sea.

Geological setting

The Aegean is a tectonically complex area of continental crust. The northern part, bounded by the Vardar and North Anatolian faults, lies on the Eurasian plate with the southern part obducting over the African plate (De Boer 1989). A mantle diapir underlies most of the Aegean (Makris 1977) and asthenospheric material is believed to flow over the subducted slab of the African plate causing uplift (Angelier *et al.* 1982). Geothermal areas are found in the coastal regions and islands around the Aegean: the regions around the Antemus, Volvi-Langada, Strymon, Nestos-Xanthi and Alexandroupoli basins in the north, Lesbos in the east, the Gulf of Maliachos and Sperkios basins in the west and Sousaki, Methana, Milos, Santorini, Kos, Yali and Nisiros in the volcanic arc to the south, (Fig. 1) (Dominico & Papastamatoki 1975; Fytikas *et al.* 1989; Minissale *et al.* 1989).

The Hellenic volcanic arc stretches from the Turkish coast, east of Kos and Nisiros, to Thebes in the northwest on the Greek mainland. Earthquake epicentres decrease in depth with distance northwards from the trenches south of Crete and the width of the arc may be defined by the zone where the epicentre depth is between 120 and 180 km (Ninkovitch & Hays 1972). Volcanic activity in the Hellenic volcanic arc started during the middle to late Pliocene as a consequence of the subduction of the African plate beneath the Aegean microplate (Fytikas *et al.* 1984). The Milos archipelago was volcanically active from 3.5 to 0.08 Ma (Fytikas *et al.* 1976) and Santorini was active from 1.6 Ma until the present (Ferrara *et al.* 1980).

Sampling procedures

At vent sites down to 10 m water depth observations and sample collections were routinely made by scuba divers. Sediment samples were collected in acrylic core tubes, 9.4 cm i.d., to a depth of 20 cm. In Paleohori Bay, Milos, along the 10 m water depth contour, samples were additionally collected using a Smith–McIntyre grab, which sampled an area of 0.1 m². Sediment samples collected for macrofauna were sieved on a 500 μm square mesh and fixed in 10% buffered formalin solution. Core samples for meiofauna studies were collected in 5.0 cm i.d. acrylic tubes. The sediment was extruded and fixed in 5–10% *v/v* neutralized formalin. Meiofauna were extracted from the fixed samples by elutriation (Uhlig *et al.* 1973).

Temperatures were measured in collected sediment samples using a thermocouple probe and *in situ* using sealed 10–50 cm long probe thermometers. Hydrogen sulphide concentrations in interstitial or vent water samples were determined by the methods of Cline (1969) or Newton *et al.* (1981). Sodium concentrations in vent and interstitial water were measured on-site using a portable sodium ion meter (Horiba), ammonia was measured by the method of Dal Pont *et al.* (1974), manganese was determined by the method of Grasshof *et al.* (1983), silicate using the method of Parsons *et al.* (1984) and sulphate by ion chromatography (Dando *et al.* 1991). Sulphate reduction rates were measured by a radiotracer method (Dando *et al.* 1994) with incubation for six hours at the temperature of the sediment.

Vents below 12 m depth were investigated using a Benthos MiniRover remote operated vehicle (ROV), equipped with a video camera, deployed from the RV *Philia*.

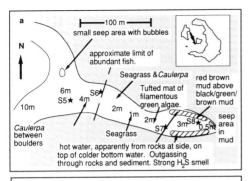

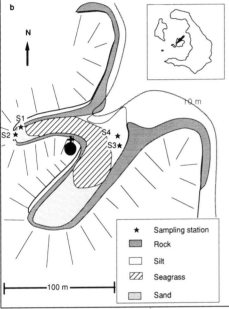

Fig. 2. Stations sampled (★) in bays of the channel between the islands in the centre of the Santorini caldera: (**a**) Nea Kameni; (**b**) Palea Kameni. The fauna obtained in core samples from these stations are listed in Table 1.

Biological investigations at sites in the Aegean

Santorini

The island of Santorini is probably the best known site of volcanic activity in the Aegean, with the principal hydrothermal areas situated around the small volcanic islands of Palea Kameni and Nea Kameni, within the ancient caldera (Holm, 1987; Varnavas *et al.* 1990). Much of the recent activity is confined to the channel between the islands and to the adjacent embayments. Fluxes of metals from the vents

are partly controlled by changes in sea level, with falling water levels resulting in increased venting (Varnavas *et al.* 1990). Extensive reddish brown gel-like sediments are found in the area: these almost entirely consist of iron oxide–hydroxide deposits attached to the stalks of the iron bacterium *Gallionella ferruginea* (Holm 1987).

Sediment samples were collected by scuba divers from areas within the caldera, between the Kameni islands, inside the 10 m depth contour, during May 1991 (Fig. 2). The species composition of the samples, which consisted almost entirely of polychaetes and oligochaetes, is listed in Table 1. Opportunist polychaetes, the spionid *Malacocerous fuliginosus* and the capitellid *Capitella capitata*, occurred at all sites with hydrothermal flows in the inlets; at station S2 the densities reached, respectively, 10 526 and 55 661 individuals m^{-2}. At the outer margins of the bay on Nea Kameni at station S5 the polychaetes *Amphiglena mediterranea* and *Ceratonereis hironicola* were also present in high densities, together with *M. fulginosus* and *C. capitata*.

Further samples were collected by Smith–McIntyre grab. In the main part of the caldera, outside the channel between the islands, there was a reasonably rich and diverse infauna. Sipunculans accounted for up to 70% of the biomass at several stations. The highest biomass found in any sample was only 56 g m^{-2} (S5).

Milos

Milos island is an area which has been well investigated in terms of geothermal sites (Fytikas *et al.* 1989). A geothermal power station was opened in 1976, but subsequently closed because of crop failures and health problems resulting from hydrogen sulphide in the discharged hydrothermal effluent (Kousis 1993). The regions of submarine hydrothermal activity around Milos cover an area of approximately 34 km^2 (Fig. 3), from the central caldera of Milos Bay on the north coast, Boudia Bay on the east coast and along the southern coast to Zefiros Point in the west (Dando *et al.* in press). All these areas show extensive gas plumes as well as venting of hydrothermal fluids without exsolved gas. The flow-rates and composition of the free gas collected from these hydrothermal areas has been described previously (Dando *et al.* in press). Gas samples contained, by volume, 71–92% CO$_2$, with up to 9.7% CH$_4$, 8.1% H$_2$S and 3% H$_2$. Vented water was enriched in ammonia, phosphate, silicate and manganese (Dando *et al.* in press). The known sites of

Table 1. *Species identified in eight core samples (94 mm i.d.) taken from the two bays off the central channel between Nea Kameni and Palea Kameni, Santorini, July 1991*

	Numbers of individuals in core							
	S1	S2	S3	S4	S5	S6	S7	S8
Polychaeta								
Aricidea catherinae	–	–	1	–	–	1	–	–
Aricidea suecica	–	–	–	–	–	–	13	–
Cirrophorus lyra	–	–	–	–	2	–	–	–
Malacoceros fuliginosus	54	73	–	–	94	1	3	2
Minuspio cirrifera	–	–	3	1	–	–	–	–
Aphelochaeta sp. 1	–	–	–	–	–	–	4	–
Capitella capitata	72	386	–	1	131	3	2	13
Notomastus latericeus	–	–	2	–	–	–	–	–
Pseudoleiocapitella sp.	–	–	1	–	–	–	–	–
Pseudocapitella sp.	–	–	–	–	–	–	1	–
Neomediomastus glabrus	–	–	5	2	–	–	–	–
Mediomastus fragilis	–	–	–	–	–	–	1	–
Syllides benedicti	2	6	1	–	–	–	–	–
Exogone gemmifera	–	1	3	–	–	–	–	–
Perinereis cultrifera	–	–	–	–	1	–	–	–
Ceratonereis hironicola	–	–	–	–	60	–	–	–
Nereis diversicolor	–	–	–	–	–	–	4	–
Nereidae sp. 1	2	18	–	–	–	–	–	–
Nereidae sp. 2	–	–	1	–	–	–	–	–
Lumbrineris funchalensis	–	–	–	–	–	–	5	–
Polycirrus denticulatus	–	–	1	4	–	–	–	–
Armandia polyophthalmus	–	–	–	1	–	–	–	–
Amphiglena mediterranea	–	–	–	–	127	–	–	–
Oligochaeta								
Oligochaete sp. 1	–	–	–	–	–	–	1	–
Echinodermata								
Ophiuridea sp. juvenile	–	–	–	1	–	–	–	–
Total number of individuals	130	484	18	10	415	5	34	15

venting cover a range of depths extending from the shoreline to 115 m, although chimney-like structures with possible gas plumes have been located in water depths of up to 350 m. Observations on the main areas of activity are described in the following sections.

Milos Bay. Much of the bay is geothermally active. Areas of gas venting in water depths of 60 m were investigated in June 1992 using the ROV. The sea bed was comprised of a fine-grained mud with areas in which animal burrows were frequently observed. Gas bubble streams issued from small craters, which were often surrounded by white bacterial mats. Similar mats were observed above areas of diffuse water flow through the sediment. Mounds and chimneys of hydrothermal deposits, up to 1.5 m high, were observed. Areas of cemented pebble pavement occurred on some of the mounds. Filamentous bacterial mats were attached to rocks surrounding the vent outlets. Black echiurins, *Bonellia* cf. *viridis*, were very common under rocks near vents; their probosces extended 1–2 m away from the rocks. These may be feeding on the bacterial mats. Epifauna appeared sparse and consisted of sponges, soft corals (*Alcyonacea*), hydrozoans and serpulid polychaetes. Occasional echinoids were also observed.

Boudia Bay. A region of active outgassing in 20–30 m water depth was studied using the ROV. The bottom was comprised of fine mud with animal burrows. Slow-flowing gas vents issued from black-centred depressions surrounded by bacterial mats. Very little epifauna was seen, with the exception of a starfish, *Astropecten* sp., lying on top of one of the gas outlets. Occasional large bivalves, *Pinna nobilis*, were seen. Fish noted in the venting areas were red mullet, *Mullus surmuletus*, and rays, *Raja* spp. The common octopus, *Octopus vulgaris*, was also present.

West of Spathi Point. Active gas venting was observed on the echosounders in water depths of

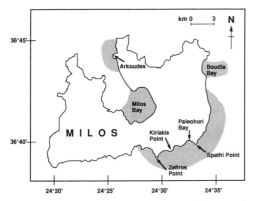

Fig. 3. Known areas of hydrothermal activity around Milos.

80–115 m. These areas were investigated using the ROV. The bottom consisted of soft mud with mounds, fissures and hydrothermal deposits, including small chimneys, in the areas of fluid venting. A rocky reef ran east – west across the area. Echiurans, *B.* cf. *viridis*, were common at the edges of the rocky patches. Large areas of the sediment were covered by white mats. These often had black sediment underneath, suggesting venting of sulphide-rich fluids. Some of the mats were covering conical mounds. In some areas reddish brown, apparently iron-rich deposits were observed surrounding the white mats of approximately 1 m diameter.

The rocks had a sparse epifauna which included sponges, soft corals, anemones, hydroids, *Aplysia* sp. and amphinomid polychaetes. Fish around the reef area were few, but included serranids and the red-band fish, *Cepola rubescens*.

Infauna from the soft iron-rich sediments collected around the vents have not been fully identified, but contained a diverse community with polychaete worms, in particular *Hyalinoecea bilineata*, dominating.

Paleohori Bay. This area of Milos was chosen for detailed studies. Hydrothermal areas in Paleohori Bay were studied in June 1991, March and June 1992, July and October 1993 and March–April 1994. Close to the shore on the western side of the bay is a 1–2 m high rocky reef, the base of which lies at 3 m water depth. Hot water and gas issued from faults running along the edge of the reef and also from small fissures in the rocks. When this area was first observed by scuba divers in June 1992 numerous active bubble streams were noted from around the reef and from cracks in the rocks. White 'smoke' issued from many of the cracks and the whole

area had reduced underwater visibility. Mats of filamentous bacteria were attached to the rocks surrounding hot water and gas outlets. These mats penetrated into the fissures around the vents and were particularly dense under rocky overhangs. The rocks were covered with hydrozoans and macroalgae, including, *Cystoseira* spp., *Anadyomene stellata*, *Dasycladus vermiculatus* and *Acetabularia acetabulum* (personal communication, F. Cinelli & M. Abbiatti, University of Pisa). In the vicinity of the vents these organisms were coated in a white bacterial and mineral film. Other epifauna noted on the reef included, sponges, bryozoans and nudibranchs. In July 1993 the outgassing from the reef area was still as extensive, but the release of 'smoke' had stopped and the algae and rocks were covered with a dark brown film, which was not sampled. Algae were still green below this coating. The following October the algae showed no brown coating, although hot water and gas outflows were still active and there appeared to be no diminution in the extent of the bacterial mats.

Close to the reef was an area of boulder-lined depressions in the otherwise sandy sea bed in which the water was boiling. Boulders close to the vents were covered in bacteria. Similar areas have been found throughout the northwest region of the bay in water depths down to 5 m.

Most of Paleohori Bay has a sea bed comprised of medium to fine sand with seagrass beds of *Cymodocea nodosa* extending from 6 m to approximately 20 m depth and *Posidonia oceanica* extending from approximately 10 to 40 m depth. Streams of gas bubbles issued from some areas of the seagrass beds. The leaves of the seagrass were occasionally covered with a white layer of bacteria and mineral precipitates. Within the seagrass were observed occasional large bivalves, *P. nobilis*, anemones and starfish.

In water depths between 5 and 13 m patches of clear sand, up to 100 m across, were often found between the stands of seagrass. These seagrass-free areas were frequently covered by white mats, which overlay warm sediment containing a hypersaline brine. The interstitial water contained up to 1065 mmol/l Na (approximately 85‰ salinity), 1 mmol/l ammonia, 2 mmol/l silicate and 1.5 mmol/l hydrogen sulphide. The sediment temperature varied from 2 to 3°C above ambient at the sediment surface and was as high as 85°C at 30 cm depth. The white mats consisted of silicates, elemental sulphur (which was sometimes attached to bacterial filaments and sometimes to mucosal strands), and globular, rod-shaped and filamentous bacteria. In places the mats had a green tinge due to gliding cyanobacteria or had a brown–orange colour due to benthic diatoms. Silicate precipitates were up to 3 cm thick in the troughs of sand ripples, especially over areas of high heat flow with little disturbance by gas venting. Yellow patches of sand coated with elemental sulphur were occasionally seen within the white mat, in areas where the sediment temperatures were above 100°C.

In the white mats the major bacterial biomass was due to the presence of a large globular, colourless sulphur bacterium, *Achromatium volutans*. The *A. volutans* cells, which were 5–50 μm in diameter, contained refractile globules of sulphur, such that areas of sand, at 10 m water depth, colonized by these bacteria, together with zones of precipitated silicates, were clearly visible from the shore at 300 m distance. These bacteria formed a 1–2 mm thick mat on the surface of the sediment overlying the brine seeps and extended at lower densities to 2 cm depth in the sand. Evidence from temperature cycles within the sediment suggests that vertical movement of the brine layer in the sand follows a diurnal cycle, resulting in an alternation of high sulphide and oxygen conditions allowing these bacteria to penetrate deeper into the sand than would normally be the case. After storm dispersal the bacterial mats visibly re-formed in 2–3 days.

Animals within the bacterial mat areas were sparse, being confined to ciliates and a gastropod, *Cyclope neritea*. *Cyclope neritea* is known to be a salinity and temperature tolerant species which is particularly common in saltmarsh areas of the Mediterranean (Sacchi 1960; Ansell & McLachlan 1980), but which, around Milos, has not been found at any distance from the hydrothermal areas. It is a scavenger with mixed nutrition, feeding on dead animals as well as on the bacterial mats (A. J. Southward, E. C. Southward & C. Kennicutt, unpublished data). This species was, however, ubiquitous on the algobacterial mat, and has never been found in the surrounding areas, except when it was observed depositing egg capsules on the seagrass fronds during March 1992 and March–April 1994. In contrast, other investigators have noted that, elsewhere in the Mediterranean, *C. neritea* deposits eggs on hard substrates such as cockle shells (Bekman 1941; Gomoiu 1964).

In April 1994 small numbers of a related nassariid gastropod, *Nassarius mutabilis*, were found, together with *C. neritea*, in some areas of bacteria mat over the main brine seeps. Both species have dense shells and are not readily displaced a great distance from the vent sites during periods of intense wave action. This, together with the observed temperature, salinity

and sulphide tolerance of *C. neritea*, may be one of the reasons it dominates the macrofauna of these shallow vents. Narcotized and dead animals, especially polychaetes, are occasionally seen on top of the sediment over the brine seep. On one occasion a large *Tonnia galea* was observed on the top of the brine.

Only a few sulphide-tolerant and apparently opportunist species, such as the nematode *Oncholaimus campyloceroides* and the polychaetes *C. capitata*, *Microspio* sp. and *Spio decoratus*, were found around the periphery of the brine seep area. Meiofauna around the periphery of the seeps consisted mainly of nematodes and ciliates. The most important nematode species present numerically, and probably in terms of biomass, was the oncholaimid *O. camplyoceroides*. This species occurs in numbers up to 3500 individuals dm^{-3} at the edge of the seep area and in the adjacent seagrass beds. *Oncholaimus camplyoceroides* is known to be a carnivorous species (Teal & Wieser 1966) and appears to be extremely sulphide-tolerant (Thiermann et al. in press). It is believed that *O. camplyoceroides* feeds on animals which are killed or narcotized by the sulphide in the brine seeps and penetrates into the fringes of the seeps for food. The less sulphide-tolerant organisms may be trapped when the brine rises towards the surface.

Other nematodes found at the periphery of the seeps are species commonly found at the oxic–anoxic interface in sediments, including *Sabateria pulchra*, *Chronadorid* spp., *Linhomoeoid* spp., *Siphonolaimus* spp. and the stillbotnematids *Eubostrichus* cf. *parasitiferus* and *Leptonemella* spp. The stillbotnematids feed on ectosymbiotic sulphur-oxidizing bacteria and are a specialist group known from sulphidic sediments (Ott et al. 1991). However, stillbotnematids have not been identified in the main part of the hydrothermal brine seeps in Paleohori Bay, only at the seep fringes.

Fauna from a grab transect along the 10 m depth contour in the bay in March 1992 varied markedly according to the temperature and sulphide conditions (Table 2). A total of 144 macrofauna species were recorded from the six grab samples taken along the 10 m depth contour, of which stations 3 and 5 (Table 2) contained a few *C. nodosa* shoots. Polychaetes represented 58% (83 species) of the total fauna, whereas crustaceans and molluscs accounted for 25% (36 species) and 13% (19 species), respectively. The polychaetes *Exogone gemmifera* and *Aricidea cerrutii* were consistently the two dominant species along the transect, with the former species having its highest densities

Table 2. *Species identified in seven, 0.1 m², Smith–McIntyre grab samples taken along the 10 m depth contour in March 1992, Paleohori Bay, Milos. Values are numbers of individuals in 0.1 m^{-2}*

Grab no:	1	2	3	4	5	6	7
Temperature (°C at 5 cm depth):	57	14	14	15.5	14	28	21.2
Species							
Cnidaria							
Anthozoa sp.indet.	–	–	2	–	3	–	2
Nemertina							
Lineidae sp. indet.	–	–	1	–	–	–	–
Nemertean sp. indet.	–	1	5	–	–	–	1
Annelida							
Scolaricia typica	–	7	19	8	8	38	28
Aricidae catherinae	–	–	–	–	3	3	–
Aricidea sp. 1	–	1	–	–	–	–	–
Aricidea cerrutii	–	5	78	32	121	27	157
Aricidea capensis bansei	–	1	1	17	27	12	10
Aricidea claudiae	–	–	2	–	–	–	–
Aricidea fragilis mediterranea	–	–	–	–	–	3	–
Cirrophorus harpagoneus	–	6	6	3	14	9	31
Cirrophorus lyra	–	–	–	7	9	2	5
Paraonidae sp. indet.	–	–	–	–	1	–	–
Spio decoratus	–	3	2	1	1	6	9
Spiophanes bombyx	–	–	1	–	1	–	–
Minuspio cirrifera	–	1	–	–	–	1	1
Prionospio malmgreni	–	–	1	6	13	2	10
Microspio mecznikowianus	–	–	4	8	5	–	13
Malacoceros vulgaris	–	–	–	–	4	–	1
Magelona equilamellae	–	1	7	–	1	1	1
Chaetozone sp.	–	–	–	3	8	7	8
Aphelochaeta sp.	–	–	1	–	–	–	–
Cirriformia tentaculata	–	–	3	–	–	–	–
Caulleriella alata	–	–	–	80	49	4	37
Caulleriella bioculatus	–	1	11	–	–	–	1
Persiella clymenoides	–	6	36	5	11	2	2
Capitella capitata	–	41	22	41	22	44	–
Paraleiocapitella sp.	–	–	2	–	–	–	–
Notomastus latericeus	–	–	–	–	1	–	–
Mastobranchus trinchesii	–	–	–	–	1	1	–
Axiothella rubrocincta	–	–	–	–	1	–	–
Axiothella sp. 2	–	–	–	–	3	–	–
Praxillella sp. 1	–	–	1	1	–	–	–
Praxillella affinis	–	–	–	–	4	–	–
Praxillella praetermissa	–	–	–	–	2	–	–
Clymenura sp. 2	–	–	–	–	12	–	–
Clymenura sp.	–	–	–	–	–	–	3
Maldaninae sp. 1	–	–	1	–	–	–	–
Euclymeninae sp. 5	–	2	1	–	–	–	–
Euclymene sp. indet.	1	–	–	–	–	–	–
Euclymeninae sp.	–	–	–	3	–	–	–
Armandia polyophthalma	–	–	–	1	1	6	–
Polyophthalmus pictus	–	–	–	1	1	–	1
Tachytrypane jeffreysii	–	–	–	–	–	–	1
Protomystides bidentata	–	1	1	–	1	–	–
Sigalion mathildae	–	3	3	–	1	3	–
Pholoe minuta	–	–	–	–	27	–	–
Adyte pellucida	–	–	–	1	1	–	–
Eunoe nodosa	–	–	–	–	–	–	1
Psammolyce arenosa	–	–	–	–	–	–	1
Microphthalmus szcelkowii	–	–	–	–	–	–	1
Langerhansia cornuta	–	–	16	–	–	–	–
Odontosyllis fulgurans	–	–	–	–	1	–	–
Sphaerosyllis hystrix	–	–	–	6	23	11	27
Sphaerosyllis tetralix	–	–	–	3	–	–	–
Exogone gemmifera	1	26	51	65	1	4	8
Exogone hebes	–	2	1	1	–	1	4
Exogone ovalis	–	–	2	1	–	1	–
Brania oculata	–	–	–	4	16	3	9
Autolytus edwarsi	–	–	–	–	10	–	–
Nereis sp.	–	–	–	–	–	–	2
Platynereis dumerilii	–	–	–	8	2	–	–
Platynereis sp. 1	–	–	2	–	–	–	–
Glycera tridactyla	–	–	1	–	10	–	–
Glycera sp. indet.	–	1	–	1	1	–	3
Nephtys assimilis	–	1	–	–	–	–	–
Micronephtys maryae	–	4	6	6	18	3	5
Hyalinoecia bilineata	–	12	20	1	3	4	8
Eunice vittata	–	–	–	1	38	–	–
Nematonereis unicornis	–	–	1	–	–	–	–
Lumbrineris emandibulata	–	3	8	–	–	2	2
Lumbrineris gracilis	–	–	4	13	1	1	17
Lumbrineris impatiens	–	–	–	–	–	4	–
Lumbrineris sp. 1	–	1	–	–	–	–	–

Grab no:	1	2	3	4	5	6	7
Temperature (°C at 5 cm depth):	57	14	14	15.5	14	28	21.2
Arabella iricolor	–	–	–	–	5	–	–
Schistomeringos neglecta	–	11	9	2	9	4	7
Schistomeringos caeca	–	2	–	–	–	2	1
Protodorvillea kefersteini	–	–	9	10	1	–	11
Owenia fusiformis	–	–	2	–	–	1	–
Myriochele heeri	–	–	–	1	2	–	–
Lanice chonchilega	–	–	–	–	1	–	1
Pista sp.	–	2	–	–	–	–	–
Polycirrus sp. 1	–	–	4	–	–	–	–
Chone filicaudata	–	–	3	–	–	3	1
Chone duneri	–	–	–	–	–	1	–
Spirorbidae sp. indet.	–	–	–	–	6	–	–
Mollusca							
Gibbula sp. indet.	–	–	1	–	1	–	3
Tricolia pulla	–	–	1	–	–	–	–
Smaragdia viridis	–	–	1	–	–	–	–
Eulima glabra	–	–	1	–	–	–	–
Gastropod sp. indet.	–	1	1	5	–	1	4
Retusa truncatula	–	–	2	–	–	1	–
Philine sp. indet.	–	–	1	–	–	–	–
Scaphopod sp. indet.	–	–	1	–	–	–	–
Solemya togata	–	2	–	–	–	–	–
Mytilacea sp. indet.	–	–	–	–	–	–	3
Loripes lacteus	–	–	2	–	–	–	–
Lucinacea sp. indet.	–	–	–	4	5	2	10
Montacutidae sp. indet.	–	–	3	–	1	–	–
Dosinia lupinus	–	–	1	–	–	–	–
Veneracea sp. indet.	–	–	–	–	–	1	–
Fabulina fabuloides	–	–	6	–	–	–	–
Arcopagia crassa	–	–	3	–	–	–	–
Tellinacea sp. indet.	–	2	5	5	2	2	7
Bivalve sp. juvenile	–	2	–	–	–	–	–
Crustacea							
Ostracod sp.	–	–	1	–	–	11	2
Bodotria pulchella	–	–	–	–	–	3	–
Bodotria scorpioides	–	–	2	2	4	–	–
Eocuma ferox	–	–	1	–	–	–	–
Eocuma sp. indet.	–	–	–	–	–	1	–
Iphinoe trispinosa	–	1	3	5	2	4	1
Leucon affinis	–	1	–	–	–	–	–
Pesudocuma longicornis	–	1	6	4	5	–	1
Pseudocuma similis	–	1	4	–	–	–	–
Apseudes latrelli mediterannea	–	1	6	–	1	–	4
Leptochelia savignyi	–	–	–	–	–	–	1
Mysid. sp.	–	1	–	–	1	1	–
Isopod sp. juv.	–	–	1	–	–	–	2
Socarnes filicornis	–	–	1	–	–	–	–
Hippomedon massiliensis	–	–	–	–	2	–	–
Bathyporeia guilliamsoniana	–	26	1	–	1	5	–
Bathyporeia megalops	–	1	1	–	–	–	–
Bathyporeia sunnivae	–	–	3	1	5	–	–
Urothoe elegans	–	1	12	8	2	11	3
Leucothoe incisa	–	1	2	–	–	–	–
Leucothoe occulta	–	–	–	1	4	–	3
Leucothoe venetiarum	–	–	1	–	–	–	–
Stenothoe monoculoides	–	–	2	–	–	–	–
Perioculodes longimanus	–	2	10	7	9	22	4
Synchelidium maculatum	–	1	–	1	1	2	1
Apherusa alacris	–	–	–	2	–	–	2
Apherusa chiereghinii	–	–	2	–	–	–	–
Atylus guttatus	–	–	–	2	4	–	2
Megaluropsis massiliensis	–	–	–	–	1	–	–
Guernea coalita	–	–	1	1	–	2	–
Aoridae sp.	–	–	–	–	1	–	2
Microdeutopus sp.	–	–	4	–	–	–	–
Ericthonius punctatus	–	–	2	–	2	–	6
Callianassa truncata	–	1	–	–	–	–	2
Paguridae sp. indet.	–	1	1	–	–	3	1
Decapod sp. juvenile	–	1	–	–	–	–	–
Echinodermata							
Acrocnidia brachiata	–	–	–	–	6	–	5
Astropecten sp. indet.	–	–	–	–	1	–	–
Echinoidea sp. juvenile	–	–	–	–	–	1	–
No. of individuals	22	1666	5064	4044	6412	2970	5962

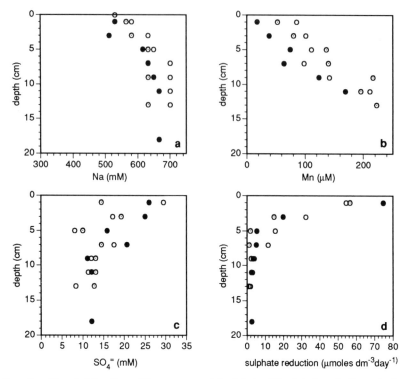

Fig. 4. Sediment profiles of: (**a**) sodium; (**b**) manganese; (**c**) sulphate; (**d**) sulphate reduction rates in cores taken around a gas vent. (○) Core taken over vent; (◉) core taken 0.5 m from vent; and (●) core taken from 1.0 m from vent.

$(65\,\mathrm{m}^{-2})$ at station 4 and the latter at station 7 $(121\,\mathrm{m}^{-2})$. Two individual polychaetes, *E. gemmifera* and a specimen of *Euclymene* sp. were the only macrofauna species to be found in the warmest sediment sample (57.4°C at 5 cm depth) from station 1.

Gas venting has an effect on the chemistry of the upper sediment layers and on the composition of the fauna. Sediment cores were collected directly over a gas vent at a depth of 12 m in a brine seep area and at distances of 0.5 and 1.0 m from it in March 1992. The sediment temperature at the seep varied from approximately 16°C at the surface (2°C above ambient) to over 30°C at 10 cm sediment depth. Depth profiles of sodium, manganese and sulphate in the interstitial water and the sulphate reduction rate are shown in Fig. 4. In the upper sediment layers the highest sodium and manganese concentrations and the lowest sulphate concentration were noted in the core taken over the vent With increasing sediment depth the three cores showed a trend towards similar interstitial concentrations of ions. This is similar to the situation which occurs around cold-seep gas vents (O'Hara *et al.* in press) due to the 'air-lift'

effect of the rising bubbles which induces an interstitial circulation around the bubble outlets. Sediment close to the vent is often oxic due to the drawdown of interstitial water to replace the water displaced by the rising gas. As the distance from the vent increases the sediment becomes less influenced by the induced circulation and becomes more reduced. Downward water flow decreases with sediment depth so that the interstitial concentrations of ions in the different cores converge with increasing sediment depth. The total organic carbon was higher at the vent station, 3.2% dry weight, compared with the cores taken further away, 1.8% dry weight, reflecting an increased microbial biomass around the vent. Water drawdown, due to the interstitial circulation induced by the rising gas bubbles (O'Hara *et al.* in press), would bring detrital material and phytoplankton into the sediment. All these changes affect the fauna; the most sulphide- and salinity- tolerant species, such as the polychaetes *C. capitata* and *Microspio* sp. and the gastropod *C. neritea*, were dominant at the vent. The number of species in the cores increased with increasing distance from the vent from six in the vent core to eight at

Table 3. *Species identified in eight core samples (94 mm i.d.) taken around a single gas vent at 12 m water depth in the centre of Paleohori Bay, Milos, June 1992. Values are numbers of individuals*

Taxon	Distance from vent (m)		
	0	0.5	1
Annelida			
Aricidea catherinae	–	–	2
Cirrophorus harpagoneus	1	1	–
Spio decoratus	1	14	9
Microspio mecznikowiani	12	1	1
Capitella capitata	31	18	8
Nephtys incisa	–	–	1
Mollusca			
Cyclope neritea	20	–	–
Crustacea			
Iphinoe serrata	–	–	5
Pseudodocuma longicornis	–	–	1
Apseudes latrelli medite	–	1	1
Perioculodes longimanus	–	–	5
Synchelidium maculatum	–	–	1
Synchelidium longidigita	–	–	1
Synchelidium haplochele	–	1	1
Isaeidae sp. *juveniles*	1	–	–
Erithonius punctatus	–	–	1
Phtisica marina	–	1	–
Callianassa truncata	–	5	3
Total no. of individuals	66	42	40

0.5 m distance and to 14 in the core taken 1.0 m from the gas vent (Table 3.). However, the total biomass of the infauna was $80 \, g \, m^{-2}$ in the vent core, decreasing to $3 \, g \, m^{-2}$, or less, in the cores taken further away. The high biomass in the vent core was almost entirely due to the presence of *C. neritea*.

Sulphate reduction rates in all three cores peaked in the upper 2 cm of sediment (Fig. 4). This shows that the major organic carbon input for the sulphate reducers comes from the surface bacterial mat and sedimenting or drawdown organic matter, rather than methane, hydrogen and other energy sources in the hydrothermal fluids.

The hydrothermal areas within the bay change with time. This was shown by the changes observed at the 3 m site, the observed growth of the bacterial mat areas into sections of seagrass bed, the change in area of actively bubbling sites, the cessation of activity at marked vents and the initiation of activity at new sites, and by the presence of recently dead seagrass in areas where low pH, sulphide- and silicate-rich interstitial water was found in the sand. Changes in the size of seagrass beds as a result of changing hydrothermal activity patterns in the bay are currently being studied.

White mats similar to those found at 10 m depth water, although of smaller area, $1–50 \, m^2$, were found at shallower depths. In shallow water areas (5 m depth) towards the west of Paleohori Bay there were river-like mats on the bottom and extensive gas bubble release. An 11.5 m long transect was made across low and high temperature areas, starting and finishing in the white mat (Fig. 5). Sediment cores were collected by scuba divers from two regions of the transect, from areas where there were sharp changes in sediment temperature and gas composition. The number of burrows of the decapod crustacean *Callianassa truncata* were counted in each $0.25 \, m^2$ quadrant along the transect line. Burrows of *C. truncata* were confined to the lower temperature region of the transect, <40°C at 10 cm sediment depth. In the highest temperature region of the transect, 57.8 and 97.3°C at 5 and 10 cm sediment depth, respectively, only four species were found (Table 4): the crustaceans *Perioculodes longimanus, Pseudocuma longicornis, Ericthonius punctatus* and *Bathyporeia* sp. These animals were almost certainly living in the cooler top layer of the sediment. The amphipod *P. longimanus* was the dominant macroinfaunal species found along the transect. Polychaetes and gastropods were few compared with the 10 m depth contour region of the bay.

It appears from the data in Fig. 5. that

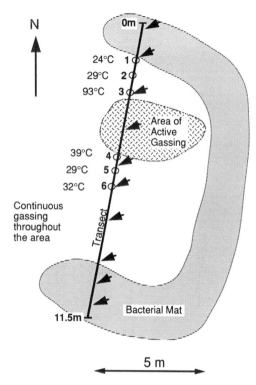

N

24°C 1
29°C 2
93°C 3

Area of
Active
Gassing

39°C 4
29°C 5
32°C 6

Continuous
gassing
throughout
the area

Transect

0m

11.5m

Bacterial Mat

5 m

Fig. 5. Transect in region of bacterial mat at 5 m water depth, Paleohori Bay, in June 1992. (○) Sediment core: (◀) gas vent. Temperatures were measured at 10 cm sediment depth. The fauna in the cores are listed in Table 4.

temperature may be the major factor influencing species distribution along the transect. This agrees with the work of Kamenev *et. al.* (1993), who found sediment temperature to be the main limiting factor in the distribution of both macro- and meiofauna around shallow water gaso- hydrothermal vents. The same workers also found motile animals, namely amphipods and decapods, in areas where sediment tempera- tures were greater than 40°C.

At 40 m depth in the bay, ROV observations showed a rocky ridge covered with sediment mounds and surrounded by soft sediment which was covered in places with large patches of dead seagrass leaves. Black areas of sediment were covered with white mats. As in previous areas, little epifauna was observed on the muddy sediment, although occasional starfish, *As- tropecten* sp., were seen. The rocks appeared to be pillow lavas encrusted with algae, sponges, hydroids and bryozoa. Filamentous bacterial mats were attached to the rocks around vent outlets. Echiurins, *B.* cf. *viridis*, were present,

lying over the rocks rather than over the surrounding sediment.

The highest infaunal biomass found in samples collected from Paleohori Bay was only 80 g m^{-2}. Biomass from the reef areas with hydrothermal venting have yet to be estimated, but do not visually appear to be unusual. During gales and storms the bacterial mats are readily dispersed, the upper 10 cm or more of sand being re-suspended, and there is an export of organic carbon and nutrients to the surrounding area. Nutrient release as a result of seismic events may also increase production in the water column (Dando *et al.* in press). These effects of bacterial and nutrient export on the productivity of the surrounding area have yet to be quantified.

Kiriakis Bay. In June 1991 Kiriakis Bay was investigated by divers. An area in a small embayment to the west of Kiriakis Bay was observed to have intense gasohydrothermal activity (Fig. 6). In this region hot water issued out of fissures in rocks and the surrounding seawater was extremely turbid. Core samples were taken within this embayment (samples K1 and K2) and along the headland, within Kiriakis Bay proper (samples K3 and K4). No fauna was found in samples K1 or K2. The fauna at stations K3 and K4 was dominated by the opportunistic polychaetes *Spio filiformis*, *C. capitata* and the crustacean *Microprotus maculata* (Table 5). In this area sporadic gassing was observed, to- gether with scattered patches of bacterial mat. Epifaunal scrapings were taken from the rocks around the area of hot water emissions. These samples were dominated by dipterid insect larvae and sabellid polychaetes.

Kiriakis Point to Zefiros Point. Rocky areas were found off both points and those in 80 to 100 m water depth were investigated by ROV deployments in July 1993. The area south of Kiriakis Point showed extensive gas venting with large pieces of bacterial mat drifting through the water. Clumps of mat had become detached and lay like 'snowballs' on the bottom. Most of the venting occurred from rock outcrops which were almost completely covered with epibionts, such as *Lithothamnion*, sponges, crinoids, anthozoa, serpulids and ophiuroids. Crayfish, *Palinurus elephas*, and echiurans, *B.* cf. *viridis* were common and the rock outcrops were surrounded by numerous fish shoals, especially of serranids. At the base of the rocks drifts of dead seagrass leaves had accumulated. These often covered the vent outlets, as shown by rising streams of gas bubbles and mounds of white bacterial mat on top of the leaves. Further away from the

Table 4. *Species identified in cores (94 mm i.d.) taken along the 11.5 m. transect at 5 m depth, Paleohori Bay, Milos*

Species	Numbers of individuals in core					
	1	2	3	4	5	6
Annelida						
Scolaricia typica	4	4	–	–	–	2
Cirrophorus harpagoneus	4	2	–	–	–	–
Spio decoratus	1	2	–	–	–	3
Spiophanes bombyx	1	–	–	–	–	–
Minuspio cirrifera	1	–	–	–	–	–
Chaetozone sp.	–	2	–	–	–	–
Caulleriella bioculatus	–	–	–	–	–	1
Capitella capitata	–	–	–	–	–	2
Sigalion squamosum	1	–	–	–	–	–
Odontosyllis ctenistoma	–	1	–	–	1	–
Micronephtys maryiae	1	–	–	–	–	–
Mollusca						
Cyclope neritea	–	–	–	–	2	–
Gastropod sp. juv.	2	–	–	–	–	–
Veneracea sp. juv.	–	–	–	–	1	–
Bivalve sp. juv.	1	–	–	–	–	–
Crustacea						
Eocuma ferox	1	1	–	–	1	–
Iphinoe serrata	1	–	–	–	–	–
Pseudocuma longicornis	1	–	–	–	–	3
Mysid spp.	1	–	–	–	–	–
Bathyporeia phaiophalma	–	–	–	–	–	1
Bathyporeia sp.	–	–	1	–	–	–
Perioculodes longimanus	2	–	1	1	3	5
Ericthonius punctatus	–	–	1	–	1	–
Isaeidae	–	–	–	–	1	–
Total no. of individuals	22	12	3	1	10	17

rocks small areas of diffuse venting occurred. These could be identified from a distance by the presence of white bacterial mats. Large bivalves, *P. nobilis*, were conspicuous, but did not appear to be associated directly with the venting areas.

Around the hot water and gas vents were thick mats of angular, filamentous bacteria, the filaments of which were surrounded by globules of elemental sulphur. Water samples collected from one vent outlet had a pH of 5.18, a hydrogen sulphide content of 500 μmol/l, ammonia of 150 μmol/l and silicate of 2 mmol/l. The bacterial mats extended up to 0.5 m from the vent outlets and on the rock past which the rising plumes flowed. Although numerous mats were seen their observed volume was insufficient to account for the material observed in the water column.

Similar epifauna covered the rocks off Zefiros Point. ROV deployment in 80 m depth showed large amounts of bacterial floc in the water column; however, no vents were seen and bubbles were not detected on sonar. No bubble plumes were detected on the echosounder to the west of Zefiros Point on the south coast of Milos.

Comparison with other hydrothermal sites

The dominant bacteria species living around the vents varied from site to site. In the Santorini caldera the iron bacterium *G. ferruginea* was the only species recorded from around the vents, forming flocculent iron-rich deposits (Holm 1987). Analogous filamentous iron-rich material has been reported from mid-ocean ridge hydrothermal sites in the Pacific (Juniper & Fouquet 1988). At the 3 m water depth site in Paleohori Bay, Milos, unidentified filamentous bacteria were found, whereas at the brine seeps in 10 m water depth the large colourless sulphur bacterium *A. volutans* dominated. *Achromatium volutans* was originally described from warm sulphidic springs in the Bay of Naples (Hinze 1903). In deeper water, 90 m depth, off Kiriakis Point a novel, unidentified type of filamentous bacterium was found. *Beggiatoa* spp., although

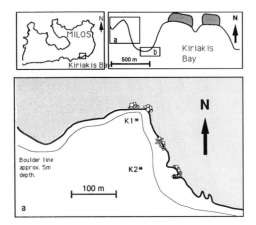

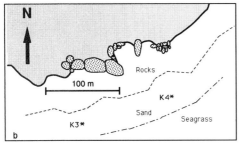

Fig. 6. Sample sites (★) in Kiriakis Bay, Milos in July 1991. The fauna obtained in core samples from these stations is listed in Table 5.

present, have yet to be identified as a dominant bacterial form at any of the Aegean sites, despite the presence of diffuse venting through soft substrates, a favoured habitat for members of this genus (Jannasch *et al.* 1989). This pattern of large differences in the composition of the dominant bacterial species between different hydrothermal sites is common elsewhere (Jannasch & Wirsen 1981; Stein 1984; Jacq *et al.* 1987; Jannasch *et al.* 1989, Tarasov *et al.* 1990; Naganuma 1991; Sorokin, 1991). As yet, there is insufficient chemical data on the water directly under the bacterial mats to specify the environmental requirements of the different bacterial species.

Novel hyperthermophiles have been isolated from nearshore vents in the Tyrrhenian Sea (Stetter 1982; Zillig *et al.* 1983; Belkin *et al.* 1986; Huber *et al.* 1986; Jannasch *et al.* 1988). However, no attempt to isolate hyperthermophilic bacteria from any of the Aegean hydrothermal vent sites has been reported.

In common with shallow hydrothermal sites

off Iceland (Fricke *et al.* 1989), California (Stein 1984; Trager & deNiro 1990) and the Kurile Islands (Tarasov *et al.* 1990) the hydrothermal areas around Santorini and Milos are inhabited by faunal species which are also found in non-vent areas. Symbiotic associations between animals and autotrophic bacteria have only, to date, been found in the stillbotnematid group of nematodes on the fringe of the brine seeps in Paleohori Bay. This group is typical of the oxic–anoxic interface in sulphidic sediments (Ott *et al.* 1991) and do not appear to be able to survive in the elevated temperatures and salinities of the seeps. The lucinid bivalves, *Loripes lacteus* and *Divaricella divaricata*, which are known to obtain nutrition from endosymbiotic sulphur-oxidizing bacteria (Le Pennec *et al.* 1987), have been found in other parts of the bay which were not conspicuously under the influence of hydrothermal activity. Lucinacean bivalves, although common world-wide in a variety of sulphide-rich habitats, except for estuaries, have not been reported from hydrothermal sites and it is possible that they do not have the temperature, salinity or metal tolerance necessary to survive under such conditions.

Echiurans of *B.* cf. *viridis* were found at all the hydrothermal rocky sites in water deeper than 35 m. Similar black boniellids have been seen on videos of the rocky areas around the hydrothermal vents in 200 m water depth at Steinahóll, at the northern end of the Reykjanes Ridge, Iceland. As these animals live with their trunks hidden among rocks, with only their proboscis extended over the sand, they are very difficult to sample.

The fauna found within the brine seeps off Milos comprise only ciliates and the gastropods *C. neritea* and *N. mutabilis*. A similar situation has been reported around hydrothermal vents off California where gastropods, *Haliotis cracherodii* and *Lottia limatula*, were similarly observed grazing on bacterial mats (Stein 1984).

Around the periphery of the brine seeps are found infaunal species such as *O. campylocercoides*, *Microspio* sp., *S. decoratus* and *C. capitata*, opportunist species relatively tolerant of sulphide and typical of disturbed habitats. *Capitella capitata* was also dominant in the vicinity of the Santorini vents. This species composition may also be a reflection on the stability of the vent habitats. There is not a sufficiently long time series of observations to know the time period over which particular sites are stable. However, we have already observed marked changes in the disposition and type of venting at Milos, as well as observing the effects of seismic activity on increasing the chemical

Table 5. *Species identified in 94 mm i.d. core samples from Kiriakis Bay, Milos, in July 1991*

	Numbers of individuals in core			
Species	K1	K2	K3	K4
Annelida				
Spio filifornis	–	–	13	9
Capitella capitata	–	–	3	27
Crustacea				
Microprotus maculata	–	–	2	25
Total no. of individuals	–	–	18	61

fluxes (Dando *et al.* in press). The biomass of the fauna both within and around these brine seep areas appears to be lower than in corresponding areas not affected by brine seepage.

Faunal biomass values at the soft-bottom hydrothermal sites in the Aegean were low, $\leqslant 80\,\mathrm{g\,m^{-2}}$, compared with reported values from other hydrothermal sites. Tarasov *et al.* (1990) report a biomass of $10\,000\,\mathrm{g\,m^{-2}}$ for the gaso-hydrothermal vents of Kraternaya Bight in the Kuriles. Values in the range $550–53\,000\,\mathrm{g\,m^{-2}}$ have been recorded for epifauna at mid-ocean ridge sites (Tunnicliffe 1991). In the shallow vents in the Bay of Plenty, Kamenev *et al.* (1993) found infaunal biomasses of $4.6–59.2\,\mathrm{g\,m^{-2}}$, decreasing with increasing temperature, in a volcanic depression, compared with a biomass of $5210\,\mathrm{g\,m^{-2}}$ for the bivalve mollusc *Tawera spissa* in a shallower region of gasohydrothermal activity. We do not, at this time, have data for rates of carbon production, or for the biomass of epifauna at the hard substrate sites at the hydrothermal vents in the Aegean. The low infaunal biomass values do not necessarily reflect the overall productivity of the area as the export of bacteria and higher organisms due to storm surges may be important.

Vestimentiferan tube worms, a group well known from mid-ocean ridge hydrothermal communities (Tunnicliffe 1991; Lutz & Kennish 1993), have been found around a hydrothermal vent at 82 m depth in Kagoshima Bay, southern Japan (Hashimoto *et al.* 1993). It was considered that the vestimentifera could have colonized the bay from nearby deep-water hydrothermal communities, although neither vestimentifera, nor other taxa typical of mid-ocean ridge communities, was observed at the Aegean sites. The Hellenic arc volcanism is relatively young: this, together with the Messinian evaporations and the lack of recent hydrothermal 'stepping stones' through the Mediterranean, could have prevented colonization from the nearest deep-water hydrothermal communities.

We acknowledge the help given to us by the following: Yvonne Leahy for organizing most of the field trips and for identification of the polychaetes; Ioanna Akoumianaki for the identification of fauna and logistic support; Mark Fitzsimons, Stewart Niven and Lesley Taylor for chemical analysis; Chris Smith for ROV operations and identification of epifauna; Marco Abbiati, Lisandro Benedetti-Cecchi, Franceso Cinelli, Inez Gamenick and Wanda Plaitis for diving assistance; and the help from the crew of RV *Philia* during the sample collections. This study was supported by CEC MAST contracts 90–0044 and 93–0058 and by a grant to PRD from the Natural Environment Research Council under the BRIDGE programme.

References

ANGELIER, J., LYBÉRIS, N., LE PICHON, X., BARRIER, E. & HUCHON, P. 1982. The tectonic development of the Hellenic arc and the Sea of Crete. *Tectonophysics*, **86**, 159–186.

ANSELL, A. D. & MCLACHLAN, A. 1980. Upper temperature tolerances of three molluscs from South African sandy beaches. *Journal of Experimental Marine Biology and Ecology*, **48**, 243–251.

BARDINTZEFF, J.-M., DALABAKIS, P., TRAINEAU, H. & BROUSSE, R. 1989. Recent explosive volcanic episodes on the Island of Kos (Greece): associated hydrothermal parageneses and geothermal area of Volcania. *Terra Research*, **1**, 75–78.

BEKMAN, M. 1941. [On the biology of the marine gastropods *Nassa reticulata v. pontica* (Mont.) and *Nassa (Cyclonassa) neritea* (L.)]. *Izvestiya Akademi nauk SSR, Serie Biologiya*, **3**, 347–352 [in Russian].

BELKIN, S., WIRSEN, C. O. & JANNASCH, H. W. 1986. A new sulfur-reducing extremely thermophilic eubacterium from a submarine thermal vemnt. *Applied and Environmental Microbiology*, **51**, 1180–1185.

CLINE, J. D. 1969. Spectrophotometric determination of hydrogen sulfide in natural waters. *Limnology and Oceanography*, **14**, 141–152.

DE BOAR, J. Z. 1989. The Greek enigma: is development of the Aegean orogene dominated by forces related to subduction or obduction? *Marine Geology*, **87**, 31–54.

DAL PONT, G., HOGAN, M. & NEWELL, B. 1974. Laboratory techniques in marine chemistry. II. Determination of ammonia in seawater and the preservation of samples for nitrate analysis. *Report of the Division of Fisheries and Oceanography , CSIRO, Australia, No. 55.*

DANDO, P. R., AUSTEN, M. C., BURKE, R. J., KENDALL, M. A., KENNICUTT, M. C., JUDD, A. G., MOORE, D. C., O'HARA, S. C. M., SCHMALJOHANN, R. & SOUTHWARD, A. J. 1991. Ecology of a North Sea pockmark with an active methane seep. *Marine Ecology Progress Series*, **70**, 49–63.

——, HUGHES, J. A., LEAHY, Y., NIVEN, S. J., TAYLOR, L. J. & SMITH, C. Rates of gas venting from submarine hydrothermal areas around the island of Milos in the Hellenic Volcanic Arc. *Continental Shelf Research*, in press.

——, ——, ——, TAYLOR, L. J. & ZIVANOVIC, S. Earthquakes increase hydrothermal venting and nutrient inputs into the Aegean. *Continental Shelf Research*, in press.

——, JENSEN, P., O'HARA, S. C. M., NIVEN, S. J., SCHMALJOHANN, R., SCHUSTER, U. & TAYLOR, L. J. 1994. The effects of methane seepage at an intertidal/shallow subtidal site on the shore of the Kattegat, Vendsyssel, Denmark. *Bulletin of the Geological Society of Denmark*, **41**, 65–79.

DOMINICO, E. & PAPASTAMATOKI, A. 1975. Characteristics of Greek geothermal waters. *In: Development and use of Geothermal Resources.* Second UN Symposium, San Francisco, 109–121.

FERGUSON, J. & LAMBERT, I. B. 1972. Volcanic exhalations and metal enrichments at Matupi Harbor, New Britain, T. P. N. G. *Economic Geology*, **67**, 25–37.

FERRARA, G., FYTIKAS, M., GIULIANI, O. & MARTINELLI, G. 1980. Age of the Formation of the Aegean Active Volcanic Arc. *In*: DOUMAS, C. (ed.) *Thera and the Aegean World II, Papers and Proceedings of the Second International Congress, Santorini, Greece.* Thera and the Aegean World, London.

FRICKE, H., GIERE, O., STETTER, K., ALFREDSSON, G. A., KRISTJANSSON, J. K., STOFFERS, P. & SVAVARSSON, J. 1989. Hydrothermal vent communities at the shallow subpolar Mid-Atlantic ridge. *Marine Biology*, **102**, 425–429.

FYTIKAS, M., GARNISH, J. D., HUTTON, V. R. S., STAROSTE, E. & WOHLENBERG, J. 1989. An integrated model for the geothermal field of Milos from geophysical experiments. *Geothermics*, **18**, 611–6211.

——, GIULIANI, O., INNOCENTI, F., MARINELLI, G. & MAZZUOLI, R. 1976. Geochronological data on recent magmatism of the Aegean Sea. *Tectonophysics*, **31**, 29–34.

——, INNOCENTI, F., MANNETTI, P., MAZZUOLI, R., PECERILLO, A. & VILLARI, L. 1984. Tertiary to Quaternary evolution of volcanism in the Aegean region. *In*: DIXON, J. E. & ROBERTSON, A. H. F. (eds) *The Geological Evolution of the Eastern Mediterranean.* Geological Society, London, Special Publications, **17**, 687–699.

GOMMOIU, M.-T. 1964. Biologisches Studium der Arten *Nassa reticulata* L. und *Cyclonassa neritea* (L.) im Schwarzen Meer (Rumanischen Kustenbereich). *Revue Roumaine de Biologie, Serie de Zoologie*, **9**, 39–49.

GRASSHOF, K., EHRHARDT, M. & KREMLING, K. 1983. *Methods of Seawater Analysis*, 2nd edn. Verlag Chemie, Weinheim, 419 pp.

HASHIMOTO, J., MIURA, T., FUJIKURA, K. & OSSAKA, J. 1993. Discovery of vestimentiferan tube-worms in the euphotic zone. *Zoological Science*, **10**, 1063–1067.

HINZE, G. 1903. *Thiophysa volutans*, ein neues Schwefelbakterium. *Berichte der deutschen botanischen Geschellschaft*, **21**, 309–316.

HOLM, N. G. 1987. Biogenic influences on the geochemistry of certain ferruginous sediments of hydrothermal origin. *Chemical Geology*, **63**, 45–57.

HUBER, R., LANGWORTHY, T. A., KÖNIG, H., WOESE, C. R., SLEYTR, U. B. & STETTER, K. O. 1986. *Thermotoga maritima* sp. nov. represents a new genus of unique extremely thermophilic eria growing up to 90°C. *Archives of Microbiology*, **144**, 324–333.

JACQ, E., GEESEY, G. & PRIEUR, D. 1987. Étude préliminaire des communautés bactériennes d'un site hydrothermal côtier (White-Point, Californe, USA). *Vie Milieu*, **37**, 59–66.

JANNASCH, H. W. & WIRSEN, C. O. 1981. Morphological survey of microbial mats near deep-sea thermal vents. *Applied and Environmental Microbiology*, **41**, 528–538.

——, HUBER, R., BELKIN, S. & STETTER, K. O. 1988. *Thermotoga neapolitana* sp. nov. of the extremely thermophilic, eubacterial genus *Thermotoga*. *Archives of Microbiology*, **150**, 103–104.

——, NELSON, D. C. & WIRSEN, C. O. 1989. Massive natural occurrence of unusually large bacteria (— sp.) at a hydrothermal deep-sea vent site. *Nature*, **342**, 834–836.

JUNIPER, K. & FOUQUET, Y. 1988. Filamentous iron-silica deposits from modern and ancient hydrothermal sites. *Canadian Mineralogist*, **26**, 859–869.

KAMENEV, G. M., FADEEV, V. I., SELIN, N. I., TARASOV, V. G. & MALAKEHOV, V. V. 1993. Composition and distribution of macro- and meiobenthos around sublittoral hydrothermal vents in the Bay of Plenty, New Zealand. *New Zealand Journal of Marine and Freshwater Research*, **27**, 407–418.

KOUSIS, M. 1993. Collective resistance and sustainable development in rural Greece. The case of geothermal energy on the island of Milos. *Sociologia Ruralis*, **33**, 3–24.

LE PENNEC, M., HERRY, A., DIOURIS, M., MORAGA, D. & DONVAL, A. 1987. Chemoautotrophie bactérienne chez le mollusque bivalve littoral *Lucinella divaricata*. *Compte Rendu Academie des Sciences, Paris, Série III*, **305**, 1–5.

LUTZ, R. A. & KENNISH, M. J. 1993. Ecology of deep-sea hydrothermal vent communities: a review. *Reviews of Geophysics*, **31**, 211–242.

MAKRIS, J. 1977. Geophysical investigations of the

Hellenides. *Hamburger Geophysikalische Einzelschriften*, **34**, 1–124.

MINISSALE, A., DUCHI, V., KOLIOS, N. & TOTARO, G. 1989. Geochemical characteristics of Greek thermal springs. *Journal of Volcanology and Geothermal Research*, **39**, 11–16.

NAGANUMA, T. 1991. Collection of chemosynthetic sulfur bacteria from a hydrothermal vent and submarine volcanic vents. *In: Technical Reports Presented at the 7th Symposium on Deepsea Research Using the Submersible 'SHINKAI 2000' System*. Japan Marine Science and Technology Center, Yokosuka, 201–219.

NEWTON, G. L., DORIAN, R. & FAHEY, R. C. 1981. Analysis of biological thiols: derivatization with monobromobimane and separation by high performance liquid chromatography. *Analytical Biochemistry*, **114**, 383–387.

NINKOVICH, D. & HAYS, J. D. 1972. Mediterranean island arcs of high potash volcanoes. *Earth and Planetary Science Letters*, **16**, 331–345.

O'HARA, S. C. M., DANDO, P. R., SCHUSTER, U., BENNIS, A., BOYLE, J. D., CHIU, T. W., HATHERELL, T. V. J., NIVEN. S. J. & TAYLOR, L. J. Gas seep induced interstitial water circulation: observations and environmental implications. *Continental Shelf Research*, in press.

ÓLAFSSON, J., THORS, K. & CANN, J. 1991. A sudden cruise off Iceland. *RIDGE Events*, **2**, 35–38.

OTT, J., NOVAK, R., SCHIEMER, F., HENTSCEL, U., NEBELSICK, M. & POLZ, M. 1991. Tackling the sulfide gradient: a novel strategy involving marine nematodes and chemoautotrophic ectosymbionts. *Marine Ecology*, **12**, 261–279.

PARSON, T. R., MAITA, Y. & LALLI, C. M. 1984. *A Manual of Chemical and Biological Methods for Seawater Analysis*. Pergamon Press, New York, 173 pp.

SACCHI, C. F. 1960. Ritmi nictemerali di fattori ecologici in microambienti acquatici e salmastri e loro significato biologico. *Delpinoa*, **2**, 99–163.

SARANO, F., MURPHY, R. C., HOUGHTON, B. F. & HEDENQUIST, J. W. 1989. Preliminary observations of submarine geothermal activity in the vicinity of White Island Volcano, Taupo Volcanic Zone, New Zealand. *Journal of the Royal Society of New Zealand*, **19**, 449–459.

SOROKIN, D. YU. 1991. [Oxidation of reduced sulphur compounds in volcanic regions in Bay of Plenty (New Zealand) and Matupy Harbour (New Britain, Papua-New Guinea).] *Proceedings of the USSR Academy of Science, Series Biological*, **3**, 376–387 [In Russian].

STEIN, J. L. 1984. Subtidal gastropods consume sulphur-oxidising bacteria: evidence from coastal hydrothermal vents. *Science*, **223**, 696–698.

STETTER, K. O. 1982. Ultrathin mycelia-forming organisms from submarine volcanic areas having an optimum growth temperature of 105°C. *Nature*, **300**, 258–260.

STÜBEN, D., BLOOMER, S. H., TAÏBI, N. E., NEUMANN, TH., BENDEL, V., PÜSCHEL, U., BARONNE, A., LANGE, A., SHIYING, W., CUIZHONG, L. & DEYU, Z. 1992. First results of study of sulphur-rich hydrothermal activity from an island-arc environment: Esmeralda Bank in the Mariana Arc. *Marine Geology*, **103**, 521–528.

TARASOV, V. G., PROPP, M. V., PROPP, L. N., ZHIRMUNSKY, A. V., NAMSARAEV, B. B., GORLENKO, V. M. & STARYNIN, D. A. 1990. Shallow-water gasohydrothermal vents of Ushishir Volcano and the ecosystem of Kraternaya Bight (The Kurile Islands). *Marine Ecology*, **11**, 1–23.

TEAL, J. M. & WIESER, W. 1966. The distribution and ecology of nematodes in a Georgia salt marsh. *Limnology and Oceanography*, **11**, 217–222.

THIERMANN, F., WINDHOFFER, R. & GIERE, O. 1994. Selected meiofauna around shallow hydrothermal vents off Milos (Greece): ecological and structural aspects. *Vie Milieu*, **44**, 215–226.

TRAGER, G. C. & DENIRO, M. J. 1990. Chemoautotrophic sulfur bacteria as a food source for mollusks at intertidal hydrothermal vents: evidence from stable isotopes. *Veliger*, **33**, 359–362.

TUNICLIFFE, V. 1991. The biology of hydrothermal vents: ecology and evolution. *Annual Review of Oceanography and Marine Biology*, **29**, 319–407.

UHLIG, G., THIEL, H. & GRAY, J. S. 1973. The quantitative separation of meiofauna. *Helgolander wissenchaftliche Meeresuntersuchungen*, **25**, 173–195.

VARNAVAS, S. 1989. Submarine hydrothermal metallogenesis associated with the collision of two plates: the Southern Aegean Sea region. *Geochimica et Cosmochimica Acta*, **53**, 43–57.

—— & CRONAN, D. S. 1991. Hydrothermal metallogenic processes off the islands of Nisiros and Kos in the Hellenic Volcanic Arc. *Marine Geology*, **99**, 109–133.

——, —— & ANDERSON, R. K. 1990. Spatial and time series analysis of Santorini hydrothermal waters. *In*: HARDY, D. A., DOUMAS, C. G., SAKELLARAKIS, J. A. & WARREN, P. M. (eds) *Thera and the Aegean World III, Proceedings of the Third International Congress, Santorini, Greece*, Vol. 2. The Thera Society, London, 312–324.

VIDAL, V. M., VIDAL, F. M. & ISAACS, J. D. 1978. Coastal submarine hydrothermal activity off northern Baja California. *Journal of Geophysical Research*, **83**, 1757–1774.

ZILLIG, W., HOLZ, I., JANEKOVIC, D., SCHÄFER, W. & REITER, W. D. 1983. The archaebacterium *Thermococcus celer* represents a novel genus within the thermophilic branch of the archaebacteriua. *Systematic and Applied Microbiology*, **4**, 88–94.

Geochemistry of the Snake Pit vent field and its implications for vent and non-vent fauna

S. M. SUDARIKOV[1] & S. V. GALKIN[2]

[1]Institute for Geology & Mineral Resources of the Ocean (VNIIOkeangeologia),
1 Angliysky pr., St Petersburg 190121, Russia

[2]Shirshov Institute of Oceanology RAN, 23 Krasikova st, Moscow 117851, Russia

Abstract: Observations made during a cruise of the R/V *Professor Logachev* at the Snake Pit hydrothermal vent field provide a large database with which to examine the relationships between the geochemical characteristics of the hydrothermal sediments and the fauna that inhabit the area. Correlations between the nature and geochemistry of sediments, the distribution of hydrothermal and peripheral fauna and the hydrothermal activity are established. Three geochemical zones in the Snake Pit area are distinguished. (i) The Central Zone, located within 10–100 m of black smoker chimneys, consisting mainly of hydroxide–sulphide sediments with Fe sulphides predominant. Barite is found only in this zone and vent-associated organisms distinctly predominate over non-vent fauna. (ii) The Intermediate Zone encompasses high-temperature springs at a distance from 50 to 150 m. Sediments are chiefly sulphide–hydroxide composition and are dominated by Fe hydroxide and sulphide minerals in an oxide envelope. The non-vent specific fauna show maximum concentrations. (iii) The Outer Zone is located within a radius of 150–500 m from the smokers. The sediment composition is characterized by a predominance of sulphide–hydroxide transitional differences with notably varying compositions of oxide and sulphide minerals. Minor sestonophages prevail in the fauna, which is typical of communities with a pronounced oligotrophic structure. From the Central Zone towards the Outer Zone, the frequency of the bacterial mats seen on photographs decreases and the hydrothermal sediments show a similar trend. Analysis of the hydrothermal sediment obtained from the cores shows: (i) Mn is absent; (ii) barite is found only in the Central Zone; and (iii) chalcopyrite and pyrrhotite concentrations decrease and opal and quartz concentrations increase with increasing distance from the nearest smoker. Comparisons between the Mid-Atlantic Ridge hydrothermal fields of TAG, Snake Pit and 15°N indicate a decrease in the hydrothermal and biological activities with latitude (TAG, Snake Pit, 15°N), which might be due to their evolutionary history.

Cruises to Mid-Atlantic Ridge (MAR) hydrothermal vent fields (Sudarikov & Ashadze 1989; Galkin & Moskalev 1990a, b; Sudarikov et al. 1990; Sudarikov & Cherkashev 1993; Batuyev et al. 1994) have provided sufficient geological and biological data to investigate the influence of MAR vent chemistry on the distribution of vent and non-vent biota (cf. ASHES vent field studies; Arquit 1990).

The distribution and composition of abyssal and hydrothermal fauna, the behaviour of the organisms and traces of their activity in the MAR were studied during several cruises using deep-sea manned submersibles (Mevel et al. 1989; Galkin & Moskalev 1990a, b; Segonzac 1992; Segonzac et al. 1993). The results of these investigations formed the basis of the interpretations presented here.

We report here the results of the biogeochemical survey during cruise 3/2 of R/V *Professor Logachev*, during which detailed studies of the Snake Pit hydrothermal field (23°N) were carried out (Fig. 1). Additionally, results obtained at the new hydrothermal field near 14°40′N MAR (seventh cruise of R/V *Professor Logachev* in 1994) and reported in Batuyev et al. (1994) and all other data from previously published work are used here for comparative purposes.

The aim of this study was to examine the relationships between the geochemical characteristics of the hydrothermal sediments and the associated fauna.

Data base and methods

The research that has been conducted is based on acoustically navigated mapping and interpretation of core and photographic data collected with the grab sampler Ocean–0.25, a 156 mm gravity corer and the deep-tow vehicle *Abyssal*.

The database developed to interpret the

From PARSON, L. M., WALKER, C. L. & DIXON, D. R. (eds), 1995, *Hydrothermal Vents and Processes*, Geological Society Special Publication No. 87, 319–327.

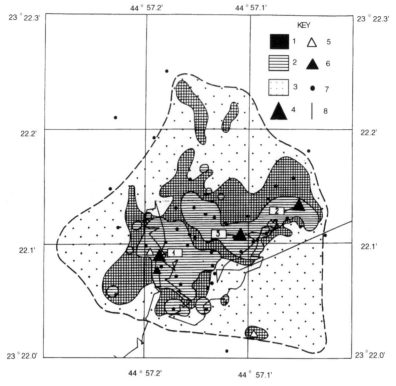

Fig. 1. Geochemical zones in the Snake Pit hydrothermal field: **1** Central Zone; **2** Intermediate Zone; **3** Outer Zone; **4** black smoker; **5** white smoker; **6** non-active vent; **7** core location; and **8** transect depicted by the cameras used in calculations.

biological and geochemical data involves the following. (i) The nature of bottom sediments (photo-profiling data): the presence of sediments and hydrothermal elements and the percentage of the bottom surface covered with hydrothermal sediment. (ii) Hydrothermal activity: the presence of particulate material in the water column (vent 'smoke') and sulphide structures (active, inactive). (iii) The presence and number of hydrothermal and peripheral fauna. (iv) The distance from the detected benthic species to the nearest high- and low-temperature vents.

Thirty-eight core samples were picked out for analysis within the 350 m zone surrounding the high-temperature vents (Fig. 1).

The sediment composition was determined according to the well-known methods of extraction of Chester & Hughes (1967) and Fomina (1975), which allow the separation of the sediment into three distinct phases (mobile forms). The first phase is characterized by carbonates, absorbed ions and evaporates of pore waters (exchange forms); the second phase

by amorphous and slightly crystallized oxides and hydroxides; and the third phase, by crystallized ('old') oxides. Combinations of Fe, Mn, Si and Al were determined in various extracts. The insoluble material which remains after extraction consists of litogenous material of diverse origin.

As part of the data processing, charts of different groups of organisms (sponges, holothurians, sea anemones, crabs, shrimps, polychaetes and bacterial mats), correlation matrixes and regression charts were compiled.

Results

Elements and mineralogy of hydrothermal sediments

The behaviour of elements accumulating within the first few hundred metres from active vents is considered.

Geochemical analysis of the sediment shows: (i) Mn is virtually absent from every type of sample or is below the detection limit ($<0.001\%$)

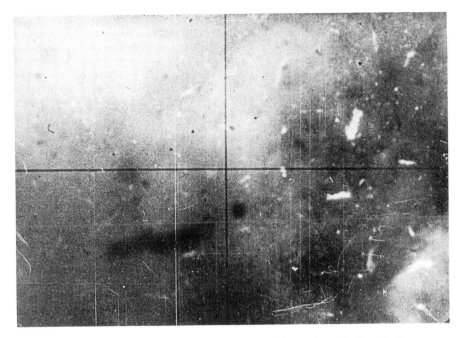

Fig. 2. Shrimps in black 'smoke' emitted from the hydrothermal chimney (vent No. 2 on Fig. 1).

of the analytical method; (ii) the content of Fe exchange forms is virtually the same in all types of sediment and easily exchangeable forms of other elements are not observed; (iii) concentrations of Fe, Si and Al obtained from the crystalline phase increase in sulphide–hydroxide sediments compared with hydroxide–sulphide sediments – amorphous forms (probably Al–Si–Fe gel) are characteristic of the hydroxide-sulphide sediments; (iv) the content of insoluble residues after leaching decreases from sulphide-rich to oxide-rich sediments; and (v) the ratio of amorphous to crystalline structures of Fe, Si and Al is higher in the sulphide-rich sediments than in the hydroxide-rich sediments.

Core analysis has been used to give an index in which the nature of the bottom sediments (colour, texture) corresponds to a typical core composition. Three groups of sediments were distinguished: (i) hydroxide–sulphide sediments (green–black, grey and black) containing Fe sulphides and barite, which are predominant in the vicinity of active vents (up to 50 m distance); (ii) sulphide–hydroxide sediments with predominant Fe hydroxides and dark brown, brick red and reddish brown sulphide minerals with surface oxides which occur within 50–250 m of the vent; and (iii) sediments (yellow, olive-yellow and olive), characterized by the transition of sulphides to hydroxides with markedly

varying compositions of oxide and sulphide minerals, occurring 100–350 m from smokers.

Iron hydroxides display a dendritic and granular appearance. Monocrystalline sulphide minerals are characterized by pyrite (marcasite), pyrrhotite, chalcopyrite and sphalerite, in order of decreasing abundance. Depletion in chalcopyrite (2.5-fold) and pyrrhotite (1.5-fold) is observed, whereas opal (2.4–11.2%) and quartz (2–5%) concentrations increase from the sulphide-rich towards the oxide-rich sediments.

Benthic fauna

Investigations on the benthic fauna from MAR hydrothermally active regions have been carried out over more than a decade. The material was mainly collected during geological expeditions. The main features of the MAR hydrothermal benthos and its structure have already been described. It is now clear that significant differences exist between Pacific and Atlantic hydrothermal fauna (Rona & Merril 1987; Mevel *et al.* 1989; Williams & Rona 1986; Fustec *et al.* 1987; Segonzac *et al.* 1993).

During the 15th and 23rd cruises of R/V *Akademik Mstislav Keldish* in 1988 and 1991, attention was focused on the spatial structure of the hydrothermal community as well as on the other fauna living in the vicinity of the venting

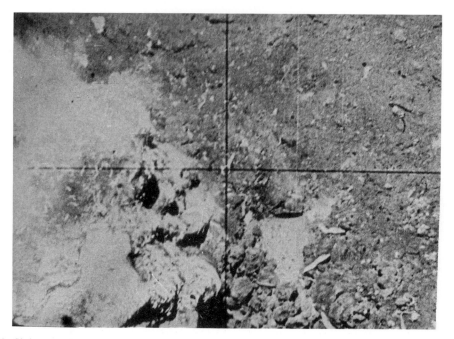

Fig. 3. Shrimps, crabs and fish near low-temperature venting zone (vent No. 2 on Fig. 1).

zone (Galkin & Moskalev 1990a). On the TAG field, faunal groups were distinguished according to oxidized and reduced ore deposits (Galkin & Moskalev 1990b). These results are used here to interpret photographic profiles obtained during the 3/2 cruise of R/V *Professor Logachev* on the Snake Pit hydrothermal field (Fig. 1).

Of the 1152 photographs studied here, benthic fauna (organisms and their remains) were observed in 252 photographs. The most typical are small sponges (Hexactinellidae), hydroids (Hydroidea), stem lilies (Crinoidea), crabs (Munidopsis and Galatheidae), sea anemones (Actiniaria), shrimps (*Alvinocaris markensis*), polychaetes (Serpulidae and Chaetopteridae), bivalve (Mytilidae), gorgonians (Elisella), holoturians (Chiridotidae), fishes (Macrouridae – *Coriphaenoides armatus*) plus bacterial mats. This short list demonstrates the predominance of minor sestonophages, which is typical of communities with a pronounced oligotrophic structure (Sokolova 1986).

The frequency of specific hydrothermal taxa on the photographs is lower than that recorded on the TAG field. Polychaete tubes (Chaetopteridae), mussel fragments and shrimps (*Alvinocaris markensis*) were observed (Figs 2 and 3). White patches of bacterial mats were found on many photographs and seem to be a good indicator of hydrothermal activity.

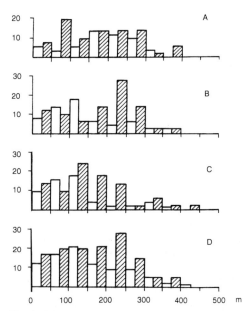

Fig. 4. Distribution of benthos (shaded columns) and nekton (open columns) versus distance from different active smokers (Fig. 1): (**A**) vent 1; (**B**) vent 2; (**C**) vent 3; and (**D**) total.

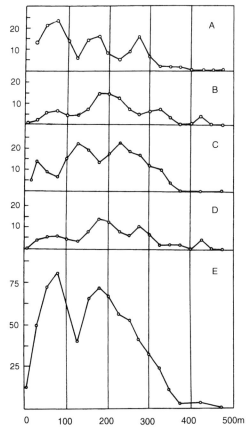

Fig. 5. Number of photographs on which organisms have been identified versus distance to the nearest black smoker: (**A**) bacterial mats; (**B**) holothurians; (**C**) sea anemones; (**D**) sponges; and (**E**) compiled data.

Relationship between biota-sediments-vents

The Snake Pit area investigated is characterized by limited sediment occurrence (coverage 0–50% of the photographed area and only 80–90% in individual photographs near hydrothermal vents). For comparison, within the TAG field, the proportion of photographs which display 100% of sediment coverage was 13–79%. The absence of bioturbated sediments indicates a poor infauna development. The faunal density indicated by the presence or absence of benthos and nekton organisms are shown on Fig 4 and 5. For the three black smokers in the vent field (Fig. 1), density generally varies with distance from the main high-temperature vents (Fig. 4A–C) This is particularly clear when looking at the compiled data (Fig. 4D). Similar trends are observed for at least two of the three chimneys investigated.

Zones where the benthos are more frequent occur between 50–100 and 200–300 m, with a maximum at 250 m from the nearest vents. The nekton distribution differs slightly. The zone of highest density occurs between 50 and 150 m with a maximum at 100–150 m. The next zone, situated between 150 and 300 m is characterized by the less frequent appearances of fish. In the interval 300–450 m, the frequency of fish is extremely low.

The frequency of the bacterial mats decreases with distance from the vent, with local maxima at 80, 180 and 280 m (Fig. 5A). These maxima may indicate the local occurrence of hot or warm venting zones. The distribution of non-vent fauna is illustrated in Fig. 5B–D, by holothurians, sea anemones and sponges. All the distributions display similar trends, where the frequency of organisms increases from 170 to 230 m and then decreases from 400 to 500 m. The cumulative data in Fig. 5E show the occurrence of two faunal maxima at distances of 80 and 180 m and a minimum between 380 and 500 m within the hydrothermal zone.

A correlation analysis was performed to determine links between the geological, geochemical and biological components of the Snake Pit hydrothermal zone and indicators of venting activity (Table 1). The frequency of the bacterial mats is positively correlated with the sediment cover, the occurrence of venting chimneys and the content of amorphous Fe and Si in sediments.

Figure 6 shows that the distribution of the bacterial mats is particularly limited in the 350–400 m band which surrounds the high-temperature chimneys and a similar trend is observed for the ratio of amorphous and crystalline forms of Fe and Si in sediments.

The density of sponges is positively correlated with distance and sediments containing Fe and Si crystalline forms and negatively correlated with the percentage of sediment cover.

Zonality of the Snake Pit field

The data show that the Snake Pit hydrothermal field is zoned according to deposit type, sediment geochemistry and benthic fauna. Three biogeochemical zones can be recognized: the Central, Intermediate and Outer Zones (Fig. 1).

The Central Zone is principally located at a distance between 50 and 100 m from the smokers. The sediment cover is densest in this area. Near the smokers the sediment is dark coloured due to high-temperature fluid precipitates and contains high levels of hydroxide–sulphide complexes, rich in amorphous forms of

Table 1. *Correlation Matrix for Biogeochemical Data*

	Distance (m)	Hydroth. sedim.	Chimneys	Vent smoke	Fe am	Fe cr	Si cr	Fe am/cr	Si am/cr	Insol. rem.
Distance (m)	1	*	*	*	*		*	*	*	*
Hydroth. sediment	-0.82	1	*	*	*		*	*	*	*
Chimneys	-0.61	0.71	1	*			*	*	*	*
Vent smoke	-0.73	0.74	0.49	1			*	*	*	*
Fe am	0.52	-0.49	-0.39	-0.41	1					
Fe cr	0.44	-0.31	-0.20	-0.18	-0.28	1	*		*	*
Si cr	0.63	-0.51	-0.61	-0.52	-0.44	0.83	1		*	*
Fe am/Fe cr	-0.50	0.81	0.72	0.61	0.29	-0.47	-0.48	1	*	*
Si am/Si cr	-0.81	0.88	0.91	0.72	0.19	-0.56	-0.66	0.72	1	*
Insoluble remains	-0.72	0.76	0.83	0.78	0.33	-0.61	-0.72	0.66	0.86	1
Bacterial mats	-0.49	0.45	0.55	0.20	0.46	-0.72	-0.69	0.77	0.81	0.79
Shrimps	-0.35	0.29	0.48	0.78	0.41	-0.29	-0.21	0.42	0.50	0.51
Crabs	-0.29	0.18	0.31	0.49	0.35	-0.18	-0.24	0.38	0.35	0.40
Polychaetes	-0.29	0.19	0.41	0.35	0.24	0.16	0.25	-0.12	0.24	0.28
Holothurians	-0.21	0.28	0.15	0.21	0.23	0.21	-0.17	0.22	0.11	0.19
Sea Anemones	-0.18	0.05	0.27	0.22	0.46	0.31	0.47	0.13	0.10	0.18
Sponges	0.54	-0.66	-0.39	-0.35	-0.51	0.48	0.51	-0.41	-0.29	-0.24
All biota	-0.38	0.43	0.25	0.28	0.24	-0.37	-0.33	0.19	0.43	0.48

Table 1. *(continued)*

	Bacterial mats	Shrimps	Crabs	Polychaetes	Holothurians	Anemones	Sponges	All biota
Distance (m)	*						*	
Hydroth. sediment							*	
Chimneys	*							
Vent smoke		*	*					
Fe am							*	
Fe cr	*							
Si cr	*						*	
Fe am/Fe cr	*							
Si am/Si cr	*	*						
Insoluble remains	*	*						
Bacterial mats	1					*		
Shrimps	0.23	1	*			*		
Crabs	0.25	0.49	1					
Polychaetes	0.34	0.41	0.39	1				
Holothurians	-0.44	-0.21	-0.25	-0.15	1		*	
Sea Anemones	0.64	0.51	0.47	0.11	-0.21	1		
Sponges	-0.39	-0.25	-0.35	-0.10	0.51	0.21	1	
All biota	0.41	0.11	0.12	-0.15	0.44	0.45	0.28	1

* Level of significance 0.001–0.1. am, amorphous; cr, crystalline.

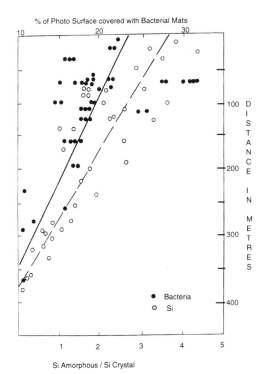

Fig. 6. Percentage of surfaces of photographs covered with bacterial mats and Si amorphous/Si crystalline ratio in the hydrothermal sediments versus distance to the nearest black smoker.

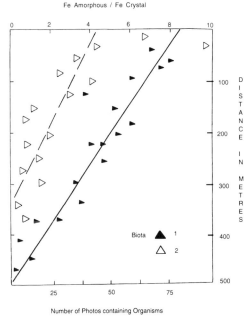

Fig. 7. Number of photographs on which organisms are seen and Fe amorphous/Fe crystalline ratio in the hydrothermal sediments versus distance to the nearest black smoker.

Fe, Si and Al (Figs 6 and 7). Within the Central Zone, the high-temperature vent area is narrow and forms a band of between 10 and 15 m around the black smokers. This area is characterized by the presence of the vent-associated organisms composed of Chaetopteridae polychaetae, Mytilidae bivalves, *Alvinocaris markensis* shrimps and bacterial mats. Up to a distance of 50–80 m from the high-temperature vents an increase is observed in the majority of non-vent organisms (Fig. 5).

The Intermediate Zone occurs at a distance from the high-temperature springs of 50–200 m. The fauna is mostly composed of hydroids, holothurians, Munidopsis ostracods and gorgonians, which display a maximum density in the interval 120–180 m. The frequency of bacterial mats is at a maximum in this zone near 180 m. Sediments are mainly composed of sulphide–hydroxide complexes in which Fe hydroxides predominate. Concentrations of amorphous and crystalline forms of Fe, Si and Al are approximately equal.

The Outer Zone is characterized by a gradual decrease in sediments of hydrothermal origin and the benthic fauna are mostly of the non-vent type. This area is located between 150 and 500 m from the smokers. The predominance of hydroxide minerals relative to sulphides in sediments as well as crystalline forms of Fe, Si, Al is typical.

Outside this hydrothermally influenced zone, the floor of the median valley graben is covered by basalts locally covered by a thin sediment film. The benthos is evidently depressed, indicating a pronounced oligotrophic ecological structure which exists under the poor productivity of the plankton community and the weakened dynamics of bottom waters.

To summarize, the occurrence of bacterial mats and hydrothermal sediments generally decreases with distance from the active vents. Bacterial mats are mainly concentrated in the Central and Intermediate zones, but can also be found in the Outer Zone in association with local venting (Fig. 6). They are markers of local gas discharge or low-temperature hydrothermal venting at the periphery of the high-temperature hydrothermal areas.

Comparison to other MAR sites

The above observations are compared with the three hydrothermal vent fields along the

Mid-Atlantic Ridge: TAG (26°N), Snake Pit (23°N) and a new active field between 14°40' and 14°48'N (Batuyev *et al.* 1994). The observations allow us to formulate a general comparison, although very few data have been obtained from the latter southernmost field (15°N).

The TAG area is the most hydrothermally active field of the three zones described. The sediment cover is thicker and has a wider distribution than in the Snake Pit area. Traces of the re-sedimentation (funnels) and bioturbation are observed at the surface of the sediments at TAG which are absent from the Snake Pit field. The biological activity is evidently high in the TAG field. Apart from these differences, the main features of the biological communities are similar.

The new hydrothermal field found near 15°N is less hydrothermally active than both the TAG and Snake Pit areas, although the dimension and distribution of the sulphide deposits are broadly equivalent (Batuyev *et al.* 1994). The extensive cover of the sediment and the absence of active chimneys on most of the sulphide mounds indicate the greater age of the structure. The characteristics of the Snake Pit hydrothermal sediments are absent and indicate differences in the composition of the sediments, which have mostly a non-hydrothermal origin. The specific hydrothermal fauna commonly found on the TAG and Snake Pit fields was not observed at 15°N, although bacterial mats are present, indicating low-temperature hydrothermal activity. The composition and distribution (such as mussel and gastropods colonies) of benthic fauna at 15°N differs from those of the TAG and Snake Pit fields.

Comparisons between the fields indicate a decrease in the hydrothermal and biological activities with latitude (TAG – Snake Pit – 15°N) and might be due to their temporal variability, as suggested in the Pacific (Fustec *et al.* 1987). The vent influence on the benthic fauna and the geochemistry of the sediments also decreases with latitude from 26° to 15°N.

Conclusion

Each field is characterized by individual faunal relationships and geochemical features and further work is necessary before a satisfactory understanding of the variation is established.

We are grateful to the RFFI Foundation for financial support in the form of a grant to the Research Institute for Geology and Mineral Resources of the Ocean for Project 94–05–1687a. Data used in this study were made available by V. Markov and A. Ashadze.

References

ARQUIT, A. M. 1990. Geological and hydrothermal controls on the distribution of megafauna in ASHES vent field, Juan de Fuca Ridge. *Journal of Geophysical Research*, **95**, 947–12 960.

BATUYEV, B. M., KROTOV, A. G., MARKOV, V. F., CHERKASHEV, G. A., KRASNOV, S. G. & LISITSYN, YE. D. 1994. Massive sulphide deposits discovered and sampled at 14°45'N, Mid-Atlantic Ridge. *BRIDGE Newsletter*, 6 April, 6–10.

CHESTER, R. & HUGHES, M. 1967. A chemical technique for separation of ferromanganese minerals, carbonate minerals and absorbed trace elements from pelagic sediments. *Chemical Geology*, **3**, 249–265.

FOMINA, L. S. 1975. *Methods of Ferromanganese Nodule Analysis*. Nauka, Moscow, 280 pp [in Russian].

FUSTEC, A., DESBRUYERES, D. & KIM, J. S. 1987. Deep-sea hydrothermal vent communities at 13°N on the East Pacific Rise: microdistribution and temporal variations. *Biological Oceanography*, **4**, 121–164.

GALKIN, S. V. & MOSKALEV, L.I. 1990*a*. Study of abyssal fauna of the North Atlantic using deep-sea inhabited apparatus. *Okeanologia*, **30**, 682–689 [in Russian].

——, & —— 1990*b*. Fauna of the Mid-Atlantic Ridge hydrothermal. *Okeanologia*, **30**, 842–847 [in Russian].

MEVEL, C., AUZENDE, J. M. & 9 others 1989. La Ride du Snake Pit (dorsale medio-atlantique, 23°22'N); resultats preliminaires de la campagne HYDROSNAKE. *Comptes Rendus de l'Academie des Sciences, Paris, II*, **308**, 545–552.

RONA, P. A. & MERRILL, G. F. 1987. Benthic invertebrates from the Mid-Atlantic Ridge. *Bulletin of Marine Science*, **28**, 371–375.

SEGONZAC, M. 1992. Les peuplement associes a l'hydrothermalisme oceanique due Snake Pit (dorsale medio-atlantique; 23°N, 3480 m): composition et microdistribution de la megafaune. *Comptes Rendus de l'Academie des Sciences, Paris, III*, **314**, 593–600.

——, DE SAINT LAURENT, M. & CASANOVA, B. 1993. L'enigme du comportement trophique des crevettes Alvinocarididae des sites hydrothermaux de la dorsale medio-atlantique. *Cahiers de Biologie Marine*, **34**, 535–571.

SOKOLOVA, M. N. 1986. *Nourishment and Trophic Structure of the Deep-sea Benthos*. Nauka, Moscow, 208 pp [in Russian].

SUDARIKOV, S. M. & ASHADZE, A. M. 1989. Distribution of metals in hydrothermal plumes of Atlantic Ocean according to ICP data. *Doklady Akademii Nauk*, **308**, 452–456 [in Russian].

—— & CHERKASHEV, G. A. 1993. Structure of hydrothermal plumes of the Pacific and Atlantic Oceans. *Doklady Akademii Nauk*, **330**, 757–759 [in Russian].

——, ASHADZE, A. M., STEPANOVA, T. V., NESHERETOV, A. B., VOROBIOV, P. V., KASABOV, R. V. & GUTNIKOV, A. S. 1990. Hydrothermal activity and ore-forming in the rift zone of the Mid-Atlantic

Ridge (new data). *Doklady Akademii Nauk*, **311**, 440–445 [in Russian].

WILLIAMS, A. B. & RONA, P. A. 1986. Two new caridean shrimps (Bresiliidae) from hydrothermal field on the Mid-Atlantic Ridge. *Journal of Crustacean Biology*, **6,** 446–462.

Lipid characteristics of hydrothermal vent organisms from 9°N, East Pacific Rise

GARETH RIELEY[1], CINDY L. VAN DOVER[2], DAVID B. HEDRICK[3],
DAVID C. WHITE[3] & GEOFFREY EGLINTON[1,4]

[1]*Organic Geochemistry Unit, School of Chemistry, University of Bristol, Bristol BS8 1TS, UK*

[2]*Department of Chemistry, Woods Hole Oceanographic Institute, Woods Hole, MA 02543, USA*

[3]*Center for Environmental Biotechnology, Knoxville, TN 37932-2567, USA*

[4]*Biogeochemistry Research Centre, University of Bristol, Bristol BS8 1TS, UK*

Abstract: Lipid compositions are reported for three distinctive deep-sea hydrothermal vent invertebrate species collected around 9°N East Pacific Rise: *Riftia pachyptila* Jones, a vestimentiferan tubeworm; *Bathymodiolus thermophilus* Kenk and Wilson, a mussel; and *Halice hesmonectes* Martin *et al.*, an amphipod crustacean. The lipid compositions of all these organisms were dominated by components characteristic of diets based on bacteria, with only very minor contributions from carbon derived from the oceanic photic zone. In all the organisms studied, large abundances of n-7 fatty acids, polyunsaturated fatty acids with unsaturations separated by more than one methylene bond, and sterol distributions dominated by cholesterol were observed. Branched fatty acids were generally of low abundance, whereas polyunsaturated fatty acids separated by single methylene groups were either absent, as in the case of *R. pachyptila*, or in very low abundance, as in the case of *B. thermophilus* and *H. hesmonectes*. Monounsaturated fatty acids were the most abundant component of *R. pachyptila* lipids, whereas non-methylene interrupted fatty acids were particularly abundant in the lipids of *B. thermophilus* (up to 45% of total fatty acids). The lipids of *H. hesmonectes* were dominated by storage lipids (e.g. wax esters). Stable carbon isotope analyses of individual sterols from the organisms examined allow specific sources to be proposed for these biochemicals. The $\delta^{13}C$ values of sterols from *R. pachyptila* were consistent with *de novo* biosynthesis, whereas that of cholesterol from *B. thermophilus* corresponded to that from marine phytoplankton. The $\delta^{13}C$ values of sterols from *H. hesmonectes* fell into two different groups and suggest that at least two distinct sources of sterols are available to these crustacea in the vent ecosystem, one of which derives from phytoplankton. Overall, the combination of the interpretation of lipid structure and distribution with compound specific isotope analyses can lead to valuable insights into trophic relationships within the deep-sea hydrothermal ecosystem.

Lipid structures and distributions, especially those of fatty acids and sterols, can be used to gain information about the ecological conditions and diets of various types of marine organisms (e.g. Sargent & Whittle 1981; Bradshaw *et al.* 1990). The vast majority of studies on marine invertebrate lipids related to diet have been on coastal and estuarine species which depend on carbon derived from photosynthetic organisms; as such, the lipids of these invertebrates are generally dominated by structures which can be related directly or indirectly to the corresponding structures in the dietary lipids (Bradshaw 1988). Investigations on organisms which derive part of their diet from bacterial sources, such as those which harbour endosymbiotic bacteria,

demonstrate that corresponding structural similarities can be observed between host and symbionts (Conway & McDowell Capuzzo 1991; Ben-Mligh *et al.* 1992; Fang *et al.* 1993).

Primary production by chemosynthetic bacteria forms the base of food webs within deep-sea hydrothermal vents and therefore a bacterial signature is expected in the lipids of vent organisms, though carbon derived directly from the photic zone may play a significant part (Dixon *et al.* this volume). Ecologically and biochemically interesting cases of eukaryotic dependence on bacterial primary production are found in associations between endosymbiotic chemoautotrophic bacteria and invertebrate hosts, with vestimentiferan tubeworms [e.g.

From PARSON, L. M., WALKER, C. L. & DIXON, D. R. (eds), 1995, *Hydrothermal Vents and Processes*, Geological Society Special Publication No. 87, 329–342.

Table 1. *Sample key to hydrothermal vent organisms studied, by site*

Sites	Samples	Latitude	Longitude
Site A	A123	9° 48.40′	104° 17.16
	A187		
	Add		
Site B	A158	9° 49.86′	104°17.41
Site C	A160	9° 50.56′	104°17.49
	A162		
	Vest		
	Troph		
Site D	A200	9°30.88′	104°14.67
	A202		
	A214		
	A216		
	Gill		
	Foot		

Axxx indicate amphipod (*Halice hesmonectes*) samples; Foot = mussel (*Bathymodiolus thermophilus*) foot; Add = mussel adductor; Gill = mussel gill; Vest = tubeworm (*Riftia pachyptila*) vestimentum; Troph = tubeworm trophosome.

Riftia pachyptila, the giant tubeworm from hydrothermal vents on the East Pacific Rise (EPR)] providing the most extreme example of this kind of association. *Riftia pachyptila* and other species within the vestimentifera have lost all vestiges of a digestive system in the adult and rely almost exclusively on their endosymbionts for nutrition, although some uptake of dissolved organic material from the surrounding seawater may make a contribution to the diet of these worms. In bathymodiolid mussels (e.g. *Bathymodiolus thermophilus* from eastern Pacific hydrothermal vents), a functional gut is present and nutrition from endosymbiotic bacteria in the gills is apparently supplemented by suspension feeding (Le Pennec & Prieur 1984). Of equal interest to host/symbiont associations are the myriad species of grazing invertebrates at hydrothermal vents (small crustaceans, polychaetes, molluscs, etc.) that are ultimately dependent on chemoautotrophic production. From the same region where *R. pachyptila* and *B. thermophilus* specimens were abundant, we sampled several populations of an amphipod crustacean, *Halice hesmonectes* (Martin *et al.* 1993), which swarms in the outflow area of warm water vents, most often above mussels or tubeworms but also above areas of bare rock (Van Dover *et al.* 1992; Kaartvedt *et al.* 1994).

Halice hesmonectes has been presumed to ingest suspended micro-organisms flowing out from subsurface reservoirs (Van Dover *et al.* 1992), though ingestion of detritus, as is common for pelagic amphipods, is also likely to be

important. Therefore, in addition to looking for bacterial signatures in the lipids of the amphipods, we were interested in identifying lipids that might be derived from vent organisms or from carbon derived from the oceanic photic zone. Amphipods, and crustaceans in general, are held to be unable to synthesize sterols *de novo* (Goad 1978; Kerr & Baker 1991), and as *H. hesmonectes* seem to lack symbiotic bacteria (Martin *et al.* 1993), they must assimilate sterols either directly from their diet or from modification of dietary sterols. To help assign sources of individual sterols, compound specific isotope analyses (CSIA) have been employed (e.g. Hayes *et al.* 1990; Rieley *et al.* 1991). Stable carbon isotope compositions ($\delta^{13}C$ values) are useful because of the distinctive fractionations imparted due to carbon source and metabolic effects (Hayes 1993).

Experimental

Sampling

The organisms in this study were collected at active venting sites along the EPR at 9°N during December 1991 using the DSV *Alvin* (support ship RV *Atlantis II*, chief scientists Mullineaux & Van Dover). Vestimentiferum tubeworm (*R. pachyptila*) and mussel (*B. thermophilus*) samples were collected using the manipulator arm of the submersible. Amphipods (*H. hesmonectes*) were collected using a vacuum pump system. On board ship, the amphipod samples were individually sorted and oven-dried at 60°C for 12 hours. On reaching the surface the remaining samples were dissected on board ship and either dried (add and vest) as for the amphipods or stored at −70°C until further analysis on land (gill, foot and troph). A key to the samples examined is given in Table 1.

Samples

Sites. Samples were collected from four geographically distinct locations along the EPR and designated sites A to D as indicated in Table 1; a guide to the organisms sampled is given in the following sections.

Tubeworms. The tubeworm samples were of *R. pachyptila* Jones (Phylum Vestimentifera; Class Axonobranchia). The tissues analysed were vestimentum (which harbours no symbionts) and trophosome (where sulphur-oxidizing bacteria live). All tissues were from single organisms.

Mussels. The mussel samples were all of *B. thermophilus* Kenk and Wilson (Phylum Mollusca; Class Bivalvia). The tissues examined were foot and adductor muscles, both of which are symbiont free, and gill which harbours sulphur-oxidizing bacteria. All tissues were from single organisms.

Amphipods. The amphipods were all of *H. hesmonectes* Martin *et al.* (1993) (Phylum Arthropoda; Class Crustacea). For analyses, approximately 10–30 organisms were pooled (individual organisms being 1–3 mm length) and so amphipod analyses represent 'snapshots' of the feeding behaviour of a larger population.

Lipid extraction

Two main isolation procedures were used for the preparation of lipids, the first being the preparation of total fatty acids and total sterols. A second procedure was used so that fatty acids could be separated into biochemically distinct classes, i.e. those bound as neutral (storage) and polar (structural/membrane) lipids.

Total lipid (TL) samples were extracted by sonication of dried samples in 2:1 dichloromethane/methanol; TL extracts were subsequently saponified in saturated KOH/methanol following the method of Bradshaw *et al.* (1990). Total lipid fatty acids (TLFA) and total lipid neutrals were separated using aminpropyl phase cartridges (Bond Elut; Analytichem), eluting total lipid extracts with 2:1 dichloromethane/isopropanol for neutral lipid fractions (including sterols) and 2% acetic acid in ethyl ether for fatty acid fractions. Before gas chromatography (GC) analyses TL fractions were derivatized using BSTFA.

For the separation of structural and storage lipids, dried samples were extracted using a modified Bligh/Dyer method (Bligh & Dyer 1959; Rieley 1993). The resultant extract was separated into the required lipid extracts using silicic acid column chromatography, sequentially eluting with chloroform, acetone and methanol to give neutral lipid (NL) fractions from the chloroform elution step and polar lipid (PL) fractions from the methanol elution step (Hedrick *et al.* 1991).

Neutral lipids. Neutral lipid fractions were saponified as for TL samples, though instead of derivatizing free fatty acids with BSTFA, the acids were converted to neutral lipid fatty acid (NLFA) methyl esters using BF_3/methanol (Bradshaw *et al.* 1990).

Polar lipids. Fatty acids were released from the PLs by heating PL fractions at 100°C for one hour in 0.05 M KOH in methanol, giving polar lipid fatty acid (PLFA) methyl esters (Hedrick *et al.* 1991).

Lipid characterization

Gas chromatography/gas chromatography – mass spectrometry. Lipids were analysed by GC using a 50 m bonded methyl silicone phase capillary column (CP Sil 5CB; Chrompack), 0.12 μm thickness, 0.32 mm internal column diameter. Fatty acid samples were also analysed on a Supelco Omegawax 320, 30 m × 0.32 mm i.d., 0.3 μm phase thickness capillary column. Hydrogen was used as carrier gas. Lipids were identified using GC–mass spectrometry (GC–MS), using a GC system as above, except using helium as the carrier gas. The gas chromatograph was attached to a Finnigan MAT 4500 mass spectrometer (70 eV ionization voltage, scan time one second). The temperature programme for the methyl silicone column was 40–150°C at 10°C min^{-1}, 150–300°C at 4°C min^{-1}, isothermal for 25 min. The temperature programme for the Omegawax 320 column was 40–150°C at 10°C min^{-1}, 150–240°C at 4° min^{-1}, isothermal for 10 min.

Positions of fatty acid unsaturation were determined by the formation of dimethyl disulphide (DMDS) adducts of fatty acid methyl esters, followed by GC–MS identification (Yruela *et al.* 1990).

Gas chromatography – isotope ratio mass spectrometry. Isotopic compositions of sterols were determined using a Finnigan MAT delta-S isotope ratio mass spectrometer as described by Rieley *et al.* (1993). Values reported are corrected for the addition of trimethyl silyl groups (*cf.* Rieley 1994).

Results

Fatty acid compositions

Fatty acid nomenclature. Fatty acids are conventionally written as a:b(n-c), where a is the carbon number of the fatty acid, b the number of double bonds and c the number of carbons to the first double bond from the terminal methyl. Unless otherwise noted, all double bonds are separated by one methylene group. The prefixes *i* and *a* indicate methyl branching at positions 2 and 3, respectively, from the terminal methyl.

Tubeworm (Riftia pachyptila). The fatty acid compositions of the tubeworm samples were

G. RIELEY *ET AL.*

Table 2. *Total fatty acid concentrations, in* µg g^{-1} *total extract, and relative concentrations for different classes of fatty acid, as % of total fatty acids identified, for total lipid (TLFA), neutral lipid (NLFA) and polar lipid fatty acid (PLFA) fractions of the organisms studied. Sites refer to geographical localities given in Table 1*

Sample:	A158	A162	A160	A202	A214	A187	Add	Vest
Site:	B	C	C	D	D	A	A	C
TLFA								
Total	41 761	40 253	28 455	35 534	31 022	25 972	9728	21 897
Sats	21.4	17.7	16.9	17.4	21.3	16.9	30.3	17.0
Unsats	78.6	82.3	83.1	82.6	78.7	83.1	69.7	80.1
Polyuns	9.2	12.6	8.9	17.2	6.5	9.1	40.0	3.8
Monouns	69.3	69.7	74.2	65.4	72.2	74.0	29.7	76.3
n-7	60.6	60.3	66.5	60.6	58.9	66.6	38.1	78.7
n-9	3.4	3.9	9.2	12.0	8.9	8.9	4.8	0.7
NMI	8.9	12.2	8.2	17.0	6.5	9.1	40.0	3.8
Branched	1.1	2.6	2.0	1.2	1.7	2.0	1.0	0.4

Sample:	A158	A162	A160	A202	A214	Gill	Foot	Troph
Site:	B	C	C	D	D	D	D	C
NLFA								
Total	9208	22 259	13 487	21 498	21 718	1453	1861	11 280
Sats	31.3	3.7	20.4	18.4	12.4	21.9	36.2	10.7
Unsats	68.7	96.3	79.7	81.6	87.6	78.1	63.8	86.2
Polyuns	14.1	4.4	24.8	22.7	18.9	42.4	31.5	3.0
Monouns	54.7	91.9	54.9	58.9	68.7	35.7	32.3	83.2
n-7	43.5	89.0	36.8	34.5	46.8	37.0	25.8	84.4
n-9	9.71	3.4	19.1	32.0	22.5	7.6	6.2	0.5
NMI	14.0	4.0	22.5	22.5	17.0	42.4	31.5	3.0
Branched	0.8	1.1	5.1	1.7	1.6	1.9	1.1	0.5

Sample:	A158	A162	A160	A202	A214	Gill	Foot	Troph
Site:	B	C	C	D	D	D	D	C
PLFA								
Total	8497	6413	8147	9562	5040	1091	2074	2957
Sats	34.3	21.2	16.0	20.6	19.1	21.0	20.7	17.9
Unsats	65.7	78.9	84.0	79.4	80.9	79.0	79.3	82.1
Polyuns	5.7	10.5	12.5	13.9	11.7	44.5	45.2	7.6
Monouns	60.0	68.4	71.5	65.5	69.2	34.5	34.1	74.5
n-7	52.7	62.1	57.5	54.7	56.4	42.1	44.8	73.2
n-9	2.2	3.0	7.7	8.5	4.9	13.0	11.2	2.4
NMI	5.3	9.6	10.7	11.9	10.2	43.8	44.5	4.4
Branched	0.7	0.3	0.7	0.5	0.4	3.8	1.3	0.1

Sats = saturated; unsats = unsaturated; polyuns = polyunsaturated; monouns = monounsaturated; NMI = non-methylene interrupted.

characterized by very high abundances of (n-7) fatty acids with up to 84% of the total fatty acids of the tubeworm trophosome storage lipids having unsaturations seven carbons from the terminal methyl (Tables 2–5). The major (n-7) fatty acids identified in all lipid extracts were 16:1(n-7) and 18:1(n-7) (Tables 3–5). Only minor amounts of (n-9) fatty acids were observed, with a maximum of 2% of trophosome structural lipids being of this type (Table 2). Small amounts of non-methylene interrupted (NMI) dienoic fatty acids were observed in all the tubeworm samples, at around 3–4% of total fatty acids (Tables 2–5). No single-methylene interrupted (SMI) polyunsaturated fatty acids were detected in the tubeworm samples. Only very small abundances of branched fatty acids (0.1–0.5% of total fatty acids; Table 2) were

Table 3. *Relative concentrations of fatty acids, as % of total lipid fatty acid extract (identified as fatty acid silyl esters). Unidentified isomers are given in order of elution (CP Sil 5CB, 50 m). Samples and sites as in Table 1*

GC*	Fatty acid	A158 B	A162 C	A160 C	A202 D	A214 D	A187 A	Add A	Vest C
–	14:0	0.86	6.39	0.17	1.01	0.97	0.78	0.65	3.23
1	15:0	–	–	0.23	0.07	–	0.23	–	–
3	16:0	14.11	7.49	12.1	10.74	15.49	12.07	15.10	12.35
6	17:0	–	0.67	–	0.15	0.13	–	–	–
12	18:0	2.49	0.33	1.75	1.86	2.31	1.77	12.35	1.03
–	19:0	–	–	–	–	0.70	–	–	–
22	20:0	2.85	0.20	–	2.36	0.03	–	0.31	–
29	22:0	–	–	–	–	–	–	0.97	–
–	i14:0	–	1.67	–	–	–	–	–	–
4	i17:0	0.27	–	0.46	0.22	0.36	0.47	0.46	–
–	a17:0	–	–	–	–	0.25	–	–	–
14	i19:0	0.87	0.94	1.12	1.02	0.43	1.14	0.96	0.38
14	a19:0	–	–	0.42	–	0.62	0.43	–	–
–	i21:0	–	–	–	–	–	–	–	–
–	14:1(n-7)	0.51	–	0.32	0.38	· 0.29	0.36	–	–
2	16:1(n-7)	15.74	33.50	23.3	19.27	25.99	23.35	5.31	54.27
5	17:1(n-7)	0.78	0.13	–	0.49	–	–	0.19	–
–	18:1(n-5)	1.10	3.93	1.15	–	0.70	1.15	–	–
11	18:1(n-7)	34.24	21.81	32.16	23.91	25.95	32.06	2.91	20.19
10	18:1(n-9)	–	2.67	6.79	6.37	6.00	6.42	0.67	0.74
10	18:1(n-13)	4.71	2.35	0.92	0.95	–	0.40	1.04	1.03
15	19:1(n-7)	0.14	–	–	–	–	–	3.53	–
21	20:1(n-7)	5.34	3.07	5.36	5.00	5.61	5.43	4.05	0.45
20	20:1(n-9)	1.84	0.34	1.57	4.06	2.39	2.70	2.02	–
20	20:1(n-13)	4.41	1.92	2.65	4.98	5.32	–	7.61	0.62
–	21:1(n-7)	–	–	–	–	–	–	0.49	–
28	22:1(n-7)	0.50	–	–	–	–	–	0.63	–
–	22:1(n-9)	–	–	–	–	–	–	1.21	–
9	18:2(n-7,13)	0.15	–	–	–	0.41	–	–	0.81
9	18:2(n-7,15)	–	–	0.85	1.98	–	0.87	0.32	–
8	18:2(n-9,15)	0.87	–	0.71	0.21	0.45	0.72	2.10	–
19	20:2(n-7,13)	–	–	–	0.23	–	–	1.02	–
19	20:2(n-7,15)	0.75	1.13	3.86	1.80	0.56	3.94	10.16	2.04
18	20:2(n-9,15)	0.85	–	0.77	2.99	2.23	0.79	2.98	–
24	21:2(n-7,14)	1.81	0.24	–	3.56	0.03	–	1.48	–
24	21:2(n-7,16)	–	–	–	–	–	–	1.32	–
–	21:2(n-?)	–	–	–	–	–	–	0.69	–
27	22:2(n-7,15)	1.11	0.38	0.61	1.21	0.10	0.62	7.28	0.95
27	22:2(n-9,15)	0.72	0.93	0.18	1.39	0.10	0.19	0.03	–
–	24:2(n-?)	0.69	0.29	–	2.74	0.02	–	–	–
7	18:3	1.04	9.19	0.94	0.67	0.25	0.96	8.38	–
–	18:3	–	–	–	–	–	–	–	–
17	20:3	0.94	–	1.01	0.20	0.51	1.03	3.43	–
25	22:3	–	–	–	–	1.78	–	0.35	–
26	22:3	–	–	–	–	–	–	0.17	–
16	20:4(n-6)	0.29	0.39	–	0.20	0.04	–	–	–

* Numbers refer to GC peaks in Fig. 1.

observed in the lipids of all the tubeworm tissues examined, the most abundant component generally being *i*19:0 (Tables 3–5).

Mussel (Bathymodiolus thermophilus). The fatty acids of the mussel samples were characterized by very large abundances of NMI polyunsaturated fatty acids, as can be observed on the gas chromatogram displayed in Fig. 1. This class of fatty acid represented up to 45% of the total fatty acids identified in the mussel tissues (Tables 2–5). The most abundant components observed were 20:2(n-7, n-15) and 22:2(n-7, n-15). Fatty acids with n-7 unsaturations were by far the most dominant type observed, being up to 45% of the total identified (Table 2); this was

Table 4. *Relative concentrations of neutral lipid fatty acids, as % total neutral lipid fatty acids identified as their methyl esters. Unidentified isomers are given in order of elution (CP Sil 5CB, 50 m). Samples and sites as in Table 1*

GC*	Fatty acid	A158 B	A162 C	A160 C	A202 D	A214 D	Gill D	Foot D	Troph C
–	14:0	–	–	–	–	–	0.61	0.66	–
1	15:0	–	–	–	–	–	0.11	0.39	–
3	16:0	19.40	1.95	5.43	9.49	8.29	6.99	16.17	9.47
6	17:0	–	–	–	–	–	–	–	–
12	18:0	8.58	0.51	4.23	4.73	2.29	8.98	16.69	0.80
–	19:0	–	–	–	–	–	0.14	0.10	–
22	20:0	2.44	0.17	5.62	2.47	0.19	2.17	0.34	–
29	22:0	–	–	–	–	–	0.95	0.74	–
–	i14:0	–	–	–	–	–	–	–	–
4	i17:0	0.28	0.26	0.31	0.09	0.25	0.47	–	–
–	a17:0	–	–	–	–	–	0.44	0.42	–
14	i19:0	0.23	0.16	1.17	–	0.32	0.97	0.37	0.45
14	a19:0	0.06	0.58	1.80	0.68	0.94	–	–	–
–	i21:0	–	–	–	–	–	0.31	0.36	–
–	14:1(n-7)	–	–	–	–	–	–	–	–
2	16:1(n-7)	9.43	76.84	13.02	8.01	11.83	4.15	4.49	61.69
5	17:1(n-7)	0.25	0.10	1.79	0.94	0.10	–	–	–
–	18:1(n-5)	–	–	–	–	–	–	–	–
11	18:1(n-7)	26.94	10.49	15.39	20.95	28.85	2.32	2.01	19.71
10	18:1(n-9)	3.64	2.29	9.12	8.71	8.94	0.76	0.76	0.50
10	18:1(n-13)	0.91	0.13	0.62	0.45	0.49	1.56	1.71	0.30
15	19:1(n-7)	–	–	–	–	–	5.14	2.75	–
21	20:1(n-7)	9.76	0.97	6.34	5.21	5.97	9.20	8.53	0.45
20	20:1(n-9)	1.22	0.55	5.76	10.35	7.72	2.63	2.20	–
20	20:1(n-13)	2.76	0.57	4.61	5.21	4.94	6.41	7.77	0.57
–	21:1(n-7)	–	–	–	–	–	0.71	0.48	–
28	22:1(n-7)	–	–	–	–	–	0.19	0.09	–
–	22:1(n-9)	–	–	–	–	–	2.18	1.13	–
9	18:2(n-7,13)	0.22	1.01	0.71	0.12	1.12	0.14	0.32	0.80
9	18:2(n-7,15)	0.49	0.13	0.93	–	0.46	–	–	–
8	18:2(n-9,15)	2.90	0.26	1.13	0.90	1.43	0.51	0.34	–
19	20:2(n-7,13)	2.90	0.06	0.35	0.36	1.20	–	–	–
19	20:2(n-7,15)	–	0.03	0.55	0.28	1.50	11.79	7.40	1.39
18	20:2(n-9,15)	1.52	0.18	1.29	10.35	2.23	3.69	2.84	–
24	21:2(n-7,14)	2.28	0.30	3.68	3.81	0.25	1.59	1.40	–
24	21:2(n-7,16)	–	–	–	–	–	2.12	1.02	–
–	21:2(n-?)	–	–	–	–	–	0.45	0.78	–
27	22:2(n-7,15)	1.25	0.23	2.15	0.93	1.61	8.5	5.53	0.83
27	22:2(n-9,15)	0.43	0.16	1.84	1.70	2.14	–	–	–
–	24:2(n-?)	1.06	0.48	1.53	2.85	0.55	–	–	–
7	18:3	–	0.05	0.77	0.12	0.45	7.26	7.53	–
–	18:3	0.96	1.15	6.87	1.10	2.64	–	–	–
17	20:3	–	0.03	0.69	–	1.36	5.32	3.31	–
25	22:3	–	–	–	–	–	0.80	0.44	–
26	22:3	–	–	–	–	–	0.16	0.53	–
16	20:4(n-6)	0.08	0.36	2.30	0.17	1.93	–	–	–

* Numbers refer to GC peaks in Fig. 1.

in contrast to the low abundance of n-9 fatty acids (around 5–13%; Table 2). Small amounts of the SMI polyunsaturated fatty acid arachidonic acid 20:4(n-6) were detected (0.7% of structural PLFA; Table 5).

Minor abundances of branched fatty acids were observed (around 1–3%; Table 2), the most abundant component being i19:0 (Tables 3–5). Greater abundances of branched fatty acids were observed in the gill tissues than in the foot and adductor tissues (Table 2).

Table 5. *Relative concentrations of polar lipid fatty acids, as % total polar lipid fatty acids, identified as their methyl esters. Unidentified isomers are given in order of elution (CP Sil 5CB, 50 m). Samples and sites as in Table 1*

GC*	Fatty acid	A158 B	A162 C	A160 C	A202 D	A214 D	Gill D	Foot D	Troph C
–	14:0	–	–	–	–	–	–	–	–
1	15:0	0.17	0.07	0.06	0.06	0.06	0.08	0.13	0.02
3	16:0	24.30	17.19	11.52	15.24	16.19	5.82	9.53	16.30
6	17:0	–	–	–	–	–	–	–	–
12	18:0	5.91	2.55	2.75	3.05	1.75	9.99	9.29	1.37
–	19:0	0.02	0.05	0.03	0.02	0.02	0.15	0.07	–
22	20:0	2.89	0.79	0.78	1.24	0.56	0.91	0.46	0.14
29	22:0	0.12	0.04	0.14	0.36	0.06	0.32	0.26	0.04
–	i14:0	–	–	–	–	–	–	–	–
4	i17:0	0.42	0.19	0.22	0.24	0.22	0.15	0.21	0.02
–	a17:0	–	–	–	–	–	–	–	–
14	i19:0	0.12	0.08	0.49	0.29	0.19	3.40	0.70	0.03
14	a19:0	0.19	0.04	–	0.02	–	0.30	0.42	0.04
–	i21:0	–	–	–	–	–	–	–	–
–	14:1(n-7)	–	–	–	–	–	–	–	–
2	16:1(n-7)	9.93	15.51	8.97	13.59	14.21	4.81	9.52	42.76
5	17:1(n-7)	0.73	0.39	0.22	0.32	0.23	0.30	0.24	0.01
–	18:1(n-5)	1.03	0.43	0.46	0.23	0.18	0.16	–	0.12
11	18:1(n-7)	29.08	34.83	34.65	25.98	29.18	1.72	1.83	27.98
10	18:1(n-9)	1.11	0.70	0.61	0.59	0.63	0.07	0.15	0.03
10	18:1(n-13)	5.28	8.22	10.91	8.75	11.91	2.69	2.10	1.60
15	19:1(n-7)	0.19	0.57	1.37	1.42	1.27	3.43	1.95	–
21	20:1(n-7)	9.92	4.26	4.09	5.32	3.59	7.04	5.26	0.45
20	20:1(n-9)	0.48	0.57	5.00	3.03	2.27	2.24	1.25	0.01
20	20:1(n-13)	1.61	2.64	4.72	6.19	5.51	9.72	10.28	1.36
–	21:1(n-7)	–	–	–	–	–	–	–	–
28	22:1(n-7)	0.71	0.18	0.12	0.12	0.06	0.10	0.11	0.03
–	22:1(n-9)	–	–	–	–	–	–	–	–
9	18:2(n-7,13)	0.59	0.80	0.76	0.63	1.06	0.02	0.08	0.83
9	18:2(n-7,15)	0.58	1.20	0.64	1.73	0.41	0.34	0.32	0.04
8	18:2(n-9,15)	1.68	1.46	0.73	0.76	0.64	0.37	0.36	0.08
19	20:2(n-7,13)	0.02	0.20	0.83	0.55	0.60	4.92	5.46	–
19	20:2(n-7,15)	0.99	3.75	5.20	3.15	4.84	15.18	13.14	2.24
18	20:2(n-9,15)	0.03	0.81	1.06	2.55	1.08	5.37	3.90	0.09
24	21:2(n-7,14)	0.07	0.03	0.09	–	0.11	0.50	0.78	–
24	21:2(n-7,16)	–	0.05	0.18	–	0.15	1.31	0.88	–
–	21:2(n-?)	–	–	–	–	–	–	–	–
27	22:2(n-7,15)	0.22	0.72	1.13	0.90	0.99	6.54	9.96	1.02
27	22:2(n-9,15)	–	–	0.07	1.61	0.27	0.08	0.27	0.02
–	24:2(n-?)	–	–	–	–	–	–	–	–
7	18:3	0.92	0.18	0.43	1.51	0.31	8.00	8.00	0.01
–	18:3	–	–	–	–	–	–	–	–
17	20:3	0.03	0.19	0.25	0.06	0.36	0.14	0.10	–
25	22:3	0.10	0.08	0.05	0.01	0.02	0.14	0.44	0.10
26	22:3	0.06	0.14	0.11	0.20	0.11	0.85	0.75	0.02
16	20:4(n-6)	0.40	0.89	1.00	0.25	0.71	0.72	0.73	3.16

* Numbers refer to GC peaks in Fig. 1.

Amphipods (Halice hesmonectes). The fatty acids of *H. hesmonectes* were characterized by high abundances of 'storage' fatty acids (NLFA; Table 2), indicating high concentrations of wax esters in these organisms as observed in many deep-sea crustacea (Sargent *et al.* 1976). As with the mussel samples, NMI fatty acids were observed in relatively high abundance (up to 25%; Tables 2–5). The NMI fatty acids were present in high abundance in the structural fatty acids (PLFA) of the amphipod samples (up to 11% of total fatty acids; Table 2). In general,

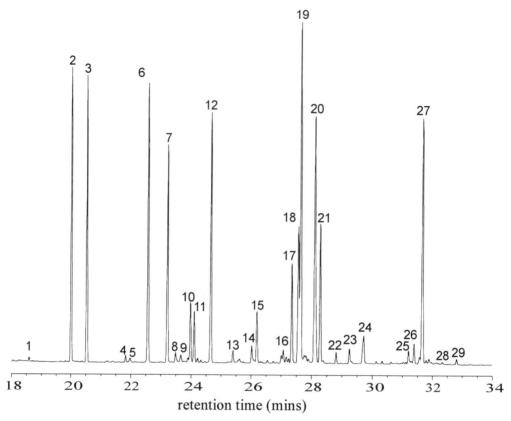

Fig. 1. Gas chromatogram of the fatty acid methyl esters of the polar lipids isolated from the hydrothermal vent mussel *B. Thermophilus*. Column: 50 m CPSil 5 CB, 0.32 mm i.d., 0.1 μm film, temperature programme 40–150°C at 12°C min^{-1}, 150–240 °C at 4°C min^{-1}. Peak numbers refer to those given in Tables 2–5.

greater abundances of NMI fatty acids were identified in the amphipod storage lipids (NLFA) than in the structural fatty acids (Table 2). As for the mussel samples, minor amounts of the 20:4(n-6) SMI fatty acids were detected in the amphipod samples (up to 1% of the total; Tables 3–5). Branched fatty acids constituted up to 5% of the total fatty acids (Table 2), $i17:0$, $i19:0$ and $a19:0$ were the most abundant components (Tables 3–5). In general, branched fatty acids were in greater abundance in the storage NLFA lipids than in the structural PLFA (Table 2).

Sterol compositions

Tubeworm (Riftia pachyptila). The major sterols identified in the *R. pachyptila* tissues were cholesterol, cholesta-5,24-dienol, and 24-methylcholesta-5,24(28)-dienol (Table 6); the major tubeworm sterols had δ^{13}C values of −15.1‰, −15.4‰ and −16.4‰ for cholesterol,

cholesta-5,24-dienol, and 24-methylcholesta-5,24(28)-dienol, respectively, compared with a total tissue value of −11‰ (Table 7), a difference of 4–5‰, which is well within that observed for lipids isolated from autotrophic organisms (Abelson & Hoering 1961; Park & Epstein 1961).

Mussel (Bathymodiolus thermophilus). A variety of sterols was observed in the mussel samples (Table 6), the most abundant component being cholesterol (68% of total sterols; Table 6). As well as cholesterol, a wide variety of minor sterols was observed (Table 6), such as cholesta-5,22-dienol, 5α-cholestanol and cholest-7-enol. The δ^{13}C value of cholesterol from *B. thermophilus* was −26.5‰ compared with a bulk tissue δ^{13}C value of −33‰ (Table 7).

Amphipods (Halice hesmonectes). A large variety of sterols were also observed in the amphipod samples (Table 6), their distribution

Table 6. *Relative abundances of sterols, as % of total identified. Total concentrations given are μg total sterols/g sample. Sites and samples as in Table 1*

Sterol	A158 B	A162 C	A160 C	A202 D	A214 D	A187 A	Add A	Vest C
C27Δ5,22[1]	0.52	–	–	–	0.59	0.45	11.55	–
C27Δ22[2]	0.61	–	–	–	–	0.53	–	–
C27Δ5[3]	80.22	40.04	94.16	92.43	90.91	90.14	67.53	39.73
C27Δ0[4]	0.14	32.37	1.06	–	2.22	1.91	8.66	–
C27Δ5,24[5]	1.04	8.49	0.81	0.60	0.62	1.55	1.18	37.06
C27Δ24[6]	–	6.41	0.08	0.03	0.09	–	0.49	–
C27Δ7[7]	4.68	1.13	0.81	1.20	1.13	0.68	6.22	1.02
C28Δ5,24[8]	–	4.42	–	–	1.12	0.23	–	15.48
C28Δ24[9]	–	5.24	–	–	–	–	–	–
C28Δ5[10]	4.21	1.93	1.01	1.20	0.35	1.86	1.58	2.21
C28Δ0[11]	–	–	0.89	0.10	–	–	0.91	–
C28Δ5,22[12]	–	–	–	0.33	–	0.37	–	–
C29Δ5[13]	6.66	–	1.00	2.34	1.86	1.31	1.48	3.50
C29Δ5,24[14]	1.92	–	0.19	1.76	1.11	0.35	0.41	1.01
Total	2944	1291	7833	6914	5272	4096	225	5605

[1]Cholesta-5,22-dienol; [2]cholest-22-enol; [3] cholesterol; [4]5α-cholestanol; [5]cholesta-5,24-dienol; [6]cholest-24-enol; [7]cholest-7-enol; [8]24-methylcholesta-5,24(28)-dienol; [9]24-methylcholest-24(28)-enol; [10]24-methylcholesterol; [11]24-methyl-5α-cholestanol; [12]24-methylcholesta-5,22-dienol; [13]24-ethylcholesterol; and [14]24-ethylcholesta-5,24(28)-dienol.

being similar to those of the mussel samples (Table 6). The sterol distributions were dominated by cholesterol in accord with distributions observed in other marine crustacea (Kerr & Baker 1991). High abundances of 5α-cholestanol were observed in one of the amphipod samples (A160; Table 7).

$\delta^{13}C$ values of the amphipod cholesterol fell into two categories, one around $-16‰$ and the other around $-21‰$ (Table 7), these categories corresponding to two sets of bulk tissue $\delta^{13}C$ values ($-14‰$ and $-24‰$; Table 7).

Discussion

Fatty acid compositions

Tubeworm (Riftia pachyptila). Vestementiferan tubeworms provide a striking example of eukaryotic dependence on carbon fixed by chemoautotrophic bacteria. Adults have no digestive systems and rely almost exclusively on their endosymbionts for nutrition, though uptake of carbon gained from dissolved organic carbon may also make a minor contribution (Childress & Fisher 1992). *Riftia pachyptila*, the giant tubeworm associated with East Pacific vent sites, is representative of this dependence, being fixed at sites of expulsion of warm vent water rich in dissolved sulphides.

The dependence of *R. pachyptila* on carbon fixed by endosymbiotic bacteria is apparent in the lipid compositions identified in this study, and especially in the fatty acid compositions. Very high abundances of (n-7) fatty acids (Table 2), and correspondingly minor amounts of (n-9) fatty acids were observed (Table 2). Bacteria generally have high abundances of (n-7) monounsaturated fatty acids, and studies of marine organisms largely dependent on bacteria for food reflect the assimilation of this class of fatty acid (Conway & McDowell Capuzzo 1991; Ben-Mligh *et al.* 1992; Fang *et al.* 1993). Small amounts of NMI dienoic fatty acids were observed in all the tubeworm samples, at around 3–4% of total fatty acids. No SMI polyunsaturated fatty acids were detected, which indicates the lack of filter-feeding as a food source for vestimentiferan tubeworms. Only very small amounts of branched fatty acids, usually considered as bacterial markers, were observed in the lipids of all the tubeworm tissues examined, indicating that branched fatty acids are only minor components of the sulphur-oxidizing symbiotic bacteria associated with *R. pachyptila* (*cf.* Conway & McDowell Capuzzo 1991, who did not identify branched fatty acids in a thiotrophic bacterium isolated from a deep-sea hydrothermal vent).

Mussel (Bathymodiolus thermophilus). The fatty acids of the mussel samples were characterized by very large abundances of NMI polyunsaturated fatty acids (Table 2). In the

Table 7. Corrected $\delta^{13}C$ values for sterols isolated from hydrothermal vent organisms (9°N East Pacific Rise). Values are corrected for the addition of derivative carbon (trimethyl silyl groups). Values in parentheses are standard deviations of uncorrected $\delta^{13}C$ values. Samples and sites as in Table 1

	A123 A	A187 A	A158 B	A160 C	A162 C	A200 D	A202 D	A216 D	Add A	Vest C
Bulk $\delta^{13}C$	−23.5	−23.1	−13.6	−23.7	−14.2	−25.1	−25.0	−25.8	−33	−11
Sterols										
C27Δ5[1]	−21.5(0.1)	−21.5(0.6)	−16.3(0)	−22.7(0.3)	−16.6(0)	−19.8(0.3)	−20.7(0.3)	−21.6(0.8)	−26.5(0.6)	−15.1(0.2)
C27Δ5,24[2]										−15.4(0.3)
C27Δ7[3]										−12.8(0.5)
C28Δ5,24[4]										−16.4(0.3)

[1]Cholesterol; [2]cholesta-5,24-dienol; [3]cholest-7-enol; and [4]24-methylcholesta-5,24(28)-dienol.

hydrothermal vent mussels examined this large abundance of NMI fatty acids apparently replaces SMI polyunsaturated fatty acids which are observed in similarly large abundances in coastal and estuarine bivalves (Joseph 1982; *cf.* Fang *et al.* 1993), e.g. arachidonic acid 20:4(n-6). The significance of these observations is that SMI fatty acids are characteristic of marine algae and the abundance of these acids in marine invertebrates is indicative of a diet based primarily on carbon fixed by phytotrophs (e.g. Bradshaw 1988). In contrast, SMI fatty acids are very rare in bacteria and therefore invertebrates which depend on a diet based on bacteria appear to desaturate bacterially derived monounsaturated (n-7) fatty acids which are extended and desaturated to produce NMI polyunsaturated fatty acids (*cf.* Zhukova 1986; Fang *et al.* 1993). The high abundance of NMI fatty acids identified in this study are in accord with the findings of Fang *et al.* (1993), who reported NMI fatty acid abundances of 14–22% of total fatty acids in bivalves containing endosymbiotic bacteria living around hydrocarbon seeps in the Gulf of Mexico. Indeed, the occurrence of NMI fatty acids is widely reported for bivalves which contain endosymbiotic bacteria, though the abundance of these lipids observed in this study is much larger than previously observed (e.g. Klingensmith 1982; Berg *et al.* 1985; Zhukova & Svetashev 1986; Zhukova *et al.* 1992). The presence of the SMI polyunsaturated fatty acid 20:4(n-6) in the mussel samples (Tables 3– 5), albeit in very low abundance, illustrates the assimilation of phytoplankton- derived carbon in the deep-sea vent ecosystem.

Branched fatty acids are indicative of a wide variety of bacteria and have been proposed by Zhukova *et al.* (1992) to be markers for symbiotic bacteria in bivalves. In this study minor abundances of this lipid type were observed (around 1–3 %; Table 2) and indicate the direct incorporation of bacterial lipids by the bivalves examined. Greater abundances of branched fatty acids were observed in the gill tissues than in the foot and adductor tissues (Table 2); this observation may be due to a combination of two factors – the gills harbour the endosymbiotic bacteria associated with *B. thermophilus* and also filter particulate bacteria from the water column (Cavanaugh 1985; Le Pennec & Prieur 1984; Page *et al.* 1991).

Amphipods (Halice hesmonectes). The fatty acid compositions of all the *H. hesmonectes* samples were remarkably consistent given that they were all taken from geographically distinct

Neutral Lipid Fatty Acids   Polar Lipid Fatty Acids

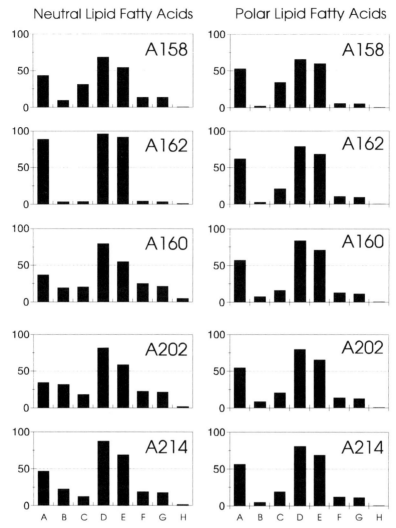

Fig. 2. Relative abundances of different amphipod lipid classes as a percentage of total fatty acids identified for neutral lipid fatty acids and polar lipids fatty acids. **A** n-7 fatty acids; **B** n-9 fatty acids; **C** saturated fatty acids; **D** unsaturated fatty acids; **E** monounsaturated fatty acids; **F** polyunsaturated fatty acids; **G** non-methylene interrupted polyenoic fatty acids; **H** branched fatty acids. Fatty acids were identified as fatty acid methyl esters.

vent sites within the Venture hydrothermal fields (9°N EPR; Tables 1 and 2; Fig. 2), which suggests similar food sources for this crustacean in all sites examined. The fatty acids of *H. hesmonectes* were characterized by high abundances of 'storage' fatty acids (NLFA; Table 2), an indication of high concentrations of wax esters as observed in many deep-sea crustacea (Sargent *et al.* 1976). Wax esters have been proposed to provide both buoyancy and energy storage in such organisms (Sargent *et al.* 1976). As with the mussel and tubeworm samples, NMI fatty acids were observed in relatively high

abundance (up to 25%) and may indicate a trophic link between the pelagic amphipod species and the benthic mussel species, i.e. detrital feeding by the amphipod, although *de novo* biosynthesis may be a factor. The generally greater abundance of NMI fatty acids in storage lipids (NLFA) than in the structural fatty acids (Table 2) reinforces the hypothesis that these acids are predominantly derived from diet (fatty acids surplus to membrane structural requirements being stored). The presence of NMI fatty acids in high abundance in the structural fatty acids (PLFA) of the amphipod samples (up to

11% of total fatty acids; Table 2) demonstrates that this class of fatty acid plays a vital structural part in the amphipod cellular membranes. As for the mussel samples, SMI fatty acids were of low abundance in the amphipod samples (Tables 3–5).

Sterol compositions

Tubeworm (Riftia pachyptila). The major sterols identified in the *R. pachyptila* tissues were the sterols cholesterol, cholesta-5,24-dienol, and 24-methylcholesta-5,24(28)-dienol (Table 6). The presence of these sterols suggests *de novo* biosynthesis by the tubeworm symbiosis (most likely biosynthesis by the tubeworm proper rather than thiotrophic endosymbiotic bacteria), the latter sterols being intermediates in the conversion of the sterol precursor lanosterol to cholesterol (Lehninger 1978; Kerr & Baker 1991). Stable carbon isotopic analyses of the *R. pachyptila* sterols provides strong evidence that this tubeworm species does indeed biosynthesize sterols *de novo*. The $\delta^{13}C$ values of the major tubeworm sterols were 4–5‰ lower than that of the total tissue value of −11‰ (Table 7), in the range for that observed for lipids isolated from autotrophic organisms (Abelson & Hoering 1961; Park & Epstein 1961; Hayes 1993). If the tubeworm sterols were derived from marine phytodetritus (or from phytoplankton derived organic matter), a value of around −27‰ may be expected (Abelson & Hoering 1961). Minor sterols were also observed, such as 24-ethylcholesterol (Table 6), which have known phytoplankton sources (Kerr & Baker 1991) and may indicate the assimilation by the tubeworm of a small amount of dissolved organic carbon.

Mussel (Bathymodiolus thermophilus). The most abundant sterol identified in the mussel samples was cholesterol (Table 6), in line with observations on other bivalves collected from coastal environments (Goad 1978; Conway & McDowell Capuzzo 1991). As well as cholesterol, a wide variety of minor sterols was observed (Table 6), including a number of sterols which are indicative of algae, such as cholesta-5,22-dienol (Goad 1978; Kerr & Baker 1991). Biosynthesis of sterols by bivalves is controversial (Conway & McDowell Capuzzo 1991), and in the case of cold hydrocarbon seep mussels from the Gulf of Mexico a product/ precursor relationship between bacterially produced 4-methyl sterols and 4-methyl sterols observed in the mussels was proposed (Fang *et al.* 1992). However, only methylotrophic

bacteria have been demonstrated to have the ability to biosynthesize sterols and since *B. thermophilus* has thiotrophic symbiotic bacteria, additional evidence other than sterol distribution and structure is needed to help assign particular sources of sterols in these organisms, such as stable carbon isotopic analyses. *Solemya velum*, a coastal clam, containing symbiotic bacteria, was reported to contain sterols with $\delta^{13}C$ values around −38.5‰ compared with a tissue $\delta^{13}C$ value of −33‰ (Conway & McDowell Capuzzo 1991) and as discussed earlier this is within the range expected if *de novo* biosynthesis is being undertaken. However, cholesterol isolated from *B. thermophilus* adductor tissues in this study had a $\delta^{13}C$ value of −27‰ compared with a tissue $\delta^{13}C$ value of −34‰ (Table 7). The 7‰ higher $\delta^{13}C$ value for the mussel cholesterol compared with total tissue rules out *de novo* biosynthesis as a major source and thus requires that the mussel studied derive sterols from filter feeding (Goad 1978). Dixon *et al.* (this volume) have demonstrated that phytoplankton are a significant proportion of the detritus associated with deep-sea hydrothermal vents.

Furthermore, electron microscopic studies on the gut of a mytilid from 12°N EPR showed that phytoplankton are present, and a source for the *B. thermophilus* sterols from phytodetritus is proposed, this proposal being reinforced by published $\delta^{13}C$ values for marine phytoplankton (Abelson & Hoering 1961; Park & Epstein 1961) which are in the range of −23‰ (Abelson & Hoering 1961; Park & Epstein 1961).

Amphipods (Halice hesmonectes). The large variety of sterols observed in the amphipod samples (Table 6) had a distribution similar to those of the mussel samples and to those of marine crustacea from coastal and estuarine habitats (Goad 1978; Bradshaw *et al.* 1990; Kerr & Baker 1991). The significance of the sterol distributions in a trophic context is that crustacea are held to be unable to biosynthesize sterols *de novo* and must assimilate these vital biochemicals from their diet (Goad 1978; Kerr & Baker 1991). In more conventional marine settings, crustacean sterols derive from algae which are rich in a variety of sterols (Goad 1978; Kerr & Baker 1991). However, in deep-sea hydrothermal vents phytoplankton are unable to exist and bacteria replace them as the base of the food chain. Bacteria are generally devoid of sterols and therefore the sterol distributions and the stable carbon isotopic composition of those sterols in vent crustacea are of particular interest.

$\delta^{13}C$ values of the amphipod cholesterol fell into two categories, one around $-16‰$ and the other around $-21‰$ (Table 7), these categories corresponding to two sets of bulk tissue $\delta^{13}C$ values ($-14‰$ and $-24‰$; Table 7). The separation of the amphipod bulk tissue and cholesterol $\delta^{13}C$ values into two classes is presumably determined by diet and suggests that at least two sources of dietary sterols are available to the amphipods, one which is similar to that observed in the tubeworm sample (i.e. most likely to derive from vent organisms) and one similar to that observed in the mussel sample (i.e. most likely derived from phytoplankton). Different proportions of these dietary sources will lead to the $\delta^{13}C$ values observed in this study.

Overview

Overall, of all lipid types fatty acids give the greatest indication as to the main diet of hydrothermal vent organisms. In this study a diet based primarily on bacteria is associated with high abundances of (n-7) fatty acids, low abundances of (n-9) fatty acids, high abundances of NMIP fatty acids and low abundances of SMIP fatty acids. These findings are in accord with similar studies on cold hydrocarbon seep organisms and studies on molluscs which contain symbiotic bacteria. The relative abundance of NMI to SMI in marine organisms gives a measure of the relative contributions of bacteria and phytoplankton in their diet. The deep-sea hydrothermal vent organisms analysed in this study from 9°N EPR are an extreme example of almost total dependence on bacterially fixed carbons which is reflected by very high abundances of NMI fatty acids and correspondingly low abundances of SMI fatty acids.

In contrast with the fatty acids, sterols, due to their absence in most bacteria and due to limitations on *de novo* biosynthesis in certain invertebrates, provide an indication of assimilation of carbon from organisms from the photic zone of the ocean. Stable carbon isotopic analyses of individual compounds can provide additional evidence so that the source of individual lipids can be assigned. For example, cholesterol has been demonstrated to have at least two sources in this vent ecosystem, one possibly deriving ultimately from the photic zone of the ocean.

This work was supported by funds from the Natural Environment Research Council (Grant numbers GR3/2951, GR3/3748 and GR3/7731) and by funds from the National Science Foundation (USA).

References

ABELSON, P. H. & HOERING, T. C. 1961. Carbon isotope fractionation in formation of amino acids by photosynthetic organisms. *Proceedings of the National Academy of Sciences*, **47**, 623–627.

BEN-MLIGH, F., MARTY, J.-C. & FIALA-MEDIONI, A. 1992. Fatty acid composition in deep hydrothermal vent symbiotic bivalves. *Journal of Lipid Research*, **33**, 1797–1806.

BERG, C. J., KRYZYNOWEK, J., ALATALO, P. & WIGGIN, K. 1985. Sterol and fatty acid composition of the clam, *Codakia orbicularis*, with chemoautotrophic symbionts. *Lipids*, **20**, 116–120.

BLIGH, E. G. & DYER, W. J. 1959. A rapid method of total lipid extraction and purification. *Canadian Journal of Biochemistry and Physiology*, **37**, 911–917.

BRADSHAW, S. A. 1988. *The biogeochemistry of lipids in model marine food chains*. PhD Thesis, University of Bristol, Bristol.

——, O'HARA, S. C. M., CORNER, E. D. S. & EGLINTON, G. 1990. Changes in lipids during simulated herbivourous feeding by the marine crustacean *Neomysis integer*. *Journal of the Marine Biological Association*, **70**, 225–243.

CAVANAUGH, C. C. 1985. Symbioses of chemoautotrophic bacteria and marine invertebrates from hydrothermal vents and reducing sediments. *Biological Society of Washington Bulletin*, **6**, 373–388.

CHILDRESS, J. J. & FISHER, C. R. 1992. The biology of hydrothermal vent animals: physiology, biochemistry, and autotrophic symbioses. *Oceanography and Marine Biology Annual Review*, **30**, 337–441.

CONWAY, N. & McDOWELL CAPUZZO, J. 1991. Incorporation and utilization of bacterial lipids in the Solemya velum symbiosis. *Marine Biology*, **103**, 277–291.

DIXON, D. R., DIXON, L. R. J., JOLLIVET, D. A. S. B., NOTT, J. A. & HOLLAND, P. W. H. 1994. Molecular identification of early life-history stages of hydrothermal vent organisms. This volume.

FANG, J., COMET, P. A., BROOKS, J. M. & WADE, T. L. 1993. Nonmethylene – interrupted fatty acids of hydrocarbon seep mussels: ocurrence and significance. *Comparative Biochemistry and Physiology*, **104B**, 287–291.

——, ——, WADE, T. L. & BROOKS, J. M. 1992. Gulf of Mexico hydrocarbon seep communities – IX. Sterol biosynthesis of seep mussels and its implications for host-symbiont association. *Organic Geochemistry*, **18**, 861–867.

GOAD, L. J. 1978. Marine Sterols. *In*: SCHEUER, P. J. (ed.) *Marine Natural Products*. Academic Press, London, 176–172.

HAYES, J. M. 1993. Factors controlling ^{13}C contents of sedimentary organic compounds: principles and evidence. *Marine Geology*, **113**, 111–125.

——, FREEMAN, K. H., POPP, B. N. & HOHAM, C. H. 1990. Compound-specific isotopic analyses: a novel tool for reconstruction of ancient biogeochemical processes. *Organic Geochemistry*, **16**, 1115–1128.

HEDRICK, D. B., GUCKERT, J. B. & WHITE, D. C. 1991. Archaebacterial ether lipid diversity analyzed by supercritical fluid chromatography: integration with a bacterial lipid protocol. *Journal of Lipid Research*, **32**, 659–666.

JOSEPH, J. D. 1982. Lipid composition of marine and esturarine invertebrates. Part II: Mollusca. *Progress in Lipid Research*, **22**, 109–153.

KAARTVEDT, S., VAN DOVER, C. L., MULLINEAUX, L. S., WIEBE, P. H. & BOLLENS, S. M. 1994. Amphipods on a deep-sea hydrothermal treadmill. *Deep-Sea Research*, **41**, 179–195.

KERR, R. G. & BAKER, B. J. 1991. Marine sterols. *Natural Products Reports*, **8**, 465–497.

KLINGENSMITH, J. S. 1982. Distribution of methylene and nonmethylene-interrupted dienoic fatty acids in polar lipids and triacyglycerols of selected tissues of the hardshell clam (*Mercenaria mercenaria*). *Lipids*, **17**, 976–981.

LEHNINGER, A. L. 1978. *Biochemistry*, 2nd edn. Worth, New York.

MARTIN, J. W., FRANCE, S. C. & VAN DOVER C. L. 1993. *Halice hesmonectes*, a new species of pardaliscid amphipod (Crustacea. Peracarida) from hydrothemal vents in the eastern Pacific. *Canadian Journal of Zoology*, **71**, 1724–1732.

PAGE, H. M, FIALA-MÉDIONI, A., FISHER, C. & CHILDRESS J. J. 1991. Experimental evidence for filter-feeding by the hydrothermal vent mussel, *Bathymodiolus thermophilus*. *Deep-Sea Research*, **38**, 1455–1461.

PARK, R. & EPSTEIN, S. 1961. Metabolic fractionation of ^{13}C and ^{12}C in plants. *Plant Physiology*, **36**, 133–138.

LE PENNEC, M. & PRIEUR, D. 1984. Biologie marine – observations sur la nutrition d'un mytilidae d'un site hydrothermal actif de la dorsale du Pacifique orental. *Comptes Rendu Academies des Sciences de Paris Série III*, **17**, 493–498.

RIELEY, G. 1993. *Molecular and isotopic studies of natural environments*. PhD Thesis, University of Bristol, Bristol.

—— 1994. Derivatization of organic compounds prior to gas chromatographic – combustion – isotope ratio mass spectrometric analysis: identification of isotope fractionation processes. *Analyst*, **119**, 915–919.

——, COLLIER, R. J., JONES, D. M., EGLINTON, G., EAKIN, P. A. & FALLICK, A. E. 1991. Sources of sedimentary lipids deduced from stable carbon-isotope analyses of individual compounds. *Nature*, **352**, 425–427.

——, COLLISTER, J. W., STERN, B. & EGLINTON, G. 1993. Gas chromatography – isotope ratio mass spectrometry of leaf wax *n*-alkanes from plants of differing carbon dioxide metabolisms. *Rapid Communications in Mass Spectrometry*, **7**, 488–491.

SARGENT, J. R. & WHITTLE, K. J. 1981. Chemistry of Marine Waxes. *In*: LONGHURST, A. R. (ed.) *Analysis of marine ecosystems*. Academic Press, London, 491–533.

——, LEE, R. F. & NEVENZEL, J. C. 1976. *In*: KOLATTAKUDY, P. E. (ed.) *Chemistry and Biochemistry of Natural Waxes*. Elsevier, Amsterdam, 50–91.

VAN DOVER, C. L., KAARTVEDT, S., BOLLENS, S. M., WIEBE, P. H., MARTIN, J. W. & FRANCE, S. C. 1992. Deep-sea amphipod swarms. *Nature*, **358**, 25–26.

YRUELA, I., BARBE, A. & GRIMALT, J. O. 1990. Determination of double bond position and geometry in linear and highly branched hydrocarbons and fatty acids from gas chromatography – mass spectrometry of epoxides and diols generated by stereospecific resin hydration. *Journal of Chomatographic Science*, **28**, 421–427.

ZHUKOVA, N. M. 1986. Biosynthesis of non-methylene-interrupted dienoic fatty acids from [^{14}C] acetate in molluscs. *Biochimica Biophysica Acta*, **878**, 131–133.

—— & SVETASHEV, V. I. 1986. Non-methylene-interrupted dienoic fatty acids in molluscs from the Sea of Japan. *Comparative Biochemistry and Physiology*, **83B**, 643–646.

——, KHARLAMENKO, V. I., SVETASHEV, V. I. & RODIONOV, I. A. 1992. Fatty acids as markers of bacterial symbionts of marine bivalve molluscs. *Journal of Experimental Marine Biology and Ecology*, **162**, 253–263.

The molecular identification of early life-history stages of hydrothermal vent organisms.

D. R. DIXON[1], D. A. S. B. JOLLIVET[1], L. R. J. DIXON[2], J. A. NOTT[1]
& P. W. H. HOLLAND[3,4]

[1]Plymouth Marine Laboratory, Citadel Hill, Plymouth PL1 2PB, UK
[2]Marine Biological Association, Citadel Hill, Plymouth PL1 2PB, UK
[3]Department of Zoology, University of Oxford, South Parks Road, Oxford OX1 3PS, UK
[4]Present address: Department of Pure and Applied Zoology, School of Animal and
Microbial Sciences, University of Reading, Whiteknights, Reading RG6 2AJ, UK

Abstract: Amplification of diagnostic genomic DNA sequences using the highly sensitive polymerase chain reaction (PCR) technique provides a fast, sensitive and relatively inexpensive approach to species identification where there is a lack of diagnostic morphological characters. This applies particularly to the early life-history stages of marine invertebrates. Using PCR primers designed to amplify diagnostic length variants within phylogenetically widespread genes (in this case an expansion segment within the 28S rRNA gene), a first attempt is presented to produce a DNA database for use in the identification of hydrothermal vent larvae. In addition, a scanning electron microscopy study of particulates recovered from the neutrally buoyant plumes of hydrothermal vents on the Mid-Atlantic Ridge revealed evidence of biological material derived both from the vent environment and from the sea surface as marine 'snow'. This investigation represents the first stage in the development of a bottom-mounted recorder to study the spatial and temporal aspects of larval dispersal in the hydrothermal vent environment. Larval dispersal processes are fundamental to the biogeography, genetics and evolution of the hydrothermal vent fauna.

A large number of coastal and deep-sea marine invertebrates have planktonic larvae which spend varying amounts of time in the water column, ranging from a few days to many months depending on the species (Thorson 1950). Larval dispersal undoubtedly plays a major part in generating and maintaining biogeographical patterns on the global scale. On a more local scale, larval dispersal maintains the gene flow between fragmented populations and thus prevents the formation of genetic bottlenecks, while preventing extinctions brought about by changes in environmental conditions (Scheltema 1971; Avise 1994). In many instances, larval identification to the species level is not possible based solely on morphological characters (Newell & Newell 1966; Hu et al. 1992). There is a need, therefore, for new molecular approaches to larval identification to enable investigations into this important area of 'supply side' ecology.

The hydrothermal vent environment is characterized by a high degree of temporal and spatial instability linked to variations in magmatic heat convection and tectonic activity along the oceanic ridges (Fustec et al. 1987; Tunnicliffe 1988; 1991). To exploit the rich source of

chemical energy represented by vent emissions, the specialized vent fauna has to cope with this highly unstable 'and patchy environment (Johnson et al. 1988; Chevaldonne et al. 1991). Regarding dispersal, representatives of several major groups (i.e. gastropods, bivalves, polychaetes, crustaceans and vestimentiferans) are known or thought to have pelagic larvae, which spend weeks or even months in the plankton (e.g. Lutz et al. 1980, 1988; Cary et al. 1989; Southward 1988; Jones & Gardiner 1989; Gustafson & Lutz 1994).

Hydrothermal vent communities exhibit biogeographical variation on a spatial scale measured in thousands of kilometres (Tunnicliffe 1991); locally there is a temporal succession of species linked with the age of the vent (Jollivet 1993). Between these two extremes is the dynamic patchwork of individual species distributions, which reflects variations in vent chemistry, the physical environment and biological factors such as reproduction. To understand the processes responsible for generating and maintaining these biogeographical patterns requires a knowledge of larval biology and dispersal in particular. It is now recognized that there is no single reproductive strategy

From PARSON, L. M., WALKER, C. L. & DIXON, D. R. (eds), 1995, *Hydrothermal Vents and Processes*, Geological Society Special Publication No. 87, 343–350.

characteristic of life at the vents. Instead, vent species exhibit a range of different reproductive strategies dictated, it appears, mainly by phylogenetic constraints and other limitations imposed by their fundamental biology (e.g. Gustafson & Lutz 1994), which share much in common with benthic invertebrates elsewhere (e.g. Turner *et al.* 1985; McHugh 1989). Dispersal and gene flow are closely linked and the levels of genetic differentiation exhibited by populations is commonly a good indicator of their dispersal potential and type of reproduction (e.g. Endler 1992), but due to strong selection pressures this relationship may not always hold for hydrothermal vent species (Jollivet 1993; Scheltema 1994).

We have developed a molecular approach to larval identification and for studying their population genetics. Our first attempts using a polymerase chain reaction (PCR)-based strategy have proved successful in allowing the analysis of DNA sequence length variation within gene introns (i.e. non-coding and potentially neutral DNA sequences) in single larvae of the shallow-water mussel *Mytilus edulis*. This has enabled us to detect evidence of restricted gene flow between mussel populations on the west and northeast coasts of Britain, and for local effects around the Thames estuary (Corte-Real *et al.* 1994*a, b*).

Currently, we are extending these investigations to include the larvae of deep-sea hydrothermal vent organisms. In this paper we present our first attempt at generating a DNA database for use in species identification of hydrothermal vent organisms. In addition, we report on a scanning electron microscopy (SEM) investigation to detect biological material on filters used to sample chemical particulates from the neutrally buoyant plumes of hydrothermal vents on the Mid-Atlantic Ridge (MAR).

Source of animals and identification

The Atlantic specimens were collected at hydrothermal vents on the MAR during the Atlantis II cruise (4 June–4 July 1993) to TAG (26°20′N), Snakepit (23°20′N) and Broken Spur (29°10′N), using the tele-manipulated claw of the deep-sea manned submersible *Alvin*; the Pacific specimens came from the northeast (Juan de Fuca Ridge), mainly 13°N, and southeast (East Pacific Rise) Pacific Ridges, and were collected by Professor Verena Tunnicliffe (University of Victoria) and scientists taking part in IFREMER cruises co-ordinated by Dr Daniel Desbruyères, using the *Alvin* and *Nautile* submersibles, respectively. The species identifications were carried out by expert scientists taking part in those expeditions or who later received the material for analysis.

Sample preservation, DNA extraction and PCR

The tissue samples used for molecular analysis were either stored deep frozen at −70°C , or stored in BLB (5% sodium dodecylsulphate, 250 mM EDTA, 50 mM TRIS·Cl pH 8). The BLB-preserved specimens were stored in the laboratory at 4°C prior to DNA extraction. High molecular weight total genomic DNA was extracted from all specimens by a standard proteinase K method (e.g. Dixon *et al.* 1992) and resuspended in 50 μl of TE (10 mM TRIS pH 8, 1 mM Na$_2$EDTA pH 8). In the case of an unidentified chaetopterid polychaete collected by the Atlantis II cruise at TAG (dive 2609: 6 October 1993) only empty, proteinaceous tubes were available, but these yielded sufficient high-quality genomic DNA for PCR amplification following standard proteinase K digestion. Great care was taken throughout all the DNA manipulations to avoid cross-contamination of samples.

Two oligonucleotide primers, MT3 (positive strand primer) and MT4 (negative strand primer) were used for PCR amplification of an expansion segment of the 28S rRNA gene in the ribosomal DNA. These were designed as 'universally applicable' PCR primers, with the potential to amplify the 28S expansion segment of any eukaryote, average fragment size 500 nucleotides (Williams *et al.* 1993).

The primer sequences are:

MT3 (5′ to 3′):
    AAAGGATCCGATAGYSRACAAGTACCG
MT4 (5′ to 3′):
    CCCAAGCTTGGTCCGTGTTTCAAGAC

Amplification reactions were performed in a volume of 10–15μl using standard conditions (Holland, 1993) with the addition that the mixes were illuminated with ultraviolet light on a trans-illuminator for four minutes before the addition of the Taq polymerase to reduce the risk of contamination. The cycling parameters used with each sample were varied until a clean product was seen by agarose gel electrophoresis. The final parameters were: 94°C, 2 min; then 35 cycles of 94°C for 45 s, 52–53°C for 1 min, 72°C for 1 min 15 s; and finally 52–53°C for 1 min, followed by 72°C for 2 min. DNA amplifications were carried out on Techne PHC-2 and Hybaid 'Omnigene' thermal cyclers. The PCR products were visualized on 2% agarose gels by ethidium bromide staining.

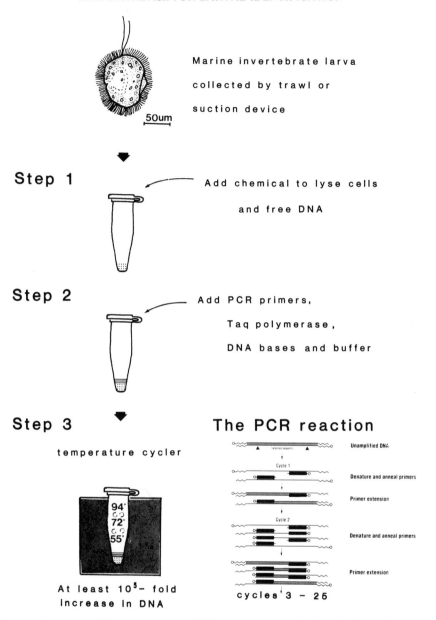

Fig. 1. Steps involved in amplifying target genomic DNA sequences from single marine invertebrate larvae using the polymerase chain reaction (PCR).

Analysis of filter samples taken from the neutrally buoyant plumes of hydrothermal vents on the MAR

Small pieces of polycarbonate filter, containing particulate matter recovered for chemical analysis from 300 m above the sea bed (i.e. at a depth of approximately 2700 m) in the neutrally buoyant hydrothermal vent plumes at the TAG, MARK and Broken Spur sites on the MAR, were analysed for signs of biology. Shortly after collection the filter pieces were rinsed with

LANES

M   1 2 3 4 5 6 7   8 910111213141516171819 20   –   24   M

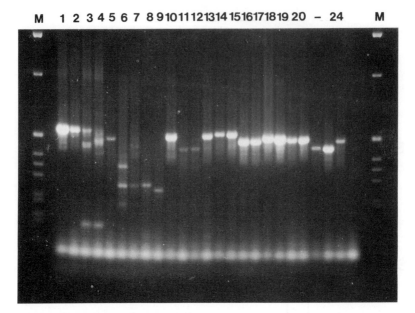

Fig. 2. Polymerase chain reaction amplification products of genomic DNA from hydrothermal vent species. Lanes 1 and 2, *Alvinella pompejana*; 3 and 4, *A. caudata*; 5, *Paralvinella sulfincola*; 6 and 7, *P. grasslei*; 8 and 9, *P. palmiformis*; 10, *P. pandorae*; 11 and 12, chaetopterid worm tubes from TAG; 13, *Escarpia laminata*; 14 and 15, *Riftia pachyptila*; 16 and 17, *Bathymodiolus* sp. from Broken Spur; 18 and 19, *Calyptogena* sp.; 20 and 21, *Lepetodrilus* sp.; 22 and 23, brachyuran crab from Broken Spur; 24, Anemone from Broken Spur; M, 1 kb molecular weight marker (Gibco-BRL). Seen by ultraviolet light after ethidium bromide staining.

deionized water to remove surface salt contamination, prior to storage in a refrigerator at 4°C. Approximately 12 weeks after collection (at the time when the samples were received for biological analysis) the filter pieces were transfered to a −20°C freezer to halt any further deterioration of the organic material. It is appreciated that any soft-bodied (i.e. protoplasmic) remains would have been largely decomposed by the time the samples were received. Consequently, the results are unavoidably biased towards those organisms which have hard skeletal structures. The filter pieces were treated with a dilute solution of rose bengal, a protein-specific stain. The material on the filters was examined under a scanning electron microscope (Jeol JSM-35C) which, with the attached solid state X-ray detector, gave an elemental analysis of the particles. Small quantities of material were dried down on graphite specimen stubs, which eliminated all spurious peaks (Nott 1993).

DNA database

Figure 1 shows the steps involved in extracting and amplifying target DNA sequences using PCR from single invertebrate larvae. PCR is an *in vitro* method for the enzymatic synthesis of specific segments of DNA, using two complementary oligonucleotide primers that hybridize to opposite strands and flank the region of interest in the target DNA. A repeated series of cycles involving template denaturation, primer annealing and the extension of the annealed primers by DNA Taq polymerase results in the exponential accumulation of the target sequence (e.g. Erlich 1989). Since its introduction in the mid-1980s (Saiki *et al.* 1988), PCR has revolutionized the study of molecular biology. PCR is so sensitive that a single DNA molecule can be amplified from a mixture of genomic sequences, yielding a sufficient amount of the target molecule for visualization on an agarose gel in only a few hours and at relatively little expense. Olson *et al.* (1991) were the first to demonstrate the feasibility of applying PCR to marine invertebrate larvae for

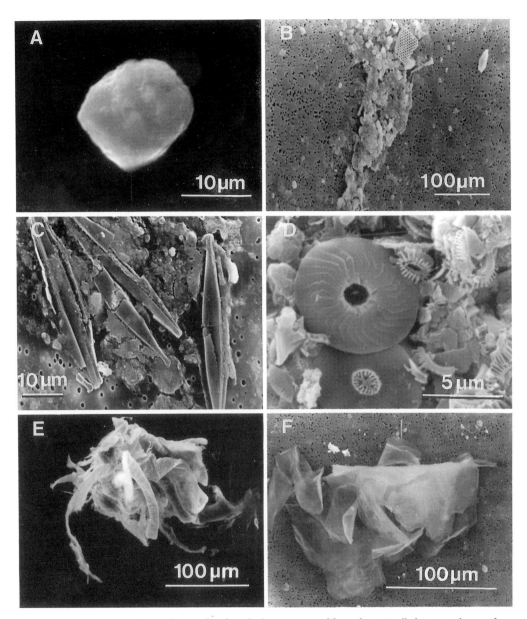

Fig. 3. Scanning electron photomicrographs of particulates recovered from the neutrally buoyant plumes of hydrothermal vents at TAG, MARK and Broken Spur (MAR): (**A**) iron–sulphur particle; (**B**) disrupted copepod faecal pellet containing diatom and coccilithophore remains; (**C**) *Nitzchia bilobata* frustules (diatom); (**D**) *Calcidiscus leptoporus* (coccilithophorid) surrounded by the remains of other coccilithophores and diatoms; (**E**) arthropod exoskeleton; and (**F**) membraneous sheet of unidentified organic matter.

species identification, but they used relatively large late stages in which adult morphology was developing. We extended this approach to include the detection of nuclear DNA polymorphism in single early bivalve larvae, based on length variation in gene introns, using these as potentially neutral markers for population genetic studies (Corte-Real *et al.* 1994*a, b*). We found that a single *Mytilus* larva (approximately 60 μm in diameter) yielded sufficient DNA for

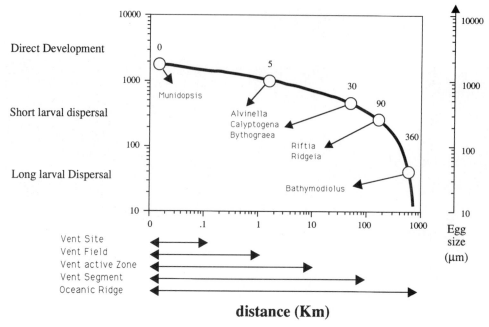

distance (Km)

Fig. 4. Relationship between length of the planktonic larval stage of hydrothermal vent organisms and the potential distance travelled in one generation, assuming a residual current speed of 2–3 m s^{-1}.

more than one type of analysis (Corte-Real *et al.* 1994*b*). This demonstrates the feasibility of coupling species identification with an investigation of the population genetics using a different primer set.

Figure 2 shows the PCR products for a range of different vent organisms. Where possible, more than one individual was analysed for each species to check the reproducability of the method. In most cases, there was good agreement between replicates both in the size and number of bands which were generated. Of particular interest was the fact that in most instances significant differences were observed between species, indicating the potential of the method for species and, because of its sensitivity, larval identification. We have since found that cutting the PCR products with restriction enzymes and/or running these (after radiolabelling) on a sequencing gel significantly increases the resolving power of the method, particularly where there is little difference in fragment size between species (Holland and Dixon, unpublished observation).

The alvinellid polychaetes *Paralvinella grasslei* and *P. palmiformis* (lanes 6, 7, 8 and 9) both showed significant levels of inter-individual variation. As these polychaetes are known to have only a short planktonic larval phase (less than five days), these differences may be a

reflection of an overall increase in genetic heterogeneity stemming from reduced levels of gene flow between breeding groups (see also Williams *et al.* 1993, p. 443, for evidence of sequence polymorphism in vestimentiferans using the same PCR primers). In a previous study of allozyme variation, high levels of genetic heterogeneity and polymorphism have been recorded for these same polychaete species (Tunnicliffe *et al.* 1993; Jollivet 1993). It is worth noting that the two specimens of *P. grasslei* used in the present study came from geographically separated vents (11° and 13°N).

Analysis of filters for biological evidence

All three filters from the TAG, MARK and Broken Spur sites contained amorphous, protein-containing masses in the size range 0.1–0.5 mm. Several of the larger objects were identified under a low power dissection microscope as copepod faecal pellets.

From the SEM results, there appeared to be no significant difference between the three filters (i.e. vent plumes) with respect to the types of particles found. In all three there was a background of small, amorphous inorganic grains, mainly silica, manganese, manganese plus iron, silica plus iron, iron (oxide), or iron plus sulphur (Fig. 3A), in the size range

5–20 µm, which is consistent with a hydrothermal origin. This was interspersed with biological particles, in the size range 10 to >1000 µm, the largest of which were identified as copepod faecal pellets (Fig. 3B). Because of their plant content and the species composition, these faceal pellets must have originated close to the sea surface during a period of phytoplankton bloom. The skeletal remains in the faecal pellets were those of diatoms, radiolarians and coccolithophores (Si and Ca elemental signatures, respectively) (Figs 3C, and D). As these skeletal remains were in a good state of preservation it seems likely that they were trapped on the filters during their descent phase and not following resuspension from the sea bed by the acidic water released by the vent. In addition to these were arthropod exhuviae (Ca signature) (Fig. 3E), plus a small amount of amorphous organic matter of unknown identity (mucus or soft-bodied larvae?), but whose elemental signature (Mg, K, S, Cl, P and Ca) was clearly indicative of protoplasm (Fig. 3F).

Thus the majority of biological material on the filters originated at or near the sea surface and is part of the marine 'snow' which links the energy pathways at the sea surface with those of the deep sea (Lampitt 1985; Alldredge & Gotschalk 1988). A small (<1%) proportion could, however, have originated locally. Given the dense swarms of bresilid shrimps (*Alvinocaris*, *Rimicaris* and *Chorocaris*) which are the major faunistic component at hydrothermal vents on the Mid-Atlantic Ridge (Segonzac *et al.* 1994), at least some of the arthropod remains may have been of this type (e.g. Fig. 4E). Confirmation of this must, however, await the analysis of fresh exhuviae using the PCR method described above.

Figure 4 shows the relationship between the time a pelagic larva spends in the plankton and the distance it could travel assuming an average current speed of 2–3 cm/s, as has been recorded along the axial valleys of the Galapagos Rift and EPR (Lonsdale 1977; A. Khripounof pers comm.). The open circles refer to development times of 0, 5, 30, 90 and 365 days, respectively. It is clear that vent species with a planktonic larval phase lasting weeks or months (this may apply to some molluscs and vestimentiferans) have the potential to disperse over large distances, greater than between adjacent vent segments; perhaps, in some cases, between oceanic ridges separated by thousands of kilometres. The low temperature conditions of the deep sea (average bottom temperature 2°C) will enhance the dispersal potential of both feeding and non-feeding larvae alike by reducing their metabolic rate and thus increasing their survival time. However, entrainment by hydrothermal vent plumes and physical and hydrographic barriers will have a modifying influence on the planktonic dispersal potential (Gustafson & Lutz 1994; Mullineaux 1994) by restricting vent larvae to particular ridge segments or even to particular vents. It is this interaction between different environmental and biological parameters which leads to the patterning of hydrothermal vent communities on different spatial scales. This study shows the feasibility of using a PCR-based strategy to identify the microscopic larvae and other life-history stages of hydrothermal vent organisms. Our plan now is to build a stand-alone pump device for sampling planktonic larvae for PCR investigation.

We express our grateful thanks to C. German (IOSDL) and M. Palmer (Bristol University) for providing samples of filters for SEM analysis; to J. Green for identifying the phytoplankton remains; and to L. Mavin and D. Nicholson for their skilled photographic assistance. B. Murton (IOSDL) kindly donated material collected by him on the June 1993 cruise to Broken Spur. This study was supported in part by NERC grant GR3/8470 awarded to A. J. Southward (L.R.J.D) and by a training bursary from the EC Human Capital and Mobility Programme (D.A.S.B.J.).

References

ALLDREDGE, A. L. & GOTSCHALK, C. 1988. In situ settling behaviour of marine snow. *Limnology and Oceanography*, **33**, 339–351.

AVISE, J. C. 1994. *Molecular Markers, Natural History and Evolution*. Chapman and Hall, New York, London, 511 pp.

CARY, S. C., FELBECK, H. & HOLLAND, N. D. 1989. Observations on the reproductive biology of the hydrothermal vent tube worm *Riftia pachyptila*. *Marine Ecology Progress Series*, **52**, 89–94.

CHEVALDONNE, P., DESBRUYERES, D. & LE HAITRE, M. 1991. Time-series of temperature from three deep-sea hydrothermal vent sites. *Deep-Sea Research*, **38**, 1417–1430.

CORTE-REAL, H. B. S. M., DIXON, D. R. & HOLLAND, P. W. H. 1994a Intron-targeted PCR: a new approach to survey neutral DNA polymorphism in bivalve populations. *Marine Biology*, **120**, 407–413.

——, HOLLAND, P. W. H. & DIXON, D. R. 1994b. Inheritance of a nuclear DNA polymorphism assayed in single bivalve larvae. *Marine Biology*, **120**, 415–420.

DIXON, D. R., SIMPSON-WHITE, R. & DIXON, L. R. J. 1992. Evidence for thermal stability of ribosomal DNA sequences in hydrothermal vent organisms. *Journal of the Marine Biological Association, UK*, **72**, 519–527.

ENDLER, J. A. 1992. Genetic heterogeneity and ecology. *In*: BERRY, R. J., CRAWFORD, T. J. & HEWITT, G. M. (eds) *Genes in Ecology*. Blackwell Scientific, Oxford, 315–334.

ERLICH, H. A. 1989. *PCR Technology: Principles and Applications for DNA Amplification*. Stockton Press, New York, London, 246 pp.

FUSTEC, A., DESBRUYERES, D. & LAUBIER, L. 1987. Deep-sea hydrothermal vent communities at 13°N on the East Pacific Rise: microdistribution and temporal variations. *Biological Oceanography*, **4**, 121–164.

GUSTAFSON, R. G. & LUTZ, R. A. 1994. Molluscan life history traits at deep-sea hydrothermal vents and cold methane/sulfide seeps. *In*: YOUNG, C. M. & ECKELBARGER, K. J. (eds) *Reproduction, Larval Biology, and Recruitment of the Deep-Sea Benthos*. Columbia University Press, New York, 76–97.

HOLLAND, P. W. H. 1993. Cloning genes using the polymerase chain reaction. *In*: STERN, C. D. & HOLLAND, P. W. H. (eds) *Essential Developmental Biology: a Practical Approach*. IRL Press/Oxford University Press, Oxford, 243–255.

HU, Y.-P., LUTZ, R. A. & VRIJENHOEK, R. C. 1992. Electrophoretic identification and genetic analysis of bivalve larvae. *Marine Biology*, **113**, 227–230.

JOHNSON, K. S., CHILDRESS, J. J. & BEEHLER, C. L. 1988. Short-term temperature variability in the Rose Garden hydrothermal vent field, Galapagos spreading center. *Deep-Sea Research*, **35**, 1723–1744.

JOLLIVET, D. 1993. *Distribution et evolution de la faune associee aux sources hydrothermales profondes a 13°N sur la dorsale du Pacifique oriental: le cas particulier des polychetes Alvinellidae*. These de Doctorat nouveau regime, Universite de Bretagne Occidentale, 357 pp.

JONES, M. L. & GARDINER, S. L. 1989. On the early development of the vestimentiferan tube worm *Ridgeia* sp. and observations on the nervous system and trophosome of *Ridgeia* sp. and *Riftia pachyptila*. *Biological Bulletin*, **177**, 154–176.

LAMPITT, R. S. 1985. Evidence for the seasonal deposition of detritus to the deep-sea floor and its subsequent resuspension. *Deep-Sea Research*, **32**, 885–897.

LONSDALE, P. 1977. Clustering of suspension-feeding macrobenthos near abyssal hydrothermal vents at oceanic spreading centres. *Deep-Sea Research*, **24**, 857–863.

LUTZ, R. A. 1988. Dispersal of organisms at deep-sea hydrothermal vents: a review. *Oceanology Acta, Special Volume No. 8*, 23–30.

——, JABLONSKI, D., RHOADS, D. C. & TURNER, R.D. 1980. Larval dispersal of a deep-sea hydrothermal vent bivalve from the Galapagos Rift. *Marine Biology*, **57**, 127–133.

MCHUGH, D. 1989. Population structure and reproductive biology of *Paralvinella pandorae* Desbruyeres and Laubier and *Paralvinella palmiformis* Desbruyeres and Laubier, two sympatric hydrothermal vent polychaetes. *Marine Biology*, **103**, 95–106.

MULLINEAUX, L. S. 1994. Implications of mesoscale

flows for dispersal of deep-sea larvae. *In*: *Reproduction, Larval Biology, and Recruitment of the Deep-Sea Benthos*. Columbia University Press, New York, 201–222.

NEWELL, G. E. & NEWELL, R. C. 1966. *Marine Plankton – A Practical Guide*. Hutchinson Educational, 221 pp.

NOTT, J. A. 1993. X-ray microanalysis in pollution studies. *In*: SIGEE, D. C., MORGAN, A. J., SUMNER, A. T. & WARLEY, A. (eds) *X-ray Microanalysis in Biology*. Cambridge University Press, Cambridge, 257–281.

OLSON, R. R., RUNSSTADLER, J. A. and KOCHER, T.D. 1991. Whose larvae? *Nature*, **351**, 357–358.

SAIKI, R. K., GELFAND, D. H. & 6 others 1988. Primer-directed enzymatic amplification of DNA with a thermostable DNA polymerase. *Science*, **239**, 487–491.

SCHELTEMA, R. S. 1971. Larval dispersal as a means of genetic exchange between geographically separated populations of shallow-water benthic marine gastropods. *Biological Bulletin*, **140**, 284–322.

—— 1994. Adaptations for reproduction among deep-sea benthic molluscs: an appraisal of the existing evidence. *In*: YOUNG, C. M. & ECKELBARGER, K. J. (eds) *Reproduction, Larval Biology, and Recruitment of the Deep-sea Benthos*. Columbia University Press, New York, 44–75.

SEGONZAC, M., DE SAINT LAURENT, M. & CASANOVA, B. 1994. L'enigme du comportement trophique des crevettes Alvinocarididae des sites hydrothermaux de la dorsale medio-atlantique. *Cahiers de Biologie Marine*, **34**, 535–571.

SOUTHWARD, E. C. 1988. Development of the gut and segmentation of newly settled stages of *Ridgeia* (Vestimentifera): implications for relationship between Vestimentifera and Pogonophora. *Journal of the Marine Biological Association of the United Kingdom*, **68**, 465–487.

THORSON, G. 1950. Reproductive and larval ecology of marine bottom invertebrates. *Biological Reviews*, **25**, 1–45.

TUNNICLIFFE, V. 1988. Biogeography and evolution of hydrothermal-vent fauna in the eastern Pacific Ocean. *Proceedings of the Royal Society London, Series B*, **233**, 347–366.

—— 1991. The biology of hydrothermal vents: ecology and evolution. *Oceanography and Marine Biology Annual Reviews*, **29**, 319–407.

——, DESBRUYERES, D., JOLLIVET, D. & LAUBIER, L. 1993. Systematic and ecological characteristics of *Paralvinella sulfincola* Desbruyeres and Laubier, a new polychaete (family Alvinellidae) from Northeast Pacific hydrothermal vents. *Canadian Journal of Zoology*, **71**, 286–297.

TURNER, R. D., LUTZ, R. A. & JABLONSKI, D. 1985. Modes of molluscan larval development at deep-sea hydrothermal vents. *Biological Society of Washington Bulletin*, **6**, 167–184.

WILLIAMS, N. A., DIXON, D. R., SOUTHWARD, E .C. & Holland, P. W. H. 1993. Molecular evolution and diversification of the vestimentiferan tube worms. *Journal of the Marine Biological Association, UK*, **73**, 437–452.

Hyperthermophilic enzymes: biochemistry and biotechnology

DON A. COWAN

Department of Biochemistry and Molecular Biology, University College London,
Gower Street, London WC1E 6BT, UK

Abstract: Deep submarine hydrothermal vent habitats have already proved to be rich sources of novel hyperthermophilic micro-organisms, most of which belong to the third super-kingdom, the Archaea. These organisms, many of which survive at temperatures at or above 100°C, contain novel macromolecules and metabolic systems which represent a vast resource for fundamental molecular and physiological studies, and for potential exploitation in biotechnology. The one guaranteed property of enzymes isolated from extremely thermophilic micro-organisms is their thermostability. Resistance to heat denaturation also ensures resistance to a number of other denaturing influences (detergents, organic solvents). This characteristic of hyperthermophilic enzymes is the most likely basis for the development of new biotechnological applications. The ability of thermostable enzymes to function in non-aqueous and mixed-phase solvent systems is a particularly useful property . Only a limited number of hyperthermophilic enzymes have found application in specialist biotechnological applications. Hyperthermophilic DNA polymerases and other DNA-modifying enzymes have already contributed significantly to the development of recombinant DNA technology and gene-based diagnostics. Other hyperthermophilic enzymes have obvious potential in the growing area of biotechnology, whereas more distant but no less exciting possibilities lie in those archaeal enzymes which are wholly novel to these organisms.

In the late 1960s when the first organisms growing optimally at above 70°C were discovered (Brock 1969), it became convenient to redefine the classification of organisms (particularly micro-organisms) on the basis of their cardinal growth temperatures (Table 1). Although this division is wholly arbitrary, it is a convenient reference structure into which the different types of high-temperature micro-organisms can be positioned.

Since the discovery in the early 1980s of micro-organisms capable of growing optimally at hyperthermophilic temperatures (i.e. above 85°C), there has been an explosion of interest in their physiology, biochemistry and genetics (see reviews by Woese & Wolfe 1985; Kandler & Zillig 1986; Stetter 1986; Woese 1987; Danson 1988; Cowan 1992a; Adams 1993). Much of the driving force for these studies has been the expectation that results will possess a high level of novelty. The combination of new microbial sources (the hyperthermophilic Archaea) and the extremely unusual environment suggests that fundamentally different biochemical strategies may await discovery.

Researchers in this field have not been disappointed. The intrinsic stability properties of macromolecules have indeed provided a means of probing some of the fundamental properties of protein structure and function, of nucleic acid structure and processing and of membrane function. The biochemistry of many metabolically important small molecules (co-factors, reaction intermediates) has been re-assessed in the light of their intrinsic chemical instability. Finally, but no less importantly, the molecular properties of the Archaea have led to a fundamental revision of the phylogenetic origins of all living organisms (Woese & Fox 1977).

There is a general assumption that the properties, particularly the stability, of thermophilic proteins must inevitably lead to valuable contributions to the field of biotechnology. Although this assumption is undoubtedly over-optimistic, it is interesting to consider the specific properties of these enzymes which are amenable to applications in biotechnology. A review of the current applications of thermophilic enzymes provides an insight into the willingness of industry to invest in this new technology. With an awareness of the developing trends in biotechnology, it may also be possible to predict the fields of biotechnology in which thermostable biocatalysts have the potential for commercial or industrial application within the foreseeable future.

From PARSON, L. M., WALKER, C. L. & DIXON, D. R. (eds), 1995, *Hydrothermal Vents and Processes,* 351
Geological Society Special Publication No. 87, 351–363.

Table 1. *Classification of organisms on the basis of growth temperature*

	Growth minimum	Growth optimum	Growth maximum
Thermophile	>30°C	>50°C	> 60°C
Extreme thermophile	>40°C	>65°C	> 70°C
Hyperthermophile	>70°C	>90°C	<115°C

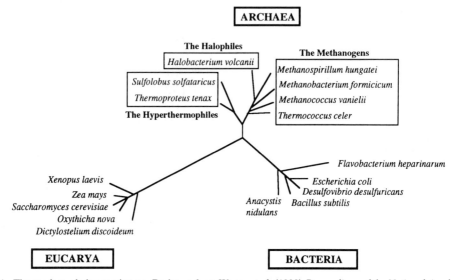

Fig. 1. The modern phylogenetic tree. Redrawn from Woese *et al.* (1990) *Proceedings of the National Academy of Sciences*, **87**, 4576, with kind permission of the authors. Only a limited number of representative species have been positioned within each domain. The branch lengths between species are a measure of the degree of evolutionary relatedness.

Microbial phylogeny

The development of molecular phylogenetic analysis methods led Woese and Fox to propose a restructuring of the fundamental phylogeny of life (Woese & Fox 1977). Based on detailed comparisons of 16S and 18S rRNA sequences, the two traditional primary kingdoms (the Prokaryotes and the Eukaryotes) were partitioned into three discrete domains, named as the Eukaryotes, the Eubacteria and the Archaebacteria (Fig. 1). A revised system of nomenclature which emphasized the fundamental evolutionary differences between the bacteria and the archaebacteria has been proposed (Woese *et al.* 1990), in which the three kingdoms (termed domains) are renamed the Bacteria, the Eucarya and the Archaea.

The domain Archaea can be further subdivided along lines which reflect phenotypic differences. Three distinct phyla, comprising the hyperthermophiles, the methanogens and the

extremely halophilic Archaea are recognized, although recent analyses of RNA extracted directly from thermal habitats suggests that these distinctions may be artificial (Barns *et al.* 1994).

Hyperthermophilic environment

Habitats suited to the survival and growth of thermophilic micro-organisms (e.g. a temperature range of 40–70°C) are widespread. For example, desert soil, compost, industrial waste water and domestic hot water supplies offer environments combining nutrient, temperature and pH conditions suitable for thermophilic growth. The majority of organisms in these thermophilic 'biotopes' are Bacteria or, at the lower temperatures, Eucarya. Higher temperature biotopes (i.e. >75°C) are less common and are typically of geothermal origin, associated with tectonically or volcanically active zones. These are widely, if sparsely, distributed over

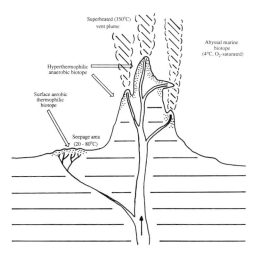

Fig. 2. Schematic representation of deep-sea hydrothermal environments. Note that the thermophilic aerobic biotope will exist in surface sediment layers (possibly only a few millimetres or centimetres in depth) where oxygen diffusion from above and thermal fluid from below combine.

the Earth's surface. These so-called hyperthermophilic biotopes are colonized by both Bacteria and Archaea, the latter being more prevalent at the upper end of the biotic temperature range (95–110°C).

Terrestrial geothermally heated waters are found in the form of thermal springs, hot pools or geysers. These are typical of many countries, but major thermal 'fields' are found in Yellowstone National Park (USA), Rotorua (New Zealand) and in many regions of the Kamchatka Peninsula (Russia), Chile, Iceland and Japan. The surface emission of magma-heated groundwater at atmospheric pressure limits the water temperature to 100°C (or less at elevated altitudes). The emission temperature, nutrient status and pH is also strongly influenced by mixing in upwelling and surface waters.

Terrestrial thermal waters typically fall in the acidophilic (pH 1.0–2.5) or neutrophilic (pH 6.0–8.5) pH ranges. These two ranges reflect the primary buffering salts: sulphuric acid ($pK_a = 1.8$) and sodium carbonate/bicarbonate ($pK_a = 6.3$ and 10.2). The alkalophilic (pH 9–10.5) thermal sites of the African Rift Valley represent a very rare and interesting biotope.

The mineralization and nutrient status of the various thermal biotopes is strongly influenced by the source and acidity of the water flow. Acid thermal waters are often S- and Fe-rich, but otherwise poorly mineralized, the lack of mineralization reflecting the fact that the hydrological cycle is brief and water penetration is shallow. The acid (primarily H_2SO_4) is derived from H_2S, carried to the surface from geological sulphide deposits by geothermal steam. There is also increasing evidence for the biogenic origin of subterranean H_2S. At or near the surface, H_2S is oxidized to S on contact with O_2 and further oxidation by S-oxidizing bacteria generates H_2SO_4. This is leached to pooling areas by surface or shallow ground water, where it may be heated by geothermal steam. Such acid thermal sites often have a low water flow and are subject to substantial temperature shifts in response to changes in both groundwater, flow and steam supply.

Neutrophilic and slightly alkaline thermal sites are usually highly mineralized and the basins in which they are sited are often heavily encrusted with silicates. This reflects the deep penetration of the source groundwater and the reactivity of superheated water. The effect of input from shallow acid groundwater is minimized because of the continual and substantial flow of neutral or alkaline waters from depth. For the same reason, alkaline thermal pools are also much more stable with respect to temperature, pH and volume.

The marine equivalent of the alkaline thermal spring differs only in that the water erupting is saline and temperatures of above 100°C can be reached because of the added hydrostatic pressure. The thermal habitat is much more localized because of the absence of a pooling area.

A specialized version of this biotope is found at great depths (1000–4000 m) in the abyssal rifts located in many of the world's oceans. Because of the effects of pressure on the boiling point of water, liquid emission temperatures in the region of 350°C have been recorded in the vent waters. The very rapid mixing of the anaerobic superheated emission water with the cold oxygenated abyssal waters generates very steep temperature and oxygen gradients (Fig. 2). The thermophilic environments are largely anaerobic. However, interface regions between the heated (anaerobic) and cold (aerobic) waters must generate some (possibly restricted) thermophilic aerobic habitats. Because of the reactivity of superheated saline water, with the solubilization of deep basalts and subsequent reductive chemical changes (Jannasch & Mottl 1985), the waters issuing from these deep-sea hydrothermal vents are very highly mineralized. The precipitation of insoluble metal salts resulting from the interactions between reduced mineralized vent water and the oxygen-rich seawater has given rise to the term 'black smokers'.

Table 2. *Representative archaeal and bacterial genera inhabiting different thermophilic biotopes*

Thermophilic habitat	Temperature (°C)	pH range	Representative genera	Domain
Coal/ore pile	50–60	Acidic	*Thermoplasma*	Archaea
			Sulfobacillus	Bacteria
			Sulfolobus	Archaea
Neutral silica pool	Up to 100	5.0–9.5	*Thermus*	Bacteria
			Bacillus stearothermophiles	Bacteria
			Thermotoga	Bacteria
			Pyrobaculum	Archaea
			Thermofilum	Archaea
			Thermoproteus	Archaea
			Desulfurococcus	Archaea
			Methanothermus	Archaea
Acid thermal pool	Up to 100	1.0–3.5	*Bacillus acidocaldarius*	Bacteria
			Sulfolobus	Archaea
			Acidianus	Archaea
			Metallosphaera	Archaea
			Stygioglobus	Archaea
			Desulfurolobus	Archaea
Shallow marine vents	Up to 103	5.0–7.0	*Thermotoga*	Bacteria
Deep-sea thermal vents	80–120*	5.0–7.0	*Aquifex*	Bacteria
			Thermococcus	Archaea
			Pyrodictium	Archaea
			Pyrococcus	Archaea
			Staphylothermus	Archaea
			Archaeoglobus	Archaea
			Thermodiscus	Archaea
			Hyperthermus	Archaea
			Methanothermus	Archaea
			Methanococcus	Archaea
			Methanopyrus	Archaea
			Desulfurococcus	Archaea
Heated oil reservoirs	60–>100	Neutral	*Thermococcus*	Archaea
			Archaeoglobus	Archaea
			Thermotoga	Bacteria

* Vent emission waters at 350°C are sterile.

There is a growing belief that an important, potentially vast, thermophilic biosphere exists below the Earth's surface. The existence of this deep subsurface biosphere has been implicated both by the discovery of hyperthermophilic Archaea in Alaskan oil reservoirs (Stetter *et al.* 1993) and by estimates of microbial biomass in deep drilling samples (Szewzyk *et al.* 1994).

Hyperthermophilic micro-organisms

Each of the thermal biotopes described in the preceding section have representative microbial genera (Table 2), although these are by no means exclusive. For example, genera identified in oil reservoirs are closely related to isolates from terrestrial hot springs and shallow water hydrothermal fields. Most micro-organisms inhabiting the hyperthermophilic environments

are archaeal, the bacterial representatives being restricted to the genera *Thermotoga* and *Aquifex*. Archaeal biomass levels in thermal waters can be surprisingly high. Cell densities in free solution are relatively low, but large numbers of cells (typically 10^3–10^4 per g) are found adhering to silicate surfaces, particulate matter and detritus.

With the rapid increase in the isolation of novel species and genera of Archaea, many of the physiological characteristics typical of Bacteria are to be found among the Archaea (Table 3). The anaerobic Archaea typically grow either chemolithoautotrophically, utilizing CO_2 as the sole C source and deriving energy by S oxidation of molecular hydrogen, or heterotrophically by S respiration of various carbon and energy sources. The former is viewed as a primeval mode of metabolism as high temperature,

Table 3. *Physiological properties of representative hyperthermophilic archaea*

Organism	Temperature maximum (°C)	pH optimum	Oxygen status	Metabolism/energy yield
Acidianus infernus	95	2.0	Aerobic or anaerobic	Obligate autotroph, either aerobic ($S° \rightarrow SO_4^{2-}$] or anaerobic ($S° \rightarrow S^{2-}$]
Archaeoglobus fulgidus	95	6.5	Obligate anaerobe	Chemolithoautotrophic on H_2, CO_2 and $S_2O_3^-$; chemo-organotrophic on complex media; SO_4^{2-} reduction ($S°$ not reduced)
Desulfurococcus spp.	95	5.5–6.5	Obligate anaerobes	Obligate heterotrophs; anaerobic $S°$ respiration; fermentation
Hyperthermus butylicus	95–106	7.0	Obligate anaerobe	Fermentative; energy also obtained from $H_2 + S° \rightarrow H_2S$ (no $S°$ respiration)
Metallosphaera sedula	80	1–4.5	Aerobe	Facultative autotroph ($S°/S^{2-} \rightarrow SO_4^{2-}$) or heterotrophy on complex media
Methanopyrus kandleri	110	6.0	Obligate anaerobe	Methanogen; obligate autotrophs, formate, $H_2 + CO_2 \rightarrow CH_4$
Pyrobaculum spp.	103	6.0	Obligate anaerobes	Facultative autotroph; H_2/S autotrophy or heterotrophy on complex media
Pyrococcus spp.	103	6.5–7.5	Obligate anaerobes	Obligate heterotroph; $S°$ respiration; fermentation
Pyrodictium spp.	110	5.5	Obligate anaerobes	Obligate $H_2/S°$ autotrophy
Staphylothermus marinus	98	6.5	Obligate anaerobe	Obligate heterotroph with $S°$ respiration, fermentation
Sulfolobus spp.	75–87	3.0	Aerobic or microaerophilic	Facultative autotrophs; $S° \rightarrow SO_4^{2-}$, $Fe[II] \rightarrow Fe[III]$, $S^{2-} \rightarrow S°$
Thermococcus celer	93	5.8	Obligate anaerobe	Obligate heterotroph; $S°$ respiration of complex media, fermentation
Thermodiscus maritimus	98	5.5	Obligate anaerobe	Facultative autotroph; $S°$ respiration
Thermoproteus tenax	97	5.5	Obligate anaerobe	Facultative autotroph; either $H_2/S°$ autotrophy or fermentation of complex media

anaerobic, aquatic or marine environments rich in S, CO_2 and H_2 may have predominated on the primeval Earth.

Of the hyperthermophilic Archaea, only the Sulfolobales demonstrate aerobic metabolism, a physiological characteristic which may reflect their relatively recent evolutionary divergence. The prevalence of strict anaerobiosis in the majority of the Archaea probably reflects both the nature of the primeval habitat and the fact that the oxygen solubility in boiling water is so low that there has not been a strong evolutionary pressure for the acquisition of aerobic metabolic systems in very high temperature aquatic and marine habitats.

The methanogenic Archaea are also well represented in hyperthermal environments. Representatives of the genera *Methanothermus*, *Methanococcus* and *Methanopyrus* have been isolated from both shallow and deep marine thermal sources. These organisms are obligately anaerobic autotrophs, utilizing H_2 and CO_2 in the biogenesis of CH_4. The methanogenic metabolism is unique to members of the Archaea and involves novel enzymes, cofactors and pathways. The microbiology and biochemistry

Table 4. *Comparative thermostability values of proteinases*

		Stability of protease	
Organism	Organism type	Half-life (minutes)	Temperature of incubation (°C)
Bacillus subtilis	Mesophilic bacterium	10	60
B. thermoproteolyticus	Thermophilic bacterium	60	80
B. caldolyticus	Extremely thermophilic bacterium	>480	80
Thermus aquaticus	Extremely thermophilic bacterium	1800	80
Sulfolobus solfataricus	Extremely thermophilic archaea	>2880	80
Pyrococcus furiosus	Hyperthermophilic archaea	>1200	95

of methanogens are covered in depth in several excellent reviews (Jones *et al*. 1987; DiMarco *et al*. 1990).

Properties of enzymes from hyperthermophilic organisms

It is obvious that structural and functional macromolecules from high-temperature micro-organisms must be sufficiently long-lived *in vivo* to enable the organisms to survive. Suggestions that thermophiles compensate for molecular instability by accelerated synthesis have not been supported experimentally. Protein stability in these organisms now appears to be either an intrinsic property or, in a limited number of cases, the result of extrinsic factors (salts, cofactors, etc.). It can be stated with confidence that any enzyme isolated from an extreme thermophile will almost certainly be more thermostable than the homologous enzyme isolated from a less thermophilic source. This trend can be clearly demonstrated in a comparison of the thermostabilities of proteinases from various sources (Table 4) and many similar comparisons with different enzymes have now been published.

The molecular mechanisms responsible for enhanced protein thermostability have proved to be both complex and subtle (see review by Jaenicke 1991). The thermodynamic stability of a protein is a balance between large stabilizing forces (of the order of a few thousands of kilojoules per mole, derived from non-covalent intramolecular interactions such as H-bonds, hydrophobic interactions and charge–charge interactions) and large destabilizing forces, primarily chain conformational entropy. The free energy difference between the folded and unfolded states is typically only 30–60 kJ mol^{-1} and can thus be perturbed by relatively minor changes in primary structure. The ability of

apparently minor structural alterations to impart relatively large changes in protein stability is better understood when it is realized that typical intramolecular interactions (H-bonds, salt bridges and hydrophobic interactions) can contribute between 5 and 20 kJ mol^{-1} depending on location and orientation.

Detailed structural comparisons of homologous mesophilic and thermophilic proteins provide the most detailed insights into the mechanisms of protein thermostability. Some of the current thermostability 'rules' derived from such studies are listed in Table 5. The consensus view is that any or all of these mechanisms may be used in any protein and that such changes are usually achieved without significant alterations in gross structural or functional properties.

A consequence of protein thermostability appears to be a high degree of resistance to other potential denaturants: detergents, chaotropic agents such as urea and guanidinium hydrochloride, organic solvents, oxidizing agents and possibly extremes of pH. There is evidence that thermostability also imparts resistance to enzymic cleavage (proteolysis) and to chemical reactions causing irreversible protein degradation at high temperatures (Cowan 1992*a*).

The resistance of thermostable proteins to denaturation by organic solvents appears to be a general characteristic. A clear positive correlation has been demonstrated between thermostability and resistance to denaturation in biphasic organic–aqueous solvent systems, both for populations of proteins and for single purified enzymes (Owusu & Cowan 1989).

In general terms, the commonality of the relationship between protein thermostability and resistance to other denaturing influences implies that some aspects of the pathway of protein unfolding may be ubiquitous. The rate-limiting process of protein unfolding may be critically dependent on the rapid reversible conformational transitions which are a reflection

Table 5. *Thermostabilization mechanisms employed in hyperthermophilic proteins*

Increase in intramolecular packing
Loss of surface loops
Increase in helix-forming amino acids
Stabilisation of α-helix dipoles
Insertion of prolines
Reduction in Asn content
Restriction of N-terminus mobility

of the normal flexibility of the protein. Loss of tertiary structure via heat or chemically induced denaturation, proteolysis or chemical degradation only progresses from the unfolded conformer. Reducing the extent of reversible conformational transitions (e.g. by molecular stabilization) should proportionally reduce the tendency of the protein to undergo further irreversible unfolding steps. The observation that unfolded proteins are much more susceptible to proteolytic attack than native protein supports this scheme. It has also been observed that the chemical degradation of proteins, including the deamination of asparagine residues and peptide cleavage at Asn–X bonds, proceeds more rapidly in pre-denatured proteins.

Although molecular stability under a wide range of different conditions is generally assumed to be an advantageous property, another apparent consequence of thermostability is, at least for the biotechnologist, a disadvantage. A common misconception is that a consequence of thermophilicity is the ability to attain very rapid reaction rates by carrying out reactions at high temperatures (Fig. 3a). This misconception is based on a simple Arrhenius law extrapolation from the properties of mesophilic enzymes, presuming that the reaction rate of an enzyme will double for every 10°C rise in temperature. The catalytic rate of thermophilic enzyme at 90°C should thus be roughly 32 times greater than that of an equivalent mesophilic enzyme at 37°C. We now know this to be incorrect in almost all instances and, for sound physical reasons, thermophilic enzymes at their 'optimum' temperatures exhibit turnover rates of the same order of magnitude as do mesophilic enzymes at their 'optimum' temperatures (Fig. 3b). The mechanism of this apparent limitation of activity is, at least in part, the restraint in conformational flexibility which is thought to be coincident with enhanced thermostability. Evidence supporting this relationship is still sketchy and there is great potential for detailed structure/function/stability studies.

Significant enhancement of reaction rates can be achieved by using thermophilic enzymes at high temperatures only where the substrate itself is rendered more susceptible to catalysis at higher temperatures. The best examples are the hydrolytic degradation of polymeric substrates, particularly the proteolytic degradation of protein and amylolytic degradation of starch. Industry has taken full advantage of the latter, particularly in the high temperature saccharification processes currently used in most enzymic starch hydrolysis plants.

Production of hyperthermophilic proteins

The efficient and cost-effective 'large-scale' production of enzymes from hyperthermophilic bacteria and Archaea is an essential prerequisite for their successful commercial or industrial application. The biomass yields from most hyperthermophilic archaeal fermentations are extremely low (typically $0.1–1.0\,g\,l^{-1}$ wet weight of biomass compared with $5–30\,g\,l^{-1}$ biomass for common mesophilic eubacteria). This is a major impediment to any direct application of hyperthermophilic organisms, whether as whole cell biocatalysts, sources of enzymes or sources of other biomolecules. However, fermentation optimization incorporating rapid degassing and/or stirring to remove inhibitory H_2 together with the careful selection of nutrient media has increased cell yields by several orders of magnitude. *Pyrococcus furiosus*, the '*Escherichia coli*' of the hyperthermophilic Archaea, has been grown in continuous culture at cell densities of over 10^{10} cells/ml in several laboratories.

Nevertheless, it is currently a reasonable assumption that the successful application of most hyperthermophilic enzymes will follow the cloning and over-expression of the protein. Early predictions that abnormal codon usage and folding kinetics would prove a barrier to the successful expression of hyperthermophilic archaeal and bacterial enzymes in mesophilic hosts have proved to be unfounded and successful cloning and over-expression of a number of such proteins has been reported. The ability of these recombinant enzymes to fold successfully under non-native temperatures (say, 50°C below the normal *in vivo* folding temperature) provides some insight into the thermodynamics of the folding process. It is also a clear demonstration that no post-translational modifications are critical to the correct folding of the functional product, despite the existence of lysine ε-methylation in some hyperthermophilic archaeal proteins.

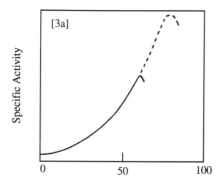
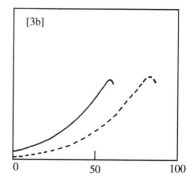

Reaction temperature (degrees C)

Fig. 3. Temperature-specific activity profiles for mesophilic and hyperthermophilic enzymes: (**a**) the perception; and (**b**) the reality.

By coincidence rather than design, the purification of recombinant hyperthermophilic enzymes has become a facile exercise. Taking advantage of the differential in heat stability between the recombinant protein and the host proteins, a single heating step (preferably above 75°C) can provide a virtually pure protein preparation. This approach to purification has been used successfully in a number of thermophile laboratories and should be amenable to industrial scale-up.

In many instances, the ability to perform enzymic reactions at greatly elevated temperatures may be of no advantage or even disadvantageous (where substrates and/or products are particularly labile, for example). In these instances, the use of moderately thermophilic enzymes below their temperature optima may still be an attractive option, as reduced reaction kinetics may be more than compensated for by very large increases in enzyme stability and hence longevity.

Current commercial applications of hyperthermophilic enzymes

The success of thermostable hyperthermophilic enzymes in biotechnology has been largely restricted to a small (in volume terms) but critically important specialist technology: the amplification of DNA sequences using thermostable DNA polymerases by what is termed the polymerase chain reaction (PCR). This technology, which provides a very rapid and efficient exponential amplification of specific sequences of DNA, has revolutionized the field of molecular genetics (see reviews by Erlich *et al.* 1991; Cha & Thilly 1993).

The PCR technique was originally developed using the Klenow fragment of *E. coli* DNA polymerase 1, but required repeated additions of this thermolabile enzyme, as all activity was lost during the high-temperature dissociation step (typically around 94°C). The automation of the PCR by the incorporation of thermostable DNA polymerase from the thermophilic eubacterium *Thermus aquaticus* was a major development as the enzyme was sufficiently stable to remain active through a number (say 20–30) of polymerization and strand-melting cycles.

The thermostability of *Taq* polymerase is marginally sufficient under normal PCR conditions and a further development has been the appearance of DNA polymerases from two of the hyperthermophilic Archaea (*Thermococcus littoralis* and *Pyrococcus furiosus*). Both these enzymes are highly thermostable and retain activity for long periods above 90°C. The commercial success of thermostable DNA polymerases has stimulated a plethora of studies on DNA polymerases from hyperthermophilic Archaea and Bacteria.

Thermostability is not the only requirement for the successful application of DNA polymerases in the PCR, sequencing and related technologies. Other factors such as processivity and proof-reading (3'–5' exonuclease) activity are also important (Table 6). There is growing interest in DNA polymerases with extended processivity. With the advent of automated DNA sequencing, high processivity polymerases provide the potential for megabase sequencing, which would be a huge advance in research fields such as the Human Genome Project.

The original PCR methodology has spawned numerous modifications, which have greatly

Table 6. *Properties of thermostable DNA polymerases (data derived from the manufacturer's technical literature)*

	Enzyme				
	Taq	Bio-X-Act™	VENT™	DEEP VENT™	Pfu™
Source	*Thermus aquaticus*	*Thermus thermophilus*	*Thermococcus littoralis*	*Pyrococcus* sp.	*Pyrococcus furiosus*
Thermostability*	1.6	>1.6	7	23	As for Deep Vent?
Processivity†	40	<40	7	n.a.‡	n.a.
3'–5' exonuclease (proofreading)	No	Yes	Yes§	Yes	Yes§
Fidelity¶	2×10^{-4}	n.a.	$1–4 \times 10^{-5}$	$1–4 \times 10^{-5}$	1.7×10^{-5}
Reverse transcriptase‖	Yes	Yes?	n.a.	n.a.	n.a.

* Half-life (minutes) at 95°C.
† Nucleotides per polymerase binding event.
‡ n.a.: data not available.
§ The cloned version of this enzyme is available with the proofreading activity deleted.
¶ Mutations per base duplication.
‖ For use in 'one-pot' reverse transcription PCR amplification.

expanded the field of application into clinical diagnostics, forensics, genomic cloning and cDNA cloning. For example, by the selection of reaction conditions, reverse transcriptase PCR enables mRNA sequences to be first transcribed and then the cDNA amplified using the same DNA polymerase. This technique has greatly simplified the cloning of eukaryotic genes, effectively circumventing the need for cDNA library preparation.

High-temperature DNA manipulation has become an accepted and valued technology and has in turn stimulated interest in other thermostable enzymes that can complement or supplement the PCR. Thermostable restriction endonucleases (for DNA cleavage) and reverse transcriptases (from hyperthermophilic viruses) have all become potential targets for commercial development, whereas a thermostable DNA ligase from *Thermus thermophilus* is already commercially available. A hyperthermophilic example has been cloned from the archaeon *Pyrococcus furiosus* and is the subject of a patent application (Mathur 1994).

Thermostable DNA ligases have attracted commercial attention because of the development (Barany 1991) of a ligase-dependent DNA amplification technique (the ligase chain reaction; LCR or LAR). This technique can be used for the identification of single nucleotide base lesions and has considerable potential as a screening assay for the gene defects that result in serious genetic disorders such as cystic fibrosis.

Future Applications of enzymes from hyperthermophiles

The field of archaeal enzymology is relatively untapped. The discovery of totally novel species or genera is currently occurring at a rate a little slower than the appearance of publications on novel enzymes. Each of these new isolates represents a tremendous resource of catalytic potential. Not only can it be assumed that the hyperthermophilic Archaea and Bacteria will contain as wide a range of different enzyme activities as any other diverse group of microorganisms, but certain activities and enzyme systems will be totally unique to these organisms, particularly the Archaea. Excellent examples must be the enzymes of the biphytanyl tetraether lipid biosynthetic pathways which are, as yet, largely uncharacterized.

In assessing the commercial potential for hyperthermophilic enzymes it is essential to consider the current biotechnological applications of enzymes. A limited range of industries use over 75% of the world tonnage of enzymes (value currently estimated at over £800 million). The detergent industry (alkalophilic proteases, cellulases and lipases), the starch industry (α-amylases, amyloglucosidases and glucose isomerases), the dairy industry (microbial and bovine milk clotting proteases) and the potable beverage industry (amylases, pectinases) are the major contributors. The question arises as to whether a hyperthermophilic enzyme is likely to

Table 7. *Potential of hyperthermophilic enzymes in the 'bulk' enzyme market*

Industrial use	Enzyme type	Thermostable enzyme in use	More thermostable enzyme required?	Critical enzyme property
Starch hydrolysis	α-Amylase	Yes	No	High stability, activity
	Glucoamylase	No	Yes	Specificity
High fructose syrup production	Glucose (xylose) isomerase	No*	Yes	Low inhibition, high activity
Detergent preparation	Protease	No	No	Broad specificity, detergent resistance, alkaliphilicity
	Cellulase	No	No	Detergent resistance, alkaliphilicity
	Lipase	No	No	Broad specificity, detergent resistance, alkaliphilicity
Clarification of beer	Pectinase	No	No	Low-temperature operation
Amino acid production	Various	No	Possibly	Specificity
Antibiotic semi-synthesis	Penicillin V acylase	No	No	Specificity

* The *Streptomyces* glucose isomerase used commercially for high-fructose syrup production is relatively stable in the immobilized form.

replace any of the enzymes used in these existing applications.

An analysis of the operational properties of some of the existing industrial enzymes (Table 7) suggests that in many cases a thermostable alternative would be very unlikely to replace the enzyme in current use. This conclusion is based on several practical considerations: (i) certain industrial biocatalytic processes are incompatible with high-temperature operation (e.g. cheese-making); (ii) some commercial trends favour low-temperature operation (e.g. domestic detergents); (iii) enzymes of sufficient thermostability are already in use (e.g. saccharification of starch); (iv) there is no obvious process advantage in increasing the reaction temperature (e.g. biosynthetic production of fine chemicals such as amino acids and penicillins); and (v) many of the large industrial biocatalyst users are committed to existing low-temperature operation through investment in specific plant and equipment. The advantages of higher temperature operation would be insufficient to offset the costs of replacing that equipment.

In summary, the conditions under which a hyperthermophilic enzyme would be a serious contender as a replacement for any 'bulk' commercial enzyme might include many (if not all) of the following: (i) available in similar (or equivalent) quantity with equally reliable supply; (ii) similar or lower price (per unit of activity); (iii) significantly better performance in many respects (not just stability); or (iv) no major alterations to existing plant and equipment (or investment in new plant) required.

Most of the 'bulk' enzymes listed in Table 7 are used as crude liquid culture supernatants of very high activity and are marketed in the cost range of £5–40l^{-1}. The biomass production of hyperthermophilic organisms is typically very low compared with many mesophilic bacteria (e.g. 0.1–1 gl^{-1} wet weight for hyperthermophiles, *cf.* 5–50 gl^{-1} for *E. coli*). Enzyme yields from fermentations parallel these values. It is therefore implicit in any successful application of hyperthermophilic enzymes that this will have been preceded by cloning of the enzyme gene and over-expression in a mesophilic host.

We must conclude that the future for hyperthermophilic enzymes will not be as replacements for existing 'bulk' enzymes. Nevertheless, the current success of thermophilic enzymes in recombinant DNA technology provides a valuable insight into one of the more likely paths of future development in biotechnology. The PCR is an excellent example of a 'niche' application, where the properties of the enzyme are complementary to the highly specific requirements of

Table 8. *Examples of enzymes in biotransformation*

Enzyme	Example Reaction	Industrial Products
Esterase		Antibiotic and Pharmaceutical intermediates
Nitrilase	$H_2C=CH-CN \Longrightarrow H_2C=CH-\underset{\underset{O}{\parallel}}{C}-NH_2$	Acrylamide
Protease	Z-L-Asp-COOH + L-Phe-OMe $\Longrightarrow$ Z-L-Asp-L-Phe-OMe	Aspartame
Lipase	Palmitate / Oleate / Palmitate + Stearic acid $\Longrightarrow$ Stearate / Oleate / Stearate	Cocoa butter substitute
Alcohol dehydrogenase		Chiral alcohols
Glucosidase		Sugar derivatives, oligosaccharide
Aldolase		Pharmaceutical intermediates
Naphthalene dioxygenase		Indigo
Steroid hydroxylase		Synthetic steroids (e.g. hydrocortisone)

a process. Based on this example, we can predict that successful applications are more likely to arise where the specific and unique characteristics of the enzymes (i.e. molecular stability and high-temperature activity) can be used to advantage.

There is growing interest in the use of enzymes in the pharmaceutical industries, where the production of chiral synthons is a vital step in the synthesis of chirally pure pharmaceutical agents (e.g. Davies *et al.* 1990; Jones 1993). In many instances the enantiospecificity of enzymes can provide a rapid and convenient route to the synthesis of chirally pure intermediates, where chemical synthesis alternatives are complex and difficult. Some examples of enzyme types which are important in industrial biotransformations are shown in Table 8.

On the 'downside', the reaction conditions necessary are not always compatible with normal enzyme stability and function. Many of the compounds involved in pharmaceutical product biosynthesis pathways are poorly soluble in aqueous solution and the presence of a significant concentration of hydrophilic miscible organic solvent can be detrimental to protein stability. This factor may be the incentive necessary to expand the role of hyperthermophilic enzymes in biotransformations. The intrinsic thermostability properties of these enzymes (see earlier) undoubtedly provides the stability characteristics necessary for their function in organic solvent media.

To date, relatively few hyperthermophilic 'biotransformation' enzymes have been identified and characterized (Table 9). This reflects the newness of the archaeal enzymology and the limited number of laboratories working in this field rather than any lack of application potential. It is further inhibited by the inability (or disinclination) of 'biotransformation' laboratories to work with organisms that require complex or difficult fermentation conditions and generate little biomass. In many cases, the successful insertion of hyperthermophilic enzymes into biotransformation processes is likely to proceed only following the commercial availability of the enzymes.

Conclusions

The hyperthermophiles are a diverse and fascinating group of organisms which can provide scientific insights into many fundamental aspects of biochemistry, genetics and even the 'origin of life'. They are valuable targets for study even in the absence of potential commercialization. Nevertheless, the biotechnological prospects for

Table 9. *Hyperthermophilic 'biotransformation' enzymes (data from Cowan 1992a; Adams 1993)*

Enzyme	Source
Protease	*Desulfurococcus* sp.
	Pyrococcus furiosus
	Thermotoga spp. and
	many others
Esterase	*Sulfolobus acidocaldarius*
	Archaeoglobus fulgidus
Alcohol dehydrogenase	*Sulfolobus solfataricus*
	Hyperthermus butylicus
Nitrilase	Not known
Aldolase	Not known
Glucosidase	*Pyrococcus furiosus*
	Sulfolobus solfataricus
Glucose isomerase	*Thermotoga littoralis*

hyperthermophilic enzymes are far from bleak. Although they may not be successful in replacing the high volume, low-cost enzymes which comprise most of 'industrial enzymology', there are a multitude of smaller, more specialized applications, particularly in the recombinant DNA and biotransformations fields, where the unique properties of hyperthermophilic enzymes may prove to be advantageous.

References

ADAMS, M. W. W. 1993. Enzymes and proteins from organisms that grow near and above 100°C. *Annual Review of Microbiology*, **47**, 627–658.

BARANY, F. 1991. Genetic-disease detection and DNA amplification using cloned thermostable ligase. *Proceedings of the National Academy of Science*, **88**, 189–193.

BARNS, S. M., FUNDYGA, R. E., JEFFRIES, M. W. & PACE, N. R. 1994. Remarkable archaeal diversity detected in Yellowstone National Park hot spring environment. *Proceedings of the National Academy of Science*, **91**, 1609–1613.

BROCK, T. D. 1969. *Thermus aquaticus* gen. n. and sp. n., a non-sporulating extreme thermophile. *Journal of Bacteriology*, **98**, 289–297.

CHA, R. S. & THILLY, W. G. 1993. Specificity, efficiency and fidelity of PCR. *PCR-Methods and Applications*, **3**, 18–29.

COWAN, D. A. 1992a. Enzymes from thermophilic archaebacteria: current and future applications in biotechnology. *Biochemical Society Symposium*, **58**, 149–169.

—— 1992b. Biochemistry and molecular biology of extremely thermophilic archaeobacteria. *In*: HERBERT, R. A. & SHARP, R. S. (eds) *Molecular Biology and Biotechnology of Extremophiles*. Blackie, USA, 1–43.

DANSON, M. J. 1988. Archaebacteria – the comparative enzymology of their central metabolic pathways. *Advances in Microbial Physiology*, **29**, 165–231.

DAVIES, H. G., GREEN, R. H., KELLY, D. R. & ROBERTS, S. M. 1990. *Biotransformations in Preparative Organic Chemistry*. Academic Press, London, 268 pp.

DIMARCO, A. A., THOMAS, A. B. & WOLFE, R. S. 1990. Unusual coenzymes of methanogenesis. *Annual Review of Biochemistry*, **59**, 335–394.

ERLICH, H. A., GELFAND, D. & SNITSKY, J. J. 1991. Recent advances in the polymerase chain-reaction. *Science*, **252**, 1647–1651.

JAENICKE, R. 1991. Protein stability and molecular adaptation to extreme conditions. *European Journal of Biochemistry*, **202**, 715–728.

JANNASCH, H. W. & MOTTL, M. J. 1985. Geomicrobiology of deep-sea hydrothermal vents. *Science*, **229**, 717.

JONES, J. B. 1993. Probing the specificity of synthetically useful enzymes. *Aldrichemica Acta*, **26**, 105–112.

JONES, W. J., NAGEL, D. P. & WHITMAN, W. B. 1987. Methanogenesis and the diversity of Archaebacteria. *Microbiological Reviews*, **51**, 135–177.

KANDLER, O. & ZILLIG, W. 1986. *Archaebacteria '85*. Gustav Fischer, Stuttgart, New York, 434 pp.

MATHUR, E. J. 1994. Purified thermostable *Pyrococcus furiosus* DNA ligase. *International Patent Application*, WO 94/02615, 83 pp.

OWUSU, R. K. & COWAN, D. A. 1989. Correlation between microbial protein thermostability and resistance to denaturation in aqueous:organic solvent two-phase systems. *Enzyme and Microbial Technology*, **11**, 568–574.

STETTER, K. O. 1986. Thermophilic Archaea. *In*: BROCK, T. D. (ed.) *Thermophiles; General, Molecular and Applied Microbiology*. Wiley, New York, 39–74.

——, HUBER, R. & 6 others 1993. Hyperthermophilic Archaea are thriving in deep North Sea sand Alaskan oil reservoirs. *Nature*, **365**, 743–745.

SZEWZYK, U., SZEWZYK, R. & STENSTROM, K. 1994. Thermophilic anaerobic bacteria isolated from a deep borehole in granite in Sweden. *Proceedings of the National Academy of Science*, **91**, 1810–1813.

WOESE, C. R. 1987. Bacterial evolution. *Microbiology Reviews*, **51**, 221–271.

—— & FOX, G. E. 1977. Phylogenetic structure of the prokaryotic domain: the primary kingdoms. *Proceedings of the National Academy of Science*, **74**, 5088–5090.

—— & WOLFE, R. S. 1985. Archaebacteria. *The Bacteria*, Vol. VIII. Academic Press, London, 581 pp.

——, KANDLER, O. & WHEELIS, M. L. 1990. Towards a natural system of organisms: proposal for the domains Archaea, Bacteria and Eucarya. *Proceedings of the National Academy of Science*, **87**, 4576–4579.

Hydrothermal fluxes of metals to the oceans: a comparison with anthropogenic discharge

CHRISTOPHER R. GERMAN & MARTIN V. ANGEL

Institute of Oceanographic Sciences Deacon Laboratory, Brook Road, Wormley, Surrey GU8 5UB, UK

Abstract: Published data for Cu, Zn, Pb, Mn, Co and Cd concentrations in hydrothermal vent fluids have been combined with best estimate vent fluid volume fluxes to generate annual gross fluxes of these dissolved metals to the oceans. These are compared with the gross annual fluxes for anthropogenic metal release to the environment, which must ultimately enter the oceans. Comparison of gross fluxes suggests that hydrothermal discharge is significant for Mn and, to a lesser extent, for Co, Zn, Cd and Cu; anthropogenic discharge is dominant for Pb. Dissolved metal fluxes from high temperature vents are modified by the precipitation of various metal sulphide and oxide phases in hydrothermal plumes. Similarly, river-borne fluxes of anthropogenic metals to the oceans are modified by the processes of dissolved metal flocculation and suspended particle coagulation and deposition in estuaries. Quantification of these processes using available published data suggests that for all metals except Mn, the net flux of metals to the oceans from hydrothermal discharge must be negligible when compared with the net anthropogenic flux.

With the first discovery of high-temperature hydrothermal vent fluids in 1979 came the realization that these fluids were chemically distinct from typical seawater. They were particularly enriched in metals such as Fe, Mn, Cu, Zn, Co, Cd and Pb and it was soon identified that hydrothermal vent fluids might represent a significant source in the global geochemical budgets of these elements (e.g. Edmond *et al.* 1979, 1982). More recently, with increasing popular concern about the industrial discharge of chemicals into the oceans, it has become of interest to consider how important metal fluxes from hydrothermal vent fluids might be with respect to anthropogenic fluxes to the oceans.

To estimate the gross flux of metals from hydrothermal vent fluids to the oceans, we first estimate a suitably representative average vent fluid composition which, combined with best estimates of high-temperature fluid volume fluxes to the oceans, yields a global gross flux of trace metals to the oceans. We then compare global estimates of total industrial metal production and total metal recovery through recycling to obtain a global estimate of metals being released into the environment in the period 1987–1992. Assuming that these metals must ultimately be transported to the oceans, presumably by riverine transport, these calculations provide a gross flux for the global anthropogenic metal input to the oceans.

It should be remembered, however, that any fluxes of dissolved metals to the oceans estimated either from rivers or from high-temperature hydrothermal fluids represent gross fluxes. The modification of dissolved metal contents of river waters during estuarine mixing with seawater are well documented (e.g. Sholkovitz, 1978). Similarly, processes active within hydrothermal plumes serve to modify strongly gross vent fluid fluxes (Mottl & McConachy 1988; German *et al.* 1991). An important consideration of the present study, therefore, is the extent of modifications of global metal fluxes to the oceans, from both hydrothermal and anthropogenic sources. Only then can a meaningful comparison of the net fluxes from these two important sources be made.

Estimation of global metal fluxes

Estimation of vent fluid compositions

Vent fluids have been identified on a wide range of tectonic settings around the global mid-ocean ridge system (German *et al.* this volume). For high-temperature vent fluids from sediment-starved mid-ocean ridges, a range of compositions has been reported. Extremes range from the metal-rich high-salinity fluids of the southern Juan de Fuca Ridge (Von Damm & Bischoff 1987) to the nearly metal-free pure vapour fluids reported from the Virgin Mound vent of Axial Seamount (also Juan de Fuca

From PARSON, L. M., WALKER, C. L. & DIXON, D. R. (eds), 1995, *Hydrothermal Vents and Processes*, Geological Society Special Publication No. 87, 365–372.

Table 1. *Vent fluid compositions*

	Mn (μmol/l)	Fe (μmol/l)	Co (nmol/l)	Cu (μmol/l)	Zn (μmol/l)	Cd (nmol/l)	Pb (nmol/l)
East Pacific Rise							
11°N	811	3583	369	36	123	25	391
13°N	2219	8370	1217	52	117	63	326
21°N	1013	1664	147	24	79	148	283
23°N	491	2180	105	14	50	100	609
Mean ± 1 SD(%)	1134 ± 67	3949 ± 77	460 ± 113	32 ± 52	92 ± 37	84 ± 63	402 ± 36

Ridge), where phase separation occurs (Massoth *et al.* 1989; Butterfield *et al.* 1990). These individual sites, however, appear to represent exceptional settings rather than the general rule. Indeed, one of the more surprising discoveries regarding the chemical composition of vent fluids over the 15 years since their discovery has been the relative stability of vent fluid compositions at any one site over 5–10 year time-scales and the general similarity of end-member vent fluid compositions from fast-spreading and slow-spreading mid-ocean ridges such as the East Pacific Rise and Mid-Atlantic Ridge (Von Damm *et al.* 1985; Bowers *et al.* 1988; Campbell *et al.* 1988*a,b*; Von Damm 1990).

For the present study we consider a mean vent fluid composition derived from data for Mn, Fe, Co, Cu, Zn, Cd and Pb concentrations for vent fluids from 11°N, 13°N and 21°N on the East Pacific Rise and from the Snakepit hydrothermal field, 23°N, Mid-Atlantic Ridge (Von Damm *et al.* 1985; Bowers *et al.* 1988; Campbell *et al.* 1988*a,b*; A. C. Campbell, unpublished data 1990). The data for each vent site and the calculated mean are presented in Table 1. Note that the variability associated with this calculation is large with respect to the calculated mean (typically ±100%, 1 SD). However, the magnitude of this variability does not prevent a useful comparison with anthropogenic fluxes from being made.

Estimation of vent fluid fluxes

Several estimates of the volume flux of high-temperature fluids associated with mid-ocean ridge hydrothermal activity have been made. Of these, arguably the most reliable is that by Palmer & Edmond (1989) based on the Sr isotope budget of global oceans. Their study estimated a global annual high-temperature fluid flux of $1.1 \pm 0.2 \times 10^{14}$ kg/a, which falls within the range of earlier estimates based on the Mg budget of the oceans which estimated

high-temperature volume fluxes of 0.7–1.5×10^{14} kg/a (Edmond *et al.* 1979; Berner & Berner 1987). Gross dissolved metal fluxes to the oceans from high temperature hydrothermal vent fluids are readily obtained by simply multiplying the mean vent fluid composition (Table 1) by the estimated global high-temperature fluid flux of Palmer & Edmond (1989). The global hydrothermal metal fluxes calculated in this way are presented in Table 2.

Estimation of anthropogenic metal fluxes

Available estimates of total global metal production for Cu, Zn, Pb, Cd, Co and Mn during the period 1988–1992 are listed in Table 3 (Data from World Bureau of Metal Statistics 1994). Also listed are reported values for recycled metal production for Cu, Zn and Pb during the same period. The difference between the two identifies the amount of new production of metals in each year for Cu, Zn and Pb. If the entire Earth system were in a steady state, this new production of metals each year would be balanced by an equal release of metals to the environment as waste. Of course, this may not necessarily be the case, because human activity may be leading to an accumulation of metals during economic growth, e.g. through stock-piling as scrap in waste dumps. Such processes might be expected to lead to a lag time between the 'new production' of waste metal and its release into the exogenic cycle as an anthropogenic flux to the oceans. However, the same processes would also increase the amount of time for which pollution would remain a problem after clean-up procedures were implemented. As a first approximation, therefore, we assume here that the calculated 'new production' of waste metals provides a reasonable integrated estimate of what must ultimately be transported by rivers to the oceans, either in solid or dissolved form. Calculations are presented here which assess whether or not these

Table 2. *Hydrothermal metal fluxes*

Element	Fluid concentration	Flux (mol/a)	Flux (kg/a)
Mn	1134 μmol/l	1.25×10^{11}	6.85×10^9
Cu	32 μmol/l	3.52×10^9	2.24×10^8
Zn	92 μmol/l	1.01×10^{10}	6.62×10^8
Pb	402 nmol/l	4.42×10^7	9.16×10^6
Co	460 nmol/l	5.06×10^7	2.98×10^6
Cd	84 nmol/l	9.24×10^6	1.04×10^6

Table 3. *Global metal production (× 1000 tonnes)*

Year	Element	Total	Scrap	New	Element	Total
1988	Cu	10 485	4307	6178	Cd	–
1989		10 878	4243	6635		21
1990		10 810	4227	6583		20
1991		10 685	4206	6479		21
1992		11 087	4267	6820		21
Mean		10 789	4250	6539		21
1988	Pb	–	–	–	Co	–
1989		5701	2722	2979		26
1990		5458	2734	2724		28
1991		5320	2651	2669		25
1992		5354	2638	2716		23
Mean		5458	2686	2772		26
1988	Zn	–	–	–	Mn	9495
1989		6787	1779	5008		10 287
1990		6724	1782	4942		10 015
1991		6884	1813	5071		8461
1992		6970	1896	5074		8365
Mean		6841	1818	5024		9325

assumptions may be reasonable and what their necessary implications must be.

For Cd, Co and Mn no data for recycling by scrap recovery are available. From consideration of the data for Cu, Zn and Pb, however, the percentage 'new' production of these metals during the period 1988–1992 appears reasonably uniform for each metal and ranges between 50 and 75% of total production. For Cd, Co and Mn, therefore, we assume the proportion of total metal production recovered as scrap to be similar and calculate the net release of these elements to the environment (and ultimately the oceans) to be 66% of their total global production. Taking the average of the annual data available for each element considered, an estimate of current anthropogenic metal fluxes to the oceans can then be obtained (Table 4).

Table 4. *Comparison of gross fluxes (kg/a)*

Element	Hydrothermal	Anthropogenic
Mn	6.85×10^9	6.22×10^9
Cu	2.24×10^8	6.54×10^9
Zn	6.62×10^8	5.02×10^9
Pb	9.16×10^6	2.77×10^9
Co	2.98×10^6	1.70×10^7
Cd	1.04×10^6	1.39×10^7

Discussion

Comparison of gross hydrothermal and anthropogenic fluxes

Estimated fluxes of metals to the oceans from hydrothermal vent fluids and from

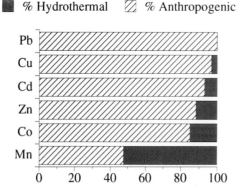

■ % Hydrothermal   ▨ % Anthropogenic

Fig. 1. Comparison of gross annual fluxes of metals to the oceans from hydrothermal and anthropogenic sources as a percentage of total (hydrothermal + anthropogenic) flux.

anthropogenic discharge are listed in Table 4. For Mn the total flux from hydrothermal vents to the oceans is comparable with that estimated for anthropogenic fluxes. Similarly, estimated hydrothermal fluxes of Cu, Zn, Co and Cd are also within roughly an order of magnitude of calculated anthropogenic fluxes. Thus, within error, the hydrothermal flux of all Mn, Co, Zn, Cd and Cu to the oceans is significant when compared with anthropogenic fluxes. By contrast, the estimated flux of Pb from anthropogenic discharge is approximately 300 times greater than the hydrothermal flux. This effect is displayed most clearly in Fig. 1, where global fluxes from hydrothermal and anthropogenic sources are plotted as a percentage of total (hydrothermal + anthropogenic) flux in order of decreasing anthropogenic input. For Pb, the calculated flux from anthropogenic sources is much greater than that calculated for hydrothermal activity, whereas only for Mn does the calculated hydrothermal flux represent greater than 50% of the total combined (hydrothermal + anthropogenic) flux.

Of course, these gross flux calculations assume a steady-state Earth in which annual waste metal production is matched by an equal riverine flux to the oceans. For an average global riverine water flux of 35×10^{15} kg/a (Palmer & Edmond 1989), these anthropogenic fluxes would require average riverine metal concentrations of 0.1–0.4 mg/kg, which are approximately 1000 times higher than previously measured uncontaminated riverine concentrations (Sholkovitz 1978). This suggests that our steady-state Earth approximation may be highly erroneous. The differences in calculated fluxes are so great, however (Table 4) that even if only 0.1% of each

year's fresh waste metal production is transported to the oceans and the other 99.9% remains land-locked and is slowly released over a much longer time-scale, gross anthropogenic fluxes of Cd, Co and Pb to the oceans would still remain significant compared with the annual hydrothermal flux. The further important implication of this calculation is that even if a global moratorium on waste metal production was effected immediately, the continuing release from one recent year's waste production would continue to significantly affect the Earth's ocean chemical cycles for about the next 1000 years.

Modification of vent fluid fluxes

High-temperature hydrothermal vent fluids erupting from the sea bed are extremely metal-rich and are compositionally very different from ambient seawater (e.g. Von Damm 1990). Upon erupting from the sea bed, however, these hot, acidic sulphide-bearing fluids mix rapidly with cold, well-oxygenated alkaline bottom waters, resulting in rapid cooling of the fluids and the precipitation of a range of trace element sulphide and oxide phases (e.g. Mottl & McConachy 1988). Therefore, any flux of dissolved trace metals to the oceans estimated from direct measurements of undiluted high-temperature vent fluids must necessarily exceed the dissolved input to the overlying water column. Instead, some large proportion of the metal flux from hydrothermal vents must be incorporated into freshly formed particulate phases and deposited to the sea bed close to the sites of active venting (Mottl & McConachy 1988; German et al. 1991).

Dissolved Fe is rapidly and quantitatively removed from solution into particulate phases during the rise time of fluids entrained within buoyant hydrothermal plumes (Rudnicki & Elderfield 1993). Therefore, if trace metal to Fe concentration ratios ([TM:Fe] ratios) in high-temperature fluids erupting from hydrothermal vents are compared with [TM:Fe] ratios in particulate samples collected from their overlying, buoyant hydrothermal plumes, the extent to which gross dissolved metal fluxes are modified by hydrothermal plume processes can be determined. For elements which are as rapidly and quantitatively removed from solution in buoyant hydrothermal plumes as Fe, [TM:Fe] ratios in plume particles should be identical to those for vent fluids. For elements which tend to persist in solution rather than precipitate with Fe, particulate [TM:Fe] ratios significantly less than vent fluid [TM:Fe] ratios would be expected, whereas for elements which were even more rapidly removed from solution

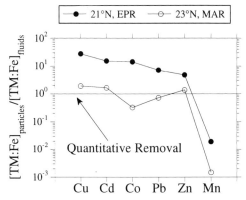

Fig. 2. Comparison of [TM : Fe] ratios for buoyant plume particles and end-member vent fluids from hydrothermally active sites at 21°N, East Pacific Rise (EPR)– and 23°N, Mid-Atlantic Ridge (MAR). Data from Von Damm *et al.* 1985; Mottl & McConachy 1988; Campbell *et al.* 1988*b*; Campbell 1991; A. C. Campbell, unpublished data 1990. [TM : Fe]$_{particle}$[TM : Fe]$_{fluids}$ = 1 indicates quantitative dissolved metal removal in hydrothermal plumes. See text for further discussion.

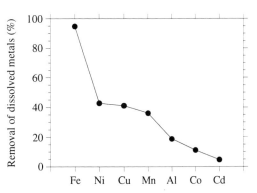

Fig. 3. Plot of percentage removal of dissolved metals from river water during estuarine mixing with seawater. Data from Sholkovitz (1978).

than dissolved Fe, plume particle [TM : Fe] ratios greater than vent fluid [TM : Fe] ratios would be expected.

In Fig. 2, particulate and vent fluid [TM : Fe] ratios are compared for Cu, Cd, Co, Zn, Pb and Mn using data from the East Pacific Rise, 21°N (fluid data from Von Damm *et al.* 1985; particulate data from Mottl & McConachy 1988) and from the Snakepit hydrothermal field, 23°N (fluid data from Campbell *et al.* 1988*b*; A. C. Campbell, unpublished data 1990; particulate data from Campbell 1991). It is immediately apparent that essentially quantitative removal of dissolved Cu, Cd, Co, Pb and Zn occurs within the buoyant hydrothermal plumes overlying hydrothermal vents on both the East Pacific Rise and Mid-Atlantic Ridge. Only dissolved Mn fluxes appear to be relatively unaffected, with evidence for ≤1% dissolved Mn removal into particulate phases. Conversely, the data allow for no more than about 1% of dissolved Cu, Co, Cd, Zn and Pb fluxes from high-temperature vent fluids to persist as dissolved metal fluxes to the overlying water column. Thus the net dissolved metal fluxes from hydrothermal vents for Cu, Co, Cd, Zn and Pb must be at least two orders of magnitude lower than the originally calculated gross fluxes listed in Table 4.

Modification of anthropogenic fluxes

Just as dissolved metal fluxes from hydrothermal vents are affected significantly by plume pro-

cesses, riverine fluxes to the oceans are modified strongly by processes which occur during mixing with seawater in estuaries. Two clearly identified processes relevant to the present study are (1) the flocculation/precipitation of dissolved metals in estuaries (Sholkovitz 1978) and (2) the estuarine deposition of suspended particle loads (Regnier & Wollast 1993). It is important to consider both of these processes because anthropogenically produced metals which escape scrap recovery and are released to the environment may be transported towards the oceans in either dissolved or suspended particulate form.

In the case of dissolved trace metals, Sholkovitz (1978) demonstrated that near quantitative removal of dissolved Fe occurs on mixing of river waters with seawater due to flocculation. The same process was also demonstrated to affect other transition metals, albeit to a lesser extent, ranging from about 35–40% removal for dissolved Cu and Mn to about 5–10% removal for Co and Cd (Fig. 3). Thus extrapolating to a global scale, dissolved riverine fluxes might be expected to be reduced by the order of about 50% during estuarine mixing. In a detailed study of the Scheldt estuary, Regnier & Wollast (1993) demonstrated that suspended particulate trace metal concentrations are also reduced significantly during estuarine mixing. In that study, suspended particulate Cu, Pb and Cd concentrations were all observed to decrease by as much as 90%, although dissolved Co data only exhibited about 50% removal (Fig. 4).

Although flocculated dissolved metals and particulate material deposited in an estuary during normal flow may subsequently be washed out in storms or dredged out by human activity, the same materials nevertheless most probably

Table 5. *Comparison of net fluxes (kg/a)*

Element	Hydrothermal	Anthropogenic (dissolved)	Anthropogenic (particulate)
Mn	6.78×10^9	3.11×10^9	6.22×10^8
Cu	2.24×10^6	3.27×10^9	6.54×10^8
Zn	6.62×10^6	2.51×10^9	5.02×10^8
Pb	9.16×10^4	1.39×10^9	2.77×10^8
Co	2.98×10^4	8.52×10^6	1.70×10^6
Cd	1.04×10^4	6.93×10^6	1.39×10^6

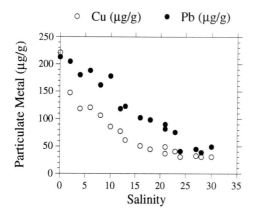

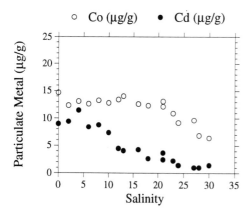

Fig. 4. Plot of decreasing suspended particulate metal concentrations with increasing salinity during estuarine mixing with seawater. Data from Regnier & Wollast (1993).

calculations are not unreasonable. Thus if the dominant mode of transport of metals in rivers is as suspended particles, the overall effect of estuarine mixing may be to reduce river-borne anthropogenic fluxes to the oceans by as much as an order of magnitude. If transport is predominantly in a dissolved form, by contrast, a reduction of only about 50% might be expected. Whichever may be the case, our calculations indicate that estuarine processes should certainly not produce as significant an effect in modifying riverine fluxes to the oceans as plume processes produce in modifying hydrothermal vent fluxes.

Comparison of net fluxes

From these discussions we calculate that the net flux of dissolved Mn from hydrothermal fluids to the oceans is best represented by about 99% of the calculated gross flux, whereas for Cu, Zn, Pb, Co and Cd the maximum possible net flux of dissolved metals to the oceans is most closely represented by about 1% of the gross flux calculated from high-temperature hydrothermal vent fluids. Thus for these elements the net flux (Table 5) is two orders of magnitude lower than the gross flux estimated previously (Table 4). For riverine fluxes, we estimate that the net fluxes of metals from anthropogenic discharge to the oceans are probably only of the order of about 50% of the gross fluxes calculated previously (assuming dissolved transport to be dominant) or about 10% of the gross flux values (assuming suspended particle transport dominates). Both sets of values are presented in Table 5.

The difference in the relative efficiencies between hydrothermal plumes and estuaries in modifying gross metal fluxes further accentuates the dominance of anthropogenic metal fluxes over hydrothermal metal fluxes when considering global ocean budgets. From Table 5 it is

become permanently incorporated into coastal sediments and are still not released to the open oceans. This is currently an area of active research but, to a first approximation, the above

■ % Hydrothermal ▨ % Anthropogenic
(dissolved)

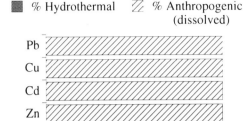

■ % Hydrothermal ▨ % Anthropogenic
(particulate)

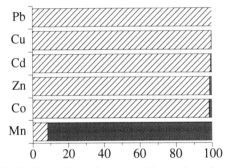

Fig. 5. Comparison of net annual fluxes of metals to the oceans from hydrothermal and anthropogenic sourdces as a percentage of the total (hydrothermal + anthropogenic) flux. Upper panel assumes dissolved riverine transport dominant; lower panel assumes suspended particulate riverine transport dominant.

clear that, of the metals considered, hydro-thermal activity only appears to remain a significant source for dissolved Mn in the modern ocean. At current levels of human activity, the net hydrothermal fluxes of Pb, Cu, Cd, Zn and Co become negligible when compared with anthropogenic discharge. These general conclusions hold whether anthropogenic metal discharge in rivers is considered to be transported predominantly in dissolved or particulate form (Fig. 5).

Conclusions

(a) The gross oceanic flux of dissolved Mn, Co, Zn, Cd and Cu from hydrothermal discharge is significant when compared with their global anthropogenic fluxes.

(b) By contrast, the global anthropogenic discharge of Pb swamps the gross hydrothermal flux for this element.

(c) For all except Mn, the dissolved metals considered tend to be quantitatively removed from solution due to the precipitation of various sulphide and oxide phases in hydrothermal plumes. This represents a significant modification of the gross flux of these dissolved metals to the oceans.

(d) For metals transported in rivers as dissolved species, similar modification of gross fluxes occurs on estuarine mixing with seawater, with metal flocculation processes resulting in up to about 50% removal of the metals considered here. For metals transported as suspended particulate phases, the predicted effect is even more pronounced and particle coagulation results in ⩽90% removal during estuarine mixing.

(e) For all except Mn, the net hydrothermal flux of metals to the oceans becomes negligible when compared with the net anthropogenic discharge.

We thank M. R. Palmer and two anonymous reviewers for their constructive criticism and acknowledge support from the NERC's Community Research Project, BRIDGE. IOSDL Contribution No. 95010.

References

BERNER, E. K. & BERNER, R. A. 1987. *The Global Water Cycle*. Prentice-Hall, Englewood Cliffs, NJ, 397 pp.

BOWERS, T. S., CAMPBELL, A. C., MEASURES, C. I., SPIVACK, A. J., KHADEM, M. & EDMOND, J. M. 1988. Chemical controls on the composition of vent-fluids at 13°–11°N and 21°N, East Pacific Rise. *Journal of Geophysical Research*, **93**, 4522–4536.

BUTTERFIELD, D. A., MASSOTH, G. J., McDUFF, R. E., LUPTON, J. E. & LILLEY, M. D. 1990. Geochemistry of hydrothermal fluids from Axial Seamount Hydrothermal Emissions Study vent field, Juan de Fuca Ridge: subseafloor boiling and subsequent fluid-rock interaction. *Journal of Geophysical Research*, **95**, 12 895–12 921.

CAMPBELL, A. C. 1991. Mineralogy and chemistry of matine particles by synchrotron X-ray spectroscopy, Mössbauer spectroscopy and plasma-mass spectrometry. *In*: HURD, D. C. & SPENCER, D. W. (eds) *Marine Particles: Analysis and Characterisation*. American Geophysical Union, Geophysical Monograph, **63**, 375–390.

CAMPBELL, A. C., BOWERS, T. S., MEASURES, C. I., FALKNER, K. K., KHADEM, M. & EDMOND, J. M. 1988a. A time series of vent fluid compositions from 21°N, East Pacific Rise (1979, 1981, 1985) and the Guaymas Basin, Gulf of California (1982, 1985). *Journal of Geophysical Research*, **93**, 4537–4549.

——, PALMER, M. R. & nine others 1988b. Chemistry of hot springs on the Mid-Atlantic Ridge. *Nature*, **335**, 514–519.

EDMOND, J. M., MEASURES, C., McDUFF, R. E., CHAN, L. H., COLLIER, R., GRANT, B., GORDON, L. J. & CORLISS, J. B. 1979. Ridge crest hydrothermal activity and the balances of the major and minor elements in the ocean: the Galapagos data. *Earth and Planetary Science Letters*, **46**, 1–18.

——, VON DAMM, K. L., McDUFF, R. E. & MEASURES, C. I. 1982. Chemistry of hot springs on the East Pacific Rise and their effluent dispersal. *Nature*, **297**, 187–191.

GERMAN, C. R., BAKER, E. T. & KLINKHAMMER, G. The regional setting of hydrothermal activity, this volume.

——, CAMPBELL, A. C. & EDMOND, J. M. 1991. Hydrothermal scavenging at the Mid-Atlantic Ridge: modification of trace element dissolved fluxes. *Earth and Planetary Science Letters*, **107**, 101–114.

MASSOTH, G. J., BUTTERFIELD, D. A., LUPTON, J. E., McDUFF, R. E., LILLEY, M. D. & JONASSON, I. R. 1989. Submarine venting of phase-separated hydrothermal fluids at Axial Volcano, Juan de Fuca Ridge. *Nature*, **340**, 702–705.

MOTTL, M. J. & McCONACHY, T. F. 1988. Chemical processes in buoyant hydrothermal plumes on the East Pacific Rise near 21°N. *Geochimica et Cosmochimica Acta*, **54**, 1911–1927.

PALMER, M. R. & EDMOND, J. M. 1989. The strontium isotope budget of the modern ocean. *Earth and Planetary Science Letters*, **92**, 11–26.

REGNIER, P. & WOLLAST, R. 1993. Distribution of trace metals in suspended matter of the Scheldt estuary. *Marine Chemistry*, **43**, 3–19.

RUDNICKI, M. D. & ELDERFIELD, H. 1993. A chemical model of the buoyant and neutrally buoyant plume above the TAG vent field, 26 degrees N, Mid-Atlantic Ridge. *Geochimica et Cosmochimica Acta*, **57**, 2939–2957.

SHOLKOVITZ, E. R. 1978. The flocculation of dissolved Fe, Mn, Al, Cu, Ni, Co, and Cd during estuarine mixing. *Earth and Planetary Science Letters*, **41**, 77–86.

VON DAMM, K. L. 1990. Seafloor hydrothermal activity: black smoker chemistry and chimneys. *Annual Review of Earth and Planetary Sciences*, **18**, 173–204.

—— & BISCHOFF, J. L. 1987. Chemistry of hydrothermal solutions from the Southern Juan de Fuca Ridge. *Journal of Geophysical Research*, **92**, 11 334–11 346.

——, EDMOND, J. M., GRANT, B., MEASURES, C. I., WALDEN, B. & WEISS, R. F. 1985. Chemistry of submarine hydrothermal solutions at 21°N, East Pacific Rise. *Geochimica et Cosmochimica Acta*, **49**, 2197–2220.

WORLD BUREAU OF METAL STATISTICS 1994. *World Metal Statistics, 1988–1992*.

Hydrothermal plumes: a review of flow and fluxes

KEVIN G. SPEER[1] & KARL R. HELFRICH[2]

[1]*Laboratoire de Physique des Oceans, IFREMER, Plouzané, France*

[2]*Department of Physical Oceanography, Woods Hole Oceanographic Institution, Woods Hole, MA 02543, USA*

Abstract: The physics of high-temperature hydrothermal venting and in particular the effects of the buoyancy flux on the background oceanic circulation are reviewed. The central theme is that the venting is dynamically active and forces oceanic flows on a variety of spatial and temporal scales. Vent fluid is not simply advected by the prevailing currents, but can actively set the scales of the flow. The cascade of scales (from fast to slow and small to large) discussed is: the buoyant plume, lateral spreading and mesoscale geostrophic vortices and the basin-scale plume. The focus is on models and observations relevant to the last two phases for which the Earth's rotation plays a significant part. Some discussion of the observational requirements for the resolution of these flows is given. The fact that small-scale buoyant convection can drive flows on scales many orders of magnitude larger can have important effects on the dispersal of vent fluid, estimates of hydrothermal heat flux and the establishment and maintenance of vent field biological communities.

The study of the circulation, both ambient and forced, in the vicinity of hydrothermal sources is important not only as a physical oceanographic problem, but also because it crosses into numerous other disciplines. For scientists interested in the heat flux along ridge crests, direct measurement of the heat flux from hydrothermal sources along a section of ridge crest is not yet feasible. Simply locating the hydrothermal sources is difficult at best, let alone the measurement of each one. Furthermore, episodic tectonic and volcanic activity rearranges the configuration of sources and has been shown to result in massive hydrothermal releases known as megaplumes (Baker *et al.* 1987; Chadwick *et al.* 1991). Although the observation of such events is an obvious goal, it is a task that requires both a fast response and luck. However, for both continuous and event-like sources, estimates of heat flux can be deduced from observations of the water column, which acts as an integrator. The key to this estimation is the availability of observations of sufficient resolution and duration for an adequate budget to be constructed or the availability of suitable models of the flow caused by hydrothermal discharge. The most familiar model of this type is the buoyant plume model of Morton *et al.* (1956), which has been used numerous times to relate observed plume characteristics such as rise height or centerline temperature to source heat flux (e.g. Little *et al.* 1987).

Knowledge of the circulation induced by venting is also important for those interested in the budgets and dispersal of chemical species associated with venting. For biologists the resolution of questions such as the establishment and maintenance of vent field communities must rely on a knowledge of the flow (Mullineaux *et al.* 1991). How do species populate new distant vent sites?

The goal of this paper is to review our understanding of hydrothermally forced circulation from the scale of the buoyant plume to that of an ocean basin. Models have an important part to play by making the connection between the ultimate forcing – the formation of ocean crust – and the oceanic manifestation – plumes – quantitatively clear. Hydrothermal venting occurs at a range of temperatures from low temperature (10–20°C above ambient) diffuse flux to high temperature (300–400°C above ambient) sources. Here we consider mainly the effects of the high-temperature venting, though it is important to point out that the integrated heat flux from the diffuse sources in a vent field may exceed the flux from the high-temperature sources (Rona & Trivett 1992; Trivett & Williams 1994).

The central idea to be discussed is that high-temperature hydrothermal venting can force circulation on a variety of spatial and temporal scales. The buoyant fluid that rises from vents is not simply advected by the ambient flow, but rather can be active in setting the scales of the local, regional and basin-scale flow. The cascade of scales initiated by a high temperature vent begins with the fast $O(1\,\text{hr})$ [where $O(.)$

From PARSON, L. M., WALKER, C. L. & DIXON, D. R. (eds), 1995, *Hydrothermal Vents and Processes*, Geological Society Special Publication No. 87. 373–385.

373

represents the order of magnitude of a quantity, e.g. time] rise of buoyant fluid from the vent to the spreading level O(100 m) above the source. This fluid then spreads laterally due to an unbalanced pressure gradient and under the influence of mean currents. After a time of O(10 hr) the rotation of the Earth acts to inhibit the lateral spreading, forcing the fluid to turn and forming a horizontal vortex of vent fluid circulating at the spreading level. An approximately geostrophic flow is established, meaning that the Coriolis force sets up a counterbalancing horizontal pressure gradient (e.g. Gill 1982). Frictional effects are normally assumed to be small away from boundaries. The horizontal scale of this flow is O(1000 m). The effect of several vent sites integrated along a ridge crest segment is to generate a density anomaly both at and below the spreading level. Information about this anomaly is eventually transmitted away from the ridge crest if conditions are suitable and a large-scale flow is established with a scale of O(1000 km).

In what follows we focus on the physics of the latter two stages, in which the Earth's rotation plays a dominant part. Throughout we try to connect the dynamic models with the available observations and also discuss the requirements of any future observational programmes.

Buoyant plume phase

Fluid emerging from an isolated hot vent rises as a turbulent plume, entraining and mixing with ambient seawater as it rises. As a result of the entrainment of cold (dense) ambient water the plume density increases with height. When the ambient environment is stratified, as is usually the case near vent sites, the density difference between the plume and the environment is eventually reduced to zero. The plume can no longer rise and begins to spread laterally. The standard model for buoyant plumes is due to Morton et al. (1956) and reviewed by Turner (1973). The model consists of conservation equations for momentum, mass and buoyancy which are derived under the assumption that the average shape of the properties (e.g. temperature) within the plume are similar at all levels. The other important assumption is that the entrainment velocity, or the rate at which ambient fluid is drawn into the plume, is proportional to the vertical velocity within the plume.

From the theory and empirical studies the height of the top of the spreading level above the source Z_N is given by

$$Z_N \approx 3.76(F_o N^{-3})^{1/4} \qquad (1)$$

where

$$F_o = Q(\frac{\rho_o - \rho_s}{\rho_o})g \qquad (2)$$

and

$$N^2 = -\frac{g}{\rho_o} \frac{\partial \rho_a}{\partial z} \qquad (3)$$

Here F_o is the source buoyancy flux, N is the background buoyancy frequency, Q is the source volume flux and ρ_s (ρ_o) is the density of the vent (background) water at the level of the source, $\rho_a(z)$ is the ambient density, z is the vertical coordinate and g is the acceleration of gravity. The coefficient of 3.76 in Equation (1) has been determined by comparison with laboratory experiments and atmospheric observations.

The plume model has been generalized to include the effects of both background temperature and salinity gradients by Speer & Rona (1989). They point out that the temperature and salinity anomalies in the spreading level are dependent on the ambient salinity and temperature gradients and can be counter-intuitive. In the Pacific, where salinity decreases with height, the spreading level consists of relatively warm salty water, whereas in the Atlantic, where salinity increases with height, the fluid in the spreading level is relatively cold and fresh. These simple plume models have been found to be consistent with the observations of both plume vertical penetration (1) and other plume characteristics (Lupton et al. 1985; Little et al. 1987; Speer & Rona 1989).

There are several important points to be made about this plume model as a precursor to the following discussion of larger scale flows. Firstly, observations of the spreading level Z_N are often used with Equation (1) to infer the heat flux from a high temperature vent. From Equation (1) the heat flux H is

$$H = \rho_s c_p Q(T_s - T_0) = \frac{\rho_0 c_p}{g\alpha}\left(\frac{Z_N}{3.76}\right)^4 N^3 \qquad (4)$$

where c_p is the specific heat of the vent fluid, α is the thermal expansion coefficient and T_s and T_0 are the temperatures of the vent and ambient fluids at the source level, respectively. From Equation (4) it is clear that estimates of H are very sensitive to errors in either Z_N or N because of the large power dependencies. Furthermore, the plume model was derived under ideal conditions and other effects such as ambient currents, local topography or, as we will see below, rotation of the Earth could act to influence the observed rise height and thus affect estimates of H obtained from Equation (4).

The effects of ambient currents have been explored in other fields with applications including industrial chimney plumes, submerged pollutant outfalls (e.g. Fischer et al. 1979; List 1982) and the dispersal of strong plumes generated during volcanic eruptions (Bursik et al. 1992). Ernst et al. (1994) have shown that weaker volcanic plumes in light cross-winds can split into two plumes, a process known as bifurcation (Fischer et al. 1979). The two plumes subsequently spread at different levels and speeds. An important effect of an ambient current is generally to reduce the total rise height of the plume. From dimensional arguments Turner (1973) shows that

$$Z_N \approx \left(\frac{F_0}{UN^2}\right)^{1/3} \qquad (5)$$

Here U is the background flow speed. The scale separating a weak effect from a strong effect is the plume velocity scale $(F_0N)^{1/4}$, of $O(10^{-1}\,\mathrm{m\,s^{-1}})$ for $F_0 = 10^{-1}\,\mathrm{m^4\,s^{-3}}$ and $N = 10^{-3}\,\mathrm{s^{-1}}$. Middleton & Thompson (1986) have considered this problem in the context of hydrothermal plumes.

Another important point regarding the plume model is that between the source and the bottom of the spreading level ambient fluid is continually entrained into the plume. This entrainment is not trivial. It leads to dilutions of $O(10^4)$ and volume fluxes of $O(100–500\,\mathrm{m^3\,s^{-1}})$ into the spreading level. In a laterally unbounded domain fluid entrained into the plume is replaced by a slow radial inflow from infinity. If the plume is isolated by vertical walls, as might occur in the North Atlantic where the axial valleys are typically deeper than the plume rise height, then eventually all the ambient water below the spreading level will be entrained into the plume. Conservation of mass then requires that this fluid is replaced by a downward vertical flow of fluid from the spreading level. This water is less dense than the original ambient fluid and once re-entrained into the plume it will lead to a higher plume rise height. Effectively the lateral boundaries result in a decreasing background stratification (decreasing N) and from Equation (1) an increasing Z_N.

This vertical circulation and increasing plume penetration is called the filling box effect and was first explored by Baines & Turner (1969) for initially homogeneous environments and extended by Manins (1979) to include diffusion, which should be important on the time-scales appropriate to venting. Cardoso & Woods (1993) added an initial stratification in the container. As we will show below, rotation acts to impose a lateral scale beyond which ambient fluid cannot be brought in and entrained. Thus either confinement or rotation can result in super-penetration of the plume. This vertical recycling of plume fluid can have serious consequences for heat flux estimates by affecting the anomalies often used to estimate fluxes, for the transport and evolution of dissolved and particulate chemical species (German & Sparks 1991) and for the dispersal of larvae.

Lateral spreading and geostrophic vortex formation

As fluid is continually delivered to the spreading level by a vent, a slow radial outflow must occur. We assume for the moment that there are no ambient currents. On a timescale f^{-1} (where f is the Coriolis frequency $\sim 10^{-4}\,\mathrm{s^{-1}}$ at 45°N) the spreading is affected by the rotation of the Earth. Parcels of fluid turn to the right (in the northern hemisphere) as they move out and an anticyclonic circulation (clockwise when viewed from above) develops. Further horizontal spreading is inhibited by rotation. Below the spreading level the entrainment into the rising limb of the plume forces the slow radial inflow of ambient fluid. Again, rotation results in fluid parcels turning to the right as they move towards the plume, thus developing cyclonic circulation (counter-clockwise when viewed from above). The net result is a baroclinic vortex pair: an anticyclonic vortex of plume fluid at the spreading level and a cyclonic companion of ambient fluid around the rising buoyant plume. Figure 1 shows a sketch of the resulting circulation.

The first model of such a flow is due to Speer (1989a). He assumed that the flow was two-dimensional (x–z instead of radial), linear and steady. The resulting equations for the rotationally dominated flow were forced by the mass flux (both entrainment and outflow) from the standard plume model. His results and dynamic scaling arguments showed that the lateral scale of a mature plume vortex l is the Rossby radius of deformation based on the plume rise height: $l = NZ_N/f$ (Gill 1982). For typical values of $Z_N = 200\,\mathrm{m}$ and $N/f = 10$, $l = 2\,\mathrm{km}$, a rather small scale. Scaling arguments show that the azimuthal velocity scale in the vortices is $u_f \sim (F_0 f)^{1/4}$. For typical values of $F_0 = 10^{-2}\,\mathrm{m^4\,s^{-3}}$ and $f = 10^{-4}\,\mathrm{s^{-1}}$, $u_f \sim 0.03\,\mathrm{m\,s^{-1}}$. This velocity scale is comparable with the observed background flows (Cannon et al. 1991) so we might expect that the plume vortex flow can exist in the presence of a background flow.

A solution from a radially symmetrical version of Speer's model is shown in Fig. 2 for some

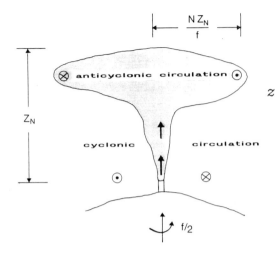

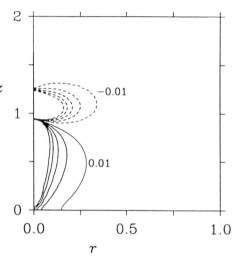

Fig. 1. Sketch of the baroclinic vortex flow forced by a buoyant plume in a rotating stratified environment. The flow is anticyclonic in the spreading layer, whereas entrainment into the rising plume forces cyclonic flow in the ambient fluid surrounding the rising plume. As a result of the rotational confinement a secondary vertical circulation is established to replace ambient fluid entrained into the rising plume.

Fig. 2. Solution for the linear steady radially symmetrical plume vortex forced by a buoyant plume with the conditions $F_o = 10^{-2}\,\text{m}^4\,\text{s}^{-3}$, $N = 10^{-3}\,\text{s}^{-1}$, $f = 10^{-4}\,\text{s}^{-1}$. The contours of azimuthal velocity (m s^{-1}) are shown as a function of height above the source z/Z_N and radius $r/(NZ_N/f)$.

typical parameter values. A consequence of the rotation is the spreading of isopycnals around the anticyclonic plume vortex. This doming results in an increase in the height of the spreading level Z_N when compared with the non-rotating situation, though the heat flux is the same [c.f., Equation (4)]. Another interesting feature of the solution is that the cyclonic flow around the rising plume extends to the bottom (Fig. 2). The length scale of the vertical shear of the horizontal velocities (the change from cyclonic to anticyclonic flow) is set by Z_N. There is a vertical recirculation that arises as a consequence of the rotation because rotation inhibits the radial inflow of ambient water from distances greater than about NZ_N/f. Thus fluid entrained into the rising plume must be replaced by a recirculation of fluid from above. In this model the assumption of steadiness required that diffusion balanced the downward advection. In a real flow diffusion and turbulent mixing may not be strong enough to achieve this balance and the plume will re-entrain light fluid brought down from the spreading level and will then rise to a higher level. Rotation then can act to place virtual walls around the plume at a scale of one Rossby radius.

Other effects can come into play that are not present in the simple model. Some of these effects lead to a different scaling and therefore a different plume structure. The fluid dynamic system responds to fluxes of properties, so changing the geometry of the source (e.g. point source to line source, Fernando & Ching 1993; Lavelle 1994) or the type of source (e.g. continuous to sudden release or thermal, Helfrich 1994) changes the scaling relationships between the source and the circulation. A sloping bottom can change the structure of the plume and make it asymmetrical (Fig. 3). The plume tends to intensify along the boundary in the direction of long topographic wave propagation (Gill 1982), or to the right-facing downslope in the northern hemisphere.

Also, the dynamics of rotating flow lead to instabilities in the plume circulation. Using laboratory experiments Helfrich & Battisti (1991) showed that a buoyant plume rising into a rotating stratified environment does not lead to the formation of a baroclinic vortex pair as described above. The upper anticyclonic lens of plume fluid had a thickness $2h$ such that Nh/f $l \sim 0.5$. They also found that as the plume grew to exceed about one Rossby radius in size the whole structure became unstable and broke up into two smaller vortex pairs. Figure 4 shows a photograph from their experiment. After the instability the daughter vortices then moved away from the source and the whole process began again. Thus a steady source resulted in the unsteady production of plume vortices. The time-scale for the break-up of a plume vortex is

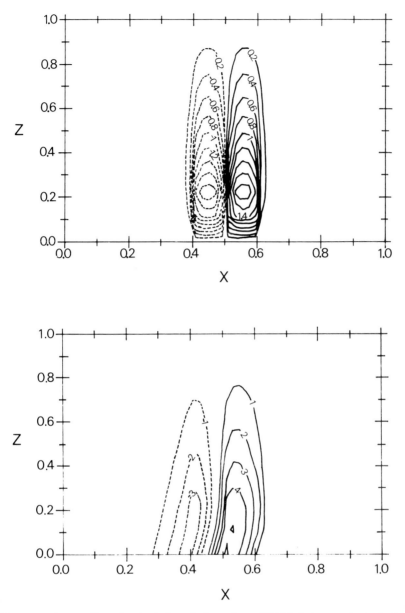

Fig. 3. Linear, steady, two-dimensional solutions similar to Speer (1989a), but over a sloping bottom, are shown. (**a**) Zero-slope reference case for $N = 7 \times 10^{-4} \, \text{s}^{-1}$, $f = 10^{-4} \, \text{s}^{-1}$, the height scale is 1 km and the width scale is 100 km. A broad, 10 km wide triangular-shaped heat source is applied at the bottom. (**b**) The bottom slope is 3.5×10^{-3} (view down slope). With a slope the plume's vertical circulation is stronger and asymmetrical.

$t_b \approx N f^{-2}$, independent of F_0. The experiments suggested that the coefficient in this relation is about 10, which for typical values of $N = 10^{-3} \, \text{s}^{-1}$ and $f = 10^{-4} \, \text{s}^{-1}$ gives $t_b \approx 2$ weeks. They also found that two vents with a separation greater than $N Z_N / f$ could interact to enhance the production of plume vortices.

These vortex pairs (or hetons; Hogg & Stommel 1985) propagate independently of any background flow and so need not simply be advected away from the vent site by background flow. The vortices also have closed streamlines and can retain anomalous properties over large distances and long times. They are similar to

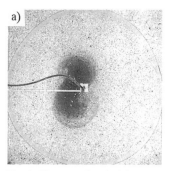 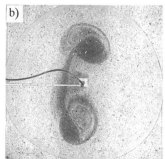 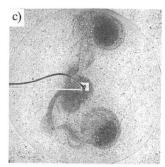

Fig. 4. Photographs of a laboratory experiment (Helfrich & Battisti 1991) showing the formation and break-up of a plume vortex in rotating stratified fluid. The view is from above. The source fluid is dyed and marks the plume fluid. Time increases from (**a**) to (**c**). A single continuous source produces one plume vortex which eventually becomes unstable and forms two smaller baroclinic vortex pairs which move away from the source.

Meddies, which are anticyclonic vortices formed from high-salinity water exiting the Strait of Gibraltar. Meddies are known to have lifetimes of one year or more (Armi *et al.* 1989). We can speculate that one mode of plume dispersal is by long-lived coherent vortices. This certainly will have important consequences for chemical and biological questions.

Speer & Marshall (pers. comm.) have begun using a non-hydrostatic numerical model to investigate the formation and stability of plume-forced vortices. The model has the advantage of removing the assumptions of linearity and steadiness of Speer's earlier study. They have been able to observe the formation of a plume vortex pair and an associated extra vertical rise due to rotational confinement. One important finding from this non-linear model is that the cyclonic circulation around the rising plume is confined to a lateral scale $(F_o f^{-3})^{1/4}$. This typically is less than the horizontal scale in the linear geostrophic models. Consequently, the vertical recirculation and azimuthal velocities around the base of the plume are larger than in the linear models. The break-up of the vortex observed in the laboratory experiments has not yet been observed in the model, in part due to the large computational resources required.

An outstanding question about the mesoscale plume circulation is whether it can exist in the presence of an ambient flow. Lavelle (1994) has undertaken a numerical study of this issue. His results show that the geostrophic flow can be established in the presence of moderate uniform background flows. Furthermore, the addition of a cross-flow enhances the vertical recirculation of plume fluid on the downstream side (Sykes *et al.* 1986; Coehlo & Hunt 1989) and gives rise to distinct offset temperature and salinity anomaly patterns. The calculations used a two-

dimensional (x–z) geometry and just a few cases of buoyancy flux and background flow strength were considered. Numerical studies of this sort need to be continued to understand fully the nature of mesoscale plume flows.

There is no conclusive observational evidence of plume vortices, primarily because of the need for measurements of sufficient duration and spatial resolution. To identify a vortex, measurements of velocity and anomaly patterns below, within and above the plume equilibrium level are a minimal level of vertical resolution. Horizontal resolution on the scale of a kilometre is also necessary. Distorting effects include tidal and topographic wave motions as well as internal waves and upper-ocean eddies. Filtering out these effects depends on information about the vertical structure of the variability. Time series of velocity over several months or more are desirable to obtain meaningful means in the presence of low-frequency motions.

Nonetheless, there are several observations which are suggestive of plume-forced mesoscale circulation. Non-uniformity in light attenuation, chemical and temperature anomaly signatures on scales suggestive of plume vortices have been observed (Baker *et al.* 1985; Baker & Massoth 1987; Rudnicki & Elderfield 1993). Cannon (pers. comm.) has observed mean vertical shear in the lower 300 m of the water column on the Juan de Fuca Ridge that has a vertical scale of the plume rise height. Franks (1992) has also observed average velocity veering with height above the bottom consistent with a change from cyclonic to anticyclonic flow near active venting on the Endeavor Ridge. Plume vortices may be difficult to find, but it does seem clear that the venting combined with rotation is helping to set the scale of the local flow.

Perhaps the best evidence to date for the

mesoscale dynamics described here are me-gaplumes, large lens-like features produced by rapid releases of large amounts of hydrothermal heat. The megaplumes rise significantly higher than those from regular venting, but the dy-namics which lead to the plume vortices de-scribed above will still come into play. D'Asaro et al. (1994) have analysed the hydrographic observations of two megaplumes over the Juan de Fuca ridge obtained by Baker et al. (1989). Their calculated geostrophic velocities show that the cores of the megaplumes are rotating anticyclonically with velocities of about $10 \, \mathrm{cm \, s}^{-1}$. The eddy shapes are also consistent with theoretical and scaling analyses; however, they were not able to identify a companion cyclonic circulation. This may be due to the difficulty in determining a level of no motion for the geostrophic velocity calculations. Lavelle & Baker (1994) have used a non-hydrostatic numerical model to study the evolution of megaplumes and their results are consistent with the observations and dynamic scaling argu-ments.

Basin-scale plume

Reid (1981) has provided a global view of the mid-depth circulation, that is to say between 2000 and 3000 m, roughly in the range of mid-ocean ridge crests (Fig. 5). It is based on horizontal density differences between two levels in the vertical and therefore represents shear, not necessarily the absolute flow. His circulation for the South Pacific shows westward flow near 15°S (assuming that the flow near 3000 m depth is relatively weak), consistent with the westward-pointing helium anomaly emanat-ing from the East Pacific Rise (Lupton & Craig 1981). Consistency with a plume tracer does not imply evidence for a hydrothermally driven large-scale circulation, however, and other studies have attempted to reconcile tracer distributions and circulation with geothermal forcing at ridge crests (Stommel 1982; Joyce & Speer 1987; Speer 1989b; Hautala & Riser, 1989, 1993). It is worth noting that similar zonal flows also exist in the North Pacific and Atlantic Oceans (Fig. 5). Talley & Johnson (1994) em-phasize this common aspect and suggest that mechanisms other than geothermal forcing also operate to drive these zonal flows. We focus here on the effect of the strong ridge-crest heat fluxes; the broad, weak conductive flux has been estimated to have an influence on the near-bottom temperature field, but not to the extent of driving a component of circulation on its own (Joyce et al. 1986).

The basic dynamic element of the steady, large-scale convectively driven flow is the β-plume (Stommel 1982). From the point of view of large-scale [O(1000 km)] circulation, venting at the ridge integrated along a segment of ridge crest produces a density anomaly in the water column over the ridge. We may think of the integrated density anomaly, or circulation anomaly, set up by a series of plumes along a finite segment of the ridge. The preferential westward group propagation of long Rossby waves (which arises due to the large-scale variations in the local Coriolis acceleration – the β-effect) away from the ridge results in a westward extending tongue of circulation. Steady flow is achieved by a balance between advection and vertical diffusion, resulting in an arrested Rossby wave. The model predicts a anticyclonic circulation extending from the ridge at the spreading level and cyclonic circulation below. Some important features of Stommel's solution are that the vertical scale of the flow is Z_N at the ridge crest and increases slowly away from the ridge crest due to diffusion. The zonal extent of the large circulation increases as the equator is approached. The inclusion of non-linear effects results in the generation of a narrower westward tongue or non-linear β-plume with sharper isopycnal displacements (Speer 1988), but otherwise the qualitative features are unchanged.

Stommel's model is for a β-plume in a resting ocean and so did not address the competition between a general thermohaline circulation represented by, for instance, the Stommel & Arons (1960) flow and the β-plume. Though the details of the general mid-depth circulation are poorly understood, the Stommel–Arons model predicts poleward and eastward flow throughout much of deep and mid-depth ocean. This is in direct competition with the westward flow forced by geothermal sources. Joyce & Speer (1987) and Speer (1989b) considered this effect and found that under the right conditions the geothermal circulation could overcome the Stommel–Arons flow, though the westward penetration was generally reduced. These studies also demonstrated that although the thermal anomaly due to vertical motion may extend westward, a passive tracer such as helium could be swept to the east. The tendency for the separation of tracers and the thermal anomaly increased as the latitude increases.

Hautala & Riser (1989) extended these modelling studies to include upper ocean wind-driven flow and topographic effects along with the Stommel–Arons abyssal flow. Their results again showed that a westward β-plume can exist

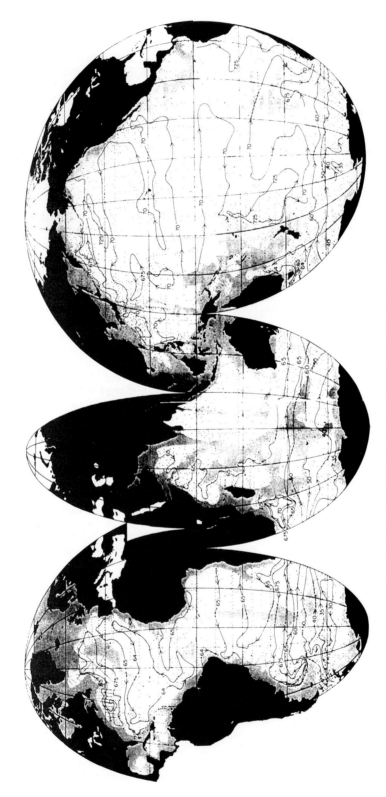

Fig. 5. Circulation at 2000 m depth relative to that at 3500 m depth (Reid 1981).

in the presence of all these additional influences. In a study of hydrographic data from the eastern Pacific Ocean near 15°S, Hautala & Riser (1993) gave a careful analysis of the various competing effects which could influence the calculated flow. They were able to determine that the data were consistent with a circulation actively driven by the ridge-crest hydrothermal flux and inconsistent with many other effects.

These studies lead to the conclusion that tracers and circulation are not only consistent with one another, but consistent also with a ridge-crest geothermal forcing. This constitutes real progress, even if a complete answer to the question 'does geothermal heating drive deep circulation in the South Pacific?' still eludes us.

Heat flux determinations

If there is one quantity which is the key to hydrothermal research in all disciplines, it is the distribution of heat flux over the seafloor. Convection in the interior of the Earth concentrates heat exchange between the solid Earth and the ocean at mid-ocean ridges and thereby determines the strength of the associated convection in the ocean. Estimates of the total hydrothermal component of heat loss at the ridge crest are typically 10^{13} W (e.g. Sclater et al. 1980), but the distribution over the crest is poorly constrained by models and direct measurements are essentially limited to the conductive component (through sediments) or a few individual vents (e.g. Converse et al. 1984). Indirect estimates of heat flux are thought to be feasible using observations of physical and chemical properties of the water column, a prospect which has motivated much fieldwork. The sampling of a highly variable, turbulent convection regime is difficult, however, and some aspects of this problem are discussed here.

Well-measured basin

One way to overcome sampling difficulties associated with calculating the net transport and divergence of mass, heat and other tracers is to exploit areas in which natural forces have created narrow channels into isolated basins. Ridge segments can have shallow (e.g. Pacific) or deep (e.g. Atlantic) axial valleys and are usually open at either end to the surrounding ocean. If active they may have mounds or volcanic peaks within the valley where the heat sources are strongest. Thus both diffuse and coherent plumes are likely to be swept up by mean currents set up by processes unrelated to geothermal activity. If circumstances create a situation in which the heat source is located within a depression or perhaps a cul de sac, such currents have less chance of overwhelming plume signals.

In conditions of no background flow, an isolated basin offers the opportunity of intensive current measurements at its sill to determine the strength of any circulation set up by heating within. Knowledge of the distribution of temperature and other properties at the sill and within the basin could allow a heat budget to be made using mass conservation (Fig. 6).

As long as the plume is confined to the basin and does not penetrate above the surrounding walls, the heat input is confined as well. In the simple case of two well-mixed layers, one of which supplies water to the plume while the other contains the spreading fluid, for inflow I at temperature T_{in} and outflow at T_{out}

$$\rho c_p I \Delta T = \mathcal{Q} \qquad (6)$$

Where $\mathcal{Q}$ is the heat source, ρ is mean density, c_p is the specific heat and ΔT is the temperature difference between layers. For a total heat source of 10^9 W and $\Delta T = 0.1°C$, the inflow $I = 2.5 \times 10^3 \, m^3 \, s^{-1}$. A sill 3 km wide with inflow roughly 100 m thick gives a mean inflow velocity of about $1 \, cm \, s^{-1}$. This is a rather weak mean flow to measure; a more concentrated flow requires a stronger source or narrower channel across the sill.

In a passage more than a few hundred meters wide, a minimal sampling strategy would have instruments on either side of the channel and three in the vertical plane to resolve the horizontal and vertical structure of the flow, for a total of six instruments. Such a minimal resolution is needed to estimate errors owing to unresolved flow. If a location rich in hydrothermal sources and with a reasonably confined inflow passage can be found, there is a chance that narrow limits can be put on the total heat input within the isolated basin (a large-scale example is the Panama Basin).

Unfortunately, the special conditions necessary for such a calculation make it impossible to generalize to large segments of the ridge crest. Thus the role of models will probably be crucial in interpreting measurements of the physical and chemical properties and in converting the information they contain into an estimate of heat flux.

Models and heat flux

Plume models have often been used to estimate heat input from various plume characteristics, such as rise height, vertical velocity or temperature statistics. Two examples, one for vents or

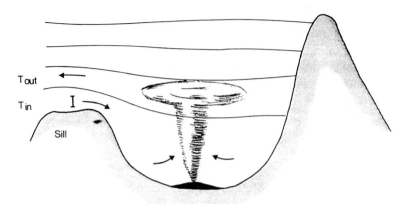

Fig. 6. A hydrothermal source in an isolated basin pumps water from a bottom layer fed by flow over a sill to a higher layer which is open to the surrounding water.

point sources and one for diffuse sources, illustrate the errors inherent in sampling a turbulent plume field. Little *et al.* (1987) carefully applied a plume model to measurements of temperature and velocity above a vent system on the East Pacific Rise near 11°N. A source strength of $3.7 \pm 0.8\,\mathrm{MW}$ was inferred from their data using a one-dimensional plume model and non-linear equation of state. The standard deviation of their maximum heat flow values, 1.2 MW, suggests an error of roughly 40%. Aside from direct heat flow measurements in individual vents, this is among the most accurate estimates to date of the heat flux from a vent system.

Diffuse sources ranging from pure conduction to broad fields of weak point sources, can be modelled in a similar fashion. Trivett & Williams (1994) and Rona & Trivett (1992) describe field measurements to which analytical and numerical models were applied to determine the strength of the diffuse sources of heat. Their results were given in terms of the total heat input by diffuse sources and imply errors of 60–70%.

In Trivett's calculations the initial buoyancy anomaly (b_0) of the warm water percolating through the seafloor into the water column is the fundamental parameter on which rise height depends ($Z \approx b_0/N^2$). This is reasonable as we can measure such anomalies near the bottom, whereas a flux is more difficult to estimate. However, the underlying hydrothermal system is driven by a flux of heat from the magma as it cools and condenses; the physics of heat conduction across the freezing lava crust may help to constrain this flux. In practice, the heat input per unit of seafloor area might be a useful, less variable quantity to estimate to compare sites.

In both point source and diffuse source cases, sources of error stem to a large extent from the difficulty of integrating turbulent fields in the deep ocean over sufficient space and time-scales to arrive at steady statistics. The physics itself imposes difficulties as well: knowing the rise height, for example, to within 20% still leads to roughly 100% errors in the total heat source strength because of the fourth power in the rise height formula. The direct solution to this problem involves deploying large arrays with high spatial resolution around source areas. Acoustic methods (Catapovic pers. comm.) look promising because of their natural integrating capabilities.

A totally different technique for determining heat source strength involves exploiting the natural time-scale of radioactive decay in certain tracers present in plume fluids. Rosenberg *et al.* (1988) presented radon (^{222}Rn) budgets in a Juan de Fuca Ridge plume and, assuming a steady state, inferred a supply of 8×10^{12} atoms per minute to balance radioactive decay (half-life of 3.85 days). They converted this mass transport to heat flux by dividing by the number of atoms per joule of anomalous specific heat content in the equilibrium layer plume, $c_p\Delta T$, where ΔT is the equilibrium temperature anomaly. However, this heat flux is not, as they assume, due directly to the vent heat source but rather to the equilibrium level heat flux (Speer & Rona 1989; see also McDougall 1990). Nevertheless, the principal interest of the method is that it provides an independent means of determining heat flux. Sampling strategies which integrate over entire ridge segments (Bougault *et al.* 1990) might be particularly helpful in the application of this method.

The difference between the vent heat source and the equilibrium level heat flux, which differs by the factor N_s^2/N^2 for zero vent salinity anomaly, deserves emphasis. This factor is the ratio of the buoyancy gradient owing to salinity alone to the total buoyancy gradient and may take on a wide range of values over the world ocean. Examples of 0.5 on the Juan de Fuca Ridge and −0.3 (negative values are associated with a destabilizing vertical salinity gradient) on the Mid-Atlantic Ridge are plausible. Much larger values, as well as near-zero values, are possible in other regions.

Methods of heat flux determination which attempt to relate a temperature anomaly to heating must be careful to distinguish between an anomaly directly due to heating and one due to the salinity effect. Obviously, hydrographic anomalies thought to be generated by hydrothermal sources must be interpreted in terms of the larger scale water mass distribution, which in this case means the ambient vertical stratification in temperature and salinity. Large-scale horizontal gradients of temperature and salinity can also introduce variability because of stirring by ocean eddies. These remarks can be generalized to other tracers of plume fluid with source and background variations, such as radon or silica. By taking rigorous sampling and modelling considerations into account for a suite of plume tracers, however, it may be possible to put tighter bounds on heat flux than by any single method alone. The problem of determining a net flux of tracer reduces to determining a time-scale for renewal, given the distribution. The radioactive tracer and isolated basin inflow, as well as plume models, are examples of means to obtain this time-scale. More direct means for measuring fluxes in the future could include techniques which use neutrally buoyant drifters to measure dispersion, or a deliberate tracer release experiment.

Discussion

It is clear that localized high-temperature hydrothermal venting at ridge crests is capable of forcing circulations on scales many orders of magnitude larger than the vent field size, or the plume rise height. What is lacking is a thorough observational confirmation of some aspects of the flows. This, however, is hindered by both the necessity for rather well-resolved measurements on the mesoscale and the problems associated with measuring relatively slow large-scale flows. Some aspects of these flows must be resolved if we are ultimately to achieve an understanding of

issues such as the dispersal and maintenance of local vent field biota.

Another major shortcoming in the field of ridge–ocean interaction is a good estimate of the heat flux over an entire ridge segment. Adding up individual and diffusive sources, plus conductive heat flow over a portion of the ridge, inside and outside the axial valley, is a difficult problem. To arrive at an answer may require estimating bounds from geophysical, chemical and physical constraints. Moreover, it is important to know over what time-scale an estimate is likely to hold; a strong flux for one year may have a lesser basin-wide influence than a weak flux for 100 years.

A companion issue to the strength of geothermal forcing is how to parameterize this forcing in models of the large-scale circulation. Given the heat input, a key quantity to estimate is the vertical velocity over the ridge segment. Presumably the localized sources such as vents and fissures contribute dominantly to the vertical transport; the diffusive sources, on the other hand, evolve more like turbulent mixed layers and the experience gained in sea surface mixed layer modelling can be applied to reproduce their characteristics. The point of transition from a distributed diffuse flux to an effective point source, and the interaction between diffuse, point-source venting and line-source venting needs to be studied.

Time-dependent processes have been dramatically illustrated in the discovery of megaplume events. However, geological events are not necessary to produce time-dependent flow. The dynamics of an expanding lens of plume fluid at the equilibrium level naturally leads to the unsteady production of hydrothermal eddies, which tend to move away from the source under their own propulsion. Their small size (5 km) makes it unlikely that they manifest a westward β-plume-like influence. Their detailed trajectories would more likely depend on interactions between individual eddies as well as the ambient flow. Though they may be described as contributing to the turbulent nature of the flow field around ridge crests, they have a very real mean flow effect because of their net vorticity, or circulation, at a given level. The circulation around a ridge segment is the sum of the circulation around all the vortices and this circulation can be carried away from their origin by the eddies, like a tracer. Precisely how the mesoscale circulation, and an ensemble of plume vortices in particular, interacts with and forces large-scale circulation remains an open question. We need also to know how this process should be parameterized in large-scale models.

Do plume vortices contribute to the basin-scale signature of hydrothermal venting like Meddies contribute to the Mediterranean salt tongue? High resolution float experiments may be one way to resolve this issue.

K.R.H. is partially supported by a NSF Ridge Program grant (OCE 92-16628). K.G.S. received support from the Centre National de la Recherche Scientifique. Interest and support from BRIDGE and RIDGE colleagues have helped to stimulate our research on hydrothermal plumes. We thank G. Ernst for contributing references to plume studies in related fields and to both he and D. Rudnicki for useful detailed comments.

References

ARMI, L., HEBERT, D., OAKLEY, N., PRICE, J. F., RICHARDSON, P. L., ROSSBY, H. T. & RUDDICK, B. 1989. Two years in the life of a Mediterranean salt lens. *Journal of Physical Oceanography*, **19**, 354–370.

BAINES, W. D. & TURNER, J. S. 1969. Turbulent buoyant convection from a source in a confined region. *Journal of Fluid Mechanics*, **37**, 51–80.

BAKER, E. T. & MASSOTH, G. L. 1987. Hydrothermal plume measurements: a regional perspective. *Science*, **234**, 980–982.

——, LAVELLE, J. W., FEELY, R. A., MASSOTH, G. L. & WALKER, S. L. 1989. Episodic venting of hydrothermal fluids from the Juan de Fuca Ridge. *Journal of Geophysical Research*, **94**, 9237–9250.

——, —— & MASSOTH, G. L. 1985. Hydrothermal particle plumes over the southern Juan de Fuca Ridge. *Nature*, **316**, 342–344.

——, MASSOTH, G. L. & FEELY, R. A. 1987. Cataclysmic hydrothermal venting on the Juan de Fuca Ridge. *Nature*, **329**, 149–151.

BOUGAULT, H., CHARLOU, J. L., FOUQUET, Y. & NEEDHAM, H. D. 1990. Activité hydrothermale et structure axiale des dorsales est-Pacifique et médio-Atlantique. *Oceanologica Acta, volume spécial 10, Actes du Colloque Tour du Monde Jean Charcot*, 199–207.

BURSIK, M. I., CAREY, S. N. & SPARKS, R. S. J. 1992. A gravity current model for the May 18, 1980 Mount St. Helens plume. *Journal of Geophysical Research*, **19**, 1663–1666.

CANNON, G. A., PASHINSKI, D. J. & LEMON, M. R. 1991. Mid-depth flow near hydrothermal venting sites on the southern Juan de Fuca Ridge. *Journal of Geophysical Research*, **96**, 12 815–12 831.

CARDOSO, S. S. S. & WOODS, A. W. 1993. Mixing by a turbulent plume in a confined stratified region. *Journal of Fluid Mechanics*, **250**, 277–305.

CHADWICK, W. W., EMBLY, R. W. & FOX, C. G. 1991. Evidence for volcanic eruption on the southern Juan de Fuca between 1981 and 1987. *Nature*, **350**, 416–418.

COEHLO, S. L. V. & HUNT, J. C. R. 1989. The dynamic of the near field of strong jets in crossflows. *Journal of Fluid Mechanics*, **250**, 95–120.

CONVERSE, D. R., HOLLAND, H. D. & EDMOND, J. M. 1984. Flow rates in the axial hot springs of the East Pacific Rise (21°N): implications for the heat budget and the formation of massive sulfide deposits. *Earth and Planetary Science Letters*, **69**, 159–175.

D'ASARO, E., WALKER, S. & BAKER, E. 1994. Structure of two hydrothermal megaplumes. *Journal of Geophysical Research*, **99**, 20361–20373.

ERNST, G. G. J., DAVIS, J. P. & SPARKS, R. S. J. 1994. Bifurcation of volcanic plumes in a crosswind. *Bulletin of Volcanology*, **56**, 159–169.

FERNANDO, H. J. S. & CHING, C. Y. 1993. Effects of background rotation on turbulent line plumes. *Journal of Physical Oceanography*, **23**, 2125–2129.

FISCHER, H. B. 1979. Turbulent jets and plumes. *In*: FISCHER, H. B., LIST, E. J., KOH, R. C. Y., IMBERGER, J. & BROOKS, N. H. (eds) *Mixing in Inland and Coastal Waters*. Academic Press, New York, 315–389.

FRANKS, S. E. R. 1992. *Temporal and spatial variability in the Endeavour Ridge neutrally buoyant hydrothermal plume: patterns, forcing mechanisms and biogeochemical implications*. PhD Thesis, Oregon State University.

GERMAN, C. R. & SPARKS, R. S. J. Particle recycling in the TAG hydrothermal plume. *Earth and Planetary Science Letters*, **116**, 129–134.

GILL, A. E. 1982. *Atmosphere–Ocean Dynamics*. Academic Press, New York, 662 pp.

HAUTALA, S. L. & RISER, S. C. 1989. A simple model of abyssal circulation, including effects of wind, buoyancy and topography. *Journal of Physical Oceanography*, **19**, 596–611.

—— & —— 1993. A nonconservative β-spiral determination of the deep circulation in the eastern South Pacific. *Journal of Physical Oceanography*, **23**, 1975–2000.

HELFRICH, K. R. 1994. Thermals with background rotation and stratification. *Journal of Fluid Mechanics*, **259**, 265–280.

—— & BATTISTI, T. 1991. Experiments on baroclinic vortex shedding from hydrothermal plumes. *Journal of Geophysical Research*, **96**, 12 511–12 518.

HOGG, N. G. & STOMMEL, H. M. 1985. The heton, an elementary interaction between discrete geostrophic vortices, and its implications concerning eddy heat-flow. *Proceedings of the Royal Society of London, Series A*, **397**, 1–20.

JOYCE, T. M. & SPEER, K. G. 1987. Modeling the large-scale influence of geothermal sources no abyssal flow. *Journal of Geophysical Research*, **92**, 2843–2950.

——, WARREN, B. A. & TALLEY, L. 1986. The geothermal heating of the abyssal subartic Pacific Ocean. *Deep-Sea Research*, **33**, 1003–1015.

LAVELLE, J. W. 1994. *A convection model for hydrothermal plumes in a cross flow*. NOAA Technical Memorandum, ERL PMEL-102, National Information Service, Springfield, VA.

—— & BAKER, E. T. 1994. A numerical study of local

convection in the benthic ocean induced by episodic hydrothermal discharges. *Journal of Geophysical Research*, **99**, 16065–16080.

LIST, E. J. 1982. Turbulent plumes and jets. *Annual Reviews in Fluid Mechanics*, **14**, 189–212.

LITTLE, S. A., STOLTZENBACH, K. D. & VON HERZEN, R. P. 1987. Measurements of plume flow from a hydrothermal vent field. *Journal of Geophysical Research*, **92**, 2587–2596.

LUPTON, J. E. & CRAIG, H. 1981. A major helium-3 source at 15°S on the East Pacific Rise. *Science*, **214**, 13–18.

——, DELANEY, J. R., JOHNSON, H. P. & TIVEY, M. K. 1985. Entrainment and vertical transport of deep-ocean water by buoyant hydrothermal plumes. *Nature*, **316**, 621–623.

MANINS, P. C. 1979. Turbulent buoyant convection from a source in a confined region. *Journal of Fluid Mechanics*, **91**, 765–781.

McDOUGALL, T. J. 1990. Bulk properties of 'hot smoker' plumes. *Earth and Planetary Science Letters*, **99**, 185–194.

MIDDLETON, J. M. & THOMPSON, R. E. 1986. *Modelling the Rise of Hydrothermal Plumes*. Canadian Technical Report on Hydrography and Ocean Science, 69. Department of Fisheries and Oceans, Sidney, BC, 18 pp.

MORTON, B. R., TAYLOR, G. J. & TURNER, J. S. 1956. Turbulent gravitational convection from maintained and instantaneous sources. *Proceedings of the Royal Society of London, Series A*, **234**, 1–13.

MULLINEAUX, L. S., WIEBE, P. H. & BAKER, E. T. 1991. Hydrothermal vent plumes: larval highways in the deep sea? *Oceanus*, **34**, 64–68.

REID, J. 1981. On the mid-depth circulation of the world ocean. *In*: WARREN, B. A. & WUNSCH, C. (eds) *Evolution of Physical Oceanography, Scientific Surveys in Honor of Henry Stommel*. The MIT Press, Cambridge, 623 pp.

RONA, P. & TRIVETT, A. 1992. Discrete and diffuse heat transfer at ASHES vent field, Axial Volcano, Juan de Fuca Ridge. *Earth and Planetary Science Letters*, **109**, 57–71.

ROSENBERG, N. D., LUPTON, J. E., KADKO, D., COLLIER, R., LILLEY, M. D. & PAK, H. 1988. Estimation of heat and chemical fluxes from a seafloor hydrothermal vent field using radon measurements. *Nature*, **334**, 604–607.

RUDNICKI, M. D. & ELDERFIELD, H. 1993. A chemical model of the buoyant and neutrally buoyant plume above the TAG vent field, 26 degrees N, Mid-Atlantic Ridge. *Geochimica et Cosmochimica Acta*, **57**, 2939–2957.

SCLATER, J. G., JAUPART, C. & GALSON, D. 1980. The heat flow through the oceanic and continental crust and the heat loss of the earth. *Reviews of Geophysics and Space Physics*, **18**, 269–311.

SPEER, K. G. 1988. *The influence of geothermal sources on deep temperature, salinity and flow fields*. PhD Thesis, MIT-WHOI Joint Program in Oceanography, Woods Hole Oceanographic Institution, Woods Hole, 146 pp.

—— 1989a. A forced baroclinic vortex around a hydrothermal plume. *Geophysical Research Letters*, **16**, 461–464.

—— 1989b. The Stommel and Arons model and geothermal heating in the South Pacific. *Earth and Planetary Science Letters*, **95**, 359–366.

—— & RONA, P. A. 1989. A model of an Atlantic and Pacific hydrothermal plume. *Journal of Geophysical Research*, **94**, 6213–6220.

STOMMEL, H. M. 1982. Is the South Pacific helium-3 plume dynamically active. *Earth and Planetary Science Letters*, **61**, 63–67.

STOMMEL, H. M. & ARONS, A. B. 1960. On the abyssal circulation of the world ocean. Part II. An idealized model of the circulation pattern and amplitude in oceanic basins. *Deep-Sea Research*, **6**, 217–233.

SYKES, R. I., LEWELLEN, W. S. & PARKER, S. F. 1986. On the vorticity dynamics of a turbulent jet in crossflow. *Journal of Fluid Mechanics*, **168**, 393–413.

TALLEY, L. D. & JOHNSON, G. C. 1994. Deep, zonal and subequatorial currents. *Science*, **263**, 1125–1128.

TRIVETT, D. A. & WILLIAMS, A. J. Effluent from diffuse hydrothermal venting, part II: measurements of plumes from diffuse hydrothermal vents at the southern Juan de Fuca Ridge. *Journal of Geophysical Research*, **99**, 18417–18432.

TURNER, S. 1973. *Buoyancy Effects in Fluids*. Cambridge University Press, Cambridge.

Particle formation, fallout and cycling within the buoyant and non- buoyant plume above the TAG vent field

MARK D. RUDNICKI

College of Oceanic and Atmospheric Sciences, Oregon State University,
Oceanography Administration Building 104, Corvallis, OR 97331–5503, USA

Abstract: The relationship between nephelometer and transmissometer measurements in the buoyant and non-buoyant plume above the TAG vent field have been investigated by calculating a phase angle between these two instruments, denoted α_{N-Tr}. This method indicates three distinct particle populations, corresponding to buoyant plume water ($\alpha_{N-Tr} < -20°$), ambient seawater ($-20° < \alpha_{N-Tr} < -10°$) and non-buoyant plume water ($-6° < \alpha_{N-Tr} < -5°$). There is a dramatic change in the nature of hydrothermally produced particles in the buoyant plume, which may be due to the formation and fallout of large sulphide particles. There is no apparent evolution of the non-buoyant plume particle population at distances of several kilometres from the vent site. Although particle processes in the buoyant plume are characterized by formation and fallout, particles settling out at the top of the buoyant plume may be recaptured and recycled. Plume-driven recirculation leads to an increase in the residence time of particles within the plume system. The residence time of particles recycled at the top of the buoyant plume has been estimated from thorium isotope distributions as *c.* 50–60 days. This model assumes that there is diffusion of Th into the plume system and the thickness of the diffusive interface has been calculated from the balance between the scavenging flux and the diffusive flux as *c.* 7 m.

The black smoker complex that tops the TAG hydrothermal mound is the largest source of high-temperature vent fluid in terms of heat and fluid output yet discovered. Although considered anomalous in this respect, TAG provides an opportunity to examine the effects of hydrothermal venting from a large steady-state system, with the hope that such processes will be more obvious to detect. Thus the TAG vent field has proved an ideal site to study particle formation kinetics in the buoyant plume (Rudnicki and Elderfield 1993) and the interaction between these particles and seawater chemistry in the non-buoyant plume (Nelsen *et al.* 1986–7; Trocine and Trefry 1988; German *et al.* 1990, 1991*a, b*; German and Sparks 1993; Rudnicki & Elderfield 1993; Mitra *et al.*; Rudnicki *et al.*). This body of work has shown the precipitation of iron sulphides and oxides in the buoyant plume, together with the coprecipitation and scavenging of seawater derived trace elements. This paper is a review of further processes which may affect the interpretation of hydrothermal particle chemistry in the TAG buoyant and neutrally buoyant plume.

Firstly, new data on combined nephelometer–transmissometer measurements from both the buoyant and non-buoyant plume are presented and used to suggest the location of particulate sulphide fallout from the TAG plume. Next, processes that lead to particle re-entrainment and recycling will be described and particulate Th isotope data (German *et al.* 1991*b*) will be reviewed in the context of these processes. A residence time for particles at the top of the TAG buoyant plume and the fraction of particles recycled will be calculated and an assessment made of the thickness of the diffusive boundary layer for Th at the margins of the non-buoyant plume. Finally, these results will be integrated into an updated model of the TAG buoyant plume.

Plume nephelometer–transmissometer relationships

Detection of hydrothermally produced particles in seawater has been of considerable importance in the discovery of black smoker venting (e.g. Nelsen *et al.* 1986–7). Two *in situ* methods for the measurement of the concentration of particles in seawater are optical backscatter (via a nephelometer) and optical transmission (via a transmissometer). If these two methods responded solely to particle concentration, then a plot of one measurement versus the other would result in a line of constant slope – a mixing line from low to high particle concentration. However, if changes in the slope of this line are seen,

From PARSON, L. M., WALKER, C. L. & DIXON, D. R. (eds), 1995, *Hydrothermal Vents and Processes*, Geological Society Special Publication No. 87, 387–396.

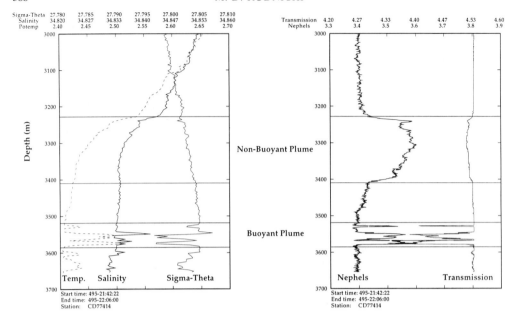

Fig. 1. CTD–transmissometer–nephelometer plots for downcast 1 of station CD77414, RRS *Charles Darwin* cruise 77, 1994. Buoyant and non-buoyant plume signatures are indicated. Transmission and nephels are reported as raw volts.

then this must be due to each instrument responding to different particle properties and thus the slope of the line could be used to discriminate differing particle populations. In this case, the word 'population' is used to describe groups of particles of different sizes and shapes so that, for example, if one sample was dominated by Fe sulphides and another by Fe oxides, then they would be considered to have differing particle populations.

In fact, the transmissometer and nephelometer do appear to respond to differing particle properties (data presented in the following). For example, it is thought that to a first approximation, nephelometry is sensitive to particle numbers and transmissometry sensitive to particle size, although there are obviously other factors which may contribute to one measurement rather than another, e.g. shape. As both devices are also sensitive to particle concentrations, then the ratio between the two measurements is used below to indicate changes in the particle populations in the TAG plume. However, as our current understanding of nephelometer–transmissometer relationships is limited, the discussion is restricted to situations where there are clear differences in the signals, without being reliant on the reason for any differences.

During RRS *Charles Darwin* cruise 77, a time series observation was conducted of the hydrothermal plume above the TAG vent field during which 24 profiles were recorded over a nine hour period (Rudnicki *et al.* 1994). On several of these profiles, both the buoyant and non-buoyant plume were recorded. The profile for downcast 1 is plotted in Fig. 1. Three water types can be discriminated. Plume water shows a positive particle anomaly (recorded by both the nephelometer and transmissometer) and can be subdivided into buoyant plume water, which shows a negative anomaly of potential density (σ_Θ), largely due to a measured temperature anomaly of up to 0.1°C, and non-buoyant plume water which does not. Ambient seawater is recognized by the lack of a particle anomaly. Additionally, buoyant warm water with no particle anomaly is recorded >3600 m, below the depth of the black smoker complex, indicating the presence of diffuse flow on the mound. An obvious difference can be seen in the response of the nephelometer and transmissometer in the buoyant plume and the non-buoyant plume. On the scale of the figure, comparable shifts in the nephelometer and transmissometer are recorded in the buoyant plume, whereas the shift in the transmissometer in the non-buoyant plume is much less than that recorded by the nephelometer.

Downcast 1 was chosen as it records buoyant

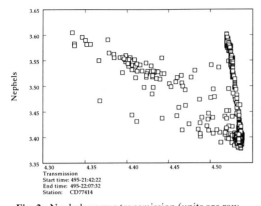

Nephels

3.65
3.60
3.55
3.50
3.45
3.40
3.35
4.30   4.35   4.40   4.45   4.50
Transmission
Start time: 495-21:42:22
End time: 495-22:07:32
Station: CD77414

Fig. 2. Nephels versus transmission (units are raw instrument volts). Data from TAG time series (Rudnicki *et al.* in press), downcast 1 (>3000 m). α_{N-Tr} is measured in degrees clockwise from a vertical line through (4.536, 3.364) – and thus produced negative values.

plume water at the depth of the TAG vent field, and thus provides an opportunity to study nephel–transmission (N–Tr) relationships in both the buoyant and neutrally buoyant plume. Figure 2 clearly shows two groups of differing N–Tr relationships representing mixing between seawater and buoyant plume water (lower slope) and between seawater and non-buoyant plume water (steeper slope). A $\Delta N/\Delta Tr$ ratio (anomalies referred to background) can be calculated, but instead of using this ratio directly, the data have been transformed by calculating the angle of a line between each data point and the point of maximum transmission–lowest nephelometer voltage, corresponding to the clearest recorded water local to the vent field, with the line of constant background transmission defined as 0°. As indicated above, changes in the slope of the N–Tr relationship (and hence in the calculated angle) indicate differing particle populations. Using this method, the data are restricted to the range $-90°$ to $0°$ and large ratios due to small ΔTr values are avoided, which is useful when gridding such data sets. This has been done for the entire TAG time series data set and the results are presented in Fig. 3a. In this figure, the three water types identified above are clearly distinguished by their N–Tr angles, denoted α_{N-Tr}: buoyant plume water ($\alpha_{N-Tr} < -20°$), ambient seawater ($-20° < \alpha_{N-Tr} < -10°$), and non-buoyant plume water ($-6° < \alpha_{N-Tr} < -5°$). Seawater with a higher α_{N-Tr} ($> -5°$) is located immediately above and below the non-buoyant plume particle maximum at the start of the

record, but is due to a slightly higher nephelometer background voltage for the first downcasts and upcasts compared with the remainder of the record. It is not possible from this data set alone to determine whether this is due to a slight change in the ambient particle population over the first *c.* two hours of measurement, or whether this is due to instrument cooling and baseline settling. However, sections measured at other vent sites with the same instrument show the same effect, indicating that for detailed plume work the Chelsea Instruments nephelometer should be held for several hours at ambient temperatures to allow the baseline to settle.

The α_{N-Tr} of particles in the non-buoyant plume is tightly constrained and distinct from buoyant plume water (Fig. 3b). There is no systematic difference resolved by this method between non-buoyant plume water immediately overlying the vent field at the end of the record and plume water further afield, recorded towards the middle of the record. This may be interpreted as an indication that the nature of particles is fairly constant in the near-field non-buoyant plume at TAG. By contrast, buoyant plume water has a distinctly lower (more negative) α_{N-Tr} which, given the simplistic interpretation of the differences between the nephelometer and transmissometer records presented above, may indicate a greater proportion of large particles in the buoyant plume. The transition between these two end-member situations must therefore occur before the level of neutral plume buoyancy.

One line of evidence for a distinct difference between particle populations comes from a comparison of total particulate + dissolved Fe/Mn ratios at the base and the top of the buoyant plume. The Fe/Mn ratio of TAG vent fluid is about 10, whereas that of the non-buoyant plume samples with the highest recorded particulate Fe and dissolved Mn, is about 5 (German *et al.* 1990, 1991*b*; Elderfield *et al.* 1993). This indicates fallout of perhaps 50% of iron bearing particles during the rise of the buoyant plume. The early fallout of freshly formed buoyant plume sulphides has been strikingly demonstrated for particles collected from the Monolith vent, North Cleft segment, Juan de Fuca Ridge (Feely *et al.* 1994). The workers found large enrichments in particulate Cu/Fe and Zn/Fe ratios within a few metres of venting, followed by a dramatic reduction thereafter due to fallout. Thus the observed differences between buoyant plume α_{N-Tr} and non-buoyant plume α_{N-Tr} may reflect a change in the proportions of Fe sulphides and Fe oxyhydroxides.

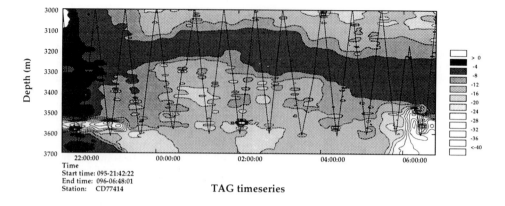

Time
Start time: 095-21:42:22
End time: 096-06:48:01
Station: CD77414

TAG timeseries

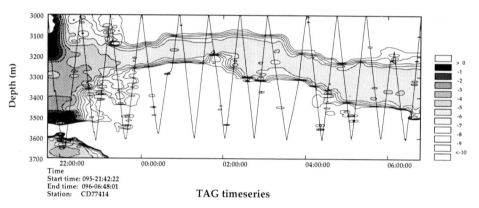

Time
Start time: 095-21:42:22
End time: 096-06:48:01
Station: CD77414

TAG timeseries

Fig. 3. TAG time series data sections of (**a**) α_{N-Tr} $-40°$ to $0°$ showing buoyant plume water (light shading), ambient seawater (medium shading) and non-buoyant plume water (dark shading). The start and end of the record represent very nearfield measurements, with distance increasing towards the centre of the record. (**b**) α_{N-Tr} $-10°$ to $0°$, showing the lack of features within the non-buoyant plume. Values of $<5°$ within the first two hours of the record are due to a settling of the nephelometer baseline.

Particle recycling

Re-entrainment and recycling of particles at the top of the buoyant plume can be accomplished by two mechanisms depending on the hydrological environment of the plume. These are (a) passive recycling, described by German *et al.* (1991*b*), Sparks *et al.* (1991) and German and Sparks (1993) and (b) active recycling, proposed by Speer (1989). Passive recycling is the process whereby particles falling out of the non-buoyant plume are re-entrained into the buoyant plume due to the horizontal inflow caused by turbulent entrainment. In contrast, active recycling is due to the vertical and horizontal inflow of water into the buoyant plume, again driven by entrainment.

Passive recycling

In quiescent environments, a particle settling out from the non-buoyant plume at a given radius from the source will be subjected to a horizontal velocity directed towards the buoyant plume due to plume entrainment. If the settling particle is drawn towards the plume sufficiently so that it reaches the buoyant plume again, it will be re-entrained. There is a critical distance at which, if the particle begins to fall out from the non-buoyant plume, it will fall just short of the buoyant plume and will not be re-entrained. This is known as the critical radius (r_c) (Sparks *et al.* 1991). German and Sparks (1993) investigated this scenario for the TAG plume and concluded that the proportion of particles re-entrained increases as plume strength increases, and that this is independent of particle size. As these workers point out, this model has several challenges in the environment of the TAG vent field – it requires a quiescent water column with no mean advective flow, but also it requires a method for aggregating the submicron sized Fe oxyhydroxide plume particles so that settling velocities are not excessively long (4.8 years for a 1 μm particle settling 350 m). If there is aggregation of fresh plume particulates leading to fallout, then the critical radius will be reduced to less that a kilometre, and we might reasonably expect to find a change in α_{N-Tr} in the non-buoyant plume, which is not seen in Fig. 3b.

Active recycling

Active recycling is caused by the imposition of rotational (e.g. coriolis) forces on the plume system. To conserve angular momentum, vertical motion of water is required in addition to horizontal inflow to provide the seawater taken into the buoyant plume due to entrainment. This inflow can include non-buoyant plume water. Recirculation of non-buoyant plume water back into the buoyant plume is one method for increasing the residence time of particles near the top of the buoyant plume.

Unlike active recycling, passive recycling results in a decoupling of particles from nascent plume water, which is a mechanism by which total (dissolved + particulate) element concentrations can be moved above the theoretical mixing line between vent fluid and ambient seawater. This situation has been found for long-lived isotopes of Th, a reactive trace element that is rapidly scavenged onto plume particles, from samples with high particulate Fe concentrations collected above the TAG vent field (German et al. 1991b). Alternatively, it has been proposed that since scavenging onto particles leaves plume water deficient in dissolved trace metals compared with ambient seawater, then diffusion may play a part in maintaining a source of dissolved trace metals to the non-buoyant plume for particle scavenging to continue (Kadko 1994; Kadko et al. 1994). If this is the case, then the measured dissolved concentration of particle-reactive trace elements in the non-buoyant plume should reflect the balance between scavenging and diffusion.

In summary, the chemistry of the particles at the top of the TAG buoyant plume requires a process which decouples the particles from nascent plume water. This may be accomplished either by passive recycling or by the active recirculation of plume waters. This last model is

investigated in the following sections. Thorium isotope data are used to calculate the particle residence time in an active recycling scenario and the balance between scavenging uptake and diffusion used to calculate the thickness of the boundary layer between the non-buoyant plume and ambient seawater.

Particle residence times in the buoyant plume

The concentration of ^{228}Th, ^{230}Th and ^{234}Th measured on particles collected from the TAG non-buoyant plume (German et al. 1991b) show several features which may be explained by particle recycling. Firstly, the total concentration (particulate + dissolved) of the long-lived isotopes (^{228}Th and ^{230}Th) are in excess over background seawater (by 20–40%) and, secondly, the particulate ^{234}Th/^{230}Th ratio (^{234}Th half-life 24 days) is less than the dissolved ratio of ambient seawater, suggesting that freshly produced hydrothermal particulates with seawater Th isotope ratios (zero age) are being supplemented by older particles with less ^{234}Th due to decay. A scavenging model has been applied to this data set (Rudnicki and Elderfield 1993) so that the evolution of particulate Th isotope concentrations on hydrothermal particles can be modelled. In that work, the difference between the measured plume top ^{228}Th/Fe ratio and the calculated primary precipitated ^{228}Th/Fe ratio was treated as being due to rapid scavenging. The idea described in the following is that the particulate Th isotope concentrations measured at the top of the TAG buoyant plume reflect freshly formed precipitates supplemented by 'older' recycled particles. Ageing of fresh particles allows two processes to occur: decay of the short-lived radionuclides such as ^{234}Th and the uptake of particle-reactive trace elements such as Th and the rare earth elements (REEs) by particle scavenging. By comparing Th isotope distributions, two parameters can be calculated. These are the fraction of re-entrained particles (f), the average 'age' of the particles (t) and thus the particle residence time at the top of the buoyant plume.

Equation (6) of Rudnicki and Elderfield (1993) describes how the $[^{230}$Th]/[Fe] ratio of freshly precipitated particles will change with time due to the scavenging of particle-reactive trace elements in the neutrally buoyant plume. The equation is rearranged below, assuming concentrations have been background corrected

$$\frac{[^{230}\text{Th}]}{[\text{Fe}]} = \left\{ \frac{[^{230}\text{Th}]_0}{[\text{Fe}]_0} + \frac{k_1}{[\text{Fe}]_0}(1 - e^{-\lambda_P t}) \right\} \quad (1)$$

Here, a subscript 0 represents (background corrected) concentrations from the top of the buoyant plume (as represented by sample TAG32T with $[Fe]_0 = 212\,nmol\,kg^{-1}$; German et al. 1991b), a subscript sw represents seawater concentrations, λ_p is the plume dilution rate constant $(0.017\,day^{-1})$ and t is time. k_1 is the model-fit neutrally buoyant plume scavenging rate and is related to the scavenging rate constant k by

$$k_1 = \frac{k[^{230}Th]_{sw}[Fe]_0}{\lambda_p} \qquad (2)$$

^{230}Th has a long half-life compared with plume processes (c. 75 000 years) and so its decay is ignored here. For ^{228}Th and ^{234}Th, the equation is modified to take radioactive decay into account [Equation (12) Rudnicki and Elderfield 1993]

$$\frac{[M]}{[Fe]} = \left\{ \frac{k_2}{[Fe]_0} + \left[\frac{[M]_0}{[Fe]_0} - \frac{k_2}{[Fe]_0} \right] e^{-\lambda_p t} \right\} e^{-\lambda^{Th} t} \qquad (3)$$

$$[M] = [^{228}Th] \text{ or } [^{234}Th]$$

The additional symbols are λ^{Th}, the decay constant for ^{228}Th $(9.99 \times 10^{-4}\,day^{-1})$ or ^{234}Th $(0.0289\,day^{-1})$, and k_2, given by

$$k_2 = \frac{k[M]_{sw}[Fe]_0}{\lambda_p + \lambda^{Th}} \qquad (4)$$

The scavenging rate constant, k, has been calculated as 1.9×10^{-9} $(nmol/kg)^{-1}s^{-1}$ for TAG non-buoyant plume particulates (Rudnicki and Elderfield 1993). The Th isotope to Fe ratios of particles at the top of the buoyant plume will reflect the precipitation of fresh particles, mixing with older particles which have scavenged Th from seawater. As time-scales in the buoyant plume are short (less than one hour) compared with the neutrally buoyant plume, scavenging in the buoyant plume will be neglected. Thus for the Th isotopes, the ratio $[Th]_0/[Fe]_0$ at the top of the buoyant plume will be given by

$$\frac{[Th]_0}{[Fe]_0} = (1-f)\left[\frac{[Th]}{[Fe]} \right]_{fresh} + f\left[\frac{[Th]}{[Fe]} \right]_{recycled} \qquad (5)$$

where the recycled ratio is given by Equation (1) for ^{230}Th and by Equation (3) for ^{228}Th and ^{234}Th. The freshly precipitated $[M]/[Fe]$ ratios can be calculated assuming quantitative removal from a vent fluid–seawater solution representative of an entrainment ratio of 570, but with only half the end-member Fe. The values used below are from Rudnicki and Elderfield (1993), and

assume that Th has a seawater source only, i.e. there is no end-member vent fluid source. Work by Mitra et al. (1994) on the concentration of REEs on particles collected from the TAG buoyant plume has shown that the approximation that only half the vent fluid Fe is available for oxide formation is good for the middle REEs (Nd, Gd: calculated Fe involved in oxide formation c. 46%), but also that the coprecipitation of the REEs with Fe is increasingly non-quantitative at atomic numbers away from theses elements. Such fractionation has been shown for REE scavenging onto Fe oxides in the neutrally buoyant plume (Rudnicki and Elderfield 1993).

Calculation of f, particle age, and particle residence time

The values of f and particle age can be obtained by comparing plots of f versus particle age for ^{234}Th with the longer lived isotopes. The parameters used to generate the curves in Fig. 4 are given in Table 1. In Fig. 4a, the general evolution of f versus time is shown. All lines converge at time 0 as a direct consequence of Equations (1) and (5) as the particles trapped at the top of the buoyant plume can only be made up of 100% recycled particles of zero age, i.e. there is no fresh input and no time is spent in the recycling process, thus no decay and no scavenging. Older particles tend towards a constant Th/Fe ratio, reflecting a balance between scavenging and decay, and so f tends to a constant. Differences in the behavior of ^{234}Th are highlighted in Fig. 4b for ages <25 days. The ^{234}Th curve produces a consistent solution with ^{228}Th at c. nine days and clearly diverges from the longer lived isotope curves at c. 12 days. Thus, on average, particles re-entrained into the TAG plume have aged by 9–12 days. These values indicate that at the top of the buoyant plume the particle population consists of 79–82% recycled particles, so that the residence time within this system, τ_{plume} $[age/(1-f)]$ is c. 50–60 days. During this time, the dissolved Th concentration in the non-buoyant plume must be maintained for scavenging to proceed. In the following section, the thickness of the diffusive boundary layer between the non-buoyant plume and ambient seawater is calculated as an order of magnitude check that the dissolved Th concentration is consistent with diffusive input to the non-buoyant plume.

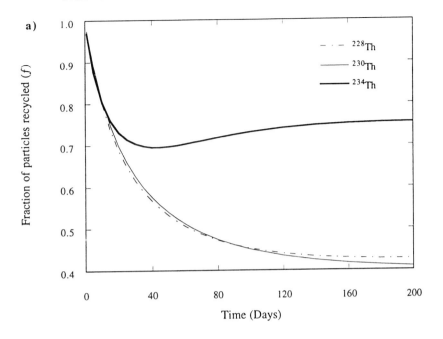

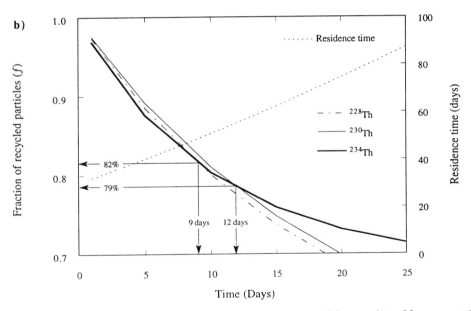

Fig. 4. Thorium isotope recycling model curves; (**a**) general behaviour; and (**b**) comparison of *f* versus age <25 days for ²²⁸Th, ²³⁰Th and ²³⁴Th. The residence time calculated from the curve for ²³⁴Th is also plotted.

Diffusion of Th into the TAG non-buoyant plume

Diffusion of particle-reactive radionuclides has been shown to be important for reducing the residence times of ²¹⁰Pb, ²¹⁰Po and ²³⁴Th with respect to scavenging onto non-buoyant plume particulates compared with deep ocean values, from studies conducted at the Endeavour and North Cleft segments of the Juan de Fuca Ridge

Table 1. *Thorium recycling model parameters*

	$\left[\dfrac{[\text{Th}]_0}{[\text{Fe}]_0}\right]^*$	$\left[\dfrac{[\text{Th}]}{[\text{Fe}]}\right]^*_{\text{fresh}}$	Model scavenging constant†
^{228}Th	1.06×10^{-2}	3.36×10^{-4}	$k_2 = 3.42$
^{230}Th	3.35×10^{-3}	$9.5 \ \times 10^{-5}$	$k_1 = 1.03$
^{234}Th	$6.6 \ \times 10^{-3}$	$4.5 \ \times 10^{-4}$	$k_2 = 1.82$

* Units are dpm/1000 kg for ^{228}Th, ^{230}Th; dpm/kg for ^{234}Th.
† Units are concentration.

(Kadko 1994; Kadko *et al.* 1994). The rate at which dissolved Th diffuses into the non-buoyant plume is given by the Reynolds' flux equation

$$[\text{Th}]_{\text{diff}} = K_Z \frac{d[\text{Th}]}{dz} \qquad (6)$$

where $d[\text{Th}]/dz$ is the diffusive gradient. Equation (6) is identical to that of Fick's law for molecular diffusion, except that here K_Z is the vertical eddy diffusivity. The units of $[\text{Th}]_{\text{diff}}$ are $\text{dpm m}^{-2}\text{s}^{-1}$. The ratio K_Z/dz can be calculated by assuming that the diffusive flux into the non-buoyant plume is balanced by the scavenging uptake of dissolved Th onto Fe oxyhydroxide particles

$$\frac{1}{2} \ hk[\text{Fe}]_0[\text{Th}]_{\text{plume}} = \frac{K_Z}{dz} \ ([\text{Th}]_{\text{sw}} - [\text{Th}]_{\text{plume}}) \quad (7)$$

The left-hand side of the equation represents the Th uptake rate of a column of non-buoyant plume water of thickness 0.5 h, where $h =$ the thickness of the plume ($\approx$100 m). The factor of 0.5 reflects diffusion occurring at both the top and bottom of the non-buoyant plume layer. k is the scavenging rate constant for Th presented earlier $[1.9 \times 10^{-9} \ (\text{nmol/kg})^{-1}\text{s}^{-1}]$, and the relevant Fe and Th concentrations are $[\text{Fe}]_0 = 212 \text{ nmol kg}^{-1}$, $[^{228}\text{Th}]_{\text{sw}} = 1.77 \text{ dpm m}^{-3}$, $[^{228}\text{Th}]_{\text{plume}} = 0.72 \text{ dpm m}^{-3}$, $[^{230}\text{Th}]_{\text{sw}} = 0.5 \text{ dpm m}^{-3}$, $[^{230}\text{Th}]_{\text{plume}} = 0.2 \text{ dpm m}^{-3}$ (values are from German *et al.* 1991*a*). The calculated ratio $K_Z/dz = 1.38 \times 10^{-5} \text{ m}^2\text{s}^{-2}$ (for ^{228}Th) and $1.33 \times 10^{-5} \text{ m}^2\text{s}^{-2}$ (for ^{230}Th) and hence there is good agreement between the isotopes. As K_Z in the oceans is generally taken to be c. $10^{-4}\text{ m}^2\text{s}^{-1}$ (Pond and Pickard 1983), this result suggests a diffusive boundary layer thickness of c. 7 m, in other words, the depletion of dissolved Th due to scavenging in the non-buoyant plume should extend outside the particle plume by about c. 7% of the plume thickness. This seems to be a reasonable value, but one which will have to be checked by detailed plume boundary sampling in future expeditions.

Buoyant plume processes above the TAG vent field

The two plume processes (recycling–diffusion) presented above can be integrated with the chemical model of Rudnicki and Elderfield (1993) to give the schematic representation of plume processes presented in Fig. 5. Evidence from combined transmissometer–nephelometer measurements suggests there is a dramatic change in the particle population between the time of early particle formation soon after venting and the top of the buoyant plume. It appears likely that this is due to the fallout of early formed Fe sulphides. Fe/Mn ratios and buoyant plume REE/Fe ratios indicate that perhaps c. 50% of the Fe is removed in this way, with the remaining Fe precipitating rapidly as Fe oxides and hydroxides. During the time of Fe oxide formation, reactive trace elements are also stripped from the entrained seawater.

In the original model it was necessary to invoke scavenging rates a thousand times greater for buoyant plume particles than for non-buoyant plume particles to obtain the particulate element to Fe ratios measured on samples with high Fe concentrations ($>$200 nmol kg^{-1}) from the top of the buoyant plume. In the present work, enhanced scavenging is replaced by the process of particle recycling. Recycling leads to an increase in the particle residence time at the top of the buoyant plume, with diffusion maintaining a supply of dissolved trace metals for scavenging to take total (particulate + dissolved) concentrations above the vent fluid–seawater mixing line. This process may also be in operation for the REEs, but as measured scavenging rate constants for the REEs are approximately two orders of magnitude lower than for Th (Rudnicki and Elderfield 1993), this hypothesis would be difficult to test.

Conclusions

The relationship between nephelometer and transmissometer measurements in the buoyant

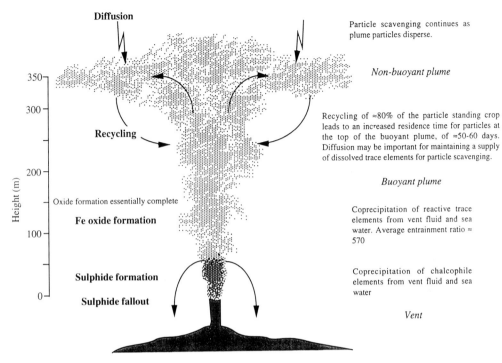

Fig. 5. An updated scheme for the chemical and physical dynamics of the TAG buoyant plume. The final chemistry of particles and fluid at the top of the buoyant plume is a function of the chemistry of the reacting vent fluid–seawater mixture, coprecipitation, particle fallout, particle recycling, diffusive input and scavenging of particle reactive trace elements.

and neutrally buoyant plume above the TAG vent field indicates three distinct particle populations. These correspond to (a) buoyant plume water, (b) non-buoyant plume water and (c) background seawater. There is thus a change in the nature of hydrothermally produced particles in the buoyant plume, which may be due to the formation and fallout of large sulphide particles. There is no apparent evolution of the non-buoyant plume particle population at distances of several kilometres from the vent site.

Several mechanisms can lead to re-entrainment of non-buoyant plume particles into the buoyant plume system. By considering the evolution of thorium isotopes scavenged onto particles in an active recycling system, the residence time of particles in this system has been estimated as *c.* 50–60 days. If the scavenging rate of dissolved Th onto non-buoyant plume particulates is balanced by the diffusive flux of Th into the plume, then dissolved Th anomalies should extend for *c.* 7 m at the top and bottom boundaries of the plume.

The work and ideas in this paper arise through close collaboration with the hydrothermal geochemistry

group, past and present, at Cambridge, notably H. Elderfield, C. R. German, M. J. Greaves, R. A. Mills and R. H. James. Many thanks to S. Sparks (Bristol) and D. Kadko (Miami) for helpful reviews. The support of NERC is recognized. This is Cambridge Earth Sciences Series Contribution 4143.

References

ELDERFIELD, H., MILLS, R. A. & RUDNICKI, M. D. 1993. Geochemical and thermal fluxes, high-temperature venting and diffuse flow from mid-ocean ridge hydrothermal systems: the TAG hydrothermal field, Mid-Atlantic Ridge 26°N. *In*: PRITCHARD, H. M., ALABASTER, T., HARRIS, N. B. W. & NEARY, C. R. (eds) *Magmatic Processes and Plate Tectonics*. Geological Society, London, Special Publication, **76**, 295–308.

FEELY, R. A., MASSOTH, G. J., TREFRY, J. H., BAKER, E. T., PAULSON, A. J. & LEBON, G. T. 1994. Composition and sedimentation of hydrothermal plume particles from North Cleft segment, Juan de Fuca Ridge. *Journal of Geophysical Research*, **99**, 4985–5006.

GERMAN, C. R. & SPARKS, R. S. J. 1993. Particle recycling in the TAG hydrothermal plume. *Earth and Planetary Science Letters*, **116**, 129–134.

——, CAMPBELL, A. C. & EDMOND, J. M. 1991a. Hydrothermal scavenging at the Mid-Atlantic

Ridge: modification of trace element dissolved fluxes. *Earth and Planetary Science Letters*, **107**, 101–114.

——, FLEER, A. P., BACON, M. P. & EDMOND, J. M. 1991*b*. Hydrothermal scavenging at the Mid-Atlantic Ridge: radionuclide distributions. *Earth and Planetary Science Letters*, **105**, 170–181.

——, KLINKHAMMER, G. P., EDMOND, J. M., MITRA, A. & ELDERFIELD, H. 1990. Hydrothermal scavenging of rare earth elements in the oceans. *Nature*, **345**, 516–518.

KADKO, D. 1994. An assessment of the effect of chemical scavenging within submarine hydrothermal plumes upon ocean chemistry. *Earth and Planetary Science Letters*, **120**, 361–374.

——, FEELY, R. A. & MASSOTH, G. J. 1994. Scavenging of ^{234}Th and phosphorus removal from the hydrothermal effluent plume over the North Cleft segment of the Juan de Fuca Ridge. *Journal of Geophysical Research*, **99**, 5017–5024.

MITRA, A., ELDERFIELD, H. & GREAVES, M. J. 1994. Rare earth elements in submarine hydrothermal fluids and plumes from the Mid-Atlantic Ridge. *Marine Chemistry*, **46**, 217–235.

NELSEN, T. A., KLINKHAMMER, G. P., TREFRY, J. H. & TROCINE, R. P. 1986–7. Real-time observation of dispersed hydrothermal plumes using nephelometry: examples from the Mid-Atlantic Ridge. *Earth and Planetary Science Letters*, **81**, 245–252.

POND, S. & PICKARD, G. L. 1983. *Introductory Dynamical Oceanography*. Pergamon Press, Oxford, 1–329 pp.

RUDNICKI, M. D. & ELDERFIELD, H. 1993. A chemical model of the buoyant and neutrally buoyant plume above the TAG vent field, 26°N, Mid-Atlantic Ridge. *Geochimica et Cosmochimica Acta*, **57**, 2939–2957.

——, JAMES, R. H. & ELDERFIELD, H. 1994. Near-field variability of the TAG non-buoyant plume, 26°N, Mid-Atlantic Ridge. *Earth and Planetary Science Letters*, **127**, 1–10.

SPARKS, R. S. J., CAREY, S. N. & SIGURDSSON, H. 1991. Sedimentation from gravity currents generated by turbulent plumes. *Sedimentology*, **38**, 839–856.

SPEER, K. G. 1989. A forced baroclinic vortex around a hydrothermal plume. *Geophysical Research Letters*, **16**, 461–464.

TROCINE, R. P. & TREFRY, J. H. 1988. Distribution and chemistry of suspended particles from an active hydrothermal vent site on the Mid-Atlantic Ridge at 26°N. *Earth and Planetary Science Letters*, **88**, 1–15.

Index

Acetabularia acetabulum, Aegean Sea, 307
Achromatium volutans, Aegean Sea, 308, 313
Acidianus infernus, 354, 355
acidophilic thermal waters, 353, 354
Actiniaria, Snake Pit, 322
Active Mound, TAG, metalliferous sediments, 223, 224, 225
active recycling, 390, 391
activity dating, 133–4
advection–diffusion equation, 162
advective heat flow within sulphide structures, 148
Aegean Sea, 305, 307
    biological communities, 303–15
aerobiosis, Archaeal, 355
African Rift Valley, 353
'air-lift' effect, 310
akaganeite, Broken Spur, 178, 184
alcohol dehydrogenase, 361, 362
Alcyonacea, Aegean Sea, 306
aldolase, 361, 362
Aleutian Arc, 3
algae, Aegean Sea, 307, 312, 338
alkalinity, MAR, 78
alkalophilic thermal waters, 353
allozyme variation, 347
alteration pipes, Pitharokhoma, 164
altered volcanic glass, Broken Spur, 52
aluminium
    Lau Basin, 236, 237
    Snake Pit, 320, 321, 325
(aluminium + iron + manganese)/aluminium ratios
        TAG, 127
aluminium oxide, 200, 203, 214, 215
Alvin mound, 123, 128, 129, 259, 260, 261
alvinellid polychaetes, 257, 285, 347
Alvinocaris, Broken Spur, 39
Alvinocaris lusca, Pacific Ocean, 259, 282, 283
Alvinocaris markensis, MAR, 259, 263, 265, 266, 283, 287, 322, 325
Alvinocaris muricola, MAR, 259
Alvinocaris stactophila, MAR, 259
amino acid production, enzymes for, 359, 360
ammonia, Aegean Sea, 307, 313
amorphous iron oxide, TAG, 125
amorphous silica
    Galapagos Rift, 210, 211, 213, 218
    Mid-Atlantic Ridge
        Broken Spur, 176, 178, 180, 181, 182, 183, 184, 186
        MARK, 49, 50
        Snake Pit, 323, 324
        TAG, 52, 53, 56, 57, 58, 60, 124, 125
    see also silica
ampharetid polychaetes, MAR, 261, 262, 263, 270, 271
Amphiglena mediterranea, Aegean Sea, 305, 306
amphinomid polychaetes, Aegean Sea, 307
amphipods
    Aegean Sea, 311

East Pacific Rise, lipid characteristics, 330, 331, 335–6, 338–40
    Mid-Atlantic Ridge, 270, 278
Anadyomene stellata, Aegean Sea, 307
anaerobiosis, Archaeal, 355
andesites, Western Pacific, 193, 195–6
anemones
    Aegean Sea, 307
    Mid-Atlantic Ridge, 261, 262, 263, 271, 286
        Broken Spur, 320, 322, 323, 324
        TAG, 124, 257, 263
anhydrite, 141, 185, 186
    Juan de Fuca Ridge, 66
    Manus Basin, 194
    Mid-Atlantic Ridge, 52, 61, 124, 125
        Broken Spur, 37, 176, 177, 178, 179, 180, 181, 182, 183, 185, 186, 187, 188
        Lucky Strike, 89, 94
anhydrite chimneys, MAR, 51, 78
annelida, Aegean Sea, 309, 311, 314
anthozoa, Aegean Sea, 312
anthropogenic discharge of metals, 365, 366–8, 369–71
antibiotic semi-synthesis, enzymes for, 360, 361
antimony
    Galapagos Rift, 214
    Manus Basin, 195
    Mid-Atlantic Ridge, 47, 48, 49, 52, 54, 56, 59, 61
    Woodlark Basin, 196, 199, 200, 202
Aphelochaeta sp., Aegean Sea, 306
Aplysia sp., Aegean Sea, 307
Aquifex, 354
arachidonic acid in vent organisms, 334, 338
aragonite, MAR, 56, 126, 176, 178, 181, 182, 183
Archaea, 351, 352, 353, 354–6, 357, 359, 362
archaeal enzymology, 359–62
Archaeglobus fulgidus, 355
Arctic Ocean fauna, 277
argon concentrations, EPR, 135, 136–7, 138–40
Aricidea catherinae, Aegean Sea, 306
Aricidea cerrutii, Aegean Sea, 308
Aricidea suecica, Aegean Sea, 306
Armandia polyophthalmus, Aegean Sea, 306
arsenic
    dissolved, Western Pacific, 9
    Galapagos Rift, 212, 213, 214, 215
    Manus Basin, 195
    Mid-Atlantic Ridge, 47, 48, 49, 52, 56, 59, 61
    Woodlark Basin, 196, 199, 200, 202
arthropod exhuviae, 349
ASHES vent field, 10, 11, 12, 68
asparagine residue deamination, 357
Astropecten, Aegean Sea, 306, 312
asymmetrical plumes, 376, 377
atacamite, MAR, 49, 52, 56, 58, 60, 126, 128
Atlantic vent fauna, seeding, 277
Atlantis Fracture Zone, 33
Austinograea alayseae, Pacific Ocean, 259
Austinograea williamsi, Pacific Ocean, 259

autotrophic bacteria, 314, 329, 355
autotrophy, 284
Axial Graben, 135, 138, 139, 140
Axial Volcano seamount, 10, 105, 199, 365
'azoic zones', 81
Azores Triple Junction, 10

Bacillus caldolyticus, 356
Bacillus subtilis, 356
Bacillus thermoproteolyticus, 356
back-arc basins, EPR, 9–10
bacteria, 297, 329
    Aegean Sea, 307, 308, 312, 313, 314
    chemautotrophic, 281, 284, 329, 330, 337, 340
    East Pacific Rise, 329, 330, 340
    ectosymbiotic, 308
    endosymbiotic, 273, 284, 314, 329, 337, 338, 340
    epibiotic, 280, 281, 283–4, 285
    hyperthermophilic, 272, 314, 352–60, 357
    iron/manganese, 81, 254, 255, 305, 313
    methanogenic, 272, 352, 355
    methylotrophic, 273, 284, 340
    Mid-Atlantic Ridge, 38, 42, 81, 182, 272, 273, 280,
        281, 283–4
    sulphide-oxidizing, 280, 284, 285
    sulphur-oxidizing, 308, 314, 330, 337, 353
    thermophilic, 115
    thiotrophic, 340
Bacteria (Eubacteria), 352, 353
bacterial mats
    Aegean Sea, 306, 307, 308, 311, 312, 313
    Mid-Atlantic Ridge, 268, 272, 281
        Broken Spur, 37, 182, 186
        Snake Pit, 320, 322, 323, 324, 325, 326
        TAG, 58
bacterial oxidation of hydrogen and methane, 116
bacterial synthesis of gases, 115, 119
bacterial transformation of metals, 254, 255
barite
    Galapagos Rift, 210, 211, 213, 215, 217
    Manus Basin, 193, 194
    Mid-Atlantic Ridge, 60, 89, 125, 176, 178, 321
    Woodlark, Basin, 198, 199, 200, 201
barite–silica chimneys, Woodlark Basin, 202
barium
    Galapagos Rift, 214, 215
    Lucky Strike, 91
    Manus Basin, 194, 195
    Woodlark Basin, 196, 199, 200
barnacles, MAR, 270
baroclinic vortex pair, 11, 375, 376, 377, 378
basalt–seawater reactions, 200
basalts, Lau Basin, 235, 275
basin-scale plume, 379, 381
bathymodiolid mussels, 257, 268, 272, 278, 284, 286
    lipid characteristics, EPR, 330, 331, 333–5, 336,
        337–8, 340
Bathymodiolus, Broken Spur, 39, 346
Bathymodiolus puteoserpentis, MAR, 262, 263, 265,
    266, 267, 286
Bathymodiolus sp., MAR, 262, 263, 346
Bathymodiolus thermophilus, 267, 285, 286
    lipid characteristics, EPR, 330, 331, 333–5, 336,
        337–8, 340

Bathyporeia sp., Aegean Sea, 311
Bauer Deep, sedimentation, 242, 246
Bay of Naples, biological communities, 313
Bay of Plenty, biological communities, 303, 315
'beehive' diffusers, 37, 175, 176, 177, 179–80, 185,
    186, 188, 270
Beehive mound, 263, 264
Beggiatoa sp., MAR, 281
Beggiatoa spp., Aegean Sea, 313
benthic diatoms, Aegean Sea, 307
benthos distribution, Snake Pit, 322, 323, 325
bifurcation, 375
biocatalytic processes, industrial, 360
biogenic carbon, 285
biogenic carbonate, Lau Basin, 232, 236
biogeographical barriers, 276, 277, 278–9
biogeographical variation of vent communities, 343
biogeography, MAR, 258, 276, 277–9, 288
biological activities and latitude, 326
biotransformations, enzymes for, 361, 362
biphytanyltetraether biosynthetic pathways, 359
'birds foot' texture, 176, 178, 180
bismuth, 47, 48, 56, 195, 196
bivalves, 272, 306, 312, 315, 322, 325
    endosymbiotic bacteria, 338
    larvae, 343
    sterol content, 340
black body radiation, 286
black smoker vent fields, diffuse flow modelling,
    159–72
black smokers, 11–12, 159, 161, 164, 172
    East Pacific Rise, 3–5, 8, 135
    faunal density and distance from, Snake Pit, 322,
        323–5
    heat supply, 68
    Mid-Atlantic Ridge, 58, 62
        Broken Spur, 35, 175, 177, 271
        MARK, 46, 82
        TAG, 51, 52, 78, 123, 124–5, 261, 387
    origin, 353
Bonellia cf. *viridis*, Aegean Sea, 306, 312, 314
boniellids, 314
bornite
    Galapagos Rift, 210, 213
    Manus Basin, 194
    Mid-Atlantic Ridge
        Broken Spur, 176, 178, 181, 182, 183, 185, 186
        MARK, 50
        Snake Pit, 264
        TAG, 52, 53, 56, 60, 125
bornite–chalcopyrite, MAR, 49, 50
boron, 73, 78, 80
boron isotope systematics, MAR, 79, 83–4
botryoidal silica, Woodlark Basin, 199
Boudia Bay, biological communities, 306
brachyuran crabs, 346
    Mid-Atlantic Ridge, 38, 39, 261, 262, 263, 266, 346
branched fatty acids in vent organisms, 332, 334, 336,
    337, 338
Branchipolynoe puteoserpentis, MAR, 267
Branchipolynoe seepensis, MAR, 267
bresiliid shrimps, 349
    Mid-Atlantic Ridge, 38, 39, 40, 257, 259, 261, 268,
        269

bresiliid shrimps—*cont'd*
nutrient sources, 281, 283, 284
sensory adaptations, 286–8
trophic studies, 280–8, 285–6
see also shrimps
brine, dilute phase, altered seawater mixing plane
model, 81, 84–5
brine filling of fractures, 74
Broken Spur field, 8, 9, 33–7, 39–40, 77, 97, 122, 123,
175
ecology, 37–40, 257, 259, 261, 270–1, 273, 277
larval identification, 344, 346, 347, 348
heat flux, 98, 156
plume characteristics, 98–108
sulphides, 175–84, 185, 187–8
vent fluid compositions, 103–8
bryozoa, Aegean Sea, 307, 312
bubble plumes, Steinahóll, 114, 118, 119
buoyancy anomaly, 382
buoyancy gradients, 383
buoyancy pressure, 160
buoyant plumes, 5, 11, 121, 159, 373, 374–84
models, 373, 374–84
particle processes, 4, 108, 387, 388–92, 394–5
Bythograea intermedia, Pacific Ocean, 259
Bythograea microps, Pacific Ocean, 259
Bythograea thermydron, Pacific Ocean, 259
bythograeid crabs, MAR, 257, 259, 262, 263, 268,
270, 281

cadmium
dissolved, gross fluxes, 365–71
Galapagos Rift, 213, 214, 215
Manus Basin, 195
Mid-Atlantic Ridge, 60
Broken Spur, 103, 105, 106, 107, 108, 184
MARK, 47, 48, 49
TAG, 48, 54, 56, 57
Woodlark Basin, 196
caesium, MAR, 78, 79, 80, 81
calcareous turbidites, Lau Basin, 234
calcium, 74
East Pacific Rise, 4, 215
Manus Basin, 195
Mid-Atlantic Ridge, 48, 52, 56, 61, 78, 81
Woodlark Basin, 196, 201
calcium carbonate, TAG, 57, 58
calcium oxide, 200, 203, 214
calcium/strontium ratio and plagioclase formation, 92
Callianassa subterranea, Aegean Sea, 311
Calvin–Benson carbon dioxide fixation, 284
Capitella capitata, Aegean Sea, 305, 306, 308, 310,
311, 312, 314, 315
carbon, organic, 310, 337
carbon assimilation by vent organisms, 285, 337, 338
carbon dioxide, Aegean Sea, 303, 305
carbon dioxide bubble plumes, Steinahóll, 114, 118,
119
carbon dioxide fixation, 201, 280, 281, 284, 354
carbon dioxide/helium ratios, Loihi Seamount, 133
carbon fixation, 337, 338, 341
carbon isotope equilibrium temperature, 117
carbonate ooze, TAG, 126, 261
carbonates, MAR, 59

cascade of scales, 373–4
catalytic rate of enzymes, 357
Central Atlantic, dispersal of fauna, 277
Central Lau Spreading Centre, CLSC, 244, 245–6
sedimentation related to propagation, 241, 243
sediments, 232, 233, 234
centralized magma feeding model, 27, 28
Cepola rubescens, Aegean Sea, 307
Ceratonereis hironicola, Aegean Sea, 305, 306
Cerianthus sp., MAR, 272
cerium anomaly, MAR, 223
chaetopterid polychaetes, MAR, 261, 262, 263, 286,
322, 325
chalcocite, MAR, 49, 56, 58, 60, 176
chalcopyrite
East Pacific Rise, 135
Galapagos Rift, 210, 211, 212, 213, 215, 216, 217
Manus Basin, 194
Mid-Atlantic Ridge, 58, 89, 124, 125, 321
Broken Spur, 176, 177, 178, 180, 181, 182, 183,
186, 187
MARK, 50
TAG, 52, 53, 54, 56, 60
Woodlark Basin, 199
chalcopyrite chimneys, Broken Spur, 181, 184, 186
chalcopyrite disease, Broken Spur, 182
chalcopyrite–anhydrite chimneys, EPR, 135
chalcopyrite–bornite aggregates, Broken Spur, 185
chalcopyrite–isocubanite, EPR, 135
chalcopyrite–pyrite, TAG, 52, 54
chalcopyrite–pyrite mineralization, Galapagos Rift,
217
chalcopyrite–sphalerite, TAG, 54
chalcopyrite–sphalerite–pyrite, MARK, 49, 61
Charlie Gibbs Transform Fault, 278, 279
chelipeds, MAR, 281, 283–4
chemoautotrophic bacteria, 281, 284, 329, 337, 340
chemoautotrophic production, 330
chemolithoautotrophic growth, 354
chemoreception, 287
chemosynthetic production, 273, 280, 281, 329
Cheshire Seamount, 195, 197
chimney formation, 125, 186, 188, 231
chiral synthon production, 362
Chiridotidae, Snake Pit, 322
chitin microfibrils, 297–8, 299, 300, 301
chitin production by tubeworms, 295, 298
chitin–protein complexes, tubeworms, 295, 296, 301
chloride–oxide aggregate, MARK, 49
chlorine and chloride concentrations, 74
East Pacific Rise, 71, 72, 73, 214
Mid-Atlantic Ridge, 78, 81, 82
chlorine/bromine ratios, EPR, 139
cholest-7-enol, in vent organisms, 336
cholest-22-enol, in vent organisms, 337
cholest-24-enol, in vent organisms, 337
cholesta-5,22-dienol in vent organisms, 336, 337, 340
cholesta-5,24-dienol in vent organisms, 336, 337, 340
5α-cholestanol in vent organisms, 336, 337
cholesterol in vent organisms, 336, 337, 340–1
Chorocaris chacei, MAR, 261, 265, 266, 280, 282,
283, 287
Chorocaris sp., MAR, 259, 262, 263, 264, 265, 269,
277, 282, 283

Chorocaris vandoverae
  East Pacific Rise, 259, 282, 287
  Mid-Atlantic Ridge, 261
Chronadorid spp., Aegean Sea, 308
chronic plumes, 65, 71–2
  Juan de Fuca Ridge, 67, 68, 69, 70, 71, 72, 73
ciliates, Aegean Sea, 308, 314
circulation forcing, 373, 374, 375, 378, 379, 381, 383
circulation within sulphide structures, 145–6, 154–7
  boundary conditions, 146–8
  predictions, 148–54
Cirrophorus lyra, Aegean Sea, 306
clams, 194, 257, 258, 278
clay, Lau Basin, 235
Cleft segment, JDFR, 6, 103, 133, 179
  heat flux from diffuse flow, 12, 159
  magmatic intrusions, 65, 66–7, 68, 71, 72, 73–4
  particle processes, 389, 393
'climax' communities, 276
cloning of genes, 359 360
cloning of proteins, 357
Cnidaria, Aegean Sea, 309
coal/ore pile, Archaeal bacteria, 354
CoAxial segment, JDFR, magmatic intrusions, 65, 66, 67, 70, 71, 74
cobalt
  dissolved, gross fluxes, 365–71
  Galapagos Rift, 212, 213, 214, 215
  Lau Basin, 236, 237
  Manus Basin, 195
  Mid-Atlantic Ridge
    Broken Spur, 103, 105, 106, 107, 108
    MARK, 47, 48
    TAG, 48, 52, 54, 56, 59, 60, 62
  Woodlark Basin, 200
coccolithophores, 347, 349
coelenterates, MAR, 268
colonization of vents, 12, 276–7, 295
  prevention, Aegean Sea, 315
community structure, 274–7
competitive exclusion, 276, 277
conduction of heat in fluids, 163
conductive cooling, 218
conductive heat flow within sulphide structures, 148, 156, 157
confinement of plumes, 375, 378
conservation of thermal energy, equation, 146
containment of plumes, MAR, 275
convecting systems, pipe models, 160–72
convective heat flow within sulphide structures, 149, 156
copepods
  faecal pellets, 347, 349
  Mid-Atlantic Ridge, 261, 266, 277
copper, 217
  dissolved, 10, 17
    gross fluxes, 365–71
  East Pacific Rise, 4, 17, 18, 20, 21
    Galapagos Rift, 207, 212, 213, 214, 215, 217, 220
  Lau Basin, 232, 236, 237
  Manus Basin, 194, 195
  Mid-Atlantic Ridge, 17, 223
    Broken Spur, 103, 105, 106, 107, 108, 184, 185
    MARK, 47, 48, 61, 62, 78, 226, 227

    TAG, 48, 52, 54, 56, 57, 58, 59, 60, 78, 81, 124, 224, 261
    Woodlark Basin, 196, 200, 201, 202
copper sulphides, Broken Spur, 177, 180, 185, 186, 187, 188
copper/iron ratios in plumes, Juan de Fuca Ridge, 389
copper/zinc ratios, effects on, 61–2
copper/zinc/lead ratios, Manus Basin, 195
corals, 272, 306, 307
Coriolis effect, 374, 379, 391
Coriphaenoides armatus, Snake Pit, 322
Corymorpha groenlandica, MAR, 272
covellite
  Galapagos Rift, 210, 211, 212, 213, 217
  Mid-Atlantic Ridge, 49, 50, 56, 176, 178, 181, 182, 186
crabs
  Mid-Atlantic Ridge, 257
    Broken Spur, 38, 39, 259, 270
    Lucky Strike, 259, 268, 270
    Snake Pit, 259, 266, 320, 322, 324
    TAG, 258, 259, 261, 262, 263
  Pacific Ocean, 194, 259, 281
crayfish, Aegean Sea, 312
Crinoidea, 312, 322
critical radius for entrainment of particles, 390
crustaceans, 227, 257, 263, 309, 311, 315, 330
  larvae, 343
  wax esters in, 339
  see also specific examples
crustal accretion and hydrothermal input, ELSC, 237, 241
crustal evolution, sedimentation and, Lau Basin, 237–46
crustal permeability, 28–9, 31, 39, 118
Cyanagraea praedator, Pacific Ocean, 259
cyanobacteria, Aegean Sea, 307
Cyclope neritea, Aegean Sea, 308, 310, 314
cyclopoid copepods, MAR, 261
Cymodocea nodosa, Aegean Sea, 307, 308
Cyprus-type ore bodies, 81
Cystoseira ssp., Aegean Sea, 307

dacite, 193, 194
dactyls, MAR, 281
Dasycladus vermiculatus, Aegean Sea, 307
de novo biosynthesis, 339, 340, 341
decapods, 257, 311, 312
degassing, magmatic, 138–40
denaturation resistance, 356–7
density anomalies, 374, 379
depth, in control of fauna, MAR, 272–3
depth of circulation, 164
desmosomatid isopods, MAR, 261
Desulfurococcus spp., 354, 355
detergent production, enzymes for, 359, 360
diatoms, 347, 349
diet, of vent fauna, 329, 330, 338, 341
  see also food webs; nutrition
diffuse flow, 4
  heat and temperature flux, 11–12, 145, 147, 373, 382
  low temperature, MAR, 123, 124, 175
  modelling, black smoker vent fields, 159–72
  oxide precipitation, Broken Spur, 186

digenite, Broken Spur, 176, 178, 181, 182, 183, 185, 186
dipterid insect lava, Aegean Sea, 312
Dirichlet condition, 146
discharge (exit) temperatures, 163, 164–9, 170, 172, 179
discharge velocity of sulphide structures, 152, 156
discharge zones, 164
discharge-dominated systems, 161, 164, 166
dispersal of fauna, 277–9, 288
dispersal of larvae, 275, 343, 344, 349
dispersion of plumes, 231
Divaricella divaricata, Aegean Sea, 314
DNA
    amplification of sequences, 358–9, 360
    data base for larval identification, 343–9
    thermostable, 359
Dobu Seamount, 195, 196, 197

Earth, effects of rotation, 374, 375
East Basin, Woodlark Basin, 196, 197
East Pacific Rise (EPR), 3–6, 9, 17–23, 39, 164, 231, 274–5
    9°N field, 329–41
    13°N field, 134–41, 295–301
    15°N field, 198, 201
    21°N field, 140, 198, 201, 368
    centralized magma feeding model, 26–7, 28–9
    ecology, 259, 266, 280, 281, 288, 295–301
        larval identification, 344
    'events', 6–8, 65, 66
    heat flux, 68, 382
    lipid characteristics of vent organisms, 329–41
    magmatic alteration of vent fields, 65, 66, 69, 72, 73, 74
    sediments, 17–18, 20–2, 24, 26, 29, 127–8, 134, 201
        metal accumulation rates, 242, 246, 247
        rare earth concentrations, 93, 94
        sulphur isotopes, 187, 188
    vent fluid compositions, 5–6, 20, 22, 115–16, 141, 198, 201, 366–71
        noble gas isotopes, 134–41, 379
    see also Galapagos Rift, Galapagos Spreading Centre
Eastern Lau Spreading Centre, ELSC
    sedimentation related to propagation, 237–43, 244–5, 246
    sediments, 232, 233, 234
Eastern Pacific Spreading Centre, ecology, 284
echinodermata, Aegean Sea, 309
echinoids, Aegean Sea, 306
Echinus sp., MAR, 262, 263, 269–70
echiurans, Aegean Sea, 306, 312, 314
ectosymbiotic bacteria, 308
Eiffel Tower, rare earth concentrations, 94
electrical potential anomalies, MAR, 58
Elisella, Snake Pit, 322
encapsulated bacteria, transformation of metals, 255
end-member mixing, 219–20
    Galapagos Rift, 218
Endeavour field, JDFR, 68, 378
    diffuse flow, 12, 156, 159
    particle processes, 393

plume-forced circulation, 378
    vent fluid compositions, 116
endosymbiosis, 280, 281, 285
endosymbiotic bacteria, 273, 284, 314, 329, 330, 337, 338, 340
entrained seawater, as coolant, 148
entrainment of brine, Broken Spur, 101
entrainment of fluid within sulphide structures, 153, 155
entrainment of seawater, 164, 185, 374, 375
    in bubbles, 113, 118
    and He–S isotope systematics, 141
enzymes, hyperthermophilic *see* hyperthermophilic enzymes
epibiotic bacteria, 285
    Aegean Sea, 312
    Mid-Atlantic Ridge, 261, 280, 281, 283–4
epidotizaton, 28
Ericthonius punctatus, Aegean Sea, 311
esterase, 361, 362
estuarine mixing with seawater, 365, 369–70, 371
24-ethylcholesta-5,24(28)-dienol in vent organisms, 337
24-ethylcholesterol in vent organisms, 337, 340
Eubostrichus cf. *parasitiferus*, Aegean Sea, 308
Eucarya (Eukaryotes), 352
Euclymene sp., Aegean Sea, 308
eukaryotic dependence, 329, 337
eukaryotic gene, cloning, 359
europium anomaly, MAR, 92, 93, 223
event plumes, 65, 66–7, 70–1, 73, 74, 133, 168
'events', hydrothermal EPR, 6–8
exit temperatures (discharge temperatures), 163, 164–9, 170, 172, 179
Exogone gemmifera, Aegean Sea, 306, 308
'eyeless' shrimp, 286

FAMOUS area, methane anomalies, 11
FARA program, 88
fast-spreading ridges, 4–6, 121–2, 274–7
fatty acids, compositions of vent organisms, 329, 331–6, 337–40, 341
faults
    effects of, 39
    role of, 28–9
fauna
    absence of in high-chloride fields, 81
    Aegean Sea, 303–15
    in diffuse flow, 175
    dispersal, 277–9
    Gulf of Mexico, 257, 266, 285, 295, 296, 338, 340
    Manus Basin, 194
    Mid-Atlantic Ridge, 37–40, 257–88, 321–3, 325, 344
    see also specific examples
faunal discontinuities and transform faults, 278–9
FAZAR cruise, 88, 91–2
felsic volcanism and ore formation, 191–2, 200
ferrigenous sediments, MAR, 223
ferrihydrite, TAG, 49, 52
ferrobasalts, EPR, 26, 29
ferromanganese oxide, Lau Basin, 236
ferruginous cherts, 200, 202, 203
Fifteen-Twenty Fracture Zone, 273, 279

filamentous bacteria, 38, 281, 306, 307, 312,
    313
  on vestimentiferan tubeworms, 297
filling box effect, 375
Fir Tree mound, 264
Fischer Tropsch reaction, 11
fish
  Aegean Sea, 306, 307, 312
  Manus Basin, 194
  Mid-Atlantic Ridge, 38, 258, 261, 267, 271, 322, 323
'flickering' water, MAR, 268
flocculation of metals, 369–70, 371
flow equations, 160, 162, 163
flow patterns within sulphide structures, *see*
      circulation within sulphide structures
flow resistance pressure, 160
fluid density/temperature, variations, 163
fluid inclusions
  Franklin Seamount, 199
  noble gas isotopes, EPR, 134–41
fluid temperatures, 353
  Broken Spur, 99, 101, 104, 108, 175, 179
  of diffuse flow effluent, 147, 163, 164–9, 172
    within sulphide structures, 147–9, 151–2, 155
  East Pacific Rise, 135
  Mid-Atlantic Ridge, 37, 81, 93, 94, 117, 175, 268,
    272
  Reykjanes Ridge, 117
fluted worm-tubes, Broken Spur, 38
food webs, 285, 288, 329
foraminiferal ooze, Lau Basin, 235
fouling organisms, 281, 284
fracture zones, MAR, 11
Franklin Seamount, 196, 197, 198, 199, 200, 201, 202,
    203
frequency of eruptive events, 275

Galapagos Rift, 3, 87, 127, 201, 207–8
  larval dispersion, 349
  massive sulphide deposits, 207–8, 216–18, 220
    geochemistry, 215–16
    minerology, 208–15
    sulphur isotope geochemistry, 187, 210, 216,
      218–20
Galapagos Spreading Centre, 117, 160
  deposits, 134, 198, 201
  ecology, 259, 266, 280, 282, 283, 284, 287
Galapagos-type diffuse flow, 81
galatheid crabs, MAR, 261, 266
galatheid squat lobsters, MAR, 262, 263
Galatheidae, Snake Pit, 322
galena
  Broken Spur, 176, 178, 180
  Galapagos Rift, 210, 213, 216, 217
  Manus Basin, 194
  Woodlark Basin, 199
Gallionella ferruginea, Aegean Sea, 305, 313
gallium, Galapagos Rift, 214
gasohydrothermal activity, 303
gastropods
  Aegean Sea, 308, 310, 311, 314
  larvae, 343
  Mid-Atlantic Ridge, 58, 261, 262, 263, 267, 270,
    272, 322, 326

gene defect screening, 359
gene flow, 343, 344, 347
genes, cloning, 360
genetic differentiation of vent fauna, 344
'geochemical minimum' zone, EPR, 252, 254
geostrophic flow, 374, 378
geostrophic vortex formation, 375–9
global budget calculations of elements, 74
glucose isomerase, 362
glucosidase, 361, 362
goethite, Broken Spur, 178, 183, 184
goethite–hydrogoethite, MAR, 58
gold
  Manus Basin, 195, 199
  Mid-Atlantic Ridge, 60
    MARK, 47, 48, 49, 61, 226, 227
    TAG, 48, 52, 54, 56, 59, 62, 126
  Woodlark Basin, 193, 196, 200, 202
Gorda Ridge, 3
gorgonians, Snake Pit, 322, 325
grey smoker plumes, 159
Guaymas Basin, ecology, 257, 272, 281
Gulf of Mexico, ecology, 257, 266, 285, 295, 296, 338,
    340

haematite, Broken Spur, 58, 176, 178, 180, 185
Halice hesmonectes, EPR, 330, 331, 335–6, 338–40
Haliotis chracherodii, Aegean Sea, 314
halophilic Archaea, 352
Hawaii, 4
heat budget for confined plumes, 381
heat exchange by 'beehive' diffusers, 185
heat flow and fluxes, 11–12, 98, 101–2, 124, 381
  diffuse flow modelling, 159–72
  plume models, and, 381–3
  in vent fluids, 70–1, 73, 74, 373, 374, 376
    equation, 374
  within sulphide structures, 146, 147–9, 152, 155–7
heat sources, TAG and MARK fields, 83
heater equation, 162, 163
height of rise of plumes, 70, 373, 374–5, 378, 382
  East Pacific Rise, 5, 9
helium
  anomalies, 77, 133, 379
  in fluid inclusions, 138–9
  movement due to geothermal circulation, 379
helium isotopes
  ^{3}He
    EPR, 127, 128
    Juan de Fuca Ridge, 66, 71
  ^{3}He enriched plumes, EPR, 5
  ^{3}He/^{4}He ratios, 134, 136, 137, 141
  ^{3}He/temperature ratios, 74
    Juan de Fuca Ridge, 71, 72, 73
  He/^{36}Ar ratios, 138, 139, 141
  as tracers of MOR vent fluids, 134–5, 136, 137, 138,
    139, 141
Hellenic Arc, 315
Hess Deep, as dispersion barrier, 278
hetons (vortex pairs), 375, 376, 377, 378
Hexactinellidae, 268, 322
Hiatella arctica, MAR, 272
high-chloride vent fields, EPR, 81
high-temperature on-axis discharge, 12

high-temperature venting, circulation forcing
    by, 373–84
high-temperature vents, 11, 12, 159
    East Pacific Rise, 3–4, 8–9, 11–12, 17, 139
    heat flux, 373, 374
    Mid-Atlantic Ridge, 8–9, 23, 25, 36–7, 92, 122, 134,
      135, 261, 268
    as source of iron, 17
holothurians, 272, 320, 322, 323, 324, 325
Hyalinoecea bilineata, Aegean Sea, 307
hydrogen
    Aegean Sea, 305
    bacterial production, 115, 119
    dissolved, Steinahóll plume, 111–12, 113–19
    oxidation, 354
hydrogen enrichments, MAR, 268
hydrogen sulphide
    Aegean Sea, 305, 307, 313
    East Pacific Rise, 4, 201, 219
    Lau Basin, 201
    Mid-Atlantic Ridge, 78, 81, 124, 187, 188
    oxidation, 353
    production, 200
    Reykjanes Peninsula, 117
    Woodlark Basin, 199, 201
hydrogen/methane ratios, 115–16, 117, 118
hydrogoethite, TAG, 49
hydrohematite, TAG, 52
hydroids, 270, 272, 307, 312, 322, 325
HYDROSNAKE mission, 283
hydrothermal oxide factor, Lau Basin, 236, 238, 239
hydrothermal oxide phase, Lau Basin, 246
hydroxide–sulphide sediments, Snake Pit, 321
hydrozoans, Aegean Sea, 306, 307
hyperthermophilic bacteria, 272, 314, 352–60
    enzyme production from, 357
hyperthermophilic biotopes, 353
hyperthermophilic environments, 352–4
hyperthermophilic enzymes
    biochemistry, 351–8
    biotechnology, 358–62
hyperthermophilic micro-organisms, 354–7
Hyperthermus butylicus, 354, 355

idaite, 56, 210, 211, 212, 213
illite, TAG, 52
industrial enzymology, 359–62
invertebrate–bacterial symbioses, 284
invertebrate–endosymbiont systems, 285
iron
    Aegean Sea, 307, 313
    chloro-complexing, 81
    dispersion patterns, 245
    dissolved
      gross fluxes, 365–6, 368–9
      Mid-Atlantic Ridge, 17, 98, 102, 103, 104
    distribution and transformation, 249–55
    East Pacific Rise, 4, 5, 17–18, 22, 26, 249–53
      Galapagos Rift, 213, 214, 215
    Juan de Fuca Ridge, 66, 71, 103
    Lau Basin, 232, 236, 237, 239
      accumulation rates, 236, 240, 241–2, 243, 245,
        246
    Manus Basin, 195

Matupi Harbour, 254, 255
Mid-Atlantic Ridge, 17, 223, 227, 228
    Broken Spur, 103, 105, 106, 107, 108, 184
    Lucky Strike, 92, 93
    MARK, 47, 48, 78, 226, 227
    Snake Pit, 320, 321, 323, 324, 325
    TAG, 48, 51, 52, 56, 57, 58, 78, 81, 124, 127,
      129, 224, 225
    Woodlark Basin, 196, 200, 201
iron bacterium 81, 255, 305, 313
iron disulphides, 183
    Broken Spur, 176, 177, 182, 187
    Galapagos Rift, 210, 211
iron hydroxide sediments, Snake Pit, 321, 325
iron oxide–hydroxide deposits, Aegean Sea, 305
iron oxides
    Galapagos Rift, 210
    Manus Basin, 194
    Mid-Atlantic Ridge, 124, 176, 180, 181, 186, 387,
      394
    Woodlark Basin, 200, 203
iron oxyhydroxides, 121–2
    Galapagos Rift, 208, 210, 213, 217
    Lau Basin, 234, 236
    Matupi Harbour, 255
    Mid-Atlantic Ridge, 227
      Broken Spur, 98, 176, 177, 181, 182, 183, 184,
        186, 188
      TAG, 389, 391
iron sulphides
    East Pacific Rise, 135, 208
    Mid-Atlantic Ridge
      Broken Spur, 176, 178, 181, 184, 185, 188
      isotopically light, 187
      Snake Pit, 321
      TAG, 124, 387, 389, 394
iron–copper sulphide deposits, Galapagos Rift, 207,
    210, 212, 213, 215, 217, 218
iron–copper sulphide mounds, EPR, 135
iron–manganese oxyhydroxides, EPR, 135
iron–silicon–manganese oxyhydroxides
    Galapagos Rift, 207–8
    Woodlark Basin, 193, 196, 198, 199, 200–1, 202,
      203
iron–titanium andesites, Woodlark Basin, 195
iron–zinc sulphide mounds, EPR, 135
iron–zinc–copper chimney, EPR, 135
iron–zinc–lead–barium mineralization, Galapagos
    Rift, 210, 211, 213
iron–zinc–lead–barium–silicon mineralization,
    Galapagos Rift, 218
iron/hydrogen sulphide/chloride ratios, 81, 83, 84
iron/manganese ratios in plumes, 389, 394
iron(II) sulphate, Galapagos Rift, 215
isocubanite, MAR, 49, 56, 60, 176, 178, 180, 182, 183
isopods, MAR, 270

Jade site, Okinawa Trough, metalliferous sediments,
    195, 199, 201, 202
Jan Mayen Ridge, ecology, 272
jordanite, Broken Spur, 176, 178
Juan de Fuca Ridge, 3, 6–7, 8, 10, 133, 139, 186, 199
    chronic plumes, 68
    diffuse flow, 12, 156, 157, 159

Juan de Fuca Ridge—*cont'd*
  ecology, 280, 295, 296
    larval identification, 344
  event plumes, 65, 66–7, 70–1, 74, 133
  heat flux, 12, 159, 382, 383
  light fields, 287
  magmatic alteration of vent fields, 65–74
  plume-forced circulation, 378, 379
  sulphides, 186, 187, 188, 389
  vent fluid compositions, 10–11, 94, 115–16, 139, 365
  see also Cleft segment, JDFR; Endeavour field, JDFR

Kagoshima Bay, ecology, 272, 315
KANE Transform Fault, 278, 279
Kane–Atlantis fracture zones, 8, 9
Kiriakis Bay, biological communities, 312–13
Kolbeinsey Island, ecology, 272, 303
Kos, 303, 304
Kraternaya, ecology, 272, 315
Kremlin area, TAG, 124, 261
Kurile Islands, ecology, 303, 315
Kuroko metalliferous deposits, 195, 202, 203

Lamellibrachia
  East Pacific Rise, 296, 298, 301
  Mid-Atlantic Ridge, 272
larval dispersal, 275, 343, 344
larval identification, 344–9
latitude, biological activities and, 326
Lau Basin, 3, 9, 10, 232–4, 243, 244
  ecology, 259
  sediments and sedimentation, 195, 202, 232, 234–47
  vent fluid compositions, 9–10, 198–9
Lau/Tonga arc, 232
lead
  Broken Spur, 103, 105, 106, 107, 108
  dissolved, gross fluxes, 9, 365–71
  East Pacific Rise, 4
  Lau Basin, 236, 237
  Manus Basin, 194, 195, 199
  Mid-Atlantic Ridge, 47, 48, 49, 52, 54, 56, 59, 60
  scavenging, 393
  Woodlark Basin, 196, 200, 201, 202
'leaky mound' model, 81
lepidocrocite, Broken Spur, 178, 183, 184
Leptonemella spp., Aegean Sea, 308
ligase chain reaction (LCR, LAR), 359
light fields, 287
light scattering, MAR, 92, 95
limpets, MAR, 267, 270, 277
Linhomoeoid spp., Aegean Sea, 308
lipase, 360, 361
lipid biomarkers, 284
lipid characteristics of vent organisms, 329–41
lithium, JDFR, 73, 79, 80, 201
lithoautotrophic oxidation, MAR, 280
Lithothamnion, Aegean Sea, 312
Loihi Seamount, 133, 139, 198, 201, 259
Loripes lacteus, Aegean Sea, 314
Lottia limatula, Aegean Sea, 314
low-temperature off-axis heat flow, 12

low-temperature venting 4, 12
  diffuse flow modelling, 159–72
  East Pacific Rise, 135
  Mid-Atlantic Ridge, 8–9, 23, 36–7, 134, 135, 175, 261, 268
lucinid bivalves, Aegean Sea, 314
Lucky Strike field, 25, 77, 87–8, 97, 122, 123
  ecology, 257, 259, 262, 263, 268–70, 272, 282, 285, 288
  high temperature vent sites, 9
  vent fluid compositions, 88–95, 268
Lumbrineris funchalensis, Aegean Sea, 306

mackinawite, MARK, 49
Macoma sp., MAR, 272
macroalgae, Aegean Sea, 307
Macrouridae, Snake Pit, 322
magma chambers, active, need for, 25
magma feeding model, EPR, 26–7, 28–9
magma–seawater mixing, Galapagos Rift, 220
magmatic alteration of vent fields, 133
  East Pacific Rise, 65, 66, 69, 72, 73, 74
  Juan de Fuca Ridge, 65–74
magmatic degassing, 71–2, 73, 138, 139, 140
magnesium
  East Pacific Rise, 20, 22, 26, 29
  Lau Basin, 236, 237
  Mid-Atlantic Ridge, 81, 82, 94, 185
magnesium oxide, Galapagos Rift, 215
magnesium silicates, 179, 180, 185, 196
magnetite, Broken Spur, 177, 178, 179, 180, 182, 185
Malacocerous fuliginosus, Aegean Sea, 305, 306
maldanid polychaetes, Broken Spur, 267
manganese, 74, 201
  Aegean Sea, 310
  dispersion patterns, 245
  dissolved, 11
    detection, 88, 89, 91
    East Pacific Rise, 5, 6, 17, 20, 22
    gross fluxes, 365–71
    Lau Basin, 9
    Marianas back-arc system, 9
    Mid-Atlantic Ridge, 8, 17, 54–5
      Broken Spur, 97–8, 103, 105, 106
      Lucky Strike, 89, 91, 92–3, 95
      Steinahóll plume, 111, 112, 114–15, 116, 117, 118
  distribution and transformation, 249–55
  East Pacific Rise, 4, 17, 18, 217, 249–53, 255
  Juan de Fuca Ridge, 66, 71, 73
  Lau Basin, 232, 236, 237, 239
    accumulation rates, 236, 240, 241–2, 243, 245, 246, 247
  Matupi Harbour, 254, 255
  Mid-Atlantic Ridge, 17, 60, 255
    Broken Spur, 97–8, 103, 104, 105, 106, 107, 108
    MARK, 47, 48, 78
    Snake Pit, 320
    TAG, 48, 56, 57, 58, 78, 123, 127, 129
manganese dioxide, 242
manganese oxide, 52, 194, 200, 203, 214, 215
manganese oxyhydroxides, Lau Basin, 234, 236

manganese/methane ratios
  Steinahóll plume, 115, 117
  TAG, 115
manganese/silicon ratios, Steinahóll plume, 115
manganite, Lau Basin, 242
mantle degassing, 139–40, 273
Manus Basin, 191–4
  ecology, 194
  polymetallic ore genesis, 193, 194–5, 199–200,
    202–3
marcasite, 62
  East Pacific Rise, 135
    Galapagos Rift, 210, 211, 212, 213, 216, 217, 218
  Mid-Atlantic Ridge, 60
    Broken Spur, 176, 178, 180, 182, 186
    MARK, 49, 50
    Snake Pit, 321
    TAG, 52, 56, 58, 125
marcasite–chalcopyrite, EPR, 135
marcasite–pyrite, MAR, 49, 52, 61
Marginal High, 135, 139, 140
Mariana Trough, 3
Marianas back-arc system, 9
  ecology, 259, 267, 277, 282, 286
  sulphide deposits, 187, 218
  vent fluid compositions, 9–10
marine 'snow', 349
MARK field, 23, 27–8, 43–9, 77, 83, 97
  larval identification, 344, 346, 347, 348
  massive sulphide deposits, 47–9, 59, 61, 62
  metalliferous sediments, 46–9, 93, 94, 223, 225–7
  vent fluid compositions, 78–80, 83–5
  see also Snake Pit field
mass flow, TAG, 223
mass flow rates, 163
mass flux within sulphide structures, 155
massive sulphide deposits, 202–3, 274
  East Pacific Rise, 17, 18–23, 29
    Galapagos Rift, see under Galapagos Rift
  Manus Basin, 194–5
  Mid-Atlantic Ridge, 44
    Broken Spur, 175–88
    MARK, 23–9, 45, 49, 62, 226
    TAG, 8, 23–9, 46, 51, 55, 58, 59, 61, 62, 261
  volcanic-hosted, 199, 200, 207, 219
  see also polymetallic ore genesis; sulphide mounds;
    sulphides
Matupi Harbour, bacterial transformation of metals,
    253, 254, 255
Meddies, 378, 384
Mediomastus fragilis, Aegean Sea, 306
megaplumes, 118, 139, 373, 379, 383
  Juan de Fuca Ridge, 6–7, 8, 66, 67, 72, 74, 133, 139
meiofauna, 261, 308
melanterite, Galapagos Rift, 213
melnicovite–pyrite, MAR, 50, 52
Menez Gwen, 122, 123
mercury, 48, 56, 195, 196, 200, 202
metabolism, Archaeal, 354–5
metal accumulation rates, Lau Basin, 236, 240,
    241–3, 244–6
metal deposition rates, 62
metal fluxes, Aegean Sea, 305
metal-rich fluids, detection of vent sites, 91

metalliferous sediments, 249
  East Pacific Rise, 4, 17–18, 20, 22–3, 27, 29, 224–5,
    249–55
  formation, 231–2
  Lau Basin, 234, 236–7, 239–41, 244–5, 246
  Manus Basin, 193, 194–5, 199–200, 202–3
  Mid-Atlantic Ridge, 58, 223, 227–8
    Broken Spur, 176–88
    MARK, 26, 45–9, 61, 225–7
    Snake Pit, 320–1
    TAG, 49–54, 56, 57, 59–60, 61–2, 121–30, 223,
      224–5
  North Fiji Basin, 195, 242, 246
  Okinawa Trough, 195, 199, 201, 202
  Woodlark Basin, 195–6, 198–9, 200–3
Metallosphaera sedula, 354, 355
metals
  deposition rates, 62
  dissolved
    anthropogenic input to oceans, 365, 366–8,
      369–71
    gross fluxes, 365–71
    riverine input to oceans, 366–7, 369–70
  flocculation, 369–70, 371
  see also particle processes; specific metals
methane, 91, 305
  biogenesis, 115, 355
  detection by shrimps, 288
  dissolved
    East Pacific Rise, 5, 6, 20
    Mid-Atlantic Ridge, 11, 54–5, 268
    Steinahóll plume, 111–12, 113–18, 272
  serpentization and, 273, 279
methane anomalies, 273
methane plumes, MAR, 274
methane/hydrogen ratios, 115–16, 117, 118
Methanococcus, 354, 355
methanogenic bacteria, 272, 352, 355
methanogenic metabolism, 355
methanol dehydrogenase, 284
Methanopyrus kandleri, 354, 355
Methanothermus, 354, 355
methanotrophic mussel endosymbioses, 273, 284–5,
    288
methanotrophic taxa, dispersion, 279
4-methyl sterols, 340
24-methyl-5α-cholestanol in vent organisms,
    337
24-methylcholest-24(28)-enol in vent organisms, 337
24-methylcholesta-5,22-dienol in vent organisms, 337
24-methylcholesta-5,24(28)-dienol in vent organisms,
    336, 337, 340
24-methylcholesterol in vent organisms, 337
methylotrophic bacteria, 273, 284, 340
microbial coenoses, MAR, 255
microbial phylogeny, 352
microfibrils, chitin, 297–8, 300, 301
Microprotus maculata, Aegean Sea, 312, 315
Microspio sp., Aegean Sea, 308, 310, 314
Mid-Atlantic Ridge (MAR), 8, 9, 28, 29, 97, 122, 383
  14°40′–15°N fields, 223, 226, 227, 271–2, 319, 326
  27°10′N field, 223, 227–8
  ecology, 257–88
    larval identification, 344

Mid-Atlantic Ridge (MAR)—*cont'd*
  massive sulphide deposits, *see* massive sulphide
      deposits
  metalliferous sediments, 17, 223, 226, 227–8, 255
  shallow water site ecology, 272–3
  vent fluid compositions, 6, 8, 11, 17, 54–5, 366
  see also Broken Spur field; Lucky Strike field;
      MARK field; Reykjanes Ridge; Snake Pit
      field; Steinahóll site; TAG field
mid-depth circulation, 379, 380
mid-ocean ridge basalts (MORBs), 133, 140
mid-ocean ridges (MORs), 121, 133
  fluid chemistry, 133–41
  minerals and mineralization, 121–2, 127–8
Middle Valley, sulphide oxidation, 186
Milos, 304, 305
  biological communities, 306–15
mineral alteration, within sulphide structures, 150,
      151, 155
mineralization
  at mid-ocean ridges, 121–2
  of thermal waters, 353
  Western Pacific, 191–203
Minuspio cirrifera, Aegean Sea, 306
Mir mound, 52–4, 55, 56–8, 123, 124, 259, 260
  ecology, 260–1
  metalliferous sediments, 60, 61, 62, 127, 223, 224,
      225
mixing equation, 162, 163
Mohn's Ridge, ecology, 273
molecular identification of larvae, 344–9
molluscs
  Aegean Sea, 309, 311, 315
  East Pacific Rise, 330
  Mid-Atlantic Ridge, 272
molybdenum
  Galapagos Rift, 214, 215
  Manus Basin, 195
  Mid-Atlantic Ridge, 47, 48, 52, 56, 59, 62
  Woodlark Basin, 200
Monolith vent, 73, 179
  particle concentrations, 389
Moose mound, 264
MOR volcanism, 139
Moresby 'Seamount', 195, 196, 197
mound growth, Mid-Atlantic Ridge, 125–6, 187–8
mudclast conglomerate, Lau Basin, 234, 235
Mullus surmuletus, Aegean Sea, 306
Munidopsis, Mid-Atlantic Ridge, 39, 322, 325
Munidopsis crassa, MAR, 266
mussels, 272, 278
  East Pacific Rise, 284, 330, 331
    lipid characteristics, 333–6, 337–8, 340
  Gulf of Mexico, 340
  Manus Basin, 194
  methanotrophic symbioses in, 273, 284–5, 288
  Mid-Atlantic Ridge, 257, 268, 273, 326
    Broken Spur, 38, 39, 261, 270
    Lucky Strike, 88, 266, 285, 288
    Snake Pit, 258, 261, 262, 263, 265–6, 285, 286,
        288, 322
mutualistic symbiosis, 281
Mya priapus, MAR, 272
Mytilidae bivalves, Snake Pit, 322, 325

Mytilus edulis, 344, 347

Nail mound, 264
nannofossil oozes, Lau Basin, 234, 235, 236
naphthalene dioxygenase, 361
Nassarius mutabilis, EPR, 308, 314
nekton distribution, Snake Pit, 322, 323
nematodes, 261, 267, 284, 308, 314
Nematonurus armatus, MAR, 267
nemerteans, 270, 309
Neomediomastus glabrus, EPR, 306
nephel anomalies, MAR, 91, 99, 101, 102, 104
nephelometer–transmissometer relationships, 387–9
Nereidae sp., Aegean Sea, 306
Nereis diversicolor, Aegean Sea, 306
Neumann condition, 146
neutral lipid fatty acids (NLFA) in vent organisms,
      331, 332, 334, 335, 336, 339, 341
neutrally buoyant plumes, 5, 9, 33, 103, 121, 122, 130,
      345–6
  see also non-buoyant plumes
neutrophilic thermal waters, 353, 354
'new' evolving vent sites, 6–8
nickel, 57, 58, 200, 214, 236, 237
niobium, Galapagos Rift, 214
nitrate, Broken Spur, 98, 99, 100
nitrilase, 361, 362
nitrite, Broken Spur, 99, 100
Nitzchia bilobata, 347
NOAA/VENTS Program, 68
noble gas isotopes in fluid inclusions, EPR, 133–41
non-buoyant plumes
  particle processes, 387, 388–9, 390, 391, 392,
      393–4, 395
  see also neutrally buoyant plumes
non-methylene interrupted fatty acids (NMI), 332,
      333, 335, 336, 337–8, 339, 341
nontronite, Woodlark Basin, 198, 200
Noranda, mineralization, 191, 200, 202, 203
North Fiji Basin, metalliferous sediments, 195, 242,
      246
North Valley, Woodlark Basin, 195
North Zone, TAG, metalliferous sediments, 223,
      224, 225
Notomastus latericeus, Aegean Sea, 306
nudibranchs, Aegean Sea, 307
nutrient release, Aegean Sea, 312
nutrition, 159, 285, 288, 329
  bivalves, 314
  gastropods, 308
  mussels, 284–5, 288
  shrimps, 281, 283, 284, 285–6
  see also diet; food webs; lipid characteristics of vent
      organisms

oceanic circulation, effect of venting on, 373
Oceanographer Transform Fault, 278, 279
ochres, TAG, 126
Octopus vulgaris, Aegean Sea, 306
off-axis volcanism, Lau Basin, 244
oil reservoirs, archaeal bacteria, 354
Okinawa Trough, 3
  metalliferous sediments, 195, 199, 201, 202
oligochaetes, Aegean sea, 305, 306

Oncholaimus campyloceroides, Aegean Sea, 308, 314
onion domes, MAR, 78, 79, 80–3
ophiuroids, Mid-Atlantic Ridge, 38, 39, 258, 267, 270 271, 312
'organ pipe' structures, Broken Spur, 176, 177, 179, 180, 186
organic carbon, 310, 311, 337
ostracods, MAR, 270
over-expression, 357, 360
oxide plume fallout variations, ELSC, 237
oxide sediments, 4, 126, 175, 184, 232, 237
oxyhydroxide sediments
  East Pacific Rise, 135, 207–8
  Juan de Fuca Ridge, 103
  Lau Basin, 234
  Mid-Atlantic Ridge, 36, 39, 181, 182, 193, 227
  Woodlark Basin, 193, 196, 198, 199, 200–1, 202

Pacific Ocean, 10, 115–16, 379
  see also East Pacific Rise
PACLARK site, Manus Basin, 193
PACMANUS sulphide deposits, 193, 194–5, 199–200
palaeotectonic controls on MAR biogeography, 277
palagonite, 52
Paleodictyon nodosum, TAG, 260
Paleohori Bay, biological communities, 307–12, 313, 314
Palinurus elephas, Aegean Sea, 312
Paralvinella grasslei, MAR, 346, 347
Paralvinella palmiformis, MAR, 346, 347
particle processes, 387–95
passive recycling, 390–1
Peclet number, 152–4, 162
Peggy Ridge, 232
pelecypod colonies, Broken Spur, 58, 60
peptide cleavage, 357
pereiopods, MAR, 281
peridotites, TAG, 24
Perinereis cultrifera, Aegean Sea, 306
Perioculodes longimanus, Aegean Sea, 311
permeability of mounds and fluid flow, 148
permeability of pipes, 164–5, 166, 171, 172
permeability of sulphide structures, 151–2, 153, 154, 155, 156
pH, of vent waters, 9, 78, 81, 198, 217, 313, 353
pharmaceuticals, enzymes for, 360, 361, 362
phase separation
  of helium, 138–9
  of seawater, 10, 73–4
  of vent fluids, 71–2, 73
phosphate, Broken Spur, 98, 99, 100
phosphorus, Lau Basin, 236, 237
phosphorus(V) oxide, 200, 203, 214, 215
photoreceptor, 'eyeless' shrimp, 286
Phymorrhynchus sp., MAR, 262, 263, 267
phytoplankton-derived carbon, 338, 340, 341
Pinna nobilis, Aegean Sea, 306, 307
pipe models of convecting systems, 160–72
Pitharokhoma, alteration pipes, 164
plagioclase formation, 92
plankton, Snake Pit, 325
planktonic dispersal potential, 348, 349
platinum, TAG, 52, 59, 62

β-plume, 379–80, 383
plume dispersion and fallout, Lau Basin, 237, 245, 246
plume theories, 11
plume velocity scale, 375
plume vortices, 375, 376–9, 383–4
plume-derived sediments, TAG, 129
plume-forced circulation, 373, 374, 375, 378, 379, 381
plumes
  asymmetrical, 376
  confined, heat budget, 381
  containment, MAR, 275
  dispersion, 231
  neutrally buoyant, 5, 9, 33, 103, 121, 122, 130, 354–6
  non-buoyant, 388–9, 390, 391, 392, 393–4, 395
  particle processes, 4, 105, 108, 121, 129, 388–95
  see also buoyant plumes
polar lipid fatty acids (PLFA) in vent organisms, 331, 332, 334, 336, 339
polonium scavenging 393
polychaetes
  Aegean sea, 305, 306, 307, 308, 311, 312
  DNA analysis, 347
  East Pacific Rise, 330
  Kraternaya, 272
  larvae, 343
  Mid-Atlantic Ridge, 286
    Broken Spur, 38, 39, 267, 271
    Lucky Strike, 270
    Snake Pit, 267, 268, 320, 322, 324, 325
    TAG, 261, 262, 263
Polycirrus denticulatus, Aegan Sea, 306
polymerase chain reaction (PCR), 344–5, 346–7, 349, 358–9, 360
polymetallic ore genesis, 191
  Manus Basin, 193, 194–5, 199–200, 202–3
  Woodlark Basin, 196, 198–9, 200–3
polynoid polychaetes, MAR, 267, 270
polyunsaturated fatty acids in vent organisms, 332, 333, 334, 335, 337–8, 339–40, 341
'popping rocks', 140
population genetic studies, 344
porous medium flow, 160
porous sulphide, Galapagos Rift, 210, 213
Posidonia oceanica, Aegean Sea, 307
potassium, MAR, 78, 81
potassium oxide, 200, 203, 214
potential temperatures, Broken Spur, 99, 101, 104
power supply equation, 163
pressure gradient equation for flow in pipe models, 161
propagation, sedimentation related to, Lau Basin, 237–46
protease, 361, 362
protein content of tubeworms, 295, 296, 298, 299, 300
protein stability, 356–7
Pseudocapitella sp., Aegean Sea, 306
Pseudocuma longicornis, Aegean Sea, 311
Pseudoleiocapitella sp., Aegean Sea, 306
Pseudorimula marianae Marianas back-arc basin, 267
Pseudorimula midatlantica, MAR, 267
Pual Ridge, 193–4, 200, 203

pyrite, 56, 60, 62, 159
  East Pacific Rise, 135
    Galapagos Rift, 210, 211, 212, 213, 215, 216, 217
    Manus Basin, 194
    Mid-Atlantic Ridge, 227
      Broken Spur, 176, 178, 180, 181, 182, 183, 186
      Lucky Strike, 89
      MARK, 49, 50, 280
      Snake Pit, 321
      TAG, 52, 53, 58, 124
    oxidation, MARK, 280
    Woodlark Basin, 199
'pyrite disease', Broken Spur, 182
pyrite–chalcopyrite, TAG, 59
pyrite–marcasite, TAG, 52
pyrite–marcasite–sphalerite–chalcopyrite, TAG, 59
Pyrobaculum spp., 354, 355
Pyrococcus furiosus, 356, 357, 358, 359
Pyrococcus spp., 354, 355
Pyrodictium spp., 354, 355
pyroxenites, TAG, 24
pyrrhotite
  Galapagos Rift, 212
  Mid-Atlantic Ridge, 58, 60
    Broken Spur, 176, 178, 179, 180, 181, 182, 183,
      185, 187
    MARK, 49, 50
    Snake Pit, 321
    TAG, 58
pyrrhotite–sphalerite–chalcopyrite, MARK, 49, 50

quartz, 49, 52, 56, 159, 321
quartz–hematite mineralization, MARK, 49

radiolarians, 349
radon budgets in heat flux determinations, 382
Raja spp., Aegean Sea, 306
Raleigh number and flow patterns, 145, 146, 147–52
rare earth elements, 89, 90, 91, 93–5, 198
  scavenging, 392, 394
rare earth/iron ratios, TAG, 129
Rayleigh number, 145, 146, 147–52, 153, 156
rays, Aegean Sea, 306
re-entrainment of particles in plumes, 390–1, 392,
  395
reaction rate enhancement, 357
recharge zones, 164
recharge-dominated systems, 161
recycling of particles in plumes, 390–1, 392, 394
recycling of plume fluid, 375
red mullet, Aegean Sea, 306
reproductive strategies, 344
residence time of particles, 391–2, 394, 395
resistance term for flow in pipe models, 160
resistance to flow in pipes, 161, 164–70
restriction endonucleases, 359
reversal of flow in pipes, 169–70, 171
reverse transcriptases, 359
Reykjanes Ridge, 10, 11, 112, 117
  ecology, 272, 303
  see also Steinahóll plume
Reynolds' flux equation, 394
rhyodacite, 193
rhyolites, Western Pacific, 193, 196, 200

ridge propagation and sedimentation, Lau Basin,
  244–6
Ridgeia piscesae, EPR, 296, 298
Riftia pachyptila
  East Pacific Rise, 295, 296, 298, 299, 300, 301
    lipid characteristics, 330, 331–3, 336, 337, 340
  Mid-Atlantic Ridge, 281, 285, 346
Rimicaris, Broken Spur, 39
Rimicaris exoculata
  Mid-Atlantic Ridge, 283, 284, 286, 287
    Broken Spur, 270
    Snake Pit, 264, 265, 266, 281
    TAG, 259, 261, 262, 263, 280, 281, 282
rise height of plumes, 70, 373, 374–5, 378, 382
  East Pacific Rise, 5, 9
riverine input of metals to oceans, 366–7, 369–70
^{222}Rn, Broken Spur, 98, 102, 103, 108
mRNA sequences, 359
rock alteration index (RAI), 150–1, 154
Romanche Transform Fault, 278–9
Rose Garden, ecology, 283
rotation of plumes, 375
rozenite, Galapagos Rift, 213
rubidium, MAR, 78, 79, 80, 81

Sabateria pulchra, Aegean Sea, 308
sabellid polychaetes, Aegean Sea, 312
salinity, Broken Spur, 99, 101, 104, 108
salinity anomalies, 374, 378
salinity-tolerant species, 308, 310
Santorini, biological communities, 303, 304, 305, 313,
  314
Saracen's Head, 36–7, 38, 39, 175, 179, 181–2, 270
scavenging rate constants, 392, 394
scavenging of trace elements, 387, 391–4, 395
Scotia Ridge, 277
Scypha quadrangulatum, MAR, 272
sea anemones, *see* anemones
see urchins, MAR, 88, 262, 263, 269–70
seafloor spreading, effect of, 73
seagrass beds, Aegean Sea, 307, 311
seamount volcanism and helium isotope ratios, EPR,
  134
seawater
  estuarine mixing with, 369–70, 371
  penetration of crust, 121
  phase separation, 10, 73–4
  reactions, 72, 121, 137, 138, 139, 191, 200, 217,
    231, 353
  as oxidant, 128, 183, 184, 185, 186, 188
Seep, fauna, 259
Segonzacia mesatlantica, MAR, 39, 259, 262, 263, 266
seismic alterations of vent fields, 133
seismic event monitoring, JDFR, 68–9, 70
selenium, Galapagos Rift, 212, 213, 214, 215, 217
sensory adaptation, 286–8
serpentinization reactions, 11, 273, 279
serpentinites, TAG, 24
serpulids, 268, 306, 312, 322
serranids, Aegean Sea, 312
sestonophages, Snake Pit, 322
Sevmorgeologija cruises, 17, 18, 26, 29, 43
'shadow effect', 245
shallow vent sites and phase separation, EPR, 10–11

shallow water vent biological communities, 272–3, 303–15
  Archaeal bacteria, 354
shimmering water, 159
  Mid-Atlantic Ridge
    Broken Spur, 36, 37, 175, 176, 179, 182, 185, 186
    Steinahóll plume, 115, 117
    TAG, 51–2, 124
shrimps, 277, 280
  'eyeless', 286
  Manus Basin, 194
  Mid-Atlantic Ridge, 277, 283–4
    Broken Spur, 38, 39, 40, 270, 271
    Lucky Strike, 88, 268, 269
    MARK, 46, 49
    nutrition, 281, 283, 284
    Snake Pit, 39, 257, 258, 262, 263, 264, 266, 278, 285, 286–8, 320, 322, 324, 325
    TAG, 39, 125, 258, 261, 262, 263, 276, 278, 286–7
  sensory adaptations, 286–8
  trophic studies, 280–4, 285–6
silica
  dissolved, 72, 198, 201
  Galapagos Rift, 207, 208, 214, 215
  Manus Basin, 193, 194, 195
  Mid-Atlantic Ridge, 57, 61, 79
  precipitation, 218
  Woodlark Basin, 196, 199, 200, 203
  see also amorphous silica
silica–barite associations, Galapagos Rift, 218
silicates, 99, 100, 307, 308, 313, 353
silicon, 74
  dissolved, Steinahóll plume, 112, 114–15, 117, 118
  Juan de Fuca Ridge, 66
  Mid-Atlantic Ridge
    Broken Spur, 98
    MARK, 47, 48
    Snake Pit, 320, 321, 323, 324, 325
    TAG, 48, 52, 56, 57, 58
silicon hydroxide, MAR, 78
silver
  Manus Basin, 195, 199
  Mid-Atlantic Ridge, 60
    MARK, 47, 48, 49, 61, 226, 227
    TAG, 48, 52, 54, 56, 59, 61, 62
  Woodlark Basin, 196, 199, 200, 201, 202
single-methylene interrupted fatty acids (SMI) in vent organisms, 332, 334, 336, 338, 341
Siphonolaimus spp., Aegean Sea, 308
sipunculans, Aegean Sea, 305
slow-spreading ridges, 8–9, 77, 121–2, 274–7
snails, Manus Basin, 194
Snake Pit field, 8, 9, 122, 123, 319–20, 325–6
  activity, 274
  bacterial activity, 255
  beehive structures, 179, 186
  biogeochemical zones, 319
  diffuse flow, 156
  ecology, 257, 258, 259, 261, 262, 263–8, 277, 278, 280, 281, 282, 283, 284, 285, 286–8, 321–3, 325
  light fields, 287
  metalliferous deposits, 185, 187, 188, 320–1, 323–5
  vent fluid compositions, 366, 367–9, 370–1
  see also MARK field

Society Islands, 4
sodium, 78, 307, 310
sodium oxide, Woodlark Basin, 200, 203
Solemya velum, 340
South Eastern Seamount, EPR, 135–6, 137–8, 139, 140
South Valley, Woodlark Basin, 195, 196
Southwest Indian Ridge, 277
spacial and temporal distribution of events, 274–7
Spathi Point, biological communities, 306–7
sphalerite
  Eastern Pacific Rise, 135, 210, 212, 213, 217
  Manus Basin, 194
  Mid-Atlantic Ridge, 58, 60
    Broken Spur, 176, 178, 180, 181, 182, 183, 184, 186
    Lucky Strike, 89
    MARK, 50
    Snake Pit, 321
    TAG, 52, 53, 54, 56, 125
  Woodlark Basin, 199
sphalerite–barite–silica deposition, Galapagos Rift, 217
sphalerite–chalcopyrite–pyrite, TAG, 52
sphalerite–pyrite, MARK, 49, 54
sphalerite-rich chimneys, TAG, 124
Spio decoratus, Aegean Sea, 308, 311, 314
Spio filifornis, Aegean Sea, 312, 315
Spire, the, 37, 38, 39, 175, 179–81, 188, 270
sponges
  Aegean Sea, 306, 307, 312
  Mid-Atlantic Ridge, 272, 273, 320, 322, 323, 324
spreading level, 374, 375, 376
squat lobsters, MAR, 38, 262, 263
Staphylothermus marinus, 354, 355
starch, hydrolytic enzymes, 359, 360
starfish, Aegean Sea, 306, 307, 312
Steinahóll plume, gas chemistry, 111–19, 272
Steinahóll site, 11, 97
  ecology, 272, 303, 314
stem lilies, Snake Pit, 322
steroid hydroxylase, 361
sterols in vent organisms, 329, 330, 336–7, 340–1
stillbotnematids, Aegean Sea, 308, 314
Stommel–Arons model, 379
'storage' fatty acids in vent organisms, 335, 339
stream function, 146, 149–50
stream functions, within sulphide structures, 150, 153, 156
strontium, 78, 81, 214, 215
strontium/calcium ratios and rare earth content, 93, 94, 95
structural fatty acids in vent organisms, 336
Stygiopontius pectinatus, MAR, 261, 266
Sulfolobales, 355
Sulfolobus solfataricus, 356
Sulfolobus spp., 354, 355
sulphate, reduction, Aegean Sea, 219, 220, 310–11
sulphates,TAG, 125
sulphide chimneys, 8, 36, 37, 138
sulphide depletion, TAG, 81
sulphide mounds
  East Pacific Rise, 135
  formation, 231

sulphide mounds—*cont'd*
  Mid-Atlantic Ridge, 58
    Broken Spur, 35, 36–7, 39, 186
    Snake Pit, 264, 267, 268
    'swelling', 59
    TAG, 8–9, 49, 52, 78
  see also massive sulphide deposits; sulphides
sulphide particles, formation and fallout, 387, 389,
    395
sulphide structures, destruction, 139
sulphide–hosted fluid inclusions, EPR, 134
sulphide–hydroxide sediments, Snake Pit, 321, 325
sulphide-oxidizing bacteria, 280, 284, 285
sulphide–silica mineralization, Galapagos Rift, 210
sulphide–sulphate–oxides, Broken Spur, 176, 178
sulphide-tolerant organisms, 308
sulphides
  detection by shrimps, 287–8
  dissolved, oxidation, 280
  East Pacific Rise, 4, 105, 134, 135, 137, 138
    Galapagos Rift, 217, 218, 220
  ingestion, 281
  Lau Basin, 232, 244
  Mid-Atlantic Ridge, 129, 280, 281, 326
    Broken Spur, 105, 175–88, 270, 271
    Snake Pit, 321
    TAG, 124, 125, 126, 128–9, 133–4
  oxidation, 128, 184, 186, 217, 218
  precipitation, 124, 159, 368
  transfer into sediments, 223
  Troodos, 164
  see also massive sulphide deposits; polymetallic ore
    genesis; sulphide mounds
sulphite chimney, EPR, 24
sulphur, 81, 220, 227, 307
sulphur bacterium, Aegean Sea, 308, 313
sulphur isotope geochemistry, EPR, 186, 187, 188,
    216, 218–20
sulphur respiration, 354
sulphur-oxidizing bacteria, 308, 314, 330, 337, 353
sulphuric acid formation, 353
super-penetration of plumes, 375
'supply side' ecology, 342
Syllides benedicti, Aegean Sea, 306
symbiosis, 280, 281, 284, 314
  see also, endosymbiosis; endosymbiotic bacteria
synaphobranchid fish, MAR, 261, 267

TAG field, 22–5, 27–8, 29, 43, 77–8, 97, 121, 122–6,
    129–30
  activity, 274
  ecology, 257, 258–63, 276–7, 278, 281, 326
    larval identification, 344, 346, 347, 348
    trophic studies, 280–2
  heat flux, 68, 83
    from diffuse flow, 156, 157, 160
  light scattering profiles, 92
  metalliferous deposits, 49–54, 58–9, 61, 62, 105,
    126–30, 223–5
    rare earth concentrations, 93, 94
  sulphide deposits, 8–9, 24–5, 28, 133–4
  vent fluid compositions, 78–85, 115
TAG plumes, particle processes, 108, 387–95
Tagiri fumaroles, ecology, 272

tanaids, MAR, 270
Tawera spissa, Aegean Sea, 315
tectonic activity and plume formation, 118
tectonic control, EPR, 20, 29
tellurium, Galapagos Rift, 214, 215
temperature anomalies, 6, 55, 67, 68, 69, 71, 374,
    378, 383
temperatures of vent fluids, *see* fluid temperatures
tennantite, Manus Basin, 194
Tethya aurantium, MAR, 272
Tevnia jerichonana, EPR, 295, 296, 298
Thames Estuary, restricted gene flow, 344
thermal anomalies, due to vertical motion, 379
thermal energy, conservation equation, 146
thermal waters, terrestrial, 353
Thermococcus celer, 354, 355
Thermococcus littoralis, 358, 359
Thermodiscus maritimus, 354, 355
thermophilic bacteria
  gas production by, 115, 119
  see also hyperthermophilic bacteria
thermophilic enzymes, 351, 357
  see also hyperthermophilic enzymes
Thermoproteus tenax, 354, 355
thermostable enzymes, 358
Thermotoga, 354
Thermus aquaticus, 356, 358, 359
Thermus thermophilus, 359
thiotrophic bacteria, 340
thorium scavenging, TAG, 128, 129, 391, 392, 393–4,
    395
tin, MAR, 47, 48, 56
titanium, Lau Basin, 236, 237
titanium (IV) oxide, 200, 203, 214
todorokite, Lau Basin, 242
Tonnia galea, Aegean Sea, 308
trace element scavenging, 387, 394, 395
trace metal/iron concentrations [TM/Fe], 368–9
Trans-Atlantic Geotraverse field *see* TAG field
transform faults, as dispersion barriers, 278
transmissometer and nephelometer relationships,
    387–9
transport velocity, 146
Troodos, 121, 164
Troodos ophiolite, massive sulphide deposits, 28
trophic interactions, MAR, 280–6
tubeworms, 125, 194, 261, 272, 278, 285
  see also vestimentiferan tubeworms; worms
Tyrrhenian sea, hyperthermophilic bacteria, 314

(U + Th)/He ratio, EPR, 137
unsaturated acids, in vent organisms, 332–5, 337–40,
    341
upwelling, 145
uranium, 128–9, 214
urchins, MAR, 88, 262, 263, 269–70

Valu Fa Ridge, 164, 232, 233, 244
  mineralization, 195, 198–9, 201, 202, 245
vanadium, 214, 215, 236, 237
Vance segment, JDFR, 6, 66
velocity of flow within sulphide structures,
    146
Vema Transform Fault, 278, 279

vent organisms
    larval identification, 343–9
    lipid characteristics, 329–41
    see also nutrition
vent-specific organisms, colonization by, 12
Venture fields, amphipods, 338–9
vertical recycling of plume fluid, 375
Vesicomyidae, MAR, 257, 273
vestimentiferan larvae, 343
vestimentiferan tubeworms
    Aegean Sea, 315
    composition and morphogenesis of tubes, 295–301
    East Pacific Rise, lipid characteristics, 329–30,
        331–3, 336, 337, 340
    Mid-Atlantic Ridge, 257
    see also tubeworms, worms
Vienna Woods site, metalliferous sediments, 195
Virgin Mound, 365
vitric sediments, Lau Basin, 234, 235
Volcania, Kos, 303
volcanic alteration of vent fields, 133
volcanic ash, Lau Basin, 234
volcanic-hosted massive sulphide deposits, 192, 193,
    199, 200, 207, 219
volcaniclastic detritus, Lau Basin, 232, 234, 236, 237
volcanism, EPR, 138
volume flow rate equation for pipe models, 160, 162,
    163
vortex pairs, 375, 376, 377, 378
vorticity equation, 146

Wasp's Nest, 37, 38, 39, 175, 179, 182–3, 270
water/rock ratios, EPR, 140
'waves of recruitment' of mussels, MAR, 269

wax esters, in crustacea, 339
well-measured basin, heat budget calculation, 381
West Florida Escarpment, ecology, 259
white smokers, 49, 159
    Mid-Atlantic Ridge, 45, 80, 81, 82, 123, 124–5, 261
White's Point, 272, 303
Woodlark Basin, 10, 191–2
    polymetallic ore genesis, 195–6, 198–9, 200–3
worm-tubes, MAR, 37, 38, 271
worms, 38, 39, 261, 263, 307
    see also tubeworms; vestimentiferan tubeworms
wurtzite, 176, 178, 180, 182, 183, 184, 194

Yali, 303, 304

Zephria Point, 305, 313
zero-angle photon fibre optic spectrometer
    technique, 88–9
zinc
    Axial Volcano, 105
    dissolved, 9, 17
        gross fluxes, 365–71
    East Pacific Rise, 4, 17
        Galapagos Rift, 212, 213, 214, 215
    Lau Basin, 232, 236, 237
    Manus Basin, 194, 195
    Mid-Atlantic Ridge, 17, 60, 223
        Broken Spur, 103, 105, 106, 107, 108
        MARK, 47, 48, 61, 62, 78, 226, 227
        TAG, 48, 51, 52, 54, 56, 57, 58, 59, 78, 81, 224
    Woodlark Basin, 196, 200, 201, 202
zinc sulphides, Broken Spur, 176, 181, 183, 184
zinc/iron ratios in plumes, JDFR, 389
zirconium, Galapagos Rift, 214